United States Nuclear Regulatory Commission

Protecting People and the Environment

NUREG-2113

I0482782

Environmental Impact Statement for the Proposed Fluorine Extraction Process and Depleted Uranium Deconversion Plant in Lea County, New Mexico

Final Report

Office of Federal and State Materials and Environmental Management Programs

AVAILABILITY OF REFERENCE MATERIALS
IN NRC PUBLICATIONS

United States Nuclear Regulatory Commission

Protecting People and the Environment

NUREG-2113

Environmental Impact Statement for the Proposed Fluorine Extraction Process and Depleted Uranium Deconversion Plant in Lea County, New Mexico

Final Report

Manuscript Completed: July 2012
Date Published: August 2012

Asimios Malliakos, NRC Project Manager

Office of Federal and State Materials and
Environmental Management Programs

ABSTRACT

International Isotopes Fluorine Products, Inc. (IIFP), a wholly-owned subsidiary of International Isotopes, Inc., has submitted a license application to the U.S. Nuclear Regulatory Commission (NRC) to construct, operate, and decommission Phase 1 of a fluorine extraction and depleted uranium deconversion facility in Lea County, New Mexico. The proposed facility would provide services to the uranium enrichment industry, which makes fuel for nuclear power reactors. The IIFP facility would deconvert depleted uranium hexafluoride (DUF_6) into fluoride products for commercial resale, and depleted uranium oxides for disposal. The license application for Phase 1 requests NRC to license the possession of up to 750,000 kilograms (827 tons) of depleted uranium under Title 10 "Energy" of the *U.S. Code of Federal Regulations* (10 CFR) Part 40, "Domestic Licensing of Source Material" in accordance with the *Atomic Energy Act of 1954*.

This Environmental Impact Statement (EIS) was prepared in compliance with the *National Environmental Policy Act* (NEPA) and the NRC regulations for implementing NEPA (10 CFR 51). This EIS evaluates the potential environmental impacts of the proposed action, which is to construct, operate, and decommission Phase 1 of the fluorine extraction and depleted uranium deconversion facility, and its reasonable alternatives, and describes IIFP's monitoring program and proposed mitigation measures.

Paperwork Reduction Act Statement

This NUREG contains and references information collection requirements that are subject to the Paperwork Reduction Act of 1995 (44 U.S.C. 3501 et seq.). These information collection requirements were approved by the Office of Management and Budget, approval numbers 3150-0014; 3150-0020; 3150-0021; 3150-0135; 3150-0009; and 3150-0008.

Public Protection Notification

The NRC may not conduct or sponsor, and a person is not required to respond to, a request for information or an information collection requirement unless the requesting document displays a currently valid OMB control number.

NUREG-2113 has been reproduced from the best available copy

TABLE OF CONTENTS

TABLE OF CONTENTS (CONTINUED)

TABLE OF CONTENTS (CONTINUED)

TABLE OF CONTENTS (CONTINUED)

TABLE OF CONTENTS (CONTINUED)

TABLE OF CONTENTS (CONTINUED)

TABLE OF CONTENTS (CONTINUED)

LIST OF TABLES

LIST OF TABLES (CONTINUED)

LIST OF TABLES (CONTINUED)

LIST OF TABLES (CONTINUED)

LIST OF FIGURES

EXECUTIVE SUMMARY

Background

Under the provisions of the *Atomic Energy Act* and pursuant to Title 10 of the U.S. Code of Federal Regulations Part 40 (10 CFR 40), the U.S. Nuclear Regulatory Commission (NRC) is considering whether to issue a license that would allow International Isotopes Fluorine Products, Incorporated (IIFP) to possess, use, transfer, or deliver source and byproduct materials at a proposed fluorine extraction and depleted uranium deconversion facility near Hobbs in Lea County, New Mexico. The scope of activities to be conducted under the license would include the construction, operation, and decommissioning of the proposed IIFP facility. The facility would deconvert commercially generated depleted uranium hexafluoride (DUF_6) into depleted uranium dioxide (DUO_2) for long-term stable disposal, and into fluorine products for resale. DUF_6 is the by-product of uranium enrichment. The application for the license was filed with the NRC by IIFP, on December 30, 2009. To support its licensing decision on IIFP's proposed facility, the NRC determined that the NRC's implementing regulations in 10 CFR 51 for the National Environmental Policy Act require the preparation of an Environmental Impact Statement (EIS). The EIS is used to examine the potential environmental impacts of the proposed IIFP facility and reasonable alternatives. Based on the EIS and other information, the NRC will determine whether to issue a license to IIFP for the construction, operation, and decommissioning of the proposed IIFP facility.

The Proposed Action

The proposed action considered in this EIS is for NRC to grant IIFP a license to construct, operate, and decommission a fluorine extraction and depleted uranium deconversion facility The IIFP facility would include a commercial plant to produce specialty fluoride gas products for sale and DUO_2 for disposal. IIFP would own the facility and be responsible for its operation and performance. The proposed facility, if licensed, would be 22.5 kilometers (km) (14 miles [mi]) west of Hobbs, New Mexico. The proposed tract of land (IIFP site) occupies 259 ha (640 ac), and the proposed facility would occupy an estimated 16 ha (40 ac) of the tract, not including roadways and other infrastructure improvements.

Construction of the IIFP facility is expected to begin in 2012 and operations would begin in late 2013. The proposed facility is designed to be capable of deconverting up to 3.4 million kilograms (kg) (7.5 million pounds, or 3,750 tons) per year of DUF_6. The annual capacity of approximately 3.4 million kg (3,750 tons) per year equates to about 9,300 kg/day (10.3 tons/day) on average. Following operations the facility is expected to be decommissioned following termination of the license.

Preconstruction Activities

The applicant's license application states that IIFP anticipates commencement of certain preconstruction activities on the proposed IIFP site prior to the NRC's decision on whether to issue a license for the construction, operation, and decommissioning of the proposed facility. The preconstruction activities would be considered by the NRC as a cumulative effect and not a part of the proposed action. Preconstruction could include the following activities and facilities: land clearing; site grading (excavating and/or blasting); erosion control and stormwater control measures installation; access road and parking facilities construction; and others.

Purpose and Need for the Proposed Action

Detailed information on the purpose and need for action is described in Chapter 1 of this EIS. The proposed action is intended to satisfy the need for a facility that would deconvert DUF_6 into DUO_2 for disposal. An added goal of IIFP would be to produce fluoride products for commercial resale. Without a facility such as the proposed IIFP facility, DUF_6 would continue to be stored, typically in 12.7-metric ton (14-ton) cylinders, at commercial uranium enrichment facilities in the United States. Although DUF_6 could be transferred to the U.S. Department of Energy (DOE) for a fee, DOE's existing inventory of DUF_6 is not projected to be deconverted for approximately 25 years. Further, long-term storage of DUF_6 represents a potential chemical hazard if the material is not properly managed, and deconversion to DUO_2 is preferable. The fluoride products are potentially valuable for applications in the electronic, solar panel, and semi-conductor markets, among others. In addition, anhydrous hydrogen fluoride (AHF) is a by-product of the deconversion process and is an important chemical in various industrial applications.

Alternatives

A detailed analysis of alternatives is included in Chapter 2 of this EIS. The no-action alternative is considered in this EIS as a baseline for comparison. Under the no-action alternative, NRC would not grant a license to IIFP to construct, operate, and decommission the proposed facility near Hobbs, New Mexico, to receive and process source material, and to ship products and low-level radioactive waste (LLW). However, impacts from preconstruction activities could occur under the no-action alternative. The proposed site would remain in its current or preconstruction condition. The regional economy would not be changed either positively or negatively, except by preconstruction. LLW would not be shipped to licensed disposal facilities for disposal. Fluoride products would not be manufactured and sold to end users. Planned or existing commercial enrichment facilities would not be able to send their DUF_6 to the IIFP facility for deconversion.

Four options would be open to these commercial facilities, in the event of the no-action alternative: (1) ship the DUF_6 to DOE facilities, (2) ship the DUF_6 to facilities overseas, (3) indefinitely store the DUF_6, or (4) construct their own deconversion facilities. DOE has constructed two facilities to deconvert DUF_6 to uranium oxides (different compounds than that which would be produced by IIFP's proposed facility) and hydrofluoric acid: one in Paducah, Kentucky and one in Piketon (Portsmouth), Ohio. Therefore, shipment to these DOE facilities is a viable option under the no-action alternative. Given that DOE has a backlog of 700,000 metric tons (771,618 tons) of DUF_6 (stored in approximately 57,000 cylinders) to deconvert, it may take DOE approximately 25 years to complete its mission before beginning to deconvert privately generated DUF_6. The DOE process does not produce the fluoride products, and it produces a hydrofluoric acid solution rather than the anhydrous hydrogen fluoride, which is an important chemical in various industrial applications.

IIFP conducted a site selection process to determine the best location, by IIFP criteria, for the proposed site. The NRC staff reviewed the IIFP site selection process and determined that the process was rational and objective. Accordingly, no alternate sites are evaluated in the EIS.

The NRC staff evaluated several alternative technologies, including: (1) a direct deconversion process; (2) the DOE deconversion process that is used at Paducah, Kentucky, and Piketon (Portsmouth), Ohio; and (3) a foreign (European) process. The direct deconversion, DOE deconversion, and foreign conversion alternative processes were eliminated from analysis in the

EIS because (1) the applicant owns and has expertise in a competing technology, (2) the impacts of implementing these technologies would be sufficiently similar to the proposed action, and (3) none of these processes would satisfy the goal to produce marketable fluoride by-products.

The NRC staff also considered an alternative that would ship the U.S.-generated DUF_6 to overseas facilities for deconversion. However, because of prohibitive cost of such shipments, this alternative was eliminated from consideration in the EIS. An alternative that would indefinitely store the DUF_6 was also eliminated from consideration because long-term storage of DUF_6 represents a potential chemical hazard if not properly managed, and such an alternative would not meet the underlying need for deconversion of the DUF_6. Lastly, the NRC staff considered an alternative in which the four U.S.-based enrichment companies could construct and operate their own deconversion facilities. However, because none of these firms has expressed an interest in constructing such a facility, NRC staff concluded that this alternative should be eliminated from consideration in this EIS.

Potential Environmental Impacts of the Proposed Action

In this EIS, NRC staff evaluates the existing conditions (Chapter 3) and potential environmental impacts of the proposed action (Chapter 4). A standard of significance, (see text box), has been established for assessing environmental impacts. The NRC staff has assigned each impact one of the three significance levels described in the textbox. The environmental impacts from the proposed action are SMALL or MODERATE and could be mitigated by the methods described in Chapter 5. Environmental monitoring methods are described in Chapter 6.

Summarized below are the potential environmental impacts of the proposed action on each of the resource areas considered in this EIS. Each summary is preceded by the impact significance level for the respective resource areas.

> **Determining the Significance of Potential Environmental Impacts**
>
> NRC has established a standard of significance for assessing environmental impacts. Each impact is assigned one of the following three significance levels:
>
> • **SMALL:** The environmental effects are not detectable or are so minor that they would neither destabilize nor noticeably alter any important attribute of the resource.
>
> • **MODERATE:** The environmental effects are sufficient to noticeably alter but not destabilize important attributes of the resource.
>
> • **LARGE:** The environmental effects are clearly noticeable and are sufficient to destabilize important attributes of the resource.

Land Use

SMALL. Construction activities would occur on about 16 ha (40 ac) within the 259-ha (640-ac) site. Construction of the proposed facility would alter the current land use of the entire IIFP site, a tract known as Section 27 of Township 18 South, Range 36 East, which is primarily used for cattle grazing. The transfer and conversion of the land for the facility would not conflict with any existing Federal, State, local, or Tribal Nation land use plans, or restrict current or planned mineral resource exploitation. The operation of the proposed facility would be consistent with the existing land use of the neighboring tracts, which support industrial facilities, natural gas and oil extraction and transmission infrastructure, and agriculture and open land.

Historical and Cultural Resources

SMALL. An archaeological survey of the entire 259-ha (640-ac) site failed to identify any archeological resources other than several isolated artifacts that were not considered to be eligible for the National Register of Historic Places. Consultation with Federally recognized Tribal Nations and the New Mexico State Historic Preservation Division (which serves as the State Historic Preservation Officer) did not identify any additional information on historically or culturally significant resources within the area potentially affected by the proposed facility. The preconstruction, construction, and operation of the proposed facility would not adversely affect historic resources or other cultural resources (e.g., significant archaeology sites).

Visual and Scenic Resources

SMALL. The proposed 259-ha (640-ac) site is flat and sparsely developed with a few irregularly-spaced structures for natural gas and oil extraction, and overhead transmission lines. The proposed IIFP facility would be approximately 22.5 km (14 mi) west of the nearest population center, Hobbs, New Mexico and would not be visible from Hobbs. The proposed site received the lowest scenic-quality rating using the U.S. Bureau of Land Management visual resource inventory process.

Climatology, Meteorology, and Air Quality

SMALL to MODERATE. Air concentrations of (1) criteria pollutants predicted for vehicle emissions and (2) emissions of particulate matter of less than 10 microns (PM_{10}) from fugitive dust during construction would be below the National Ambient Air Quality Standards (NAAQS) for carbon monoxide (CO) and sulfur dioxide (SO_2) emissions, and above NAAQS for nitrogen dioxide (NO_2), and particulate matter ($PM_{2.5}$ and PM_{10}) emissions. Fugitive dust emissions would be temporary and localized. During construction of the IIFP facility, carbon dioxide (CO_2) emissions are projected to be 2,110 metric tons (2,326 tons) or 0.003 percent of New Mexico's statewide output and 0.00003 percent of the projected nationwide CO_2 emissions for the same period. A National Emissions Standards for Hazardous Air Pollutants Title V permit would not be required for operations due to the low levels of estimated emissions. All stack emissions would be monitored. During any typical year of IIFP facility operation, CO_2 emissions are projected to be 5,774 metric tons (6,373 tons), approximately 0.009 percent of the New Mexico statewide output or 0.0009 percent of the nationwide emissions for calendar year 2000.

Geology, Minerals, and Soils

SMALL. Construction-related impacts on the geology, minerals, and soils would occur within the 16 ha (40 ac) of the 259-ha (640-ac) site on which the proposed facility would be built, and for the construction of the access road, which would extend roughly 1 kilometer (1/2 mi) from Arkansas Junction Road (NM 483) to the entrance of the proposed facility. The site has no prime farmland, as defined by the U.S. Department of Agriculture. The site has been explored for oil and gas and mined for caliche and, thus, it has very limited leasable, locatable, or marketable mineral resources. Therefore, the proposed facility construction activities would not result in loss of mineral resources. No impact to the underlying bedrock, mineral resources, or soils is expected during the facility operations. The site is in an area of limited seismic activity and operation of the IIFP facility is not expected to cause seismic or fault-related impacts. Any seismic risk would be mitigated by incorporation of seismic criteria in the facility design.

Water Resources

SMALL. The site has no permanent surface water and no jurisdictional wetlands. The closest source of a named ephemeral stream is more than 5 km (3 mi) from the property, and the nearest permanent surface water is more than 32 km (20 mi) from the site. The site, which overlies the Lea County Underground Water Basin, would utilize water from the Ogallala Aquifer to support construction and operation. Groundwater demand on the Ogallala Aquifer during construction would be relatively low, mainly for dust suppression. During operations, groundwater use for potable water and process water needs is estimated to be less than 38,000 liters (10,000 gallons [gal]) per day peak, averaging an estimated 13,000 liters (3,000 gal) per day. The proposed facility would use approximately 0.5 percent of the estimated additional annual 40-year planning period groundwater demand for Lea County and only 0.15 percent of the unappropriated water rights that have been assigned to Lea County.

Ecological Resources

SMALL. Approximately 16 ha (40 ac) of land would be disturbed, which represents approximately 6 percent of the site's 259 ha (640 ac). There are no wetlands or unique habitats, and no threatened or endangered species on the proposed site. Fencing around the proposed IIFP facility would restrict wildlife access to the facility. Mitigation measures proposed by the New Mexico Department of Game and Fish (Appendix B - Consultation/Correspondence) would be considered to lessen impacts.

Socioeconomics

SMALL. Construction of the IIFP facility would employ approximately 140 people. Eighty percent of this staff is expected to be current residents in the socioeconomic region of influence (ROI): Lea and Eddy Counties, New Mexico. It is expected that the other 20 percent (28 workers) would migrate into the socioeconomic ROI. Including family members, the total increase in residents to the ROI is expected to be 72 people, which would result in a 0.06 percent increase in the ROI population. During operation, the proposed IIFP facility would employ approximately 140 people, and 20 percent (28 individuals with their families) are expected to in-migrate, increasing the population in the socioeconomic ROI by 90 people, or less than 0.1 percent of the 2009 population. The impacts on the local unemployment rate, housing vacancies, schools, and public services and utilities would be minimal during operations and construction.

Environmental Justice

SMALL. The environmental justice analysis focused on census blocks and block groups in an area within 80 kilometers (50 miles) of the proposed IIFP site. The largest minority population within 80 kilometers (50 miles) of the proposed site is the Hispanic/Latino population. The nearest minority or low-income population as defined by NRC criteria is 22.5 km (14 mi) from the proposed site. The impacts of IIFP construction and operation on resources would be SMALL and, in most cases, localized. Therefore, because all impacts would be SMALL, and the identified minority and low-income populations are not in close proximity to the proposed site, impacts would not be disproportionately high and adverse for any populations in the region, including minority or low-income populations.

Noise

SMALL. Noise would come predominantly from construction equipment and traffic. Construction activities would be temporary and limited to daytime working hours. The nearest residence is approximately 2.6 km (1.6 mi) northwest of the site and there are no recreational areas within 8.0 km (5.0 mi) of the proposed site. Noise levels during operations would be within the U.S. Department of Housing and Urban Development guidelines.

Traffic and Transportation

SMALL. The potential maximum increase from construction workforce traffic would be 280 round trips per weekday, and the potential maximum increase to traffic due to construction deliveries and waste removal would be 40 round trips per weekday. The majority of the construction worker trips would use US 62/180 to access NM 483. These trips would increase traffic on NM 483 by 33.5 percent daily, but the design capacities of NM 483 and US 62/180 would not be exceeded. Statistically, the risk of an accident with injuries (risk of less than 0.8 injury crashes per year) or fatality (risk of less than 0.03 fatal crashes per year) to the construction workforce is unlikely.

The operational workforce could increase the traffic on NM 483 by 29 percent and on US 62/180 by 8 percent daily. With the predicted increased traffic volumes, the design capacities of NM 483 and US 62/180 would not be exceeded. Statistically the risk of an accident with injuries (risk of less than 0.7 injury crashes per year) or a fatality (risk of less than 0.02 fatal crashes per year) for the operations traffic is unlikely.

Operation of the IIFP facility would require shipment of full DUF_6 cylinders from commercial enrichment facilities, empty DUF_6 cylinders back to the commercial enrichment facilities, DUO to waste disposal facilities, and other miscellaneous process and LLW to waste disposal facilities. Approximately 730 radiological shipments would occur annually. The collective doses from shipments and accidents involving shipments would be comparatively low, versus natural sources of radiation (Appendix E - Transportation of Radioactive Materials).

Public and Occupational Health and Safety

SMALL. During construction, a fatality would be unlikely (the probability of fatality is less than one per year). During normal operations, based on statistical probabilities, there could be six industrial injuries per year and no fatalities. Worker radiological doses were conservatively estimated to be about 0.75 mSv/yr (75 millirem/yr) for those workers involved in the deconversion processing operations within the proposed facility. The average individual dose for workers at the cylinders yards was estimated to range from a low of 4.3 mSv/yr (430 millirem/yr) to a high of 6.9 mSv/yr (690 millirem/yr). All public radiological exposures would be significantly below the 10 CFR 20 regulatory limit of 1 mSv (100 millirem) per year. The maximally exposed member of the public would receive approximately 0.21 mSv/yr (21 millirem/yr) from the proposed facility operations. For comparison purposes, the average annual dose to a member of the public due to background radiation is estimated to be about 3.1 mSv/yr (310 millirem/yr) (see details in the body of the EIS [Section 3.12]).

The most significant possible accident consequences would be those associated with the rupture of a cylinder containing liquefied DUF_6. However, the facility emergency plan addresses this type of event, and all other high- and intermediate-consequence events. The facility design and procedures would reduce the likelihood of this type of event by requiring a robust cylinder design that maintains its integrity during credible drops, shocks, collisions, and thermal events. In addition, facility design features, which prevent release of liquid DUF_6 or rupture of cylinders during processing cycles, would be implemented. Procedures would be instituted which would minimize the possibility of an accident scenario occurring, and would provide steps to take should an accident occur. The NRC staff concludes that through the combination of facilities design, engineered controls, and administrative controls, including procedures, accidents at the facility would pose a small risk to workers, the environment, and the public.

Waste Management

SMALL. Nonhazardous waste generated from the proposed construction activities would result in a negligible increase (less than 0.0007 percent) in the waste that the Lea County landfill receives annually from all sources. Less than 0.9 metric ton/yr (1 ton/yr) of hazardous wastes would be expected from construction of the proposed IIFP facility. This would represent less than 0.00009 percent of the overall hazardous waste generated in the State.

During operations, industrial waste generated from the proposed facility would result in an increase of approximately 0.06 percent in the waste that the Lea County landfill receives annually from all sources. Hazardous waste generated during operations would also be small, resulting in an increase of less than 0.02 percent in the hazardous waste generated in the State of New Mexico. Up to 3,170 tons per year of LLW could be sent for disposal annually. There is enough existing national disposal capacity to accept the LLW that would be generated at the proposed facility.

Summary of the Costs and Benefits of the Proposed Action

The costs of construction activities is estimated to be between $100 million and $140 million (in 2009 dollars), excluding escalation, contingencies, and interest. Construction-related activities, purchases, and workforce expenditures would incur several types of taxes, including individual income taxes, gross receipts taxes, and property taxes. Approximately $554,400 of fee in lieu of property tax would be paid to the Hobbs Municipal School District and the New Mexico Junior College during the construction period.

During operations, about $56 million to $71 million (in 2009 dollars) in wages (wages account for $7.9 million to $9.1 million), benefits, goods and services would be spent annually. Construction and operation of the facility would have additional indirect economic impacts by creating additional employment and economic activity within the region of influence. Over the lifetime of operations, the low estimate of corporate income taxes and gross receipts taxes paid is $144,200,000 to the State of New Mexico. Over the lifetime of operations, the low estimate of gross receipts taxes is $6,500,000 (in 2009 dollars) to Lea County.

Comparison of Alternatives

Under the no-action alternative, NRC would not grant a license to IIFP to construct the proposed facility near Hobbs, New Mexico, to receive and process source material, and to ship products and LLW. The four planned or existing commercial enrichment facilities would not be able to send their DUF_6 to the IIFP facility for deconversion. DOE has constructed two deconversion facilities to convert DUF_6 to U_3O_8 and hydrofluoric acid: one in Paducah, Kentucky and one in Piketon (Portsmouth), Ohio. Therefore, shipment to these DOE facilities is a viable option under the no-action alternative, but the timeframe for deconversion would be much greater than what the proposed IIFP facility would provide, and goals to create commercial fluorine products would not be realized. Under the no-action alternative, the proposed site would be impacted by preconstruction, but would not be impacted by operation of the proposed facility. The no-action alternative would have cumulative impacts due to preconstruction on current land use; visual/scenic and cultural resources; air; water; ecological resources; geology, minerals and soils; socioeconomics; environmental justice; traffic and transportation; public and occupational health; and waste management. These impacts would be SMALL for all resources except for air quality, for which they would be SMALL to MODERATE.

In comparison to the no-action alternative, the proposed action would have SMALL impacts on land use; air; water; ecological resources; geology, minerals and soils; noise; traffic and transportation; public and occupational health; socioeconomics (these impacts would be SMALL and positive); environmental justice; and waste management.

ACRONYMS AND ABBREVIATIONS

$^{\circ}$C	degrees Celsius
$^{\circ}$F	degrees Fahrenheit
AADT	annual average daily traffic
ac	acre(s)
ACEC	area of critical environmental concern
ACHP	Advisory Council on Historic Preservation
AEGL	acute exposure guideline levels
AHF	anhydrous hydrogen fluoride
ALARA	as low as reasonably achievable
APE	area of potential effect
API	American Petroleum Institute
AQCR	Air Quality Control Region
B_2O_3	boron oxide
BEA	Bureau of Economic Analysis
BF_3	boron trifluoride
bgs	below ground surface
BLM	Bureau of Land Management
BMP	best management practice
Bq/L	becquerel per liter
$CaCO_3$	calcium carbonate
CaF_2	calcium fluoride
$Ca(OH_2)$	calcium hydroxide
CCS	Center for Climatic Strategies
CEDE	committed effective dose equivalent
CEQ	Council on Environmental Quality
CERCLA	Comprehensive Environmental Response Compensation and Liability Act
CFR	Code of Federal Regulations
CH_4	methane
Ci	curie
cm	centimeter
CMA	critical management areas

ACRONYMS AND ABBREVIATIONS
(continued)

CO	carbon monoxide
CO_2	carbon dioxide
CWA	Clean Water Act
dB	decibel (sound pressure level)
dB(A)	decibel, A-weighted (humanly audible frequency)
DNFSB	Defense Nuclear Facilities Safety Board
DOE	U.S. Department of Energy
DOT	U.S. Department of Transportation
DU	depleted uranium
DUF_4	depleted uranium tetrafluoride
DUF_6	depleted uranium hexafluoride
DUO_2	depleted uranium dioxide
DUO	depleted uranium oxides (general term; not a compound that exists)
DUO_2F_2	depleted uranyl dioxyfluoride
DWB	Drinking Water Bureau
EDE	effective dose equivalent
EPP	environmental protection process
EIS	environmental impact statement
EPA	U.S. Environmental Protection Agency
Eq	equivalents
ER	environmental report
ERPG	Emergency Response Planning Guideline
ESA	Endangered Species Act
FD&C	Federal Food, Drug and Cosmetic Act
FEMA	Federal Emergency Management Agency
FEP/DUP	Fluorine Extraction Process/Depleted Uranium Deconversion Plant
ft	foot or feet
FR	Federal Register
g	gravity
gal	gallon(s)
GHG	greenhouse gases
gpd	gallons per day

ACRONYMS AND ABBREVIATIONS
(continued)

gpm	gallons per minute
GPS	global positioning system
GWP	global warming potential
H_2	hydrogen
ha	hectare(s)
HAP	hazardous air pollutant
HF	hydrofluoric acid or hydrogen fluoride
HFC	hydrofluorcarbon
HPD	[New Mexico] Historic Preservation Division
HS&E	health, safety and environmental
HUD	U.S. Department of Housing and Urban Development
IIFP	International Isotopes Fluorine Products, Inc
in	inch
IPCC	Intergovernmental Panel on Climate Change
IRB	industrial revenue bond
ISA	integrated safety analysis
kBq	kilobecquerel
KF	potassium fluoride
kg	kilogram
KOH	potassium hydroxide
km	kilometer(s)
km/hr	kilometers per hour
km^2	square kilometer
kV	kilovolt
L	liter
lb	pound
LCF	latent cancer fatality
LLD	lower limit of detection
LES	Louisiana Energy Services
LLRW	low-level radioactive waste
LLW	low-level (radioactive) waste
L/min	liters per minute

ACRONYMS AND ABBREVIATIONS
(continued)

m	meter(s)
MCL	maximum contaminant levels
MDC	minimum detection concentrations
MEI	maximally exposed individual
mg	milligram
MGD	million gallons per day
mg/kg	micrograms per kilogram
mg/L	milligrams per liter
mg/m^3	milligrams per cubic meter
mi	mile(s)
mi^2	square mile
MM	modified Mercalli
mpg	miles per gallon
mph	miles per hour
mrem	millirem
mSv	millisievert
MW	megawatt
N_2O	nitrous oxides
NAAQS	National Ambient Air Quality Standards
NAGPRA	Native American Graves Protection and Repatriation Act
NCRP	National Council on Radiation Protection
NEPA	National Environmental Policy Act
NESHAP	National Emissions Standards for Hazardous Air Pollutants
NHPA	National Historic Preservation Act
NH_3	ammonia
NM	New Mexico
NMAAQS	New Mexico Ambient Air Quality Standards
NMAC	New Mexico Administrative Code
NMDOT	New Mexico Department of Transportation
NMED	New Mexico Environment Department
NMEDAQB	New Mexico Environmental Department Air Quality Bureau
NMEDHWB	New Mexico Environmental Department Hazardous Waste Bureau

ACRONYMS AND ABBREVIATIONS
(continued)

NMEDRCB	New Mexico Environmental Department Radiation Control Board
NMEDWQB	New Mexico Environmental Department Water Quality Bureau
NMGF	New Mexico Department of Game and Fish
NMOSE	New Mexico Office of the State Engineer
NMRL/CID	New Mexico Regulations and Licensing/Construction Industries Division
NMRPR	New Mexico Radiation Protection Regulations
NMRPTC	New Mexico Rare Plant Technical Council
NMSA	New Mexico Statutes Annotated
NM SHPO	New Mexico State Historic Preservation Office
NMSLO	New Mexico State Land Office
NMSS	Nuclear Materials Safety and Safeguards
NMVOC	non-methane volatile organic compound
NO_2	nitrogen dioxide
NO_x	oxides of nitrogen
NRC	U.S. Nuclear Regulatory Commission
NRHP	National Register of Historic Places
O&M	operating and maintenance
OSHA	Occupational Safety and Health Administration
pCi/L	picocurie per liter (1×10^{-12} curie/liter)
PFC	perfluorocarbon
PGA	peak [horizontal] ground acceleration
PILT	payment in lieu of taxes
$PM_{2.5}$	particulate matter with a diameter less than 2.5 microns
PM_{10}	particulate matter with a diameter less than 10 microns
PPE	personal protective equipment
ppm	parts per million
PSD	prevention of significant deterioration
psig	pounds per square inch gauge
PSTB	Petroleum Storage Tank Bureau
QA	quality assurance
RAI	Requests for Additional Information

ACRONYMS AND ABBREVIATIONS
(continued)

RCRA	Resource Conservation and Recovery Act
REMP	Radiological Environmental Monitoring Program
ROD	record of decision
ROI	region of influence
RMP	resource management plan
RMPA	resource management plan amendment
SARA	Superfund Amendment and Reauthorization Act
SF_6	sulfur hexafluoride
SiF_4	silicon tetrafluoride
SiO_2	silicon dioxide
SPCC	Spill Prevention Control and Countermeasures
SO_2	sulfur dioxide
SRP	site redress plan
s.u.	standard units
SVOC	semivolatile organic compound
SWPP	Stormwater Pollution Prevention Plan
SWU	separative work unit
TDS	total dissolved solids
TEDE	total effective dose equivalent
Tg	teragram [1×10^{12} gram]
TLD	thermo luminescent dosimeters
TSD	treatment, storage, disposal
TSP	total suspended particulate
U-234	a uranium isotope
U-235	a uranium isotope
U-236	a uranium isotope
U_3O_8	variously known as uranium oxide, triuranium octoxide, or "yellowcake"
UF_4	uranium tetrafluoride
UF_6	uranium hexafluoride
µCi/g	microcuries per gram
µg/L	micrograms per liter

ACRONYMS AND ABBREVIATIONS
(continued)

$\mu g/m^3$	microgram per cubic meter
UNFCCC	United Nations Framework Convention on Climate Change
UO_2	uranium oxide or uranium dioxide
UO_2F_2	uranyl oxyfluoride
USACE	U. S. Army Corps of Engineers
USCB	U.S. Census Bureau
USEC	U.S. Enrichment Corporation
USFWS	U.S. Fish and Wildlife Service
USGS	U.S. Geological Survey
UWB	underground water basin
VA	volt-ampere
VOC	volatile organic compound
WCS	Waste Control Specialists
WIPP	Waste Isolation Pilot Plant

1.0   INTRODUCTION

1.1   Background

Nuclear reactor fuel requires uranium with a higher proportion of the uranium-235 (U-235) isotope than is found in naturally occurring uranium (approximately 0.7 percent by weight). To increase the portion of U-235 isotopes in the fuel, an enrichment process is used. Uranium in the form of uranium hexafluoride (UF_6) is the feed for the enrichment process, and depleted uranium hexafluoride (DUF_6) is a byproduct of the process. During enrichment, the U-235 is extracted from a portion of the natural uranium in order to concentrate the U-235 into nuclear fuel. This lowers the concentration of U-235 in the remainder of the material so that its proportion is lower than the 0.7 percent by weight found in natural uranium (DOE, 2004). The UF_6 with an increased concentration of U-235 is known as "enriched uranium". The UF_6 with a reduced concentration of U-235 is referred to as DUF_6, which is primarily stored at the enrichment facilities. DUF_6 is considered source material. Source material licensees are regulated under Title 10, Part 40, of the Code of Federal Regulations (10 CFR 40), in accordance with the Atomic Energy Act of 1954, as amended.

Forecasts of operating nuclear-generating capacity suggest a continuing demand for uranium enrichment services both in the United States and abroad. Four new commercial enrichment plants in the U.S. are either in planning, construction, or start-up-phases, and the amounts of DUF_6 are projected to increase. Although there are potential beneficial uses for depleted uranium (DU), the current need for DU is low compared to the existing inventory, and the potential for significant commercial demand is considered to be low. The Defense Nuclear Facilities Safety Board (DNFSB) has reported that long-term storage of DU in the UF_6 form represents a potential chemical hazard if not properly managed, and conversion to more-stable DU oxides is preferable to continued long-term storage (NRC, 2005). Because significantly increased use of DU is not expected, this material will likely require disposal. DU can be disposed of as low level (radioactive) waste (LLW).

In 1998, Congress directed the U.S. Department of Energy (DOE) to construct DU deconversion facilities next to the existing gaseous diffusion uranium enrichment plants in Piketon (Portsmouth), Ohio and Paducah, Kentucky. The Portsmouth, Ohio facility began operating in October, 2010 and the Paducah, Kentucky facility began operating in December, 2010. With both fully operational, these plants will deconvert more than 700,000 metric tons (771,000 tons) of DUF_6 currently stored by DOE. This inventory is projected to require 25 years to deconvert, once the facilities become operational. DOE plans to dispose of the 551,000 metric tons (607,200 tons) of deconverted DU as LLW (DOE, 2004).

International Isotopes Fluorine Products, Inc. (IIFP) proposes to construct, operate, and decommission a facility for deconversion of DUF_6 (IIFP, 2009a). The deconversion process is used to convert DU to more chemically stable uranium oxide compounds, such as triuranium octoxide (U_3O_8) or uranium dioxide (UO_2), that are similar to the chemical form of natural uranium (DOE, 2004) and are generally suitable for disposal as LLW.

High-purity silicon tetrafluoride (SiF_4) and boron trifluoride (BF_3) would be manufactured in the IIFP facility from the fluorine derived from the deconversion of DUF_6. The fluoride gas products are valuable for applications in the electronic, solar panel, and semi-conductor markets. Anhydrous hydrogen fluoride (AHF), which is not produced by the DOE facilities described above, is another by-product of the deconversion process, which is used for various industrial applications (IIFP, 2009a).

The U.S. Nuclear Regulatory Commission (NRC) staff has prepared this Environmental Impact Statement (EIS) in response to an application submitted by IIFP for a license that would allow the construction, operation, and decommissioning of a commercial facility for deconversion of DUF_6 in Lea County, New Mexico.

The NRC's Office of Federal and State Materials and Environmental Management Programs has prepared this EIS as required by 10 CFR 51, "Environmental Protection Regulations for Domestic Licensing and Related Regulatory Functions." The NRC's regulations under 10 CFR 51 implement the requirements of the *National Environmental Policy Act of 1969*, as amended (NEPA) (Public Law 91-190). NEPA requires Federal agencies to prepare an EIS for every major federal action significantly affecting the quality of the human environment.

Source material licenses, such as the one requested for the IIFP facility, are regulated under 10 CFR 40. This licensing action is considered a major federal action because it may significantly affect the quality of the human environment consistent with 10 CFR 51, and must therefore meet the requirements of the NEPA for an EIS. The NRC staff has prepared this EIS to evaluate the potential environmental impacts of the proposed IIFP facility and reasonable alternatives to the proposed action.

1.2   Proposed Action

The proposed action is for the NRC to grant IIFP a license (under 10 CFR 40, "Domestic Licensing of Source Material") to construct, operate, and decommission a facility to deconvert commercially generated DUF_6 to depleted uranium dioxide (DUO_2) and other deconversion products. IIFP would own the facility and be responsible for its operation and performance. If the NRC issues a license to IIFP under the provisions of the *Atomic Energy Act of 1954*, as amended, the license would authorize IIFP to possess and use special nuclear material, source material, and byproduct material at the proposed IIFP facility for a period of 40 years in accordance with the NRC's regulations in 10 CFR 40. The scope of activities to be conducted under the license would include the construction, operation, and decommissioning of the proposed IIFP facility.

If issued a license by NRC, IIFP has proposed that the IIFP facility, comprising 16 hectares (ha) (40 acres [ac]) would be located within a 259-ha (640-ac) section in Lea County, near Hobbs, New Mexico. This parcel of land which was previously publicly-owned and comprises open range land

---

Potential Beneficial Uses of DU

- Further enrichment – DU can be used as feedstock for uranium enrichment. The low cost of uranium ore and postponed deployment of advanced enrichment technology have indefinitely delayed this application.
- Nuclear reactor fuel – DU can be mixed with plutonium oxide from decommissioned nuclear weapons to make mixed oxide fuel (typically about 6 percent plutonium oxide and 94 percent depleted uranium oxide) for commercial power reactors.
- Down-blending highly-enriched uranium – Nuclear disarmament treaties allow the down-blending of some weapons-grade highly enriched uranium with DU to make commercial reactor fuel.
- Munitions – DU metal can be used for tank armor and armor-piercing projectiles.
- Biological shielding – DU metal has a high density, which makes it suitable for shielding from x-rays or gamma rays for radiation protection.
- Counterweights – Because of its high density, DU has been used to make small but heavy counterweights.

Source: NRC, 2005

used for grazing as well as overhead transmission lines and underground petroleum pipelines, has been conveyed from the State of New Mexico to Lea County and, ultimately, to IIFP for construction and operation of the proposed facility.

The IIFP initial (Phase 1) plant would include two main chemical processes that, when integrated, will comprise the Fluorine Extraction Process and Depleted Uranium Deconversion Plant (FEP/DUP). The potential future Phase 2 facility expansion would provide additional deconversion capability.

Construction of the IIFP facility is expected to begin in 2012 and operations would begin in late 2013. The construction for the Phase 2 expansion, which is not part of the current license application but is anticipated, is expected to begin in 2015 and full operations would begin in late 2016. At the end of its useful life, the IIFP FEP/DUP plant would be decommissioned consistent with the plan developed and submitted to NRC in the IIFP License Application. IIFP expects to capture beneficial byproducts as result of the deconversion process, including SiF_4, BF_3, and AHF. IIFP's license application states that IIFP also intends to convert DUF_6 to chemically stable compounds discussed in Section 1.1 above, for disposal. Additional details, including volumes of nuclear material, are discussed in Section 1.3.

1.3 Purpose and Need for the Proposed Action

The proposed action under consideration by the NRC is a license application to construct, operate, and decommission a facility to deconvert DUF_6 into depleted uranium oxides for disposal. Additionally the process will recover fluoride products for commercial sale. With the existing inventory of stockpiled depleted uranium and four new commercial enrichment plants in the United States expected to be operating within the next few years, there is a need to deconvert the quantity of DUF_6 that exists and would be produced at these enrichment facilities. Without a deconversion facility, DUF_6 would continue to be stored, primarily at commercial uranium enrichment facilities in the United States, typically in 12.7-metric ton (14-ton) cylinders. Although DUF_6 could be transferred to DOE for deconversion for a fee, DOE's existing inventory of DUF_6 is not projected to be deconverted for 25 years. The proposed IIFP facility should be capable of deconverting up to 3.4 million kilograms (kg) (7.5 million pounds, or 3,750 tons) per year of DUF_6, (NRC, 2010a) which would be approximately one-tenth of the DUF_6 that is projected to be produced annually in the United States by commercial enrichment facilities. The annual capacity of 3.4 million kg (3,750 tons) per year equates to about 9,340 kg/day (10.3 tons/day) on average.

IIFP is proposing to perform the following activities:

- Construct, operate, maintain, and decommission the proposed facility.
- Receive full and return empty DUF_6 cylinders from various commercial enrichment facilities.
- Transport marketable deconversion byproducts to end users.
- Transport depleted UO_2 for LLW disposal or other potential disposition.

IIFP is planning, but has not formally submitted an application for, an expansion of the facility. Expansion and operation of the expanded facility (Phase 2) would be a reasonably foreseeable action and is evaluated as a cumulative impact in this EIS.

Activities that do not constitute construction under 10 CFR 40 and 51 are those that do not have a reasonable nexus to radiological health and safety or the common defense and security, and could include clearing of the facility area, grading, installation of drainage and erosion control and other environmental mitigation measures, and construction of access roads. These "preconstruction" activities are evaluated in this EIS as cumulative impacts because they are expected to occur independently of the proposed licensing action by NRC.

1.4 Scope of this Environmental Analysis

On December 30, 2009, IIFP submitted an application to the NRC (IIFP, 2009b), seeking a license to construct, operate, and decommission a facility for deconversion of DUF_6. As part of that license application, IIFP submitted an Environmental Report (ER) (IIFP, 2009a) for the proposed facility.

On February 24, 2010, the NRC accepted the IIFP application for formal review (NRC, 2010b). A safety review team and an environmental review team are conducting both safety and environmental reviews of the license application. To fulfill its responsibilities under NEPA, the NRC staff has prepared this EIS to analyze the potential environmental impacts of the proposed IIFP facility, and of reasonable alternatives to the proposed action. The scope of this EIS includes consideration of both radiological and nonradiological (including chemical) impacts associated with the proposed action and the reasonable alternatives. The EIS also addresses the potential environmental impacts of transportation. It addresses cumulative impacts to physical, biological, and socioeconomic resources. In addition, it identifies monitoring and mitigation activities. This EIS is the result of the NRC staff's review of the IIFP facility license application, the ER, information obtained from the NRC staff's independent research, and IIFP's responses to Requests for Additional Information (RAIs). This review has been closely coordinated with the NRC staff's development of the Safety Evaluation Report (SER).

Scoping

Scoping is an early and open part of the NEPA process designed to help determine the range of actions, alternatives, and potential impacts to be considered in the EIS, and to identify significant issues related to the proposed action. In addition to the public scoping process, the NRC solicits input from State, local, and other Federal agencies, and potentially affected Native American Tribes in order to focus on issues of genuine concern.

1.4.1   Scoping Process and Public Participation Activities

The NRC regulations in 10 CFR 51 contain requirements for defining the scope of an EIS and identifying issues that should be addressed in depth. The scoping process was used to solicit public and agency input to identify those issues to be discussed in the EIS in detail, and to identify those issues that are either beyond the scope of this EIS or are not directly relevant to the assessment of potential impacts from the proposed action.

As part of the NRC staff's environmental review and in compliance with 10 CFR 51.26 and 10 CFR 51.27, the scoping process was initiated on July 15, 2010, with the publication in the Federal Register (FR) of a Notice of Intent (NOI) to prepare an EIS (NRC, 2010a). The NOI summarized the NRC's plans to prepare an EIS and presented background information on the proposed IIFP facility. The NOI also invited comments on the appropriate scope of issues to be considered and announced NRC's plan to hold a public scoping meeting. The public scoping comment period ended on August 30, 2010.

On July 29, 2010, the NRC staff held a public scoping meeting in Hobbs, New Mexico, to receive oral and written comments from interested parties. The public scoping meeting began with the NRC staff providing a description of the NRC's roles, responsibilities, and mission. A brief overview of the licensing process was followed by a description of the environmental review process and a discussion of how the public can effectively participate. Most of the meeting was reserved for attendees to ask questions and make comments on the scope of the environmental review. Prior to the public scoping meeting, the NRC staff hosted an informal "open house" for those who wished to attend. The open house provided members of the public with an opportunity to speak informally with individual NRC staffers.

Scoping meeting attendees submitted oral and written comments. Additional comments were received during the scoping period via electronic and postal mail. As a result of the scoping process, the following public comments were received:

•   Expressions of general support for the IIFP facility.

•   Opposition to locating the IIFP facility, or any facility that deals with nuclear byproducts, over an aquifer and in an area with a history of earthquakes.

•   Expressions of support for the project, specifically for the jobs that would be created by construction and operation of the facility and the positive economic impact it would have on the region.

•   Support for the project as a way to use depleted uranium that will be generated at the nearby URENCO USA uranium enrichment plant, which would otherwise have to be stored or disposed of as DUF_6 waste.

•   Concern that a disposal path for waste from the IIFP facility to the Andrews County, Texas, nuclear waste disposal facility is an unsafe disposal path.

•   A statement that the EIS should include the aquifer map that has been prepared by Mesa Water Company.

•   A statement that the EIS should address the seismic hazards that have been indicated for Lea County by the U.S. Geological Survey.

Appendix A (Scoping Summary) of this EIS includes the scoping summary report that summarizes the comments received during the scoping process as required in 10 CFR 51.29(b).

The NRC staff has requested information regarding the scope of its environmental review from the New Mexico State Historic Preservation Officer (NM SHPO) and Native American Tribes identified by the NM SHPO. The NRC staff has also asked for comment from the U.S. Fish and Wildlife Service (USFWS) and the New Mexico Department of Game and Fish (NMGF) regarding threatened and endangered species. The NRC staff also sought information from the New Mexico Energy, Minerals, and Natural Resources Department. The NRC staff has not identified any cooperating agencies for the preparation of this EIS.

Information received from these agencies and potentially affected Native American Tribes was important in assessing impacts to cultural and ecological resources and determining if there were environmental justice concerns. Correspondence with the NM SHPO and potentially affected Native American Tribes (Apache Tribe of Oklahoma, Comanche Tribe of Oklahoma, Kiowa Tribe of Oklahoma, Mescalero Apache Tribe, Ysleta del Sur Pueblo, and Shawnee Tribe) is included in Appendix B (Consultation/Correspondence) of this EIS. Correspondence with the USFWS and the NMGF is also included in Appendix B of this EIS.

1.4.2   Issues Studied in Detail

In the (July 15, 2010) NOI, the NRC staff tentatively identified issues to be studied in detail as they relate to implementation of the proposed action. These issues were:

- Land Use: plans, policies, and controls.
- Historic and Cultural Resources: archaeological sites (historic and prehistoric archaeological artifacts/features and information), architectural historic resources (structures and districts), and historic properties of traditional religious significance to the Native American Tribes.
- Visual Resources: the visual setting on and near the proposed site.
- Transportation: transportation modes, routes, quantities, and risk estimates.
- Geology and Soils: physical geography, topography, geology, and soil characteristics.
- Water Resources: surface water and groundwater hydrology, water use and quality, and the potential for degradation.
- Ecology: wetlands; aquatic, and terrestrial economically or recreationally important species; and threatened and endangered species.
- Air Quality: meteorological conditions, ambient air quality, pollutant sources, and the potential for degradation.
- Socioeconomics: demography, economic base, labor pool, housing, transportation, utilities, public services and facilities, and education.
- Environmental Justice: potential disproportionately high and adverse impacts to minority or low-income populations.
- Noise: noise receptors and potential noise impacts in the vicinity of the proposed facility.

- Public and Occupational Health: potential public and occupational consequences from construction, routine operation, transportation, and credible accident scenarios (including natural events).

- Waste Management: types of wastes expected to be generated, handled, and stored.

- Cumulative Effects: impacts from past, present and reasonably foreseeable future actions at and near the site.

After completion of the scoping process, the NRC staff determined that these issues are still appropriate for detailed study in the EIS, for the following reasons: (1) the fact that the resources identified for study are present and have the potential to be impacted by the action and (2) the fact that participants in the scoping process raised many of the same issues, including perceived beneficial impacts, that were identified in the NOI. Therefore, the initial issues identified in the July 5, 2010, NOI for consideration were carried forward for further analysis. In addition, the NRC staff identified no new issues that require detailed study in the EIS.

1.4.3   Issues Beyond the Scope of the EIS

The purpose of an EIS is to assess the potential environmental impacts of a proposed federal action in order to assist in an agency's decision-making process. In this case, the NRC's decision is whether to grant the license. Some issues and concerns raised during the scoping process are not relevant to the EIS because they are not directly related to the environmental impact analysis or to the NRC's decision. The lack of an in-depth discussion in the EIS, however, does not mean that an issue or concern lacks value. Issues beyond the scope of the EIS either may not yet be at the point where they can be resolved, or are more appropriately discussed and decided in other venues. Appendix A includes a discussion of issues identified during scoping that are beyond the scope of the EIS.

Some of the issues raised during the public scoping process for the proposed facility are outside the scope of the EIS, but are analyzed in the SER. For example, health and safety issues are considered in detail in the SER prepared by the NRC staff for the proposed action and are summarized in the EIS. The EIS and the SER may cover some of the same topics and may contain similar information, but the analysis in the EIS is focused on the assessment of potential environmental impacts. In contrast, the SER deals primarily with safety evaluations and procedural requirements or license conditions to ensure the health and safety of workers and the general public.

1.5   Applicable Statutory and Regulatory Requirements

This section summarizes compliance with legal/regulatory requirements, including permits, licenses, and other authorizations, and approvals at the Federal, State, and local level, which would be necessary for the proposed IIFP facility's construction, operation, and decommissioning, should NRC grant the license.

1.5.1   State of New Mexico Laws and Regulations

Certain Federal environmental requirements, including some discussed earlier, have been delegated by the Federal agencies to State authorities for implementation, enforcement, or oversight. In addition, the State of New Mexico has its own state laws, and Lea County has its own local laws. Table 1-1 provides a list of applicable New Mexico laws, regulations, and agreements, whereas Table 1-2, includes anticipated requirements of those agency laws, regulations, and policies where federal agencies have delegated authority to the state, and those laws, regulations and policies administered under autonomous state legal authority. New Mexico Statutes Annotated (NMSA) and implementing regulations in New Mexico Administrative Code (NMAC) are listed numerically by citation (primarily by NMSA statutory Chapter, Article, and Section; or secondarily by NMAC regulation Title, Chapter, and Part).

Table 1-1.   Applicable State of New Mexico Laws, Regulations, and Agreements

Law, Regulation, or Agreement	Citation	Requirements
New Mexico Wildlife Conservation Act	NMSA, Chapter 17, Game and Fish and Outdoor Recreation, Article 2, Hunting and Fishing Regulations, and Part 3, Wildlife Conservation Act	Requires a permit and coordination if a project may disturb habitat or otherwise affect threatened or endangered species. There are no known, or anticipated (other than transient), threatened or endangered species on the proposed site.
New Mexico Raptor Protection Act	NMSA, Chapter 17, Game and Fish and Outdoor Recreation, Article 2 Part 14, Hawks, Vultures, and Owls; taking, possessing, trapping, destroying, maiming, or selling prohibited except by permit; penalties	The act makes it unlawful to take, attempt to take, possess, trap, ensnare, injure, maim, or destroy individuals of any species of hawk, owl, or vulture.
New Mexico Cultural Properties Act	NMSA, Chapter 18, Libraries and Museums, Article 6, Cultural Properties	The act defines the NM SHPO role and responsibilities, and establishes requirements to prepare an archaeological and historic survey and consult with NM SHPO. A cultural resources inventory was completed for the project. The survey for cultural resources consisted of a file search, field inventory, and inventory report. A negative declaration was prepared by the applicant and the NM SHPO concurred. NRC staff has not yet completed its consultation with the NM SHPO.
New Mexico Occupational Safety and Health	NMSA, Chapter 50, Employment Law NMAC Title 11, Labor Workers Compensation, Chapter 5, Occupational Safety and Health	The act and implementing regulations establish State requirements for assuring safe and healthful working conditions for every employee. These State regulations are being followed to ensure any additional requirements beyond the Federal Occupational Safety and Health Administration (OSHA) regulations are adequately addressed.

Table 1-1. Applicable State of New Mexico Laws, Regulations, and Agreements (Continued)

Law, Regulation, or Agreement	Citation	Requirements
New Mexico Air Quality Control Act	NMSA, Chapter 74, Environmental Improvement, Article 2, Air Pollution NMAC Title 20, Environmental Protection, Chapter 2, Air Quality	The act and implementing regulations establish air quality standards and permit requirements that must be met prior to construction or modification of an emissions source. These regulations also define requirements for an operating permit for major producers of air pollutants and impose emission standards for hazardous air pollutants.
New Mexico Radiation Protection Act	NMSA, Chapter 74, Environmental Improvement, Article 3, Radiation Control NMAC, Title 20, Environmental Protection, Chapter 3, Radiation Protection	The act and implementing regulations establish State requirements for worker protection from radiation sources. Because the facilities would be privately owned, the State will require registration of security X-ray machines.
New Mexico Hazardous Waste Act (see note below)	NMSA, Chapter 74, Environmental Improvement, Article 4, Hazardous Waste NMAC Title 20, Environmental Protection, Chapter 4, Hazardous Waste	The act and implementing regulations establish State standards for the management of hazardous wastes. The New Mexico Environmental Development (NMED) regulations imposed on a generator or on a treatment, storage, or disposal (TSD) facility, vary according to the type and quality of material or waste generated, treated, stored, or disposed. The method of treatment, storage, or disposal also impacts the extent and complexity of the requirements. The IIFP plant may generate hazardous waste during construction and operation. These hazardous wastes will be temporarily stored and shipped off site for treatment and disposal in accordance with applicable NMAC and Resource Conservation and Recovery Act (RCRA) requirements.

Note: Source, special nuclear, or by-product, material as defined by the *Atomic Energy Act of 1954* is specifically excluded from the definition of a solid waste and therefore is not a hazardous waste regulated under RCRA or NMSA, Chapter 74, Article 9, Solid Waste Act and implementing regulations at NMAC Title 20, Chapter 9, Solid Waste. The IIFP facilities would not store (other than temporarily) or dispose of hazardous waste on site. IIFP may need a permit for operation of its Environmental Protection Process under the authority of RCRA or the New Mexico Hazardous Waste Act.

Table 1-1.   Applicable State of New Mexico Laws, Regulations, and Agreements (Continued)

Law, Regulation, or Agreement	Citation	Requirements
New Mexico Radioactive and Hazardous Materials Act	NMSA, Chapter 74 Environmental Improvement,, Article 4, Article 4A, Radioactive and Hazardous Materials	The act establishes a system of assuring public health and safety with regard to safe treatment, disposal, and transportation of radioactive and hazardous materials and coordinates efficient and timely emergency response to accidents and natural disasters with a centralized and coordinated source of information.
New Mexico Hazardous Chemicals Information Act	NMSA, Chapter 74 Environmental Improvement,, Article 4E-1, Hazardous Chemicals Information Act	The act implements the hazardous chemicals information and toxic release reporting requirements of the *Emergency Planning and Community Right-to-Know Act of 1986 (Superfund Amendment and Reauthorization Act [SARA] Title III)* for facilities such as the proposed IIFP plant.
New Mexico Water Quality Act	NMSA, Chapter 74, Environmental Improvement, Article 6, Water Quality NMAC Title 20, Environmental Protection, Chapter 6, Ground and Surface Water Protection	The act and implementing regulations establish water quality standards and apply to permitting prior to construction, during operation, and decommissioning, if necessary. Generally, a permit is required for discharges that could affect surface or groundwater. Any impoundments for sewage treatment facilities, cooling water or other discharges that exceed the standards listed in 20.6.2.3103 NMAC or contain toxic constituents require a permit. No site-specific issues have been identified which would preclude permitting of needed water control and treatment facilities at the IIFP Site.
New Mexico Groundwater Protection Act	NMSA, Chapter 74,Environmental Improvement, Article 6B, Groundwater Protection NMAC Title 20, Environmental Protection, Chapter 5, Petroleum Storage Tanks	The act and implementing regulations establish State standards for protection of groundwater from leaking underground and above-ground storage tanks.

Table 1-1. Applicable State of New Mexico Laws, Regulations, and Agreements (Continued)

Law, Regulation, or Agreement	Citation	Requirements
New Mexico Night Sky Protection Act	NMSA Chapter 74, Environmental Improvement, Article 12, Night Sky Protection	The act establishes requirements to preserve and enhance the State's dark sky while promoting safety, conserving energy and preserving the environment for astronomy. These requirements will be addressed during detailed design of the IIFP facility.
Exchanges of State Trust Lands	NMAC Title 19, Natural Resources and Wildlife, Chapter 2 State Trust Lands, Part 21, Land Exchanges	The act establishes State standards and procedures for exchanges of lands held in trust, including consideration of cultural resources, natural resources, and wildlife.
New Mexico Endangered Plant Species Act	NMAC Title 19, Natural Resources and Wildlife, Chapter 21, Endangered Plants	The act establishes an endangered plant species list and rules for collection. There are no threatened or endangered plant species on the proposed IIFP site.
Registration of Tanks	NMAC, Title 20, Environmental Protection, Chapter 5, Petroleum Storage Tanks, Part 2, Registration of Tanks	The regulations establish the State standards for the regulation of petroleum storage tanks. If needed at the IIFP facility, storage tanks would be designed in accordance with State requirements and registration application made.
Drinking Water Regulations	NMSA, Chapter 74, Environmental Improvement, Article 1, General Provisions, Sections 1-8 and 1-13.1, and Article 6 Water Quality NMAC Title 20, Environmental Protection, Chapter 7, Wastewater and Water Supply Facilities, Part 10 Drinking Water	The acts require the establishment of drinking water standards for New Mexico. These standards are found at 20.7.10 NMAC. The proposed facility would use an on-site groundwater supply for all domestic water needs. Under the New Mexico drinking water regulations, the facility would be classified as a non-transient, non-community water supply system because it would regularly serve more than 25 people.
Transportation and Highway	NMAC Title 18, Transportation and Highways, Chapter 31, Classification and Design Standards, Part 6, State Highway Access Management Requirements	The regulations establish State highway access management requirements that will protect the functional integrity of and investment in, the State highway system.
Threatened and Endangered Species of New Mexico	NMAC. Title 19, Natural Resources and Wildlife, Chapter 33, Endangered and Protected Species	The regulations establish the State of New Mexico's list of threatened and endangered wildlife species. There are no threatened or endangered species on the proposed plant site

Source: IIFP, 2009a

1.5.2   Lea County and Local Laws and Regulations

Lea County requires county permits for most major construction activity, but these permits are issued in accordance with subdivision ordinances at the time when parcel subdivision is approved; thus most other parcels where subdivision is not requested are not restricted by local subdivision ordinances and do not require county permits for construction activity. In other words, building permits for foundations, structures, electrical/mechanical systems, roadways, or temporary construction-related structures are not required by local ordinance, except where subdivision regulations apply. Because subdivision is not necessary for the IIFP facility, Article 8 of Lea County's subdivision regulations (or other local regulations) do not apply.

1.5.3   Permit and Approval Status

IIFP would prepare and submit several construction and operating permit applications, and regulatory approval and/or permits would be received prior to preconstruction, construction, or facility operation. It is IIFP's responsibility to adhere to necessary permit application schedules and permit requirements prior to preconstruction, construction, or operation, as applicable. Tables 1-2 and 1-3 list the required Federal and State construction and operation permits and their status.

1.5.3.1   Permits, Licenses, Authorizations, Approvals, and Consultations Required for Preconstruction and Construction

Table 1-2 identifies the anticipated legal/regulatory requirements for site preparation and construction of the proposed IIFP facility. These include any permits, licenses, authorizations, approvals, or other regulatory entitlements required for constructing the proposed facility. Table 1-2 also identifies the status of these possible requirements.

1.5.3.2   Permits, Authorizations, Approvals and Consultations Required for Operations

Table 1-3 identifies the anticipated legal/regulatory requirements for operation of the proposed IIFP facility. Table 1-3 also identifies the status of these possible requirements.

1.6   Cooperating Agencies

No Federal, State, or local agencies or Native American Tribes have requested to be considered as cooperating agencies in the preparation of this EIS.

1.7   National Historic Preservation Act of 1966 and the Endangered Species Act of 1973 Consultations

The consultation requirements of the *National Historic Preservation Act* (NHPA) and *Endangered Species Act* (ESA) apply to the NRC with regard to the proposed IIFP facility licensing action. Consultation correspondence is provided in Appendix B (Consultation/Correspondence).

Table 1-2. Legal/Regulatory Requirements and Authorizations for Site Preparation and Construction of the Proposed IIFP Facility

Agency	Legal/Regulatory Authority	Activity Covered	Status	Permits, Licenses, Authorizations, Approvals, and Consultations Identification	Permit Dates
Federal – NRC	10 CFR 40	Domestic licensing of source material	IIFP's application accepted February 24, 2010; pending NRC review ongoing	License Application (LA-IFP-001, IFP-002, & ER-IFP-001)	Pending
Federal – EPA Region 6 and State – NMED	CWA 33 USC 1251	Construction Stormwater General Permit and NOI	3rd Qtr. 2012; Stormwater Pollution Prevention Plan and NOI (IIFP)	Pending	Pending
Federal – USACE	CWA 33 USC 1251	Fill of wetlands	Applicability of depressions onsite determined by consultation (IIFP)	Jurisdictional Determination complete (no wetlands present; no permit necessary)	No permit needed; no jurisdictional wetlands or waters of the U.S. are present
State – NMDOT District 2	NMAC Title 18, Chapter 31, Part 6, State Highway Access Management Requirements	Highway Access Management Permit for NM 483 and/or US 62/180 driveway entry and/or second (emergency) access point	3rd Qtr. 2012 (IIFP or Lea County). The District 2 permit, if issued, would stipulate any required highway safety enhancements.	Pending	Pending

Table 1-2. Legal/Regulatory Requirements and Authorizations for Site Preparation and Construction of the Proposed IIFP Facility (Continued)

Agency	Authority	Activity Covered	Status	Permits, Licenses, Authorizations, Approvals, and Consultations Identification	Permit Dates
State – NMED/AQB	Title 20 Chapter 2 Part 72, Clean Air Act Title V and NESHAP	Preconstruction and New Source Review, NESHAP Permit (if needed; pending consultation with state AQB)	Submitted 1st Qtr. 2012 (IIFP). A "Request for a No Permit Required Determination" for pre-licensing construction activities would be submitted.	Pending	Pending
State – NMRL/CID	NMAC Title 14	Building permits for foundations, structures, and electrical/mechanical systems, as well as temporary construction-related structures	Projected for 3rd Qtr. 2012 (IIFP)	Pending	Pending
State – NMED/PSTB	Above Ground Storage Tank Registration	Petroleum storage tanks (size, design specifications, and fuel type)	Projected for 2nd Qtr. 2013 (IIFP)	Pending	Pending

(IIFP, 2012) International Isotopes Fluorine Products, Inc. "Response to NC EIS Follow Up Questions on EIS Table 1-9." E-mail communication to J. Miller and J. Thomas (IIFP) from M. Bartlett (NRC). Table 1-9, Federal and State Permits and Requirements. ADAMS Accession No.ML12089A470..
AQB—Air Quality Bureau, CWA—Clean Water Act, EPA—U.S. Environmental Protection Agency, ER—environmental report, IIFP—International Isotopes Fluorine Products, Inc., LA—license application, NESHAP—National Emissions Standards for Hazardous Air Pollutants, NMAC—New Mexico Administrative Code, NMDOT—New Mexico Department of Transportation, NMED/AQB—New Mexico Environment Department/Air Quality Bureau, NOI—Notice of Intent, NRC—U.S. Nuclear Regulatory Commission, USACE—U.S. Army Corps of Engineers, NMED/PSTB—Petroleum Storage Tank Bureau, NMRL/CID—New Mexico Regulations and Licensing/Construction Industries Division.

Table 1-3.   Legal/Regulatory Requirements and Authorizations for Operation of the Proposed IIFP Facility

Agency	Authority	Activity Covered	Status	Permits, Licenses, Authorizations, Approvals, and Consultations Identification	Permit Dates
Federal – NRC	10 CFR 40	Domestic licensing of source material	IIFP's application Accepted February 24, 2010 (IIFP); pending review	License Application (LA-IIFP-001, LA-IIFP-002, & ER-IIFP-001)	Pending
Federal – NRC	10 CFR 73 and 74	Domestic licensing of physical protection and material control/accountability	IIFP's application accepted February 24, 2010; pending NRC review	License Application (LA-IIFP-001, IIFP-002, & ER-IIFP-001)	Pending
Federal – NRC	10 CFR 20	Standards for Protection Against Radiation	IIFP's application accepted February 24, 2010; pending NRC review	License Application (LA-IIFP-001, IIFP-002, & ER-IIFP-001)	Pending
Federal – EPA Region 6 and NMED	CWA 33 USC 1251	Multi-sector Industrial Stormwater General Permit	Projected for 2nd Qtr. 2013 (IIFP)	Pending Multi-sector Industrial General Permit (IIFP)	Pending
Federal – EPA Region 6 and state— NMED/HWB)	RCRA Operations Permit	EPA Operation	If necessary, IIFP would submit 2nd Qtr. 2013 (IIFP)	Pending	Pending
Federal – EPA Region 6 and NMED/HWB	RCRA	EPA Waste Activity, EPA ID Number, for storage and use of hazardous chemicals	2nd Qtr. 2013 (IIFP)	Small Quantity Generator pending	Pending

Table 1-3. Legal/Regulatory Requirements and Authorizations for Operation of the Proposed IIFP Facility (Continued)

Agency	Authority	Activity Covered	Status	Permits, Licenses, Authorizations, Approvals, and Consultations Identification	Permit Dates
State—NMED/AQB	Clean Air Act Title V and NESHAP	Air Operation Permit (if needed: pending consultation with State)	Submitted 1st Qtr. 2012 (IIFP); a separate NESHAP operating permit is required.	Pending	Pending
State—NMED/HWB	Hazardous Chemicals Information Act	Hazardous Waste Permit	Projected for 3rd Qtr. 2013 (IIFP)	Pending	Pending
State – NMED/HWB	Title 20, Chapter 4, Hazardous Waste	EPA Waste Activity EPA ID Number	Notification Form 8700-12 projected for 2nd Qtr. 2013 (IIFP)	Pending	Pending
State – NMED/RCB	NMSA, Chapter 74, Article 3, Radiation Control and Title 20, Chapter 3	Machine-Produced Radioactivity (X-Ray Inspection)	May be required for security contractor but not required for IIFP	Not required	Not required
State – NMED/DWB	Safe Drinking Water Act (40 CFR 141, 142, 143; 07/01/2009) and 20-7.10 NMAC (10/15/2008)	Drinking Water System Permit	Projected for 2nd Qtr 2013 (IIFP)	Pending	Pending

Table 1-3.   Legal/Regulatory Requirements and Authorizations for Operation of the Proposed IIFP Facility (Continued)

Agency	Authority	Activity Covered	Status	Permits, Licenses, Authorizations, Approvals, and Consultations Identification	Permit Dates
State—NMED/WQB	Title 20 Chapter 6, Groundwater	Groundwater Discharge Plan and Permit; Liquid Waste Permit	Projected for 3rd Qtr. 2013 (IIFP)	Pending	Pending
State—NM WQCC	NPDES	NPDES Industrial Stormwater Permit	Not required	not required	Not required
State—RCB	OSHA Regulations, 29 CFR Part 1910; Federal FD&C Act; New Mexico Radiation Control Act	Required for x-ray security inspection machines	Identify permit-holder by 1st Qtr. 2011 (IIFP); submit list of source equipment 2nd Qtr. 2012 (IIFP)	Machine-Produced Radiation Registration (X-Ray inspection)	Security contractor may be required to obtain this permit under a separate application
State—Utah	Low-Level Radioactive Waste Policy Act (Northwest Interstate Compact, Utah Code Title 19 Chapter 3)	LLW disposal at Nuclear Solutions (if chosen)	Arrangement is needed for LLW disposal at Energy Solutions (if chosen)		
State – Washington	Low-Level Radioactive Waste Policy Act (Northwest Interstate Compact, Compact for the State of Washington)	LLW disposal at Hanford (Richland) (if chosen)	Arrangement is needed for LLW disposal at Waste Control Specialist (if chosen)		

Table 1-3.   Legal/Regulatory Requirements and Authorizations for Operation of the Proposed IIFP Facility (Continued)

Agency	Authority	Activity Covered	Status	Permits, Licenses, Authorizations, and Consultations Identification	Permit Dates
State—Texas	Low-Level Radioactive Waste Policy Act and Texas Radiation Control Act (Texas LLW Interstate Compact with Vermont)	LLW disposal at Waste Control Specialists (if chosen)	Arrangement is needed for LLW disposal at U.S. Ecology Hanford (Richland) (if chosen). IIFP will not likely select this site but it was included as part of the impact assessment as an option. Would require approval from Rocky Mountain Compact		

(IIFP, 2012) International Isotopes Fluorine Products, Inc. "Response to NC EIS Follow Up Questions on EIS Table 1-9." E-mail communication to J. Miller and J. Thomas (IIFP) from M. Bartlett (NRC). Table 1-9, Federal and State Permits and Requirements. ADAMS Accession No. ML12089A470.CWA—Clean Water Act EPA—U.S. Environmental Protection Agency; ER—environmental report; FD&C—Federal Food, Drug, and Cosmetic Act; ID—identification number; IIFP—International Isotopes Fluorine Products, Inc.; LLW—low-level radioactive waste; NESHAP—National Emissions Standards for Hazardous Air Pollutants; NMAC—New Mexico Administrative Code; NMED/AQB—New Mexico Environment Department /Air Quality Bureau; NMED/HWB—New Mexico Environment Department/Hazardous Waste Bureau; NMED/RCB—New Mexico Environment Department/Radiological Control Bureau; NMED/GWQB—New Mexico Environment Department/Ground Water Quality Bureau; NMED/DWB—New Mexico Environment Department/Drinking Water Bureau; NMSA—New Mexico Statutes Annotated; NM WQCC—New Mexico Water Quality Control Commission; NPDES—National Pollutant Discharge Elimination System; NRC—U.S. Nuclear Regulatory Commission; OSHA—Occupational Safety and Health Administration; RCRA—Resource Conservation and Recovery Act.

1.7.1 National Historic Preservation Act of 1966 Section 106 Consultation

NRC staff initiated the NHPA Section 106 consultation process by letter dated July 2, 2010. NRC staff contacted the NM SHPO regarding information about historic sites and cultural resources that could potentially be affected by the proposed IIFP facility. In the letter, the NRC staff identified the Area of Potential Effect (APE) for the proposed project and requested information from the NM SHPO related to the proposed action's potential to affect cultural resources. Also in the letter, the NRC staff stated its intent to use the NEPA process to comply with Section 106 of the NHPA as allowed in 36 CFR 800.8. The NM SHPO replied on July 15, 2010, that the SHPO had no record of any cultural resources surveys having been conducted and outlined the process for completing a survey, undertaking tribal consultation, and completing the Section 106 consultation process. IIFP conducted an archeological reconnaissance survey of the proposed site (as explained in later chapters of this document) according to New Mexico's *Cultural Properties and Historic Preservation, Standards for Survey and Inventory* (NMHPD, 2005). By letter dated October 14, 2010, the New Mexico Commissioner of Public Lands, following his review of IIFP's cultural resources survey document, recommended "a finding of no effect/no cultural properties/no historic properties....There are no documented cultural properties within the APE when considering direct effects. Similarly, there are no registered cultural properties within the assumed, five-mile APE when considering indirect effects." The NM SHPO concurred with the New Mexico Commissioner of Public Lands determination on October 25, 2010.

Consultation under NHPA with Native American Tribes (listed below) was undertaken using a list maintained by the NM SHPO. The list is based partially on U.S. Indian Claims Commission data and also on an NM SHPO Historic Preservation Division (HPD) ethnographic study, the National Park Service's Native American Consultation Database, and Tribes that have notified NM SHPO directly that they wish to be consulted. Based on tribal information provided for Lea County, in July 2010, the NRC staff contacted the Tribes listed below and requested information on historically or culturally significant resources within the APE of the proposed facility. The NRC staff also contacted the NM SHPO tribal liaison (Appendix B). Correspondence between NRC staff and the responding tribes is provided in Appendix B.

- Apache Tribe of Oklahoma
- Comanche Tribe of Oklahoma
- Kiowa Tribe of Oklahoma
- Mescalero Apache Tribe
- Ysleta del Sur Pueblo
- Shawnee Tribe

NRC staff will consider comments received from tribes concerning this EIS. Otherwise, the coordination that has been conducted in accordance with the NHPA is complete.

1.7.2 Endangered Species Act of 1973 Section 7 Consultation

The NRC staff consulted with the USFWS to comply with the requirements of Section 7 of the *Endangered Species Act (ESA)*. On July 2, 2010, the NRC staff sent a letter to the USFWS (New Mexico Ecological Field Office) describing the proposed action and requesting a list of threatened and endangered species and critical habitats that could potentially be affected by the proposed action. The USFWS, in a letter dated August 10, 2010, provided general information about species of concern and critical habitat in New Mexico and Lea County, but made no site-specific comments. In response to a verbal inquiry from the NRC staff, the NMGF responded in a letter dated June 21, 2011, with further information about wildlife habitat on the proposed IIFP site, recommendations for avoiding impacts to wildlife, and other best management practices (Appendix B). No federally threatened or endangered species or critical habitat have been identified on the proposed IIFP site to date; therefore formal Section 7 ESA consultation is not required for the NRC action (licensing) to occur.

1.8   References

(DOE, 2004) U.S. Department of Energy. 2004. Final Environmental Impact Statement for Construction and Operation of a Depleted Uranium Hexafluoride Conversion Facility at the Paducah, Kentucky, Site, DOE/EIS-0359. June 2004. ADAMS Accession No. ML 050380331.

(IIFP, 2009a) International Isotopes Fluorine Products, Inc. 2009. Fluorine Extraction Process and Depleted Uranium De-conversion Plant (FEP/DUP) Environmental Report, Revision A, ER-IFP-001. December 27, 2009. ADAMS Accession No. ML100120758.

(IIFP, 2009b) International Isotopes Fluorine Products. Inc. 2009. Fluorine Extraction Process and Depleted Uranium De-conversion Plant (FEP/DUP) License Application, Revision A, ER-IFP-001. December 23, 2009. ADAMS Accession No. ML100630503.

(NMHPD, 2005) New Mexico Historic Preservation Division, Department of Cultural Affairs. 2005. New Mexico Register, Volume XVI, Number 15, Title 4, Cultural Properties and Historic Preservation; Chapter 10, Cultural Properties and Historic Preservation; Part 15, Standards for Survey and Inventory. Santa Fe, NM. August 15, 2005. ADAMS Accession No. ML112710497.

(NRC, 2005) U.S. Nuclear Regulatory Commission. 2005. Environmental Impact Statement for the Proposed National Enrichment Facility in Lea County, New Mexico, NUREG-1790, Vol. 1. June 2005. ADAMS Accession No. ML051730238.

(NRC, 2010a) U.S. Nuclear Regulatory Commission. 2010. Notice of Intent to Prepare an Environmental Impact Statement for the Proposed International Isotopes Uranium Processing Facility. July 8, 2010. ADAMS Accession No. ML101330539.

(NRC, 2010b) U.S. Nuclear Regulatory Commission. 2010. Letter from Matt Bartlett (NRC) to John J. Miller (IIFP), "License Application for International Isotopes Fluorine Products, Inc. Facility – Acceptance Review (TAC No's L32739 and L32740), February 23, 2010. ADAMS Accession No. ML100480302.

2.0   ALTERNATIVES

This chapter describes and compares the proposed action and its alternatives. As discussed in Section 2.1, the proposed action is for IIFP to construct, operate, and decommission a DUF_6 deconversion facility near Hobbs in Lea County, New Mexico. In this EIS the NRC staff evaluates a reasonable range of alternatives to the proposed action, including alternative sites for the IIFP facility, alternative deconversion technologies, other DUF_6 management options, and the no-action alternative. Under the no-action alternative, IIFP would not construct, operate, or decommission the proposed facility. Therefore, the no-action alternative provides a basis against which the potential environmental impacts of the proposed action are evaluated and compared.

Section 2.1 presents detailed technical descriptions of the proposed action and related actions, including descriptions of the proposed site, preconstruction and construction activities, chemical process operations within the proposed plant, and decommissioning. Disposition of DUO_2 is also discussed in Section 2.1. Section 2.2 describes alternatives to the proposed action, including the no-action alternative. The chapter concludes with a comparison of predicted potential environmental impacts of the proposed action and no-action alternative (Section 2.3) and a recommendation from the NRC staff regarding the proposed action (Section 2.4).

2.1   Proposed Action

The proposed action evaluated in this EIS is for NRC to grant IIFP a license to construct, operate, and decommission a facility (the proposed IIFP facility) in Lea County, New Mexico, for the deconversion of commercially generated DUF_6 inventories into DUO_2 and other deconversion products. The NRC would grant IIFP a license under 10 CFR 40 (Domestic Licensing of Source Material) to possess and use special nuclear material, source material, and byproduct material at the proposed IIFP facility.

If the NRC issues a license to IIFP, the license would authorize IIFP to:

- construct, operate, and decommission the proposed DUF_6 conversion facility.

- receive DUF_6 cylinders from various commercial uranium enrichment facilities.

- transport marketable deconversion byproducts to end users.

- transport DUO_2 for disposal as LLW or other potential disposition.

IIFP anticipates that the proposed project would be implemented in two phases, but the current license application is for the first phase only (Phase 1), and only the potential impacts of the first phase are evaluated in this EIS. Phase 2 would be an expansion of the facility that would use a direct conversion technology described in Section 2.2.2.2.1. Because Phase 2 is a "reasonably foreseeable future action" (as defined in 40 CFR 1508.7), impacts associated with Phase 2 are considered cumulative impacts under NEPA. Cumulative impacts are discussed in Section 4.2.2 of this EIS.

Phase 1 and 2 milestones are shown below. Phase 2 milestones are presented for information only.

IIFP submitted license application to NRC	December 30, 2009
IIFP begins construction (Phase 1)	2Q 2012
IIFP begins Phase 1 operations	4Q 2013
IIFP submits license application for plant expansion (Phase 2)	2Q 2013
IIFP begins construction of plant expansion (Phase 2)	2Q 2015
IIFP begins Phase 2 operations	late 2016

The proposed action is described in detail in Sections 2.1.1 through 2.1.8. Unless otherwise indicated, the information presented in Section 2.1 is from the IIFP's environmental report (IIFP, 2009) and responses to NRC staff requests for additional information (IIFP, 2011a).

2.1.1   Site Location and Description

The proposed IIFP site is 22.5 km (14 mi) west of Hobbs, in Section 27 of Township 18S, Range 36E, in Lea County, New Mexico. Figure 2-1 depicts the general site location in southeast New Mexico. Approximately 16 ha (40 ac) of the 259-ha (640-ac) Section would be dedicated to the deconversion facility. The remaining 243 ha (600 ac) would remain undeveloped. Figure 2-2 locates the 16-ha (40-ac) facility within the Section. The Section now consists of mostly undeveloped land that has been used in the past for cattle grazing and gas and oil production.

2.1.2   IIFP Deconversion Process

At the proposed IIFP facility, the FEP/DUP would employ three basic processes, as described in detail in the sections that follow. In summary, the DUF_6 would first be deconverted from DUF_6 to depleted uranium tetrafluoride (DUF_4), with marketable AHF produced as a byproduct. Then, DUF_4 would be processed to produce two marketable deconversion byproducts: high-purity SiF_4, and BF_3, as needed. Plant throughputs are provided in Figure 2-3. The amount of silicon and boron byproducts produced would likely outpace the demand for these byproducts if all the potentially available DUF_6 were converted using this process. Therefore, Phase 2 of the project would support a process that allows the direct conversion of DUF_6 to uranium oxide, without producing the silicon and boron compounds.

2.1.2.1   Deconversion of DUF_6 to DUF_4

As described in Chapter 1, DUF_6 results from the enrichment of natural uranium during the manufacture of nuclear reactor fuel. It is stored and transported as a solid in cylinders specifically designed for these purposes. DUF_6 is a solid at temperatures below 52°C (125°F). After receipt at the proposed IIFP facility, as the first step in the deconversion process, the cylinders would be placed in an autoclave enclosed in containment to vaporize the contents. The DUF_6 vapor would be captured in a reaction vessel where it would react with hydrogen to produce DUF_4 powder and AHF. The chemical equation for this process is as follows:

$$DUF_6 \text{ (gaseous)} + H_2 \text{ (gas)} \rightarrow DUF_4 \text{ (solid)} + 2HF \text{ (anhydrous)}$$

The DUF_4 powder would be continuously withdrawn from the bottom of the vessel and fed to the FEP for further deconversion in either the silicon separation process or the boron separation

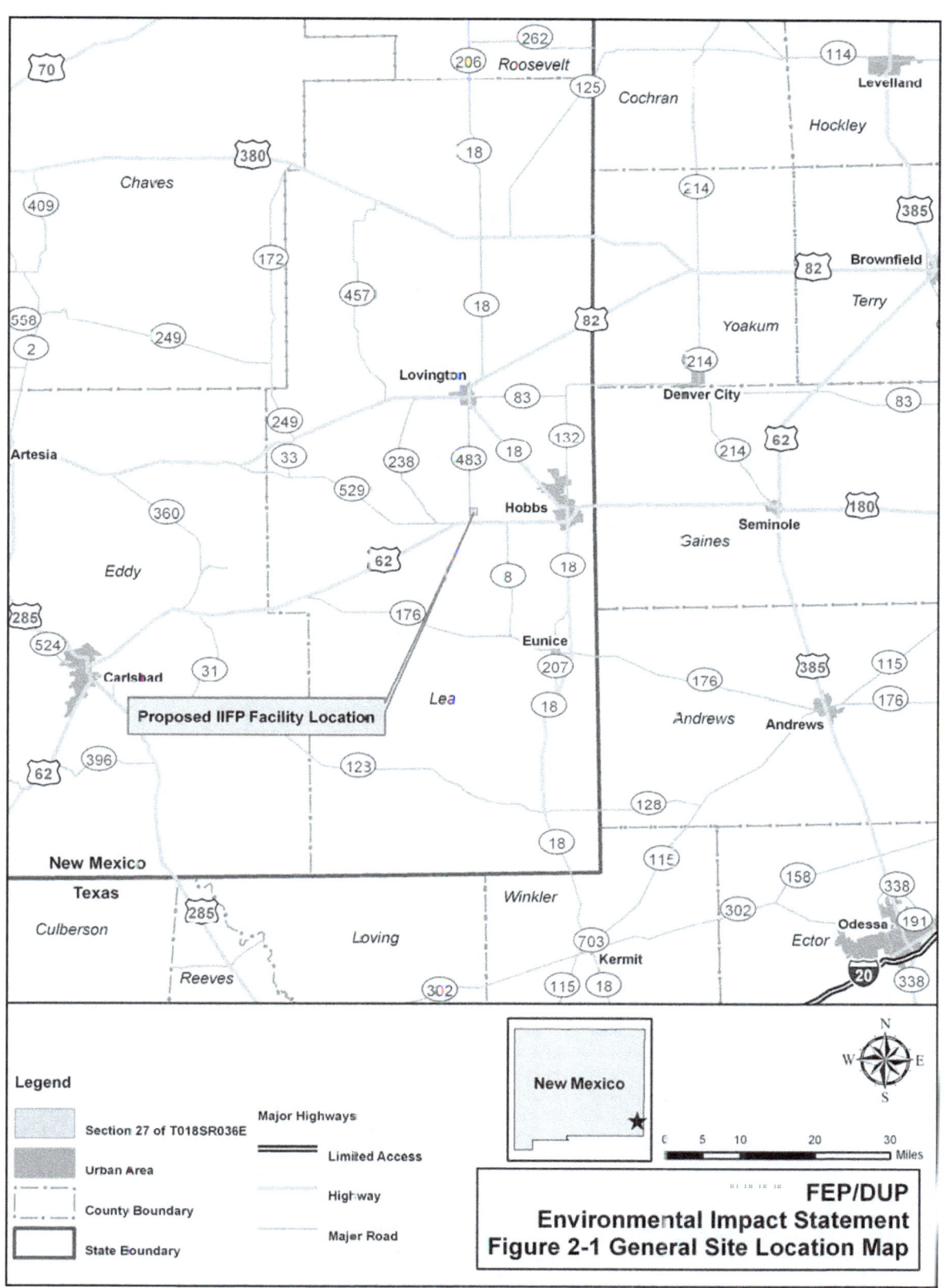

Legend

Section 27 of T018SR036E

Urban Area

County Boundary

State Boundary

Major Highways

Limited Access

Highway

Major Road

Proposed IIFP Facility Location

New Mexico

FEP/DUP
Environmental Impact Statement
Figure 2-1 General Site Location Map

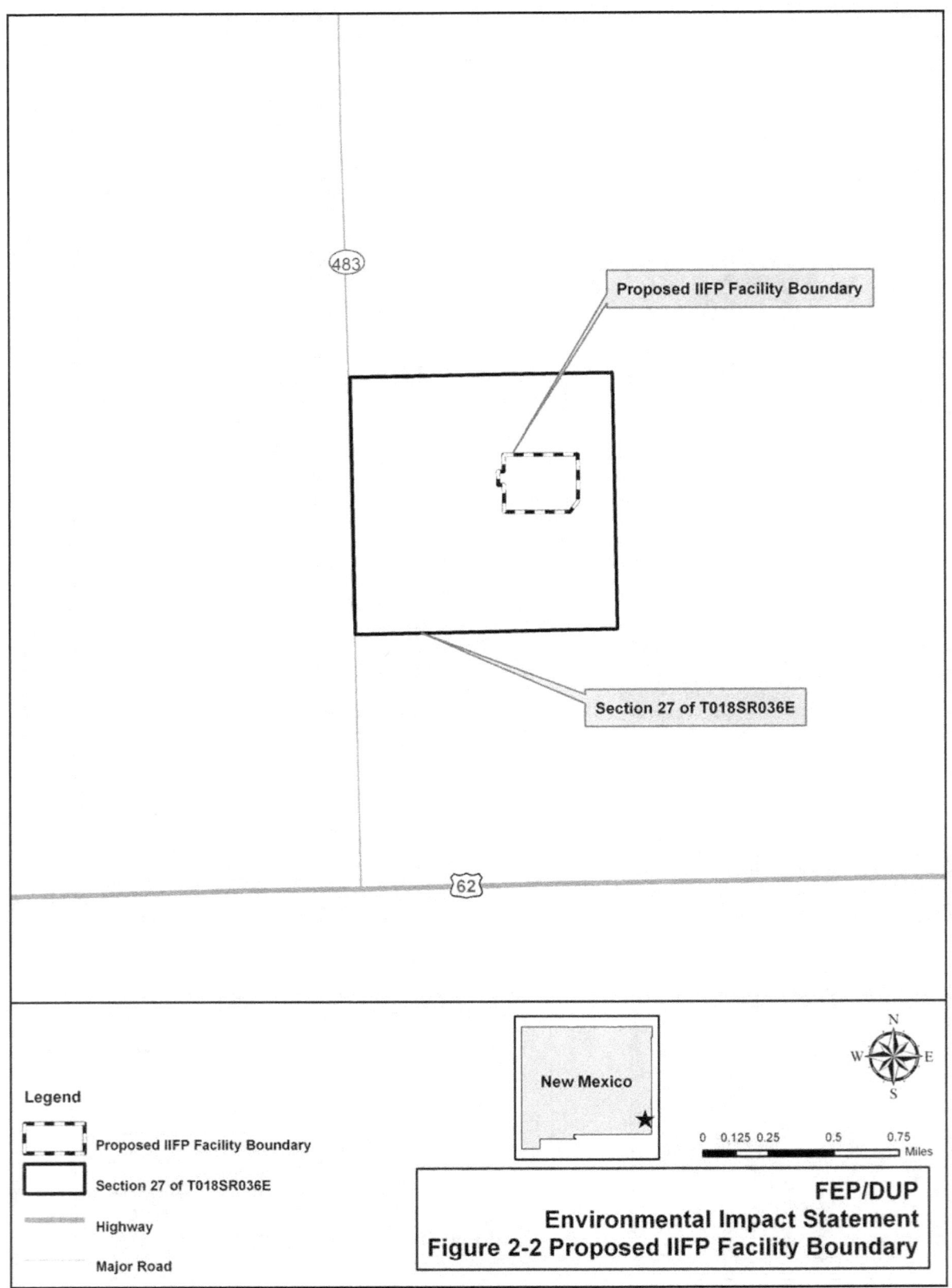

Legend

- ▭ (dashed) Proposed IIFP Facility Boundary
- ▭ (solid) Section 27 of T018SR036E
- ═══ Highway
- ──── Major Road

New Mexico

FEP/DUP
Environmental Impact Statement
Figure 2-2 Proposed IIFP Facility Boundary

Figure 2-3. Plant Throughput for the DUF$_6$ Deconversion Process

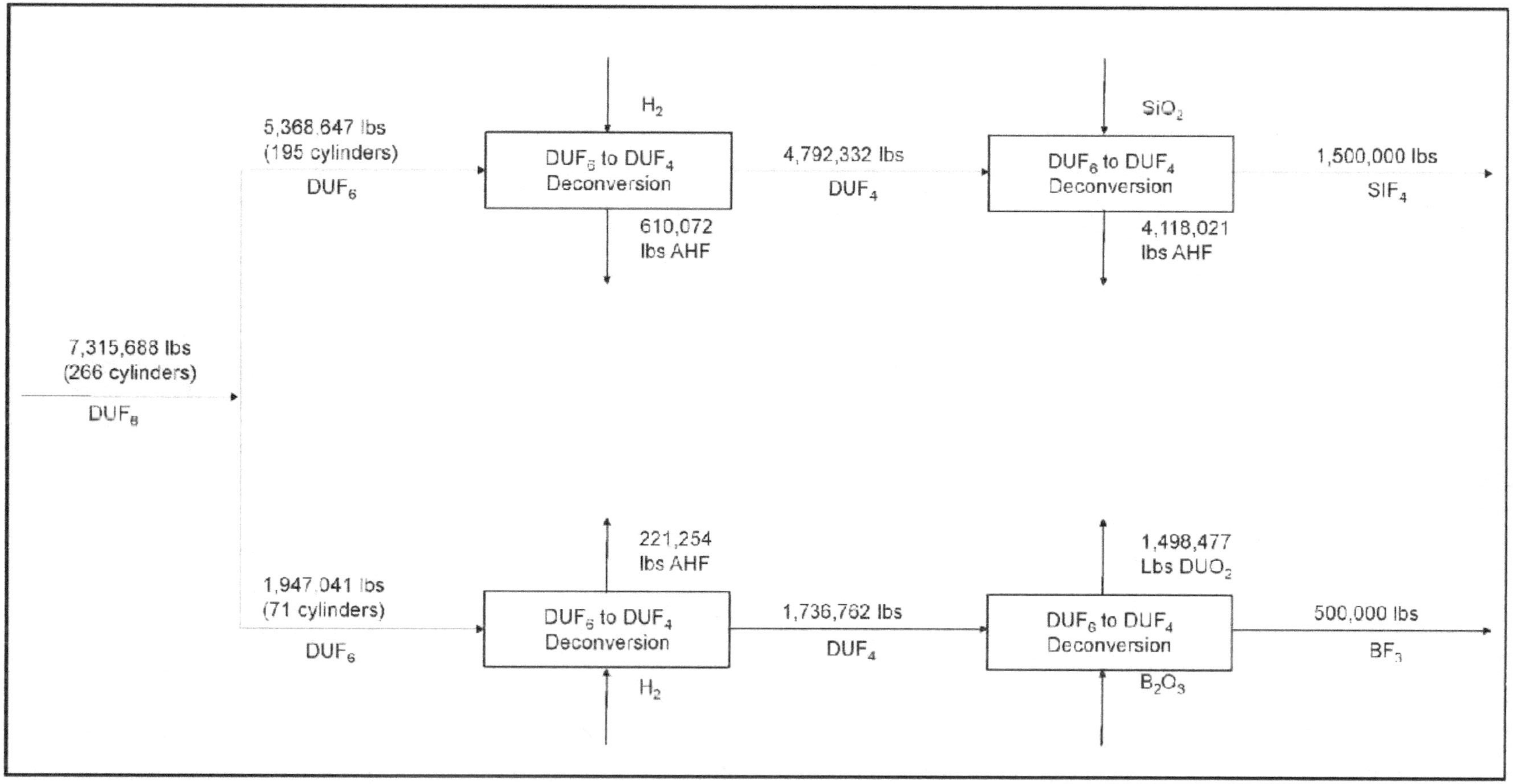

process (Sections 2.1.2.2 and 2.1.2.3, respectively). Also, hydrogen fluoride (HF) can be anhydrous (meaning pure hydrogen fluoride without water) or not. In chemical equations, hydrogen fluoride is depicted as HF, but the parenthetic expression (anhydrous) is added when appropriate. Hydrofluoric acid is another term for hydrogen fluoride combined with water. HF offgases would be filtered, and any residual DUF_6 would be trapped on carbon filters. The AHF would then be condensed to liquid form, and any entrained hydrogen burned. Offgas treatment is described in Section 2.1.6.4.1. AHF would be collected in 3,630-kg (8,000-lb) storage vessels to limit inventory should a leak occur. AHF storage vessels would be located in a building designed to contain a leak. The AHF would be loaded from this building into tanker trucks and shipped to customers. Figure 2-4 shows the process flow chart for this process.

2.1.2.2   SiF₄ Production

To produce SiF_4, the powdered DUF_4 would be mixed with powdered silicon dioxide (SiO_2) in a rotary calciner, and heated to react to form gaseous SiF_4 and solid UO_2 (U_3O_8) triuranium octoxide, sometimes referred to simply as uranium oxide or "yellowcake." The chemical equation for this process is as follows:

$$SiO_2 \text{ (solid)} + DUF_4 \text{ (solid)} \rightarrow SiF_4 \text{ (gas)} + DUO_2 \text{ (solid)}$$

The gaseous SiF_4 would be collected from the calciner, filtered to remove any particulate contamination, and cooled to condense any hydrofluoric acid or other trace gases. The purified, gaseous SiF_4 then would be collected in cold traps. The cold traps would be warmed to vaporize the SiF_4, and the gaseous SiF_4 would be stored in a vessel for subsequent packaging and shipment to customers. Offgas treatment is described in Section 2.1.6.4.1. Figure 2-5 shows the process flow chart for this process.

2.1.2.3   BF₃ Production

The BF_3 production process would be very similar to that for SiF_4, except that there would be a pretreatment step in which a feed mixture of boron oxide (B_2O_3) and DUF_4 would be heated prior to mixing in the rotary calciner (Figure 2-6). The preheating would remove moisture by reacting the water with the DUF_4, releasing gaseous (anhydrous) HF. The gaseous (anhydrous) HF would be filtered and scrubbed in the offgas system. The remainder of the process would be very nearly the same as for SiF_4 production. The chemical equation for this process is as follows:

$$2B_2O_3 \text{ (solid)} + 3DUF_4 \text{ (solid)} \rightarrow 4BF_3 \text{ (gas)} + 3DUO_2 \text{ (solid)}$$

2.1.3   Description of the Proposed Facility

The proposed facility would be typical of specialty industrial chemical facilities. The proposed 16-ha (40-ac) facility would be enclosed with a security fence with a surveillance road just inside the fence. Pole-mounted security lighting would be installed around the entire perimeter. Entry into the proposed facility would be from the west via a paved road accessed from New Mexico Highway (NM) 483 which bounds the proposed site on the west (Figure 2-2). Structures within the security fence would include process, administration, and laboratory buildings; a maintenance shop; security facilities; utilities; cylinder storage pads; and warehouses. The parking lot would be outside the security fence. The tallest building is expected to be approximately 21 meters (m) (70 feet [ft]) high, and the tallest structure is a 40 m (131 ft) meteorological tower.

Figure 2-4. Process Flow Chart for Deconversion of DUF_6 to DUF_4

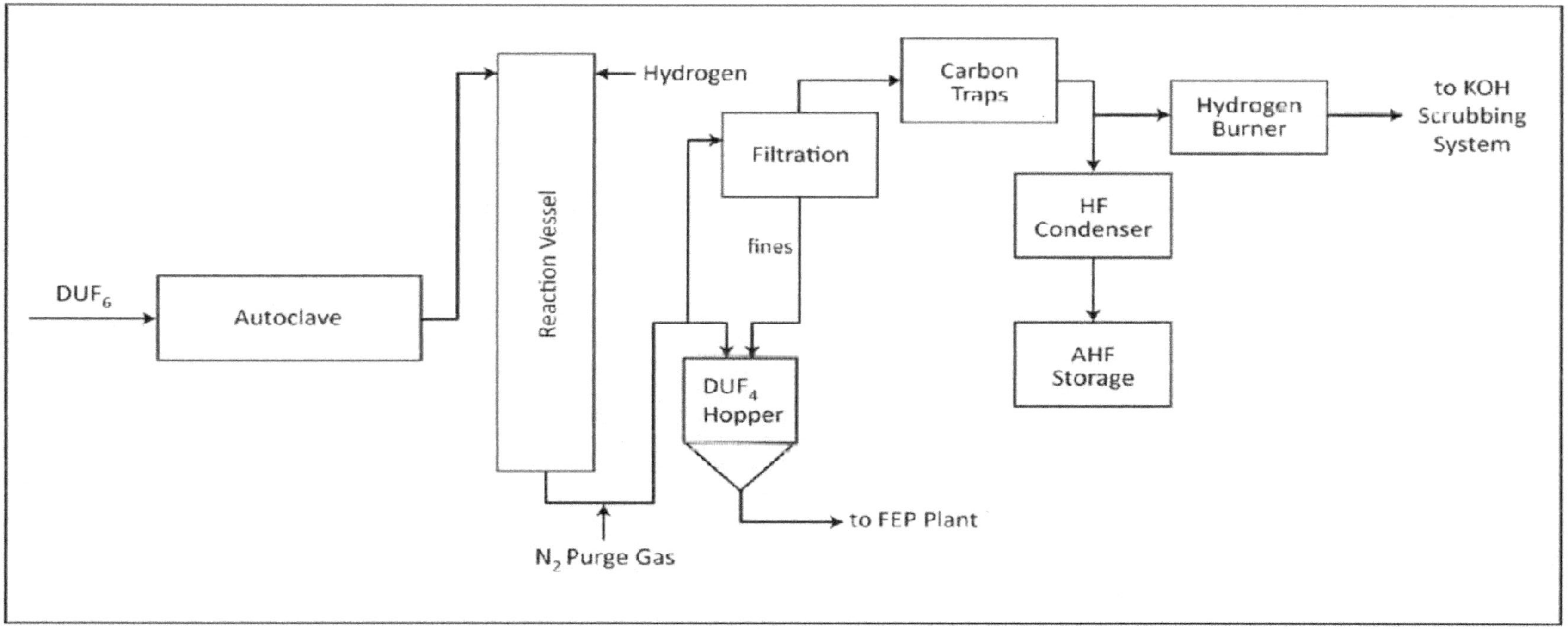

Figure 2-5. Process Flow Chart for SiF$_4$ Production

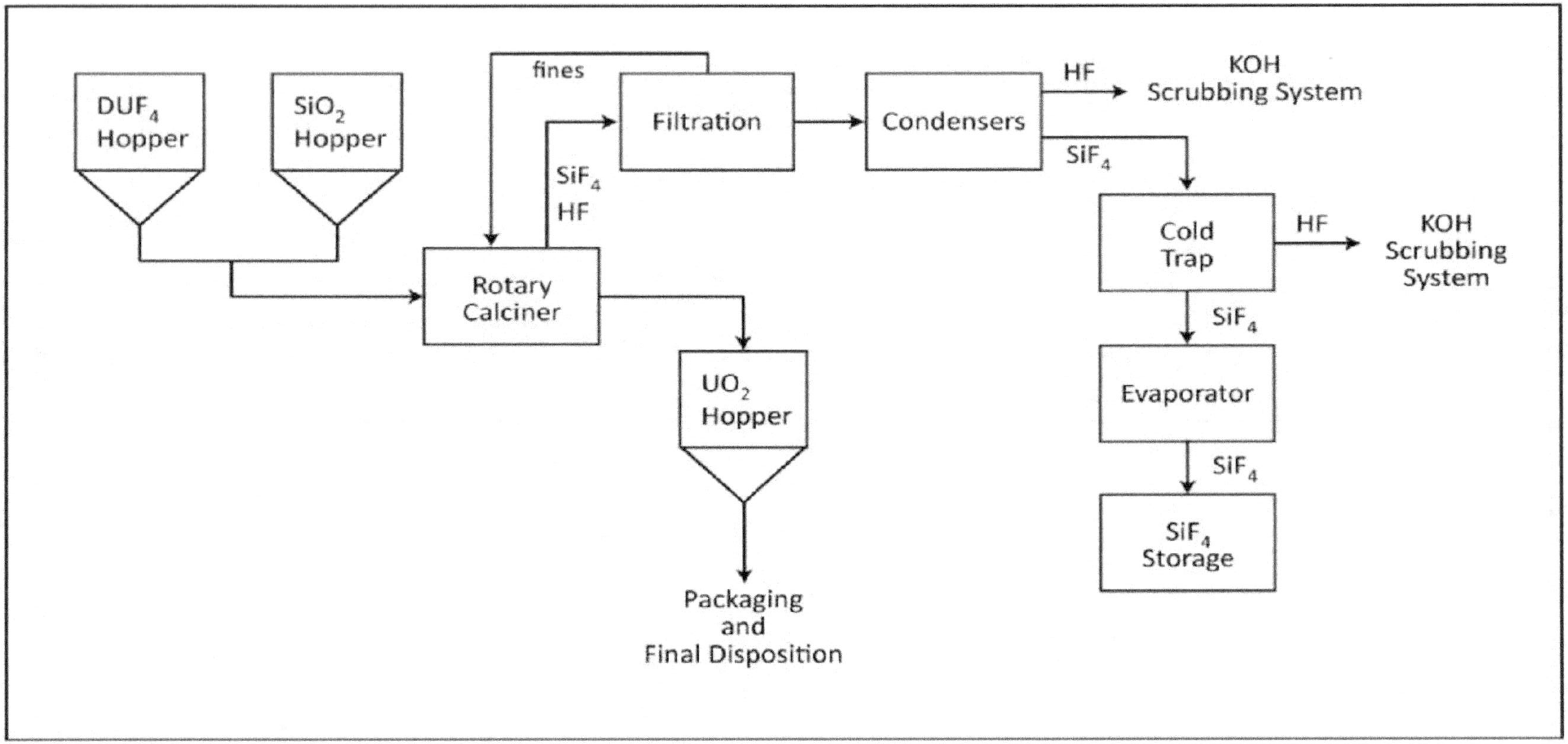

Figure 2-6. Process Flow Chart for BF₃ Production

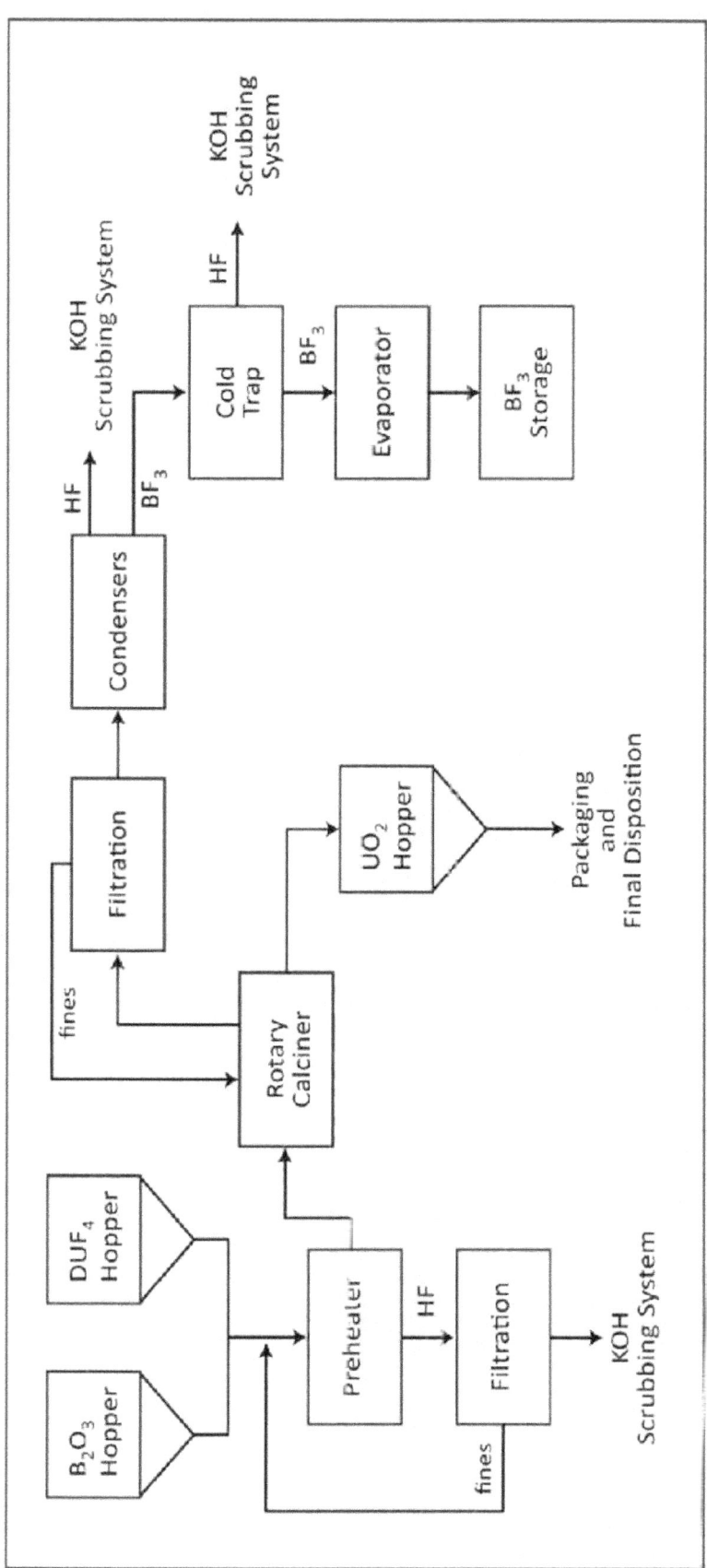

The proposed IIFP facility would have a Full DUF_6 Cylinder Storage Pad with bollards to protect the cylinders from vehicles. The Pad is designed to be 53.3 m wide by 61 m long (175 ft wide by 200 ft long) and is sized to store up to 60 full cylinders. The Pad would be curbed for stormwater collection and provided with underground drains to a stormwater retention basin south of the Pad. There would also be a 32 m by 56.3 m (105 ft by 185 ft) Empty DUF_6 Cylinder Storage Pad, with capacity for 40 empty cylinders. It would be the staging area for the shipment of empty cylinders.

The main process buildings, listed below, would be on the proposed 16-ha (40-ac) facility.

- DUF_6 Autoclave Building
- DUF_4 Process Building
- DUF_4 Container Staging Building
- Decontamination Building
- FEP Process Building
- FEP Oxide Staging Building
- DUF_4 Container Storage Building
- FEP Product Storage and Packaging Building
- AHF Staging Containment Building
- Fluoride Products Trailer Loading Building
- SiO_2 Storage Silo
- Potassium Hydroxide (KOH) Storage Tank
- FEP and DUF_4 Scrubbers and Scrubber Containment Pads

Hydrogen used as a reactant in the deconversion processes would be generated onsite from natural gas using a vendor-supplied steam reforming system. The system would provide approximately 6 to 9 pounds per hour of hydrogen at 24.7 to 29.7 pounds per square inch absolute. The natural gas requirement is approximately 18.7 pounds per hour (420 standard cubic feet per hour). Other than a small surge tank, the site would not store hydrogen gas.

All the building area aprons and areas surrounding outside equipment would have concrete curbing dikes designed to contain the largest possible spill of liquid chemicals, based on the volume of chemicals expected to be stored in each building/area. Pads for the storage of hazardous or corrosive chemicals would be coated to prevent leaks penetrating through the pads. The dikes would be equipped with pumps to transfer any spills to the Environmental Protection Process (EPP) equipment (Section 2.1.6.4.2). Radiological hand and foot monitors would be installed at exits of buildings where uranium would be handled. Fluoride and radiological detection systems, local alarms, and alarms in the control rooms would alert workers to potentially hazardous conditions.

Auxiliary buildings would generally house:

- materials
- maintenance shops

- laboratories

- steam boilers and supporting utilities

- electrical utility equipment

- sanitary water treatment equipment

- equipment for process water treatment and recycling

- personnel offices, break rooms, changing rooms, and restrooms

2.1.4   Preconstruction and Construction

Preconstruction activities include activities that would occur prior to issuance of the license and are discussed in cumulative impacts (Section 4.2.2). Preconstruction activities include site preparation activities and would not include the construction of process buildings or any safety-related structures. Preconstruction activities would include:

- clearing land

- grading the site and installing erosion controls

- building the main entrance roadbed and drainage

- setting up construction trailers

- preparing preliminary site roadways and gravel parking area

- (potentially) drilling water wells

- constructing an electrical substation

- stubbing in gas line to meter

- constructing the administrative building shell

- constructing the maintenance and storage building

- constructing the material warehouse building shell

- installing temporary fencing

- constructing facility roadbeds and gravel parking areas

- installing a geothermal heat pump loop

- installing a firewater tank

- installing a truck washing station

During preconstruction, the 16-ha (40-ac) IIFP facility site would be graded to provide an approximately level grade at elevation 1,157 meters (3,797 ft) above mean sea level. Approximately 11-ha (26-ac) on the northeast would be cut approximately 0.3 meters (1 foot) in depth, resulting in a cut of an estimated 32,400 cubic meters (42,400 cubic yards). This excavated material would be used as fill in the northwest and southwest areas of the proposed facility location, including two isolated depressions on the west side of the site (approximately 7.3-ha [18-ac]). The amount of fill required would be approximately 32,600 cubic meters (42,600 cubic yards), resulting in a deficit of 150 cubic meters (200 cubic yards) of fill needed that would be obtained onsite.

Heavy equipment that would be required for preconstruction (and construction) would include tractor/backhoes, graders, excavators, dozers, dump trucks, cranes, fuel trucks, water trucks, forklifts, and flatbeds. Additional equipment could include air compressors, concrete pumps, generators, and welding machines. During "construction," which refers to all construction activities that occur after the license is issued, the remainder of the facility, including the process buildings, would be constructed. The following activities would occur during construction:

- connecting utilities

- completing the access road and parking lot

- completing the construction of multiple structures including 13 process buildings, an administration building, laboratories, a maintenance shop, security facilities, cylinder storage pads, and warehouses

- construction of a meteorological tower

- installation of process equipment and other interior infrastructure

- construction of the wastewater management system

Construction of Phase 1 of the facility is expected to require 140 workers.

During construction, a 0.6-m (2-ft) depth of topsoil (approximately 2,400 cubic meters [3,100 cubic yards]), would be removed in the areas of buildings and adjacent pads to provide adequate bearing for concrete floors and pads. Additionally, an estimated 3,000 cubic meters (4,000 cubic yards) would be removed at an approximate 0.6-m (2-ft) depth in the areas for the (full and empty) DUF_4 cylinder pads. The material used to fill back to the foundation level would have soil compacting specifications suitable for the load bearing requirements that would be determined during the detailed engineering of the project.

Foundations and footings for buildings, tanks and equipment, and for evaporation basins and the storm and sanitary sewer systems, would require excavation of an equivalent 3,170 cubic meters (4,150 cubic yards), encompassing excavation less backfill.

The roadbed for the access road from NM 483 to the 16-ha (40-ac) site would require approximately 6,700 cubic meters (8,800 cubic yards) of fill. This fill would use most of the 8,600 cubic meters (11,250 cubic yards) of material from the excavations described above. Any excess (or unsuitable fill material) would be spread approximately 0.15-m (6 inches [in]) deep and compacted over an estimated 0.4 to 0.8-ha (1 to 2-ac) area of the 258-ha (640-ac) section. The grading and temporary preparation of a construction access road would be included in the preconstruction activities, but final construction would occur during Phase 1 activities.

2.1.5   Utilities and Other Services

The FEP/DUP plant would require the installation of electrical and natural gas service lines from existing utilities that cross the proposed site and are outside the facility boundary. It is expected that these utility connections would be installed during preconstruction. Steam and compressed air would be generated on site (Section 2.1.5.4). Nitrogen would be internally generated on site or procured from a vendor. Hydrogen would be generated on site. Water would be obtained from on-site groundwater wells.

2.1.5.1   Electrical Power

Most of the electrical power required by the proposed facility would be to operate four reaction vessels (calciners) in the FEP process building, and the refrigeration system and reaction vessel in the DUP process building. A new electrical substation and distribution line are proposed for providing electrical service to the facility. Currently 115- and 230-kilovolt transmission lines run along NM 483 and across Section 27. The local electric utility would install a 4.9 kilovolt-ampere substation and distribution lines to the facility. The substation would be within the facility fence. For some lighter loads, solar electric panels, both ground- and roof-mounted, would supplement the offsite power.

2.1.5.2   Water

The proposed facility would require relatively low volumes of process water because it would recycle process water and re-circulate cooling water. IIFP estimates that the total water supply requirement is less than 38,000 liters (L) (10,000 gallons [gal]) per day. Sanitary water requirements for showers, lavatories, drinking, toilets, and the laboratory would be 11,000 L to 17,000 L (3,000 to 4,500 gal) per day of the total. Treated sanitary waste water would be used for landscape watering. Boiler blow-down would be sent to the EPP (Section 2.1.6.4.2) for treatment, if needed, and evaporation.

No municipal water line runs near the proposed site. Therefore, it is anticipated that there would be at least one but no more than two groundwater wells to supply water for the facility. Lea County will install and provide one groundwater well as part of the land transfer to IIFP; IIFP would install another, if necessary, to obtain the desired yield for operations (of both the Phase 1 and Phase 2 facilities operations). A package treatment plant would render the groundwater acceptable for potable water use.

2.1.5.3   Natural Gas

The proposed facility would require natural gas for two gas-fired boilers that would support process steam production, the autoclave feed system, and the hydrogen production plant. Several natural gas pipelines cross Section 27. Gas would be conveyed to the facility from one of these existing pipelines via a smaller-diameter distribution pipeline.

2.1.5.4   Internal Utilities

2.1.5.4.1   Steam

Steam would be the primary heat source for vaporizing DUF_6 in the autoclave, heating some process and warehouse buildings, and warming pipes as necessary to prevent solidification of temperature-sensitive substances. Steam requirements for the facility are estimated to be 2,500 to 3,500 pounds per hour. Steam would be generated on site at 150 pounds per square inch (psig) using package boilers of about 10,000 pounds per hour capacity.

2.1.5.4.2   Compressed Air

Compressed air would be needed for operation of some instrumentation, control valves, dust collector blow-back, hopper vibrators, and miscellaneous uses. Ambient air would be filtered, compressed, and dried to deliver approximately 100 psig.

2.1.5.4.3   Nitrogen

Gaseous nitrogen would be required for purge gas and for cooling pre-condensers in the FEP process building. Liquid nitrogen would be used for the cold traps. The cold nitrogen vapor exiting the product cold traps would be used for the pre-condenser cooling. Gaseous nitrogen leaving the condensers would be collected and compressed to supply gaseous nitrogen to the parts of the facility that require a dry inert gas. The main application would be for purge and seal systems, such as the rotary calciner inlet and discharge seals. IIFP plans to conduct a cost-benefit analysis during detailed design to determine whether to make or buy the liquid nitrogen or to use another cryogenic system, such as gaseous helium. It is assumed for this EIS that liquid nitrogen would be procured from a vendor.

2.1.6   Facility Operations

2.1.6.1   Workforce

During Phase 1 operations, the continuous, fulltime workforce is expected to be approximately 140 workers.

2.1.6.2   Feedstocks

The primary raw materials used in the facility would be DUF_6, SiO_2, and B_2O_3. Annual throughputs of these materials are provided in Figure 2-3. Other materials needed would be hydrogen, nitrogen, potassium hydroxide (KOH), and lime.

2.1.6.3   Products

The finished products are fluoride products, namely AHF, SiF_4, and BF_3. The byproduct of the facility is a chemically stable DUO_2 suitable for permanent offsite disposal, if desired (Section 2.1.8). The expected annual production of these materials is provided in Figure 2-3. The design-basis inventories are provided in Table 2-1.

2.1.6.4   Waste Streams

The wastes from the FEP/DUP plant include gaseous emissions, process wastewaters, sanitary wastes, and solid wastes. These waste streams and their treatment methods are described below. Gaseous emissions rates are provided in Table 2-2.

2.1.6.4.1   Process Offgas Treatment and Stacks

The plant would have three stacks to vent treated process offgases and particulates to the atmosphere: the KOH Scrubbing System Stack, the DUF_4 Dust Collector System Stack, and the FEP Dust Collector System Stack. Prior to venting, the particulate and gas process streams would be filtered and/or scrubbed using multi-stage equipment. Additionally, one boiler vent stack would release natural gas combustion products to the atmosphere.

Offgas Treatment

Final off-gas streams from the DUF_6 to DUF_4, SiF_4, and BF_3 processes (comprised mostly of nitrogen, air, and trace fluorides) would enter the Plant KOH Scrubbing System, a three-stage

Table 2-1.   Facility Design Basis Inventories

Material	Maximum Limit Agreement with New Mexico	Projected Average Phase 1
Total depleted uranium (DUF_6, DUO_2 and DUF_4)	2,200,000 kg (4,851,000 lb)	See Note 2
DUF_6	See Note 1	15-20 full cylinders 165,000-220,000 kg (363,000-484,000 lb)
DUF_6 in process	See Note 1	19,500-30,000 kg (43,000-66,000 lb)
DUF_4	See Note 1	63,500-136,100 kg (140,000-300,000 lb)
Uranium oxides as DUO_2	See Note 1	154,200-213,200 kg (340,000-470,000 lb)
HF (aqueous)	23,300 kg (51,400 lb)	4,500-6,800 kg (10,000-15,000 lb)
AHF	45,000 kg (99,200 lb)	14,000-15,900 kg (31,000-35,000 lb)
SiF_4 (packaged + in-process)	64,700 kg (142,700 lb)	21,800-31,800 kg (48,000-70,000 lb)
BF_3 (packaged + in-process)	22,400 kg (49,400 lb)	7,800-15,000 kg (17,000-33,000 lb)
KOH	8,100 kg (17,900 lb)	6,800-7,700 kg (15,000-17,000 lb)
CaF_2 (calcium fluoride)	36,500 kg (80,500 lb)	20,400-22,700 kg (45,000-50,000 lb)

Source: IIFP, 2009
lb = pound; kg = kilogram
Note 1: The "Maximum Limit" applies to the total depleted uranium as either DUF_6 (both in cylinders and in process), DUO_2 or DUF_4.
Note 2: The "Projected Average" is provided as individual breakdowns for DUF_6 in cylinders and in process, DUO_2, and DUF_4.

Table 2-2.   Projected Annual Gaseous Emissions to the Atmosphere from Phase 1 Facility Operations

Pollutant	Emissions	Units
CO	1,200 (1.3)	kg/yr (tons/yr)
NO_2	290 (0.32)	kg/yr (tons/yr)
$PM_{2.5}$	100 (0.11)	kg/yr (tons/yr)
PM_{10}	100 (0.11)	kg/yr (tons/yr)
SO_2	18 (0.02)	kg/yr (tons/yr)

Table 2-2. Projected Annual Gaseous Emissions to the Atmosphere from Phase 1 Facility Operations (Continued)

Pollutant	Emissions	Units
SiF_4	3.7 (8.2)	kg/yr (lbs/yr)
BF_3	64 (141)	kg/yr (lbs/yr)
HF	53 (117)	kg/yr (lbs/yr)
CaF_2	3.5 (7.8)	kg/yr (lbs/yr)
$CaCO_3$	61 (134)	kg/yr (lbs/yr)
B_2O_3	4.9 (10.8)	kg/yr (lbs/yr)
U-234	5.2×10^5 (1.4×10^{-5})	becquerels (Bq)/yr curies (Ci)/yr
U-235	4.8×10^4 (1.3×10^{-6})	Bq/yr (Ci/yr)
U-238	4.4×10^7 (1.2×10^{-3})	Bq/yr (Ci/yr)

Source: IIFP, 2011a
CO = carbon monoxide; NO_2= nitrogen dioxide; $PM_{2.5}$=particulate matter less than 2.5 microns in diameter; PM_{10}=particulate matter less than 10 microns in diameter; CaF_2=calcium fluoride; $CaCO_3$ = calcium carbonate; U-234, U-235 and U-238= isotopes of uranium.

scrubber system, to remove fluoride from the offgases prior to releasing them to the atmosphere.

Two parallel systems would provide operating flexibility. The first stage of each scrubber system would consist of a primary wet venturi scrubber. The second stage would consist of a countercurrent-flow, gas-liquid packed tower scrubber. The third-stage scrubber would route gas exiting the secondary packed tower scrubber though a bed of sized coke (a cellular, carbonaceous material derived from the destructive distillation of coal or petroleum products). The coke would be wetted by an aqueous KOH solution that serves as the scrubber liquor. The aqueous KOH solution would be recycled within each of the scrubbers until the concentration of KOH needs replenishment (i.e., until the KOH no longer effectively captures the fluoride residuals, referred to as being "spent"). The KOH solution concentration in the scrubber equipment would be maintained to ensure it effectively reacts with (scrubs) the fluoride components in the gas stream.

When the KOH scrubbing liquor concentration needs replenishment, some of the spent scrubbing solution, containing potassium fluoride (KF), water, and some excess KOH, would be pumped from the scrubber recycle tanks to the EPP (described in Section 2.1.6.4.2). The Plant KOH Scrubbing System process flow is depicted in Figure 2-7 and consists of a KOH storage tank, KOH pump tank, regenerated KOH tank, two or three (installed spare) venturi scrubbers, two packed towers, and two coke boxes.

Figure 2-7. Plant KOH Scrubbing System Process Flow

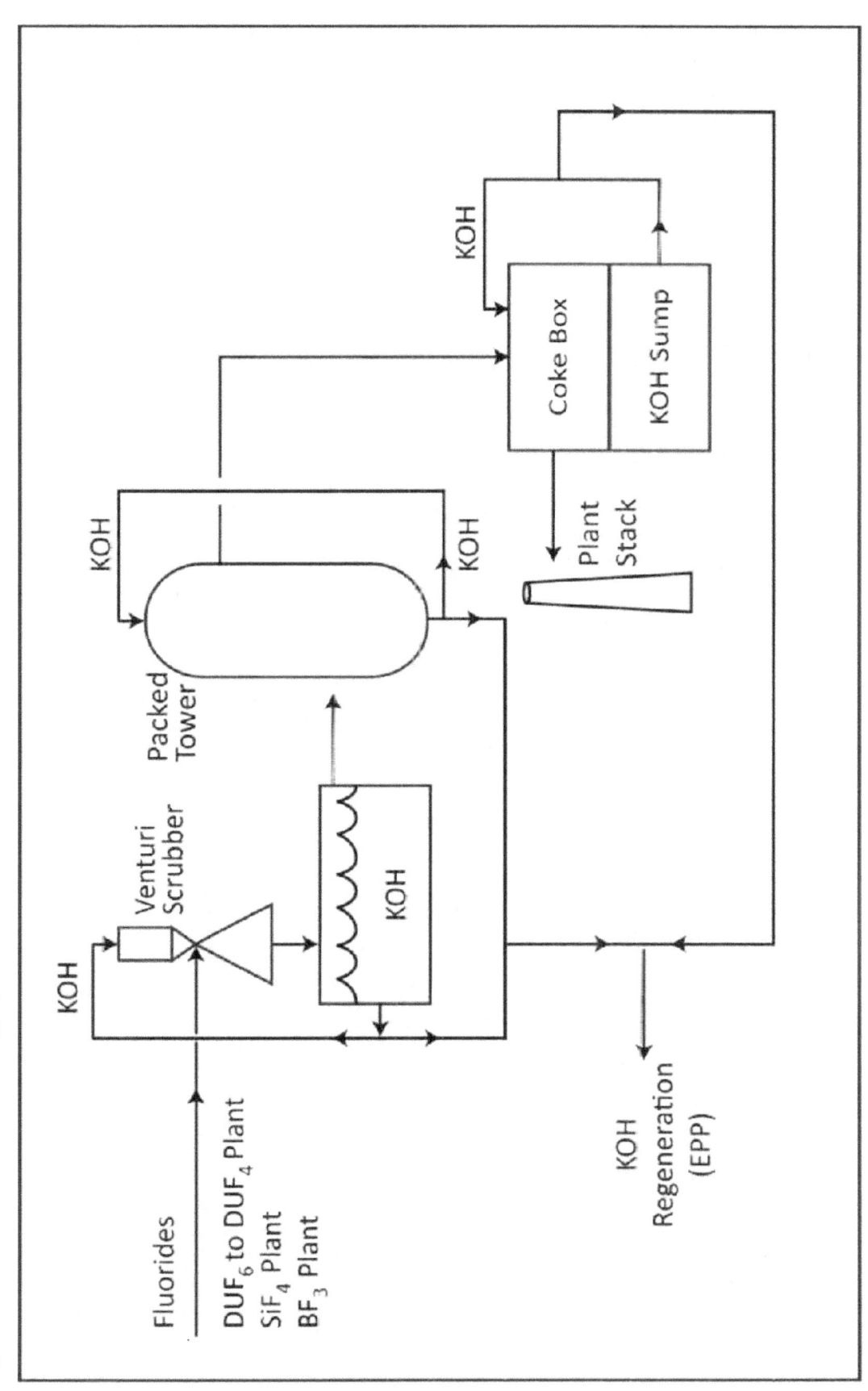

The three-stage KOH scrubbing system would be designed to remove fluoride-bearing components in the gas streams at approximate efficiencies of greater than 80 percent, 95 percent, and 99 percent for the first, second, and third stages, respectively. The overall system removal efficiency would be designed at greater than approximately 99.9 percent. The plant KOH scrubbing system stack would be continuously sampled to measure for traces of fluorides or uranium in the vent gas.

Process Dust Collection

Dust capture and collection systems would be installed in areas where depleted uranium particulates, such as DUF_4 or DUO_2, would be handled or processed. The dust collection systems would be filter-type baghouses that would remove the depleted uranium particulates prior to discharging the process gas to the outside environment through the DUF_4 Dust Collector Vent Stack.

Equipment where depleted uranium-bearing powders would be handled or stored, such as storage hoppers and enclosed drum packaging stations, would be connected to the dust collection intakes. Uranium particulates captured by the dust collection systems either would be recycled back to the respective process operations or packaged and sent to an approved off-site disposal facility. The design efficiency of baghouse dust collectors would be greater than 99.5 percent for each collector. At least two components would be used in series to ensure an overall system efficiency of greater than 99.9 percent in the collection and removal of particulate uranium from the vented process gas.

Sampling and analysis for uranium would be performed routinely on each baghouse dust collector. If an unacceptable level of uranium carryover was detected on any given dust collector, it would be removed from service for maintenance. Additionally, each baghouse would be continuously monitored for differential pressure across the filter bag sections to ensure bag design integrity was maintained.

2.1.6.4.2   Environmental Protection Process

The EPP would treat KF solutions to regenerate KOH, and neutralize weak aqueous HF. Both of these waste streams originate from offgas scrubbing systems designed to prevent air emissions, as described in Section 2.1.6.4.1.

A KF solution would be generated when KOH was used as a scrubbing medium. In the KOH regeneration process of the EPP, the KF solution, water, and excess KOH spent solution from the plant KOH scrubbing system would react with a lime slurry, producing calcium fluoride (CaF_2) and regenerated KOH solution. The regenerated KOH would be recycled and reused in the plant scrubbing process. The CaF_2 would be filtered, dried, and packaged for shipment to an approved disposal facility, to an HF producer, or to another potential user.

The other stream treated in the EPP would consist of a weak aqueous HF solution, water, or KOH solution that may contain a low concentration of fluorides. Also, small spills that could occur in spill control containment areas and require clean up and that could contain weak fluoride concentrations would be treated in the EPP like the weak HF solution. In these cases, the fluoride-bearing liquids could have too much water to send to the KOH regeneration system. The HF neutralization process would use lime slurry to react with weak HF to produce CaF_2 and water.

Figure 2-8 depicts the EPP HF Neutralization and KOH Regeneration processes. These processes are discussed below.

KOH Regeneration

Lime would be fed to an agitated mix tank and mixed with water. The lime/water slurry would be approximately 30 percent solids. Spent KOH solution (KOH solution with a weak concentration of KOH) would be transferred from a storage tank to an agitated reaction vessel that has a volume of about 22,712 L (6,000 gal). The lime slurry would be transferred from the mix tank to the reaction vessel. The solutions would remain in the reaction vessel tank for one hour or more to ensure the reaction was complete. Then the contents of the reaction vessel would be transferred to a thickening tank for settling. CaF_2 from the chemical reaction and excess lime would be transferred by a slurry pump from the bottom of the thickening tank to a rotary drum vacuum filter. Solids would be discharged from the filter to a dryer to remove excess water. Liquors would be transferred from the filter to a clarifier to allow trace solids to settle. Regenerated KOH solution would be decanted from the top of the clarifier and passed through a set of filters to the regenerated KOH storage tank. The regenerated KOH solution would be recycled to the Plant KOH Scrubbing System, as needed, for reuse by the scrubbers. Solids from the clarifier would be transferred via a slurry pump from the bottom of the clarifier to the rotary drum vacuum filter and subsequently transferred to the dryer. The dried material would be packaged and stored for sale or sent to an approved disposal facility. The primary chemical reaction is:

$$2KF + Ca(OH)_2 \text{ (calcium hydroxide)} \rightarrow CaF_2 + 2KOH$$

HF Neutralization

The HF Neutralization process would operate intermittently, as needed. A lime silo would hold an inventory of hydrated lime. The silo would include a dust collector. Lime would be fed to a mix tank and mixed with water. The slurry would be approximately 30 percent solids. Dilute HF solution would be transferred from the weak HF solution tank to an agitated acid reaction vessel with a volume of about 22,712 L (6,000 gal). The lime slurry would be transferred from the mix tank to the acid reaction vessel. The solutions would remain in the acid reaction vessel for one hour or more to ensure the reaction was complete. Then the solution from the acid reaction vessel would be transferred to a thickening tank for settling. After thickening, CaF_2 and excess lime would be transferred by a slurry pump from the bottom of the thickening tank to a rotary drum vacuum filter. Solids would be discharged from the filter to a dryer to remove excess water. Liquors from the filter would be recycled to the weak HF solution tank for recycling. After drying, the CaF_2 would be packaged for sale or disposal at an approved disposal facility. The primary chemical reaction is:

$$2HF + Ca(OH)_2 \rightarrow CaF_2 + 2H_2O \text{ (water)}$$

2.1.6.4.3   Sewer Systems

Storm Sewers and Stormwater Collection Basins

The facility storm sewer system design assumes a 100-year storm for the Hobbs, New Mexico, area of 8.9 to 10.2 centimeters (cm) (3.5 to 4 in) of rain falling in one hour. IIFP performed preliminary engineering of the drainage system size and layout to estimate costs, determine requirements, and provide information for later detailed design. The preliminary design includes

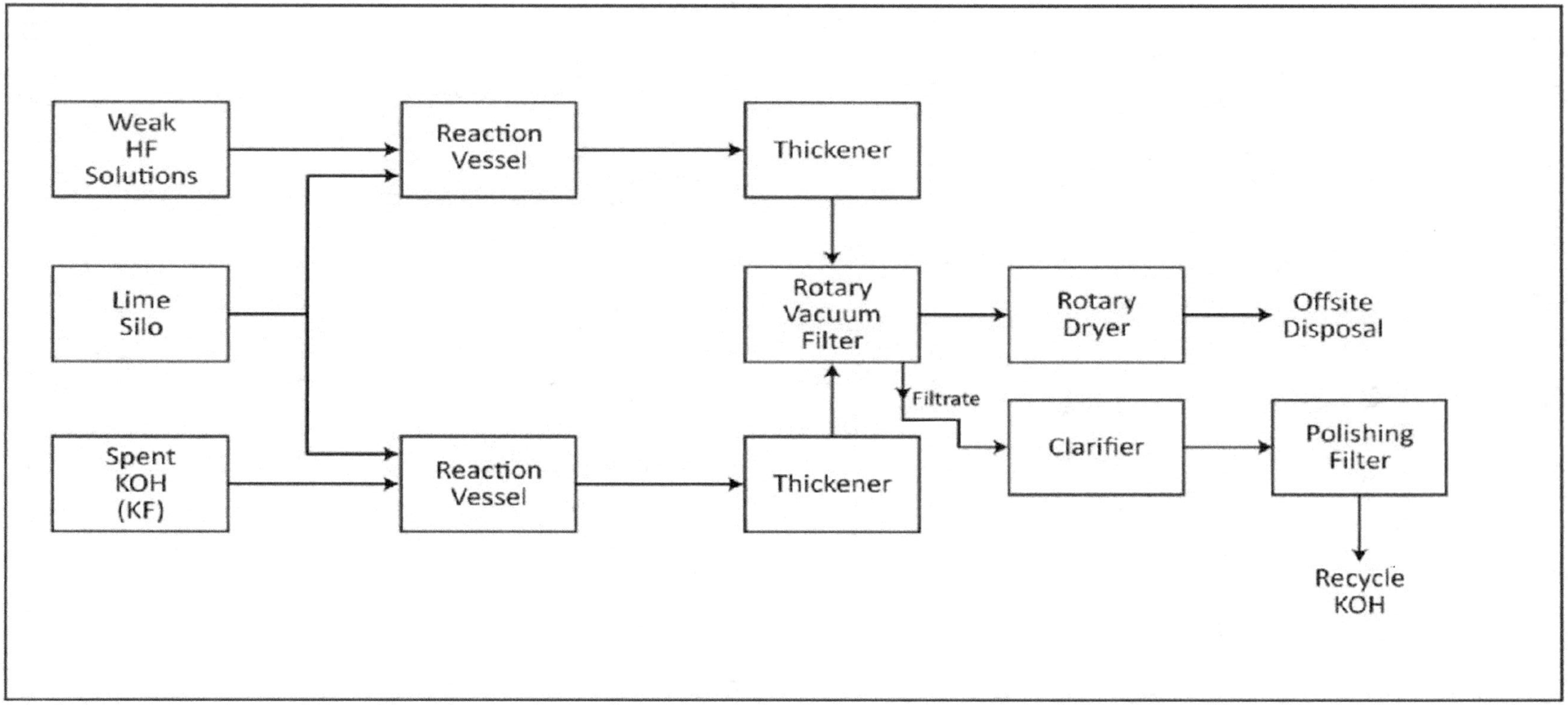

Figure 2-8. EPP HF Neutralization and KOH Regeneration Processes

the locations of the process buildings, auxiliary buildings, pads, roads, parking lot, and water treatment plant and electrical substation. All of the storm sewer systems would be inside the fenced area and would collect rainwater runoff from an estimated 8- to 10-ha (20- to 25-ac) area, including roadways, building roofs, and pads.

Two collection basins are planned to handle storm water drainage surges. One basin would serve the Full DUF$_6$ Cylinder Storage Pad. The other would be the main holding basin for the site storm sewer drainage system. Preliminary engineering calculations performed by IIFP estimate the main basin needs to be approximately 2,800,000 L (100,000 cubic ft) in volume, assuming a 20 percent freeboard above the maximum design water level. The basin would be double-lined with impervious synthetic materials typically used in these applications. IIFP's current plans are to use a sand base with a layer of geo-synthetic liner and a second layer of high density polyethylene. Detail engineering and specifications would be refined after civil engineering data are obtained from the site surveys and after discussions with the New Mexico Environment Department regarding permits.

Sanitary Sewer

Preliminary design of the sanitary sewer system provides for capability to handle hydraulic loading of about 11,356 to 17,034 L (3,000 to 4,500 gal) per day. Sanitary sewer discharge would be treated in primary and secondary package systems for digestion and activation. Tertiary treatment with disinfection, probably using ultraviolet radiation, would follow. Biomass generated by the treatment would be removed from the site by a licensed disposal contractor. The triple-treated water would be re-used as process water in the facility or for landscape irrigation.

Process Sewer

Process water and solutions, and KOH liquors would be pumped, when contaminant concentrations dictate, from process systems or the air emissions scrubbing units, respectively, to the EPP via above-ground piping. Pipes would be double-walled to prevent leakage of hazardous solutions out of the piping system if the piping cannot be routed through areas with adequate spill containment.

2.1.6.4.4   Solid Wastes

IIFP would use solid waste management systems including facilities, administrative procedures, and practices for the collection, temporary storage, and disposal of categorized solid waste. No solid waste processing is planned. The facility would generate industrial (nonhazardous), radioactive, hazardous, and mixed wastes. Radioactive and mixed waste would be segregated according to the volume of liquid that could not be readily separated from the solids. Solid radioactive wastes would be low-level (radioactive) waste (LLW) as defined in 10 CFR 61, "Licensing Requirements for Land Disposal of Radioactive Waste." Table 2-3 provides the estimated annual quantities of solid waste.

Industrial waste, including sanitary waste, miscellaneous trash, vehicle air filters, empty cutting oil cans, miscellaneous scrap metal, and paper would be shipped off site for minimization, if appropriate, and then disposed in an appropriate licensed landfill.

Table 2-3.   Phase 1 Estimated Annual Solid Waste Generation - Operations

Material	Estimated Annual Amount
Low-Level Waste	1,309,000 – 2,875,000 kg (2,885,650 – 6,337,300 lbs)
Hazardous Waste [a]	92,000 – 140,000 kg (203,200 – 308,400 lbs)
Other Solid Waste	27,510 – 41,400 kg (60,650 – 91,300 lbs)

Source: IIFP, 2011a
[a] Includes calcium fluoride which would not be hazardous waste if it is sold as a byproduct.

The DUO_2 waste from the deconversion process could be shipped to an offsite LLW disposal facility licensed to accept DUO_2 (Section 2.1.6.5.3 and Section 2.1.8). Other LLW, including dust collector bags, ion exchange resin, crushed contaminated drums, contaminated trash, contaminated coke, and carbon trap material would be collected in labeled containers in each radiological Restricted Area and transferred to the Radioactive Waste Storage Area for inspection. Waste would be volume-reduced, if appropriate, and disposed at a licensed LLW disposal facility.

Hazardous wastes and some mixed wastes would be collected at the point of generation, transferred to the Waste Storage Area, inspected, and classified. Any mixed waste that could be processed to meet land disposal requirements would be treated in its original collection container and shipped as LLW for disposal at a licensed LLW disposal facility. Hazardous wastes would be collected and packaged in approved containers and shipped by a licensed transporter to a licensed hazardous waste disposal facility. There would be no on-site disposal of any solid waste at the IIFP facility.

2.1.6.5   Product and Byproduct Packaging and Shipping

Three types of products/byproducts would be shipped: AHF, FEP products (SiF_4 and BF_3), and depleted uranium oxides (e.g., DUO_2). Given the hazards of fluoride products, especially AHF, the AHF Staging Containment Building and the FEP Products Trailer Loading Building would be equipped with an array of water-fog nozzles that would automatically activate in the event of a leak of AHF or fluoride product chemicals. Fluoride detectors would be deployed throughout the two buildings to ensure effective coverage. The detection and control system would be designed for automatically closing isolation valves at the storage tanks and at the tank trailer fill lines. The detection system would also provide automatic and manual controls for initiating the water deluge system in event of chemical leaks in either building.

2.1.6.5.1   AHF

The AHF Staging Containment Building and equipment would provide temporary storage of AHF received from the DUF_6 to DUF_4 process AHF condensers. AHF would be stored in the AHF Staging Containment Building in approximately 3,630-kg (8,000-lb) capacity tanks. Dikes around each storage tank would be sized to hold the contents of the storage tank, with a margin of safety to minimize the surface area (and evaporation rate of the AHF) in the unlikely event the tank breached and spilled liquid AHF.

When AHF inventories reach a volume suitable for shipment, the AHF would be loaded into an approved tank trailer staged in the Fluoride Products Trailer Loading Building, which would be connected to the AHF Staging and Containment Building. The transporter-owned or customer-owned tank trailer would be approved by the U.S. Department of Transportation (DOT) and of the design and type routinely used for shipping AHF nationwide (typically type DOT-412 trailer loaded to about 13,608 to 18,144 kg [30,000 to 40,000 lb] of product). The Fluoride Products Trailer Loading Building would have a truck entrance door that remained closed, sealed, and controlled except for short periods when the tank trailer would be moved in and out. A transfer line from the storage tanks in the AHF Staging Containment Building would enter the tank trailer side of the Fluoride Products Trailer Loading Building. Safety precautions, controls, and barriers would prevent the tank trailer from inadvertently being moved or from contacting the fill line.

2.1.6.5.2   FEP Products

The SiF_4 and BF_3 products awaiting shipment to customers would be stored in the FEP Product Storage and Packaging Building until packaged within the Building into customer-owned, DOT-approved shipping cylinders (typically type 3A or 3AA). The SiF_4 or BF_3 product could be packaged into DOT-approved shipping tube trailers, and in this case the product would be transferred from the storage vessels to a tube trailer parked in the Fluoride Products Trailer Loading Building.

2.1.6.5.3   Depleted Uranium Oxides

DUO_2 and all other LLW materials generated at the facility would be transported by truck in 208-(wet) or 242-(dry) L (55-gal) drums in accordance with NRC and DOT packaging and shipping regulations (10 CFR 71 and 49 CFR 171-173). Trucks would carry 20 to 25 drums per shipment. The drums would be disposed of at a licensed off-site LLW disposal facility. For purpose of analysis in this EIS, the expected disposal site is considered to be the Energy *Solutions* facility at Clive, Utah. See Section 2.1.8 for a discussion of depleted uranium oxide disposal options.

2.1.7   Decommissioning

The proposed IIFP facility would be licensed to operate for 40 years. At the end of this period, unless IIFP files a timely application for license renewal, the proposed facility would be decontaminated and decommissioned in accordance with applicable NRC license termination requirements. The FEP/DUP facility would be decommissioned such that the site and remaining facilities could be released for unrestricted use as defined in 10 CFR 20.1402. Decontamination and decommissioning would occur over three years, after the NRC operating license expires and if no application to renew the license is submitted. Decommissioning would employ 40 workers for the three-year period (IIFP, 2009).

Two possibilities exist for the facility structures and paved areas. One is to leave the structures and most (non-uranium-processing) support equipment in place after it is decontaminated to free release levels, in accordance with 10 CFR 20, for ultimate use by another industrial tenant or owner. The second scenario is to raze the structures and remove the pavement, restoring the site for use as open range land (e.g., grazing and wildlife habitat). IIFP's analytical assumption is that decommissioning would involve the removal of the internal equipment (both Phase 1 and Phase 2 expansion, if built), utilities, and products from the building(s); however, the physical structures, associated foundations, access roads, and utility lines would likely remain intact, (i.e., the first scenario). Decommissioning of the proposed IIFP facility would include decontamination and removal of uranium-processing equipment and other materials that would be shipped offsite for licensed disposal. Radioactively contaminated equipment and materials would be sent to a licensed treatment and/or disposal facility in a manner authorized by the NRC (IIFP, 2009). Prior to the expiration of the license or cessation of facilities operation, whichever comes first, IIFP would submit a detailed decommissioning plan, which would undergo additional NEPA review.

2.1.8   Depleted Uranium Disposition Options

On average, the facility would produce approximately 0.32 to 0.36 kg (0.7 to 0.8 lb) of DUO_2 for every pound of DUF_6 processed, yielding approximately 2.5 million kg (5.6 million lb) of DUO_2 annually (Figure 2-3). The DUO_2 could either be disposed as LLW or recycled. Potential reuses of depleted uranium are as aircraft and ship ballast, as ingredients in pigments and glazes, as shielding material, as forklift counterweights, in armor-piercing projectiles, in high density concrete, as material to downblend highly enriched uranium, as a component of fuel in fast breeder reactors (including the proposed variant, the traveling wave reactor), as an ingredient in mixed-oxide fuel for thermal reactors, and as shielding/absorber in waste repositories. Some of these uses are conceptual and have never been employed. Others are in little demand or use only small quantities of depleted uranium, making them unfavorable for disposition of large volumes of depleted uranium. The uranium fuel cycle as currently configured in the U.S. does not have the capacity to accept significant quantities of depleted uranium (DOE, 1999).

Depleted uranium is different from most LLW in that it consists mostly of long-lived isotopes of uranium, with small quantities of thorium-234 and protactinium-234. The Commission affirmed that depleted uranium is properly considered a form of LLW in Louisiana Energy Services, L.P. (National Enrichment Facility), CLI-05-5, 61 NRC 22 (January 18, 2005; NRC, 2005a). This means that depleted uranium could be disposed of in a licensed LLW facility if the licensing requirements for land disposal of radioactive waste in 10 CFR 61 are met. However, a specific site may place additional limits on concentration, volume, or waste form.

Disposal options, including waste form, would be determined after licensing and may change over the operating life of the facility; however, licensed LLW disposal facilities, including the U.S. Ecology site in Richland, Washington, Energy*Solutions* site in Clive, Utah, DOE's site in Area 5 of the Nevada National Security Site (formerly the Nevada Test Site), and the Waste Control Specialists (WCS) facility in Andrews, Texas are potential options, provided regulatory and contractual conditions can be satisfied. The U.S. Ecology facility is in the Pacific Northwest Compact which has an agreement with the Rocky Mountain Compact (of which New Mexico is a member) to dispose of waste, but the U.S. Ecology facility's license would need a revision in the allowable total uranium inventory. Energy*Solutions* accepts shipments from all states and is currently developing a performance assessment to establish inventory limits, if needed. Shipment to the Nevada National Security Site would require DOE to accept possession of the LLW (consistent with Section 13 of the USEC Privatization Act of 1996).

The WCS facility is 42 km (26 mi) southeast of the proposed site but is currently limited to waste from the Texas Compact and therefore, would have to establish approval mechanisms for out-of-compact waste to be disposed. Furthermore, the Rocky Mountain Compact would have to approve shipment outside the compact. The analysis in the EIS is not intended to support selection of the LLW disposal facility for the DUO_2.

2.2 Alternatives to the Proposed Action

The range of alternatives to the proposed action was determined by considering the underlying purpose and need for the proposed action and consideration of the no-action alternative. In addition, DUF_6 management options from the DOE's programmatic EIS on long-term DUF_6 management (DOE, 1999) were considered. From this evaluation, the NRC staff developed a set of reasonable alternatives. These alternatives include:

- a no-action alternative under which the proposed FEP/DUP facility would not be constructed

- deconversion of DUF_6 at DCE facilities

- alternative sites for the proposed facility

- alternative technologies available for DUF_6 deconversion

- overseas deconversion of DUF_6

- indefinite storage of DUF_6 at the uranium enrichment facilities

- deconversion of DUF_6 at the uranium enrichment facilities

2.2.1 No-Action Alternative

Under the no-action alternative, NRC would not grant a license to IIFP to construct, operate, and decommission the proposed IIFP facility near Hobbs, New Mexico. The proposed site would remain undeveloped except for preconstruction activities performed by IIFP. The regional economy would not be changed either positively or negatively. LLW would not be shipped to disposal facilities. Fluoride products would not be manufactured, sold, and shipped to end users.

The comparison of impacts between the proposed action and no-action alternative is provided in Table 2-4. Environmental impacts of the no-action will be less than the proposed action. However, the no-action alternative does not serve the purpose and need. Presently, there are four existing or planned domestic commercial enrichment facilities: URENCO USA (formerly Louisiana Energy Services) National Enrichment Facility, Eunice, New Mexico; AREVA Eagle Rock, Idaho Falls, Idaho; American Centrifuge Plant, Piketon, Ohio; and GE Global Laser Enrichment, Wilmington, North Carolina. Under the no-action alternative, the four planned or existing domestic, commercial uranium enrichment facilities would not send their DUF_6 to the IIFP facility for deconversion. Four other options would be open to them: (1) ship the DUF_6 to a DOE deconversion facility; (2) ship the DUF_6 to one of the deconversion facilities overseas; (3) indefinitely store the DUF_6; or (4) construct their own deconversion facilities. As explained in the subsequent paragraphs of this Section and in Section 2.2.2, all of these options but the first are identified in Section 2.2.2 as alternatives considered but eliminated from further consideration in this EIS.

DOE has constructed two deconversion plants to convert DUF_6 to U_3O_8 and hydrofluoric acid; one in Piketon (Portsmouth), Ohio and one in Paducah, Kentucky (DOE, 2004a; DOE, 2004b). The Ohio facility began operating in October 2010, and the Kentucky facility is slated to begin operating in 2011. Therefore, shipment to these DOE facilities is a viable option under the no-action alternative. Such shipment is allowed under the provisions of the United States Enrichment Corporation (USEC) Privatization Act of 1996.

The option to ship DUF$_6$ to the DOE deconversion facilities was considered in the National Enrichment Facility EIS (NUREG-1790; NRC, 2005b). This facility, near Eunice, New Mexico, is now known as the URENCO USA facility. As quoted in its Commission Order CLI-05-05 (NRC, 2005a), NRC stated (in CLI-04-3 regarding the LES facility) that, "an approach by LES to transfer to DOE for disposal by DOE of LES['s] depleted tails pursuant to Section 3113 of the USEC Privatization Act constitutes a 'plausible strategy' for dispositioning the LES depleted tails" if the tails could be considered LLW under 10 CFR 61.[1] Commission Order CLI-05-05 further stated that DUF$_6$ tails are a form of LLW. Accordingly, deconversion by DOE is retained as part of the no-action alternative for this EIS.

Given that DOE has a backlog of 700,000 metric tons (771,618 tons) of DUF$_6$ (DOE, undated) (in approximately 57,000 cylinders) to deconvert, it is expected to take DOE approximately 25 years to complete its mission (DOE, undated) and have the facility capacity to begin deconverting privately generated DUF$_6$. The DOE process does not produce the FEP products, and it produces hydrofluoric acid solution rather than the more useful AHF.

Table 2-4. Comparison of Impacts between Proposed Action and No-Action Alternatives

Affected Environment	Proposed Action *IIFP would construct, operate, and decommission the proposed IIFP facility in Lea County, New Mexico.*	No-Action Alternative *IIFP would perform preconstruction activities, but would not construct, operate, and decommission the proposed IIFP facility.*
Land Use	The NRC staff has determined that land use impacts resulting from construction of the facility and restricting the current land use would be SMALL due to the abundance of other nearby undeveloped land.	IIFP would obtain the proposed site, complete preconstruction of the IIFP facility, and institute restrictions on grazing and agriculture. The 16-ha (40-ac) site would be cleared and potentially reseeded. Grazing could resume on the entire 259-ha (640-ac) site. Impacts would be SMALL.
Historic and Cultural Resources	The NRC staff has determined that impacts of the construction and operations of the facility to historic resources or other cultural resources would be SMALL.	IIFP would obtain the proposed site and complete preconstruction of the IIFP facility. The site would be cleared and graded. Impacts to historic and cultural resources would be SMALL.
Visual Resources	The NRC staff has determined that the proposed facility would not affect visual resources.	IIFP would obtain the proposed site and complete preconstruction of the IIFP facility which would not adversely affect visual resources.

---

[1]See Louisiana Energy Services, L.P. (National Enrichment Facility), CLI-04-3, 59 NRC 10, 22 (2004), reprinted in 69 FR 5873, 5877 (Feb. 6, 2004).

Table 2-4.   Comparison of Impacts between Proposed Action and No-Action Alternatives (Continued)

Affected Environment	Proposed Action *IIFP would construct, operate, and decommission the proposed IIFP facility in Lea County, New Mexico.*	No-Action Alternative *IIFP would perform preconstruction activities, but would not construct, operate, and decommission the proposed IIFP facility.*
Climatology, Meteorology, and Air Quality	Small amounts of nonradioactive emissions and small quantities of uranium isotopes would be released to the atmosphere. The NRC staff concludes that impacts to air quality during construction and operation of the IIFP would be SMALL to MODERATE.	IIFP would obtain the proposed site and complete preconstruction of the IIFP facility. Smaller amounts of nonradioactive air emissions would be released than by the proposed action. Impacts would be SMALL.
Geology, Minerals, and Soil	The NRC staff has concluded that construction impacts and operation of the proposed IIFP facility to geology, minerals, seismicity, and soil would be SMALL, if proper best management practices are instituted as mitigation. Note that seismicity was a key consideration in IIFP's site evaluation process, and the proposed site is not in Seismic Zone 4 or within 48 km (30 mi) of a quaternary active fault.	IIFP would obtain the proposed site and complete preconstruction of the IIFP facility. Impacts would be SMALL.
Water Resources	The NRC staff has concluded that no impacts would occur to surface waters, and groundwater use impacts during construction and operations are expected to be SMALL.	IIFP would obtain the proposed site and complete preconstruction of the IIFP facility. Additional groundwater use may or may not occur, depending on future uses of the site. Impacts would be SMALL.
Ecological Resources	The NRC staff has concluded that direct and indirect adverse impacts to ecological resources during construction and operation of the proposed facility would be SMALL.	IIFP would obtain the proposed site and complete preconstruction of the IIFP facility. The 16-ha (40-ac) site would be cleared and potentially reseeded. Impacts would be SMALL.
Noise	The NRC staff has determined that the proposed facility would not affect ambient noise levels.	IIFP would obtain the proposed site and complete preconstruction of the IIFP facility which would not adversely affect ambient noise levels.
Traffic and Transportation	The NRC staff has concluded that impacts to traffic due to the IIPF construction and operation would be SMALL on NM 483 and US 62/180.	IIFP would obtain the proposed site and complete preconstruction of the IIFP facility. Impacts would be SMALL.
Public and Occupational	Regulated gaseous effluents would be below regulatory limits as	IIFP would obtain the proposed site and complete preconstruction of the

Table 2-4. Comparison of Impacts between Proposed Action and No-Action Alternatives (Continued)

Affected Environment	Proposed Action *IIFP would construct, operate, and decommission the proposed IIFP facility in Lea County, New Mexico.*	No-Action Alternative *IIFP would perform preconstruction activities, but would not construct, operate, and decommission the proposed IIFP facility.*
Health	specified by the New Mexico Air Quality Bureau. Radiological impacts to off-site receptors from routine combined effluent releases and direct radiation are anticipated to be SMALL. Doses to public receptors at other sites of interest are also anticipated to be SMALL. The radiation exposure of involved workers is estimated to be well within public health standards and impacts would be SMALL. The impacts to human health from occupational injuries during operation would be SMALL.	IIFP facility. Impacts would be SMALL.
Waste Management	Waste DUO_2 and LLW materials would be disposed of at a licensed LLW disposal facility. There would be no onsite disposal of any solid waste at the IIFP facility. Hazardous wastes would be shipped to a *Resource Conservation and Recovery Act* (RCRA) disposal facility. The quantity of construction and operations hazardous and nonhazardous waste material would result in SMALL impacts that could be managed effectively.	IIFP would obtain the proposed site and complete preconstruction of the IIFP facility. Impacts would be SMALL.
Socioeconomics and Environmental Justice	The NRC staff has determined that impacts of the IIFP facility on tax revenues, housing, and community services for the two-county Region of Interest (ROI), consisting of Lea and Eddy Counties, where most in-migrating construction and operations workers are likely to live, and where the majority of economic impacts	IIFP would obtain the proposed site and complete preconstruction of the IIFP facility. Impacts would be SMALL.
Socioeconomics and Environmental Justice (Continued)	would occur would be SMALL and positive; and where not positive, would still be SMALL. Decommissioning would provide short-term employment, and	IIFP would obtain the proposed site and complete preconstruction of the IIFP facility. Impacts would be SMALL.

Table 2-4. Comparison of Impacts between Proposed Action and No-Action Alternatives (Continued)

Affected Environment	Proposed Action *IIFP would construct, operate, and decommission the proposed IIFP facility in Lea County, New Mexico.*	No-Action Alternative *IIFP would perform preconstruction activities, but would not construct, operate, and decommission the proposed IIFP facility.*
	depending upon the option chosen, the facility could be used for other industry and/or the site for agriculture. All resource impacts are SMALL and the identified minority and low-income populations are not in close proximity to the proposed site, so impacts would not be considered disproportionately high and adverse for any populations in the region, including minority or low-income populations.	
Accidents	NRC regulations and IIFP's operating procedures for the proposed facility would ensure that the high and intermediate probability accident scenarios would be unlikely. Items which mitigate or prevent emergency conditions, and the implementation of emergency procedures and protective actions in accordance with the facility emergency plan, would limit the consequences and reduce the likelihood of accidents that could otherwise extend beyond the proposed facility site and property boundaries. IIFP would be required by NRC and DOT regulations to package and manage the transported waste to minimize the probability of accidental release of radioactive material. IIFP facility design, passive and active engineered controls, and administrative controls would reduce the likelihood of accidents. Therefore, the NRC staff has concluded that accident impacts would be SMALL.	IIFP would obtain the proposed site and complete preconstruction of the IIFP facility. Impacts would be SMALL.

2.2.2   Alternatives Considered But Eliminated

2.2.2.1   Alternative Sites

IIFP conducted a site selection process (IIFP, 2009) to determine a suitable location for the proposed facility. The NRC staff reviewed the IIFP process to determine whether an obviously superior site was identified. This section discusses IIFP's site-selection process, identifies the candidate sites for the proposed FEP/DUP facility, and discusses NRC staff's review of the process, screening criteria, and results used by IIFP for selecting the preferred site.

The IIFP site selection process involved (1) a solicitation of community interest to find potential sites; (2) coarse screening to identify the viable sites among those suggested; (3) fine screening to further narrow to the candidate sites based on the criteria listed in Table 2-5; and (4) final site selection based on quantitative criteria.

Table 2-5.   IIFP's Evaluation Criteria for Fine Screening

Evaluation Criteria	Project Objective	Impact Value
Local community residents must accept and support facility siting	Required	Pass/Fail
Local and state governments must support Regulatory Activities	Required	Pass/Fail
Site cannot be in Seismic Zone 4	Required	Pass/Fail
Site cannot be within 50 km of a quaternary active fault	Required	Pass/Fail
Presence of nearby activities or structures that could be exposed to a hazard by the facility (NUREG-1513)	Regulatory	0.8
Presence of nearby activities or structures that could pose a hazard to the facility (NUREG-1513)	Regulatory	0.8
Commitment of natural resources for site offered including the destruction or diminution of wildlife habitats, flora, woodlands, and marshlands	Regulatory	0.8
Presence of endangered or threatened species, or critical habitat in Endangered Species Act	Regulatory	0.8
Environmental Justice Requirements (minority and low-income populations: multiple effects to be considered)	Regulatory	0.8
Will action cause a violation of Federal, State, local, tribal laws or requirements for protection of environment (Air Quality, Water Quality, other)	Regulatory	0.8
Location of adjacent hazards or hazardous operations leading to cumulative impacts	Regulatory	0.8
State and local government financial incentives	Cost	0.4
Property tax incentive	Cost	0.8
State Income taxes	Cost	0.8
State Sales and use taxes	Cost	0.8
Transportation routes (impacts) for incoming feed material, considering distances & routes	Cost	0.8
Transportation cost to uranium oxide waste disposal site	Cost	0.8
Transportation cost to primary anhydrous HF buyers	Cost	0.8
Schedule time required to license and construct	Schedule	0.4

Table 2-5.   IIFP's Evaluation Criteria for Fine Screening (Continued)

Evaluation Criteria	Project Objective	Impact Value
Existence of chemical or radiological contamination	Regulatory	0.4
Adequate water supply and cost	Cost	0.4
Presence of special interest groups (interveners)	Regulatory	0.4
Acreage Offered (min 640-acres) and cost	Cost	0.4
Waste types generated during construction, operation and demolition, RCRA, etc.	Regulatory	0.4
Cost of construction and operation	Cost	0.4
Electrical supply and cost	Cost	0.4
Gas supply and cost	Cost	0.4
Impact on water quality or water supply (reduction)	Regulatory	0.4
Site characteristics: Geology, topography, seismic	Regulatory	0.2
Decommissioning Requirements	Regulatory	0.2
Site characteristics: depth to frost line	Regulatory	0.2
Infrastructure incentive	Cost	0.2
Contaminants	Regulatory	0.2
Training, accessibility, availability of emergency response personnel / facilities	Regulatory	0.2
Existing environmental data	Regulatory	0.2
Ambient noise levels	Regulatory	0.1
Site characteristics: climatology and meteorology	Regulatory	0.1
Sanitary wastewater treatment availability	Cost	0.1
Availability of road, rail, and airport	Cost	0.1
Buildings offered and terms	Cost	0.1
Condition of land	Cost	0.05
Unemployment insurance tax	Cost	0.05

Source: IIFP, 2009

Potential environmental impacts can be avoided or significantly reduced through proper site selection. IFFP used an approach to select a preferred site based on technical, environmental, safety, and economic considerations (IIFP, 2009). The NRC staff reviewed the site selection process used by IIFP and determined that the process is comprehensive because it takes into account all applicable criteria, structured because it follows from coarse to more fine screening process, and appropriate for identifying and evaluating the proposed site and alternative candidates.

2.2.2.1.1   Solicitation of Interest

IIFP determined that desirable locations for the plant would be proximate to existing, private, DUF_6 sources and near LLW disposal facilities that could accept DUO_2 for disposal. This resulted in IIFP soliciting site proposals from communities in the states of Texas, Idaho, and New Mexico (IIFP, 2011b). The IFP inquiry package requested information about the community and any interest or proposal for attracting and accepting a DUF$_6$ deconversion facility. As a result, six potential sites were identified: one in Texas, two in Idaho, and three in New Mexico.

2.2.2.1.2   First-Phase Screening

IIFP used the following criteria to evaluate the six potential sites:

- acceptance of the proposed facility by community

- acceptance of the proposed facility by state and local governments

- appropriate seismic qualifications (not to be in seismic zone 4)

- no environmental legacy potential liabilities

- location in proximity to customers and waste disposal sites

- availability of utilities infrastructure

The NRC staff reviewed the IIFP's first-phase screening process, elimination criteria, and results and determined that they are reasonable and appropriate because the elimination criteria consists of considerations relevant to the evaluation of potential impacts to environmental resource areas discussed in this EIS. Further, the NRC staff agrees that these elimination criteria allow IIFP to exclude from further consideration certain sites due to their potential environmental impacts.

Sites were excluded from further consideration based on the outcomes of these screening criteria when applied to each potential site: two of the six potential sites were eliminated from consideration. One New Mexico site was eliminated because it was distant from utilities and population centers, and it had no characterization data. One of the Idaho sites was eliminated because it was located on a previous radioactive materials processing site and, thus, had legacy issues that IIFP chose to avoid. As a result, one Texas site, one Idaho site, and two New Mexico sites moved on to IFFP's fine screening. The NRC staff agreed it was appropriate to eliminate the two sites, based on the first-phase screening criteria.

2.2.2.1.3   Second-Phase Screening

The second-phase screening occurred in two rounds in which IIFP evaluated the remaining four sites using used various categories of evaluation criteria. The first round evaluated the four sites on qualitative site-specific criteria and quantitative cost-benefit criteria. The qualitative criteria included public and state support, seismic characteristics, land/soil issues, land/mineral rights, aesthetics, and licensing and permits. The cost-benefit criteria included incentives, infrastructure cost, operating costs, state and local taxes, and transportation costs.

During this screening, two sites (one in New Mexico and one in Idaho) were eliminated because of site-specific features and excessive land and/or infrastructure costs. Subsequently, the communities that had offered the eliminated sites were asked to nominate a second (replacement) site, resulting in a second iteration of first-phase screening. The New Mexico replacement site was rejected in a reiteration of first-phase screening because of numerous oil wells on the site and the complexity of acquiring the land. The Idaho replacement site survived this screening (IIFP, 2011b). This left three sites to undergo the second round of IIFP's second-phase screening: one in Texas, one in Idaho, and one in New Mexico (the Hobbs site).

Table 2-5 identifies the screening criteria used by IIFP in their siting selection process and the criterion's relative importance.

The NRC staff reviewed the IIFP's second-phase screening process and determined that it is reasonable and appropriate as it consists of criteria allowing the applicant to consider potential environmental impacts as a result of the site selection process. Further, appropriate consideration was given to seismic potential, threatened and endangered species and critical habitat, economic considerations, emergency preparedness and response, air quality, climatology and meteorology, water quality and water supply, waste management, acreage, noise, nearby hazards or hazardous operations, and environmental justice. The NRC staff finds that consideration of these criteria is appropriate and comprehensive because it takes into consideration environmental resources such as wildlife habitats, potential for exposure to hazards to the facility, proximity to quaternary active faults, and applicable Federal State, local, and Tribal laws for protecting the environment. The application of the criteria allow for the identification and selection of a site that would be expected to result in reduced potential environmental impacts.

2.2.2.1.4   Final Site Selection

In the final site selection, the sites were evaluated by IIFP using the criteria listed in Table 2-5, assigning an impact value and an evaluation value to each criterion in Table 2-5 (IIFP, 2009). For each potential site the impact value of each criterion listed in Table 2-5 was multiplied by an evaluation number assigned for each potential site. The evaluation number for each criterion ranged from 1 for most favorable to 10 for least favorable potential site. The summation of the product of these multiplications produced the total score for each site. The lower the evaluation score, the more favorable the site. The Hobbs, New Mexico, site was ultimately selected because it has the lowest (best) score.

IIFP determined that the Hobbs site offers overall the most beneficial combination of technical, safety, economic, and environmental factors (IIFP, 2009). IIFP selected the Hobbs, New Mexico, site for the proposed facility in part because it is not near an active fault, there is no legacy chemical or radiological contamination, there are no air quality non-attainment areas in the vicinity, the site is sparsely populated, the availability of water, electricity and natural gas, and public and state, and local support. Consideration was also given to threatened and endangered species, critical habitats, and historic and cultural properties.

The NRC staff reviewed the IIFP process and determined that the process used by IIFP is reasonable and appropriate because the list of criteria is comprehensive and considers elements relevant to the evaluation of potential environmental impacts. It also includes regulatory requirements, and considers costs, scheduling impacts, and community support. The results concluding that the Hobbs site offers overall the most beneficial combination of environmental, technical, safety, and economic factors are reasonable. The NRC staff further concludes that none of the candidate sites is obviously superior to the IIFP preferred site.

2.2.2.2 Alternative Technologies

2.2.2.2.1 Direct Deconversion (IIFP Facility Phase 2)

In Section 2.1.2, a direct conversion process is mentioned as a possible, future Phase 2 licensing action. This technology is very similar to that of the proposed action, but it does not yield the marketable FEP products, SiF_4 and BF_3. In direct conversion, all the fluorides in the DUF_6 would be converted to AHF. As an alternative, IIFP could seek a license for the Phase 2 process without obtaining a license for Phase 1. The process, which directly converts DUF_6 to uranium oxide, mainly as U_3O_8, is described in more detail below.

In the direct conversion to oxide process, the DUF_6 feed would be vaporized in the same type of autoclave as in the proposed DUF_6 to DUF_4 process. The DUF_6 vapor would be fed to a first-stage reaction vessel where it would react with a feed of a vaporized mixture of HF and steam that has been recycled from the back end (distillation system) of the process. The reaction results in the formation of uranyl oxyfluoride (UO_2F_2) and HF.

The UO_2F_2 powder would be withdrawn from the bottom of the reaction vessel and sent to a second-stage reaction vessel where it would undergo a reaction with steam to form U_3O_8 and HF. A more concentrated HF vapor mixture and water would exit the tops of the first and second stage reaction vessels and be condensed using heat exchanger equipment. The condensed and concentrated HF would then be distilled to produce commercial grade AHF. The resulting distillation bottom material of less concentrated HF would be recycled, vaporized, and returned as feed to the first-stage reaction vessel.

U_3O_8 formed in the second-stage reaction vessel would be transferred to storage hoppers. A two-stage dust collector system would control and recycle U_3O_8 dust generated by air or gas flows associated with the solids handling equipment. The U_3O_8 in the storage hoppers would be packaged into DOT-approved shipping containers and transported to an off-site, licensed LLW disposal facility.

The potential environmental impacts of the proposed action and the alternative technology of direct conversion would be nearly the same, because the throughput of DUF_6 would be the same, both processes produce large volumes of AHF (the alternative technology somewhat larger), both produce the same quantity of chemically stabilized uranium (although in slightly different chemical forms), and the basic chemical processes are very similar. Because (1) direct conversion is analyzed as a cumulative impact in this EIS and (2) there is so little difference in the expected environmental impacts between the proposed action and the direct conversion alternative, this alternative is eliminated from consideration in this EIS as a separate alternative.

2.2.2.2.2 DOE Deconversion Technology

DOE has constructed two deconversion facilities, one at the site of the former Paducah Gaseous Diffusion Plant in Paducah, Kentucky and the other at the site of the former Portsmouth Gaseous Diffusion Plant near Piketon, Ohio. These plants were constructed to deconvert the approximately 700,000 metric tons (771,618 tons) of DUF_6 stored at the Paducah plant, the Portsmouth plant, and the East Tennessee Technology Park (formerly the K-25 Gaseous Diffusion Plant) at the Oak Ridge Reservation, Tennessee. Shipment of full DUF_6 cylinders to these plants is allowed as described in Section 2.2.1 (the no-action alternative).

As an alternative technology to the technology proposed in this EIS, IIFP could construct a plant that uses the technology of the DOE plants. The DOE deconversion process reacts DUF_6 with water (steam) and hydrogen to produce U_3O_8 and aqueous HF. The DUF_6 is directly converted to U_3O_8 in a one-stage reaction vessel. HF and water vapor exit the reaction vessel and are collected as aqueous HF. The U_3O_8 solids exit the reaction vessel, and are stored temporarily until they are shipped to a waste disposal site. DOE plans to market the aqueous HF, but any HF that is not sold may have to be treated as a waste liquid. This liquid waste would likely be reacted with lime to form CaF_2 and stored in retention basins or sold. (DOE, 2004a; DOE, 2004b). Assuming the CaF_2 can be sold, it could be used to produce AHF at an industrial AHF production plant.

This alternative is eliminated from consideration in this EIS because:

- The DOE process has already been analyzed in two DOE-prepared NEPA documents (DOE, 2004a; DOE, 2004b), and the impacts of implementing the DOE technology would be sufficiently similar to the proposed action. For this reason no value would be gained by further analyzing this technology as an alternative to the proposed action. The throughput of DUF_6 would be the same, both processes would produce the same quantity of chemically stabilized depleted uranium (although in slightly different chemical forms), and the basic chemical processes are very similar.

- The DOE alternative has greater uncertainty regarding the disposition of the aqueous HF, and could result in higher environmental consequences should the conversion of HF to CaF_2 and then AHF, be required.

2.2.2.2.3   European Deconversion Technology

Three processes can convert DUF_6 to uranium oxide; the IIFP process, the DOE process, and the European process. The European process involves reacting DUF_6 directly with steam in a first-stage reaction vessel, producing aqueous HF and depleted uranyl dioxyfluoride (DUO_2F_2). The DUO_2F_2 is processed further in a second-stage reaction vessel to form aqueous HF and depleted uranium oxide for disposal. The HF is collected in an aqueous form that can be sold or treated (IIFP, 2009).

As with the DOE technology, this alternative is eliminated for the following reasons:

- The impacts of implementing the European technology would be sufficiently similar to the proposed action such that no value would be gained by further analyzing this technology as an alternative to the proposed action. The throughput of DUF_6 would be the same, both processes would produce the same quantity of chemically stabilized uranium (although in slightly different chemical forms), and the basic chemical processes are very similar (IIFP, 2009).

- The European alternative has greater uncertainty regarding the disposition of the aqueous HF, and could result in higher environmental consequences should the conversion of HF to CaF_2 and then AHF, be required.

2.2.2.3   Overseas Shipment of DUF_6 for Deconversion

URENCO and AREVA are foreign companies that operate or are planning to operate enrichment plants in the U.S. These firms own and operate deconversion facilities overseas and could choose to ship their U.S.-generated DUF_6 to those facilities for deconversion. Also,

Russia has recently commissioned a deconversion facility and is planning another. Under this alternative, any of the four U.S.-based commercial enrichment companies could ship their DUF_6 overseas for deconversion. However, this would involve shipping DUF_6 long distances overseas and the uranium oxides would have to be shipped back to United States for licensed disposal (IIFP, 2009). The cost of such shipments would likely be significant.

In its EIS for the National Enrichment Facility (now URENCO USA), the NRC staff (NRC, 2005b) examined three foreign disposition alternatives for DUF_6: Russian re-enrichment, French deconversion or re-enrichment, and Kazakhstan deconversion. The NRC staff concluded, "Due to the costs for disposition in Russia, France, or Kazakhstan, the NRC staff does not consider these alternatives to be viable" (NRC, 2005b).

For reasons discussed above, the NRC staff concludes that overseas shipment of DUF_6 for deconversion is not a reasonable alternative. Thus, this alternative has been eliminated from further analysis.

2.2.2.4 Indefinite Storage of DUF_6

Commercial enrichment facilities in the U.S. could store their DUF_6 at their enrichment facilities, much like DOE has done for decades. As described in Section 2.2.1, No-Action Alternative, the DOE deconversion facilities could eventually (approximately 25 years in the future) take this DUF_6, making this alternative evolve over time to the no-action alternative.

The DNFSB has reported that long-term storage of DUF_6 represents a potential chemical hazard (DNFSB, 1995). DOE policy (DOE, 2000) is that alternatives for the long-term management of DUF_6 include its deconversion to a more stable uranium oxide. DOE evaluated long-term storage in its Programmatic EIS on DUF_6 management (DOE, 1999), but did not select the long-term storage option (64 FR 43358).

In addition to creating a potential chemical hazard, the alternative of indefinite storage of DUF_6 does not meet the need to deconvert this material (as discussed in Section 1.3) and has therefore been eliminated from consideration in this EIS.

2.2.2.5 Commercial Enrichment Plant Deconversion of DUF_6 at Uranium Enrichment Facilities

The four U.S.-based enrichment companies could decide to construct and operate their own deconversion facilities. The only operational commercial enrichment facility in the U.S., the URENCO USA plant near Eunice, New Mexico, has already signed an agreement with IIFP for IIFP to accept URENCO's DUF_6. Furthermore, it is expected that the potential environmental impacts of implementing this alternative at each enrichment facility would be similar to that for the proposed action. However in this event these impacts would occur at up to four locations as a result of the construction of four deconversion facilities rather than just one as would be the case for the proposed action. One deconversion facility for each U.S.-based enrichment company would have greater environmental impacts than the construction of one facility to support all the enrichment facilities, which is the proposed action. Thus, the NRC staff has concluded that this alternative offers no meaningful advantages over the proposed action, and therefore does not warrant further consideration in this EIS.

2.3   Comparison of Predicted Environmental Impacts

Chapter 4 of this EIS presents a detailed evaluation of the potential environmental impacts of the proposed action and the no-action alternative. Table 2-4 summarizes and compares these environmental impacts. A common element between the two alternatives is the occurrence of preconstruction activities. It is assumed that preconstruction activities take place under both alternatives and, therefore, the impacts associated with preconstruction activities would occur regardless of which alternative is selected. As a result, the comparison of alternatives presented in Table 2-4 is intended primarily to highlight the differences between the two alternatives after preconstruction activities have occurred. A standard of significance has been established for assessing potential environmental impacts. In its implementation of the Council on Environmental Quality's regulations on significance (40 CFR 1508.27), NRC staff has assigned each impact one of the following three significance levels, as defined in NRC (2003):

- SMALL. The environmental effects are not detectable or are so minor that they would neither destabilize nor noticeably alter any important attribute of the resource.

- MODERATE. The environmental effects are sufficient to noticeably alter but not destabilize important attributes of the resource.

- LARGE. The environmental effects are clearly noticeable and are sufficient to destabilize important attributes of the resource.

These impact levels are used in the summary and comparison of alternatives in Table 2-4.

2.4   Staff Recommendation Regarding the Proposed Action

After weighing the impacts of the proposed action in Chapter 4 and comparing the impacts of the proposed action and the no-action alternative in Table 2-4, the NRC staff, in accordance with 10 CFR 51.91(d), sets forth its NEPA recommendation regarding the proposed action.

The NRC staff recommends that, unless safety issues mandate otherwise, the proposed license be issued to IIFP. The NRC staff has concluded that potential environmental impacts are in all aspects SMALL or MODERATE, and application of the environmental monitoring program described in Chapter 6 and the proposed IIFP mitigation measures discussed in Chapter 5 would eliminate or substantially lessen any potential adverse environmental impacts associated with the proposed action.

The conclusion of the NRC staff is that the overall benefits of the proposed IIFP Facility outweigh the environmental disadvantages and costs based on consideration of the following:

- The need for a facility to deconvert the domestic stockpile of DUF_6 into more stable depleted uranium oxides; and

- The potential environmental impacts from the proposed action are in most aspects SMALL with the exception of short term construction related air quality impacts and in some cases, beneficial.

2.5   References

(DNFSB, 1995) U.S. Defense Nuclear Facilities Safety Board. 1995. Integrity of Uranium Hexafluoride Cylinders. Technical Report DNFSB/TECH-4. May 5, 1995.

(DOE, undated) U.S. Department of Energy. Undated. Transporting DOE Uranium Oxide from Portsmouth Ohio, and Paducah Kentucky. Portsmouth/Paducah Project Office, Lexington, KY. ADAMS Accession No. ML110120583.

(DOE, 1999) U.S. Department of Energy. 1999. Final Programmatic Environmental Impact Statement for Alternative Strategies for the Long-term Management and Use of Depleted Uranium Hexafluoride, DOE/EIA-0269. Office of Nuclear Energy, Science and Technology, Washington D.C. April 1999. ADAMS Accession No. ML050180302.

(DOE, 2000) U.S. Department of Energy. 2000. Assessment of Preferred Depleted Uranium Disposal Forms. ORNL/TM-2000/161, Oak Ridge, TN. June 2000. ADAMS Accession No. ML060110351.

(DOE, 2004a) U.S Department of Energy. 2004. Final Environmental Impact Statement for Construction and Operation of a Depleted Uranium Hexafluoride Conversion Facility at the Paducah, Kentucky Site, DOE/EIS-0359, Washington, D.C., June 2004. ADAMS Accession No. ML050380331.

(DOE, 2004b) U.S. Department of Energy. 2004. Final Environmental Impact Statement for Construction and Operation of a Depleted Uranium Hexafluoride Conversion Facility at the Portsmouth, Ohio Site, DOE/EIS-0360, Washington, D.C., June 2004. ADAMS Accession No. ML051030392.

(IIFP, 2009) International Isotopes Fluorine Products. 2009. Fluorine Extraction Process and Depleted Uranium De-conversion Plant (FEP/DUP) Environmental Report, Revision A, ER-IFP-001, Revision A. December 27, 2009. ADAMS Accession No. ML100120758.

(IIFP, 2011a) International Isotopes Fluorine Products, Inc. 2011. Fluorine Extraction Process and Depleted Uranium De-conversion Plant (FEP/DUP) Official Responses to Environmental Report RAIs, Revision A. March 31, 2011. ADAMS Accession No. ML110970481.

(IIFP, 2011b) International Isotopes Fluorine Products, Inc. 2011. "Site Selection Information for International Isotopes Fluorine Products, Incorporated (IIFP) Fluorine Extraction and Depleted Uranium Deconversion Facility near Hobbs in Lea County, New Mexico," E-mail, September 22, 2011. ADAMS Accession No. ML11265A322.

(NRC, 2003) U.S. Nuclear Regulatory Commission. 2003. Environmental Review Guidance for Licensing Actions Associated with NMSS Programs. Final Report. NUREG 1748, Division of Waste Management, Washington, D.C. August 2003. ADAMS Accession No. ML032450279.

(NRC, 2005a) U.S. Nuclear Regulatory Commission. 2005. In the matter of Louisiana Energy Services, L.P. (National Enrichment Facility), CLI-05-05, Memorandum and Order. Docket No. 70-3103-ML. January 18, 2005. ADAMS Accession No. ML050180295.

(NRC, 2005b) U.S. Nuclear Regulatory Commission. 2005. Environmental Impact Statement for the Proposed National Enrichment Facility in Lea County, New Mexico, NUREG-1790, Vol. 1. Washington, D.C. June 2005. Available at http://www.nrc.gov/reading-rm/doc-collections/nuregs/staff/sr1790/vl/. ADAMS Accession No. ML051730238.

3.0 AFFECTED ENVIRONMENT

This chapter describes the existing regional and local environmental conditions at and near the proposed IIFP site before any preconstruction activities begin and prior to the proposed action. This chapter presents information on land use; historic and cultural resources; visual resources; climatology, meteorology, and air quality; geology, minerals and soils; water resources; ecological resources; socioeconomic resources; traffic and transportation; noise; and public and occupational health. The data and information presented here provide a baseline against which to assess impacts (Chapter 4) of the proposed action described in Chapter 2 of this EIS.

3.1   Site Location

The proposed IIFP site would occupy Section 27, in Township 18S, Range 36E of southeastern New Mexico, in Lea County. The 259-ha (640-ac) site is approximately 22.5 kilometers (km) (14 miles [mi]) west of Hobbs, New Mexico, 27.4 km (17 mi) west of the Texas/New Mexico border, and 362 km (225 mi) southeast of Albuquerque, New Mexico. The nearest population center is Hobbs, New Mexico, which had an estimated population of 30,838 in 2009 (USCB, 2010a). The nearest important permanent surface water is the Pecos River, approximately 146 km (91 mi) west of the site. The southern boundary of Section 27 is 1.6 km (1 mi) north of U.S. Highway 62/180 (US 62/180) and the western boundary is NM 483. Figures 3-1 and 3-2 depict the 80-km (50-mi) and 10-km (6-mi) radii surrounding the site, respectively, and are referred to in subsequent analyses.

IIFP has set aside approximately 16 ha (40 ac) in the northeast quadrant of the proposed site for the deconversion facility (see Figure 2-2). The remainder of Section 27 would remain as undeveloped (IIFP, 2009). The facility's location was selected to avoid, to the extent possible, utility rights-of-way including overhead transmission lines and underground pipelines. The facility would be enclosed by a security fence with a surveillance road just inside the fence. See Section 2.1.3 for a list of the structures at the facility.

3.2   Land Use

This section includes a description of land use on and near the proposed IIFP site, including a description of offsite areas and the regional setting. For the purposes of this EIS, the Region of Influence (ROI) for land use is defined as the area within a 10-km (6-mi) radius of the center point of the proposed IIFP site (see Figure 3-2 for site location and nearby features within the ROI).

The proposed IIFP site is in Section 27, in Township 18S, Range 36E of southeastern New Mexico, in Lea County, approximately 22.5 km (14 mi) west of Hobbs, New Mexico. Property ownership in the county (which is approximately 1.1 million ha [2.8 million ac] in size) is 17 percent Federal ownership, 31 percent State ownership, and 52 percent private ownership. The Federally owned land is primarily in the southwestern portion of the county, the State-owned land is located throughout the central portion of the county, and the privately owned land primarily extends from north to south in the county's eastern portion. Large tracts of land in Lea County are privately owned by farmers, ranchers, and oil, gas, and mining companies. Urbanized areas near cities and towns include smaller tracts used for residential, municipal, and commercial purposes. Approximately 93 percent of Lea County is used as range land for grazing and approximately 4 percent is used for crop farming. Urban areas and the roadway

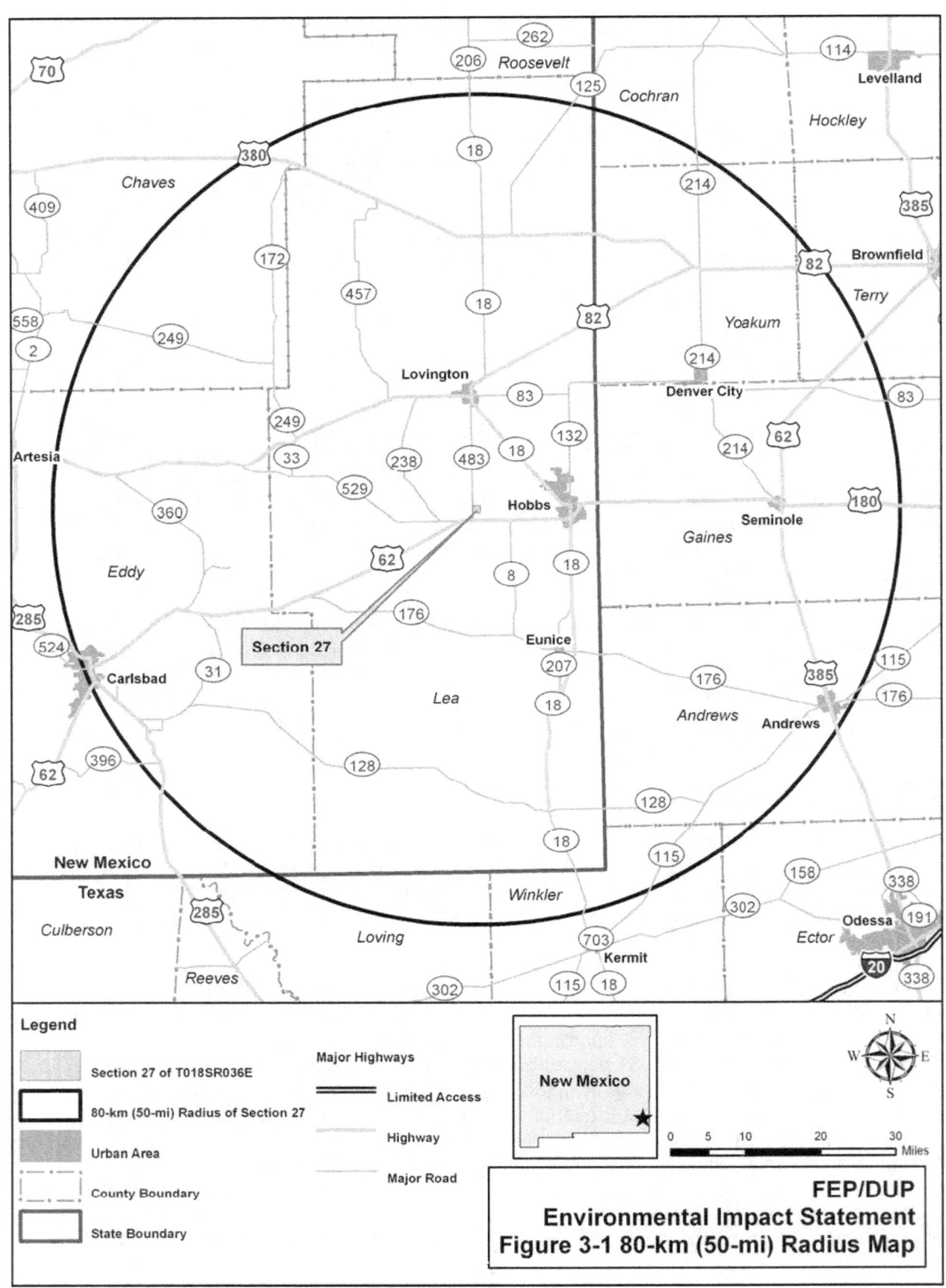

Legend

Section 27 of T018SR036E

80-km (50-mi) Radius of Section 27

Urban Area

County Boundary

State Boundary

Major Highways

━━━ Limited Access

━━━ Highway

━━━ Major Road

New Mexico

0 5 10 20 30 Miles

FEP/DUP
Environmental Impact Statement
Figure 3-1 80-km (50-mi) Radius Map

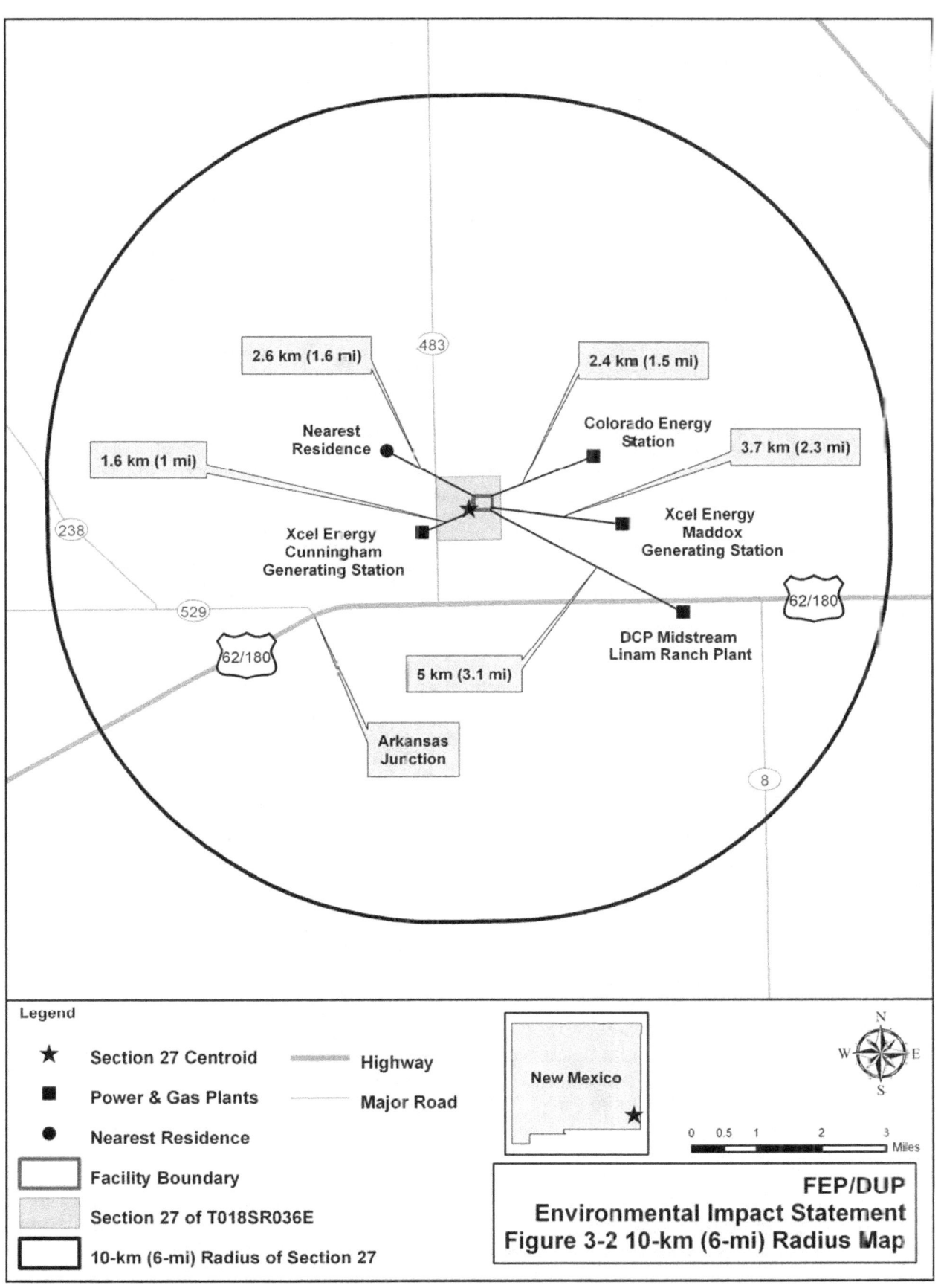

Legend

★ Section 27 Centroid

■ Power & Gas Plants

● Nearest Residence

☐ Facility Boundary

▨ Section 27 of T018SR036E

☐ 10-km (6-mi) Radius of Section 27

═══ Highway

─── Major Road

New Mexico

0 0.5 1 2 3 Miles

N
W   E
S

FEP/DUP
Environmental Impact Statement
Figure 3-2 10-km (6-mi) Radius Map

2.6 km (1.6 mi)

2.4 km (1.5 mi)

Nearest Residence

Colorado Energy Station

1.6 km (1 mi)

3.7 km (2.3 mi)

Xcel Energy Cunningham Generating Station

Xcel Energy Maddox Generating Station

DCP Midstream Linam Ranch Plant

5 km (3.1 mi)

Arkansas Junction

483

238

529

62/180

62/180

8

system account for the county's remaining land use. Most of the land actively farmed in Lea County is irrigated (Leedshill-Herkenhoff et al., 2000).

The public roadways closest to the proposed IIFP site are US 62/180 running east-west and NM 483 running north-south (Figure 3-2). These roadways mostly traverse open range land in the ROI.

The proposed 259-ha (640-ac) IIFP site and the surrounding ROI are largely undeveloped. The land has been used for cattle grazing and for gas and oil development (IIFP, 2009). There are 715 oil or gas wells within the 10-km (6-mi) ROI. Seven of these wells, all of which have been abandoned, are within 1.6 km (1.0 mi) of the site (Oil and gas wells are discussed in more detail in Section 3.6.3). Both overhead and underground utilities, and their associated rights-of-way, cross the site and the ROI. Several overhead transmission lines and underground gas/oil pipelines run generally east to west across the proposed IIFP site. Xcel Energy's Cunningham Station, just west of Section 27, has four groundwater monitoring wells in Section 27 (see Figures 3-2 and 3-14). Three other energy production facilities are within 10 km (6 mi) of the proposed facility: the Colorado Energy Station (also known as the Hobbs Generating Station) is 2.4 km (1.5 mi) northeast of the proposed IIFP site; the Xcel Energy Maddox Generating Station is 3.7 km (2.3 mi) east-southeast of the proposed site; and the DCP Midstream Linam Ranch Plant is 5 km (3.1 mi) southeast of the proposed site (Figure 3-2).

The nearest residence is west-northwest of the site approximately 2.6 km (1.6 mi) from the northern boundary of the site (Figure 3-2). There are no public recreational areas or National Register of Historic Places (NRHP)-listed historic structures or properties within 10 km (6 mi) of the proposed facility.

Other than the proposed IIFP facility (Phase 1 and Phase 2), there are no current developments or proposed developments in the ROI. The proposed IIFP site is not subject to local or county zoning, land use planning, or associated review process requirements; and there are no potential conflicts of land use plans, policies, or controls (Appendix A).

The State of New Mexico and Lea County have transferred ownership of the proposed 259-ha (640-ac) site from the State and Lea County to IIFP. The transfer to IIFP was part of an economic incentives package developed by the Economic Development Corporation of Lea County. The land transfer was carried out in accordance with the *New Mexico Economic Development Act.*

See Section 3.6.4 for a discussion of farmland protection programs in New Mexico.

3.3   Historic and Cultural Resources

This section includes a description of the potential and documented human habitation on and near the proposed IIFP site, and a discussion of significant offsite historic resources and the regional setting. The ROIs for historic and cultural resources are explained later in this section, corresponding to Areas of Potential Effect (APEs), which vary for different resources that could be directly or indirectly affected.

Southeastern New Mexico was settled by humans approximately 12,000 years ago. The cultural sequence in the region includes six chronological periods: the Paleo-Indian period (10,000 B.C. to 7,000 B.C.), the Archaic period (5,000 B.C. to A.D. 1000), the Formative period (A.D. 900 to 1500), the Protohistoric/Spanish Colonial period (1541 to 1800), the Mexican period (1828 to 1834), the Territorial period (1834 to 1912), and the Statehood period (1912 to present) (NMHPD, 2001).

While archeological sites documenting occupation during all of these periods have been identified in southeastern New Mexico, the proposed project site on the Llano Estacado is part of a flat, arid plain without permanent or even intermittent water sources. While the Paleo-Indians used the Llano Estacado for hunting when the climate was less arid (prior to 6000 B C.), it was not hospitable to more extensive human occupation (Rothman and Holder, 1998). Archaeological resource records indicate prehistoric, protohistoric, and historic human occupation in areas of southeastern New Mexico within the Llano Estacado only in areas with reliable potable water, shelter, and food. Therefore, because there is no permanent water source near the site (see Section 3.7.2), the potential for archeological resources is low at the IIFP site and any prehistoric, protohistoric, or historic activity would have been transient. Only isolated artifacts have been found in the vicinity of the proposed IIFP site (Daras, 2009).

3.3.1 Prehistoric Occupation

The initial prehistoric period in New Mexico, the Paleo-Indian period, is characterized by kill sites, camp sites, butchering sites, and lithic quarries associated with small, nomadic groups subsisting by the hunting of now-extinct large game animals such as mammoths and large bison. The Paleo-Indian period is better represented in the southeastern quadrant of the State than in any other area of New Mexico (Main, 1992). Several Paleo-Indian hunting sites have been found on the Llano Estacado although none have been found in the vicinity of the proposed IIFP site (Daras, 2009). During the Archaic period, people became more sedentary as a society, settling in small bands along major watercourses in response to drier conditions. The vicinity of the proposed IIFP site, far from permanent watercourses, would have been unattractive for hunting, gathering, or settlement (Rothman and Holder, 1998). During the Formative (or Ceramic) period tribal groups became increasingly settled and concentrated within villages and base campsites. Formative period sites also were generally located near permanent sources of water, and the proposed site's setting within the Llano Estacado - far from watercourses - remained unsuitable for increasingly sedentary and growing populations (Rothman and Holder, 1998). Therefore, the proposed IIFP site is not expected to yield significant Formative period archaeological resources.

3.3.2 Historic Indian Tribes

By the early 1540s, when Spanish explorer Vasquez de Coronado arrived in New Mexico, the southeastern quadrant of the State was dominated by small hunter-gatherer groups and small settlements along river valleys such as the Pecos (Rothman and Holder, 1998). The groups occupying the region included the Suma, the Tigua, and the Jumano (Gerald, 1974; Kelley, 1986; Hickerson, 1994). In the nineteenth century, these groups were replaced by Apache and Plains Indians, including the Kiowa and Comanche (Hickerson, 1994). Tribal testimony before the U.S. Indian Claims Commission indicates that the proposed IIFP site lies west of a large area used and/or or occupied by the Plains Apache, Comanche, and Kiowa; and east of a large area used by the Mescalero Apache (ICC, 1979). However, the proposed IIFP site was not known to be occupied, or known to have been used, other than for hunting, by tribes who occupied lands to the east (Plains Apache, Comanche, and Kiowa) and west (Mescalero Apache).

Today, the Mescalero Apache Reservation is approximately 190 km (118 mi) west of the proposed site, in northeast Otero County, New Mexico. The Kiowa, Plains Apache, and

Comanche reservations are in south central Oklahoma, approximately 570 km (354 mi) to the northeast. A remnant group of the Tigua (Ysleta del Sur Pueblo) has traditional-use areas, where activities such as hunting and gathering have traditionally occurred, in the general project area, which includes large areas appropriate for traditional cultural uses throughout the region. Therefore, the land containing the proposed IIFP site is not unique in providing traditional cultural use opportunities for Native Americans.

3.3.3 Historic Euro-American Exploration and Settlement

Historic Euro-American interests in the region began with Spanish exploration in the mid- to late-sixteenth century during the Protohistoric/Spanish Colonial Period. There is no indication that any of the Spanish expeditions during the sixteenth and seventeenth centuries ventured near the vicinity of the proposed project site, with almost all activity confined to the river valleys of the Rio Grande and Pecos River (Rothman and Holder, 1998). The Llano Estacado region was one of the last areas of New Mexico to be settled by Euro-Americans, because of the lack of surface water and semiarid climate. No settlement or significant activities took place in the vicinity of the proposed site during the Mexican Period. In the 1810s and 1820s, sheep ranchers, formerly concentrated west of and along the Rio Grande River, began to move into the eastern plains and the Pecos Valley (Merlan, 2010), but would not likely have ventured into the dry Llano Estacado region in the vicinity of the proposed site, and no archeological resources from this period were found in the vicinity of the proposed IIFP site.

During the Territorial Period, some Texas cattle ranchers drove their herds through southeastern New Mexico along the Goodnight-Loving Trail, which followed the Pecos River approximately 80 km (50 mi) west of the proposed site (Clampitt, 2008). Euro-American settlement in the area began in the late nineteenth and early twentieth centuries during the Territorial period, prior to New Mexico achieving statehood in 1912. After the American Civil War, homesteaders established ranches at Monument Springs, near Monument, New Mexico (Anderson, Undated).

The Hat Ranch, established in 1890 and variously known as "Monument Springs" and the "Monument Springs Ranch," was the largest ranch in the area. It operated on more than one million acres of purchased and leased public lands from Seminole, Texas westward to the Pecos River, and northward to the vicinity of Tatum, New Mexico, including the area of the proposed project. The ranch headquarters was established near Monument Springs, about 10 km (6 mi) south-southeast from the southern extent of the proposed IIFP site (Anderson, Undated). As one of New Mexico's first large-scale cattle ranching operations, the Hat Ranch at its peak in the early 1900s had 50,000 head of cattle, 500 saddle horses, 26 water wells, and several windmills and ranch houses (Anderson, Undated; NMMA, Undated), however, no structures associated with the Hat Ranch have been documented on the IIFP site.

During the early Statehood Period, most of the land continued to serve as pasture for cattle grazing. In the open-range tradition, no fences were used to demarcate property lines in Lea County into the 1910s and 1920s (Merlan, 2010). The ranch continues operation today on reduced acreage, and the headquarters remain at Monument Spring, where the ranch owners reside (Hat Ranch, 2010). The Hat Ranch is not listed in or determined eligible for the NRHP, but is listed on New Mexico's State list of historic resources (NMHPD, 2011).

3.3.4   Archaeological and Historic Resources at the Proposed IIFP Site

This section describes the historic and cultural resources in the vicinity of the proposed IIFP facility. Historic and cultural resources include archaeological sites, architectural resources (such as historic structures, objects, districts, or landscapes) and places of cultural importance to groups for maintaining their heritage. Cultural resources are nonrenewable; that is, once adversely altered, the information contained in cultural resources cannot be recovered.

The *National Historic Preservation Act of 1966*, as amended (NHPA), requires that all adverse effects to *National Register of Historic Places* (NRHP)-listed and eligible historic and cultural resources be considered during Federal undertakings, such as the NRC licensing activity for the proposed IIFP facility. The requirement to consider adverse effects to cultural resources takes the form of a consultation process and/or mitigation. A resource is eligible for listing on the NRHP if it meets at least one of the following four criteria (36 CFR 60.4): (1) association with an historic person, (2) association with a historic event, (3) representation of the work of a master, or (4) potential to provide information on the history or prehistory of the United States.

Section 106 of the NHPA identifies the process for considering whether a project would affect significant cultural resources, which are discussed in Chapter 4 of this EIS. The archaeological APE for the Section 106 review for the proposed IIFP facility is the 16 ha (40 ac) proposed facility site (Figure 2-2), the access road from NM 483, and an external parking lot immediately outside the security fence surrounding the facility property all of which would be directly affected by construction. The architectural resource APE includes the viewshed from any NHRP-eligible historic resources which would include the proposed facility structures. The APE also considers the potential auditory or direct physical impacts of the project to historic architectural resources A distance of 10 km (6 mi) from the project site was determined to be an appropriate APE, based on the height of the facility security fence and the facility buildings, which would generally be less than 6 m (20 ft) high.

The Section 106 process requires consultation between the lead Federal agency and the State Historic Preservation Office (SHPO), which is the custodian of information on cultural resources for the State. The Section 106 process also requires that Federally recognized Native American Tribes who have ancestral interest in the property be consulted to determine if resources important to the Tribe are present (36 CFR 800.2(4)(c)(ii)).

Information on historic and archaeological resources at the proposed IIFP site was obtained by a file review completed by the applicant's archaeological resources consultant at the Historic Preservation Division of the New Mexico Office of Cultural Affairs on April 17, 2009, and from an archaeological survey conducted by IIFP's archaeological consultant in response to the New Mexico State Land Office for the proposed project in May 2009 (Daras, 2009).

According to the Archaeological Resource Management Records Section of the New Mexico Office of Cultural Affairs, the proposed IIFP site had not been the focus of a cultural resource survey, and no archaeological or historic architectural resources were identified at the proposed site prior to the 2009 cultural resources survey (Daras, 2009) conducted by the applicant.

The New Mexico State Register of Cultural Resource Properties lists one historic resource within 10 km (6 mi) of the proposed site. This is the cluster of stone ranch houses and outbuildings that make up the Hat Ranch Headquarters (Site LA 43256, SR #162), which is approximately 9.5 km (5.9 mi) from the southern boundary of the proposed site. As noted earlier, the Hat Ranch Headquarters has not been listed in or determined eligible for the NRHP

(NMHPD, 2011); therefore, it is not an historic property subject to protection under Section 106 of the *National Historic Preservation Act.*

IIFP's archaeological consultant conducted a cultural resource survey in May 2009 on the entire 249-ha (640-ac) Section 27 tract. The survey was conducted according to New Mexico's Cultural Properties and Historic Preservation, Standards for Survey and Inventory (NMHPD, 2005 and consisted of systematic surface pedestrian coverage at 15-m (49-ft) intervals. The survey identified three isolated artifacts, but no archaeological sites. The isolated artifacts were the distal end of a San Jose (Archaic period) chert projectile point, a gray quartzite hammerstone, and three decolorized manganese glass vessel fragments. In accordance with Section 18-6-5 of the New Mexico Cultural Properties Act, the archaeological consultant completed a "negative survey report" to describe the survey and record the isolated artifact occurrences. The isolated occurrences were recorded, and the State Land Office and their contract archaeologists recommended no further archaeological studies or evaluations (Daras, 2009). The State Land Office cultural resources survey did not note the presence of any historic structures or structural remains.

Consultation with Native American Tribes was undertaken using a list maintained by the New Mexico (NM) SHPO. The NRC staff received three responses (Appendix B). On July 13, 2010, the Ysleta del Sur Pueblo stated that the Pueblo believes the project will not adversely affect traditional, religious, or culturally significant sites, but requested consultation should human remains or artifacts regulated by the *Native American Graves Protection and Repatriation Act* be discovered. On June 15, 2011, the Tribal Historic Preservation Officer for the Comanche tribe noted that he had no comments on the project. On July 13, 2011, the Shawnee Tribe's Tribal Historic Preservation Department concurred that no known historic properties would be impacted by the project, but also requested consultation if archaeological materials are encountered during construction or operation of the facility.

In a letter dated October 14, 2010, from David C. Eck, Trust Land Archaeologist, in the office of the State of New Mexico Commissioner of Public Lands, to Jan Biella, New Mexico Historic Preservation Division (Appendix B), the Commissioner's Office states that it "recommends a finding of no effect/no cultural properties /no historic properties... There are no documented cultural properties within the Area of Potential Effect (APE) when considering direct effects. Similarly, there are no registered cultural properties within the assumed, five-mile APE when considering indirect effects." The New Mexico State Historic Preservation Officer concurred with this determination (Appendix B).

3.4   Visual Resources

This section includes a description of the visual setting on and near the proposed IIFP site. For the purposes of this EIS, the ROI for visual resources is defined as the area within a 10 km (6 mi) radius of the center point of the proposed IIFP site; accounting for the view of, and view from, the proposed facility.

The 259-ha (640-ac) site is flat and sparsely developed with a few irregularly spaced structures for natural gas and oil extraction, and overhead transmission lines. The proposed IIFP facility would be approximately 22.5 km (14 mi) west of the nearest population center, Hobbs, New Mexico, and would not be visible from Hobbs. As noted in Section 3.2, four energy industry facilities are less than 10-km (6-mi) of the proposed site, and three are visible from the site. The proposed IIFP site is 1.6 km (1 mi) north of US 62/180 and is bordered on the west by NM 483.

Although the proposed IIFP facility would be located near the center of the site, it is anticipated that it would be visible from both roads.

No mountain ranges are in the site vicinity. The landscape of the site and vicinity is typical of a semi-arid climate and consists of caliche soils with Plains vegetation such as mesquite bushes and native prairie grasses.

As noted in Section 3.2, no recreational areas are within 10 km (6 mi) of the site. The closest recreation facilities are golf courses 12 km (7.5 mi) east and northeast (Hobbs Country Club, 2010; Ocotillo Park Golf Course, 2010), and a motorsports park, also 12 km (7.5 mi) northeast (Hobbs Motorsport Park, 2010). These recreational facilities are not within the visual resources ROI of the site. The nearest residence is approximately 2.6 km (1.6 mi) northwest of the site boundary. A State-listed historic site, Monument Springs, is approximately 10 km (6 mi) southeast of the site.

IIFP assessed the scenic quality of the proposed IIFP site using the U.S. Bureau of Land Management (BLM) visual resource inventory process (BLM, 2007). A visual rating is determined by evaluating potential impacts of a proposed project on the surrounding area. Classes range from Classes I and II (most valued), through Class III (moderate value), to Class IV (least valued). Based on the visual resource inventory, the proposed IIFP site was determined to be in Class IV (IIFP, 2009). This rating means that the level of change to the characteristic landscape can be high, and allows for the greatest level of landscape modification.

3.5   Climatology, Meteorology, and Air Quality

3.5.1   Climatology

The climate in the region of the proposed IIFP site is semi-arid with mild temperatures, low precipitation and humidity, and a high evaporation rate. The weather is often dominated in the winter by high-pressure systems in the central part of the U.S. and low-pressure systems in north-central Mexico. The region is typically affected by low-pressure systems located over Arizona in the summer (WRCC, 2010).

The mean monthly temperature over the last 98 years has ranged from 5.7°C (42.2°F) in January to 26.8°C (80.2°F) in July. July is the hottest month with an average maximum of 34.3°C (93.9°F) and an average minimum of 19.2°C (66.6°F). January is the coldest month with an average minimum of -2.3°C (27.9°F) and an average maximum of 13.6°C (56.4°F) (WRCC, 2010).

The average annual total rainfall in Hobbs, New Mexico, is 40.54 cm (15.96 in). Average monthly rainfall ranges from 1.14 cm (0.45 in) in January to 6.58 cm (2.59 in) in September. The mean annual snowfall in Hobbs is 12.7 cm (5.01 in) (WRCC, 2010).

Thunderstorms occur in Lea County throughout the year, though they are most common in the spring and summer, and occasionally include hail. Thunderstorms occur on average 36 days per year. Summer rains fall almost entirely during brief, but frequently intense, thunderstorms. Rain showers and thunderstorms from June through September account for more than half the annual precipitation. The general southeasterly circulation towards the Gulf of Mexico brings moisture from the Pacific into New Mexico, and strong surface heating, combined with lifting as the air moves over higher terrain, causes air currents and condensation. As storms move inland from the Pacific Coast, much of the moisture is precipitated over the coastal and inland mountain ranges of California, Nevada, Arizona, and Utah. Much of the remaining moisture falls on the western slope of the Continental Divide, which is west of the proposed site, and over northern and high-central mountain ranges. Winter is the driest season in eastern New Mexico.

On average, nine tornadoes are reported per year in New Mexico. Ninety-two tornadoes were reported in Lea County from January 1950 to May 2010. There has only been one tornado reported in Hobbs, New Mexico, in May 1997 (NCDC, 2010).

Wind speeds in New Mexico are usually moderate, although relatively strong winds often accompany occasional weather fronts during late winter and spring and sometimes occur just in advance of thunderstorms. Frontal winds may exceed 48.3 kilometers per hour (km/hr) (30 miles per hour [mph]) for several hours and reach peak speeds of more than 80.5 km/hr (50 mph) (WRCC, 2010). Winds are generally stronger in the eastern plains, to the north of the proposed facility, than in other parts of the state. Winds are generally from the southeast in summer and from the west in winter, but local surface wind directions will vary greatly because of local topography and mountain and valley breezes (WRCC, 2010).

Blowing sand may occur occasionally in the area due to the combination of strong winds, sparse vegetation, and the semi-arid climate. High winds associated with thunderstorms are frequently a source of localized blowing dust. Dust storms that cover an extensive region are rare; and those that reduce visibility to less than 1.6 km (1 mi) occur only with the strongest pressure gradients such as those associated with intense extra-tropical cyclones which occasionally form in the area during winter and early spring (DOE, 2006).

3.5.2 Greenhouse Gases

Greenhouse gases (GHGs) include those gases, such as carbon dioxide (CO_2), water vapor, nitrous oxide (N_2O), methane (CH_4), hydrofluorocarbons (HFCs), perfluorocarbons (PFCs), and sulfur hexafluoride (SF_6), that are transparent to solar (short-wave) radiation but opaque to long wave (infrared) radiation from the earth's surface. The net effect over time is a trapping of absorbed radiation and a tendency to warm the planet's surface and the boundary layer of the earth's atmosphere, which constitute the "greenhouse effect" (IPCC, 2007). Some direct GHGs[1] (CO_2, CH_4, and N_2O) are both naturally occurring and the product of industrial activities, while others, such as the hydrofluorocarbons, are man-made and are present in the atmosphere exclusively due to human activities. Each GHG has a different radiative forcing potential, which is defined as the gas' ability to affect a change in climatic conditions in the troposphere[2] (IPCC, 2007). The radiative efficiency of a GHG is directly related to its concentration in the atmosphere.

As a way to compare the radiative forcing potentials of various GHGs without directly calculating changes in their atmospheric concentrations, an index known as the Global Warming Potential (GWP) (IPCC, 2007) has been established with CO_2 as the reference point[3]. GWPs are calculated as the ratio of the radiative forcing that would result from the emission of 1 kg

---

[1] Direct GHGs are those gases that can directly affect global warming once they are released into the atmosphere.
[2] Radiative forcing potential is expressed as the amount of thermal energy [in watts] trapped by the gas per square meter of the earth's surface.
[3] Water vapor is the most abundant and most dominant greenhouse gas in the atmosphere. However, it is neither long-lived nor well mixed in the atmosphere, varying from 0 to 2 percent. CO_2 is the most abundant of GHGs released to the atmosphere after water vapor.

(2.2 lbs) of a GHG to that which would result from the emission of 1 kg (2.2 lbs) of CO_2 over a fixed period of time. GWPs represent the combined effect of the amount of time each GHG remains in the atmosphere and its ability to absorb outgoing thermal infrared radiation. As the reference point in this index, CO_2 has a GWP of 1. On the basis of a 100-year time horizon, GWPs for other key GHGs are as follows: 21 for CH_4, 310 for N_2O, 11,700 for HFC-23, and 23,900 for SF_6 (IPCC, 2007). Indirect GHGs, carbon monoxide (CO), nitrogen oxides (NO_x)[4], nonmethane volatile organic compounds (NMVOCs), and sulfur dioxide (SO_2), indirectly affect terrestrial solar radiation absorption by influencing the formation and destruction of tropospheric and stratospheric ozone or, in the case of SO_2, by affecting the absorptive characteristics of the atmosphere.

3.5.2.1 Greenhouse Gas Emissions and Sinks in the United States

The U.S. Environmental Protection Agency (EPA) is responsible for preparation and maintenance of the official U.S. Inventory of Greenhouse Gas Emissions and Sinks to comply with existing commitments under the United Nations Framework Convention on Climate Change. GHG sinks are those activities or processes that can remove GHGs from the atmosphere. GHG emissions[5] are reported in sectors, using the GWPs established in the Second Assessment Report of the Intergovernmental Panel on Climate Change (IPCC)[6]. Site preparation, construction, operation, and decommissioning of the proposed IIFP facility would result in the release of GHGs as a result of the same human activities that were identified by EPA as the sources of GHGs in the U.S. Inventory. Results of the most recent report on the U.S. Inventory of GHG Emissions and Sinks (EPA, 2010a) for direct GHGs that are most relevant to the proposed IIFP facility include:

- The primary GHG emitted by human activities in the U.S. was CO_2, representing approximately 85.1 percent of the total GHG emissions.

- In 2008, total U.S. GHG emissions were 6,957 teragrams of CO_2 equivalent (Tg CO_2 Eq), an increase of 14 percent from 1990.

- Overall emissions of GHGs fell 2.9 percent from 2007 to 2008.

- CO_2 emissions for 2008 were 5,921.2 Tg CO_2 Eq, of which 5,572.8 was the result of the combustion of fossil fuel primarily related to electricity generation (2,363.5), transportation (1,785.3), industrial applications (819.3), residential heating (342.7), and commercial applications (219.5) (this considers only fossil fuel emissions in the 50 U.S. states, not the U.S. territories).

- Fifty-three percent of the CO_2 emissions related to transportation were the result of consumption of gasoline in privately owned vehicles; the remainder was from combustion of fuels in diesel trucks and aircraft.

- Emission of CH_4 in 2008 as a result of combustion of fossil fuels in mobile sources was 2.0 Tg CO_2 Eq.

---

[4]NO_x represents all thermodynamically stable oxides of nitrogen, excluding nitrous oxide (N_2O).
[5]In keeping with the GWP convention that names CO_2 as the reference gas, assigning it a GWP of 1, GWPs of other direct GHGs are expressed as equivalents (Eq.) of CO_2, as teragrams (Tg) of CO_2 equivalent (Tg CO_2 Eq.). One Tg is equal to 10^{12} grams, or one million metric tons (1.102 million tons).
[6]IPCC assessment reports are a compilation of separate reports of the various working groups that are established by the Panel. IPCC periodically updates assessment reports to incorporate newly established data, including revisions to GWPs and radiative forcing potentials of GHGs.

- Emission of N_2O in 2008 as a result of combustion of fossil fuels in mobile sources was 26.1 Tg CO_2 Eq.

- Emission of HFCs (released from equipment) in 2008 was 113 Tg CO_2 Eq.

- Emission of SF_6 in 2008 as a result of electrical transmission and distribution[7] was 13.1 Tg CO_2 Eq.

- The primary GHG sinks functional in 2008 included carbon sequestration in forests, trees in urban areas, agricultural soils, and landfilled yard trimmings and food scraps, all of which, in aggregate, offset 13.5 percent of the total GHG emissions in 2008.

- The most significant emissions of indirect GHGs in 2008 included:

  o 13,578 Tg CO_2 Eq. of NOx, primarily from mobile fossil fuel combustion (7,441), stationary fuel combustion (5,148), and industrial processes (520).

  o 60,739 Tg CO_2 Eq. of CO, primarily from mobile fossil fuel combustion (51,533), stationary fossil fuel combustion (4,792), and industrial processes (1,682).

  o 13,254 Tg CO_2 Eq. of NMVOCs, primarily from mobile fossil fuel combustion (5,447), solvent use (3,834), industrial processes (1,804), and stationary fossil fuel combustion (1,321).

  o 10,368 Tg CO_2 Eq. of SO_2, primarily from stationary fossil fuel combustion (8,891), industrial processes (795), and mobile fossil fuel combustion (472).

As noted above, consumption of fossil fuels for electricity generation represents the single greatest source of CO_2 emissions in 2008 (2,363.5 Tg CO_2 Eq.). The total gross GHG emissions in the United States from all sectors (transportation, industrial, residential, and commercial) in 2008 were 6,957 Tg CO_2 Eq. Net emissions (considering all emissions and sinks) were 6,016.4 Tg CO_2 Eq.

3.5.2.2   Greenhouse Gas Emissions in New Mexico

A review of statewide emissions of GHGs can inform an understanding of the impact anticipated GHG emissions from the proposed IIFP facility would have in a regional context. Among the United States, New Mexico ranks 35th with respect to GHGs emissions and 35th in population, based on 2003 data (CRS, 2007). The Center for Climate Strategies[8], published a report in November 2006 on New Mexico's Greenhouse Gas Inventory and Reference Case Projections for the period 1990–2020 (CCS, 2006). Table 3-1 shows New Mexico GHG emissions by sector. Table 3-2 compares the most recent GHG inventories by sector in New Mexico with the United States as a whole in calendar year 2000.

---

[7]SF_6 is a gas at standard conditions and is used as a dielectric medium in high-voltage electrical equipment.
[8]The Center for Climate Strategies is a public-purpose, nonprofit, nonpartisan 501(c)(3) partnership organization established in 2004 to assist in climate policy development at the Federal and State levels.

Table 3-1.   New Mexico GHG Emissions, by Sector

Sector	Carbon Dioxide Equivalents (million metric tons)			
	1990[1]	2000[1]	2010[2]	2020[2]
Electricity Production	29.5	33.2	33.3	38.8
Coal	28.0	30.7	30.4	35.5
Natural Gas	1.4	2.5	2.9	3.2
Oil	0.0	0.0	0.0	0.0
Residential/Commercial/Non-Fossil Industrial (RCI)	7.0	7.3	8.5	9.9
Coal	0.1	0.2	0.2	0.2
Natural Gas	3.8	4.6	4.5	5.4
Oil	3.1	2.5	3.8	4.3
Wood (CH_4 and N_2O)	0.0	0.0	0.0	0.0
Transportation	11.0	14.2	17.6	22.3
On-road Gasoline	7.2	8.7	10.2	12.2
On-road Diesel	2.5	4.2	5.6	7.9
Natural Gas, LPG, Other	0.1	0.1	0.1	0.1
Jet Fuel and Aviation Gasoline	1.2	1.2	1.6	2.0
Fossil Fuel Industry	15.2	19.5	20.3	20.7
Natural Gas Industry	12.7	17.0	17.3	17.7
Oil Industry	2.3	2.3	2.3	2.3
Coal Mining (Methane)	0.2	0.2	0.7	0.7
Industrial Processes[3]	0.5	1.5	2.0	2.8
ODS Substitutes	0.0	0.5	1.3	2.3
PFCs in Semi-conductor Ind.	0.1	0.5	0.2	0.1
SF_6 from Electric Utilities	0.2	0.1	0.1	0.0
Cement & Other Industry	0.2	0.4	0.4	0.4
Waste Management	0.8	1.2	1.4	1.2
Solid Waste Management	0.6	1.0	1.1	0.9
Wastewater Management	0.2	0.2	0.3	0.3
Agriculture	4.5	6.0	6.4	6.7
Manure Management & Enteric Ferment (CH_4)	2.3	3.5	4.1	4.4
Agricultural Soils (N_2O)	2.2	2.4	2.3	2.3
Total Gross Emissions	68.5	82.9	89.4	102.4
Forestry and Land Use	-20.9	-20.9	-20.9	-20.9
Net Emissions (incl. forestry)	47.6	62.0	68.5	81.5

Source: CCS, 2006
[1] Historical estimates
[2] Projected estimates
[3] The proposed facility would be classified in the Industrial Processes sector.

Table 3-2.   Comparison of New Mexico vs. U.S. GHG Emissions (Percent) by Sector[1]

Sector	Percent of State Total GHG Emissions[2]	Percent of U.S. GHG Emissions[2]
Electricity	40	32
Fossil fuel industry (CH_4)	23	3
Transportation	17	26
Agriculture	7	7
Residential/commercial fuel use	5	9
Non-Fossil Industrial fuel use	4	14
Waste	2	4
Industrial processes	2	5

Source: CCS, 2006
[1] All data, calendar year 2000
[2] As shown in Table 3-1, total net CO_2 emissions for New Mexico for the year 2000 were 62 million metric tons (68 million tons) of CO_2 equivalents. For the United States for that same year, total net CO_2 emissions were 5,977 million metric tons (6,588 million tons) (EPA, 2010a)

In March 2010, the NMED published the Inventory of New Mexico Greenhouse Gas Emissions: 2000-2007 which presented estimates of historical New Mexico anthropogenic GHG emissions. This information was compiled to support efforts to address anthropogenic climate change, including those of the Climate Change Action Implementation Team, which was created by Executive Order 2006-69 – New Mexico Climate Change Action. As reported in the inventory, after a 3 percent annual GHG emissions growth rate experienced from 1990 to 2000, the total (gross) direct emissions in New Mexico remained essentially level from 2000 to 2007 despite a 6.7 percent growth in New Mexico's population over that period (NMED, 2010a).

3.5.2.3   Projected Impacts from Construction and Operation of the Proposed IIFP Facility on Carbon Dioxide and Other Greenhouse Gases

Site preparation, construction, operation, and decommissioning of the proposed IIFP facility can be expected to result in emissions of CO_2 and other GHGs through various mechanisms, primarily from combustion of fossil fuels in both mobile and stationary sources. Individual contributions of construction and operations are discussed in Chapter 4. Transportation volumes used in the following sections were established in Section 4.1.2.9 and are applied here without modification.

3.5.3   Meteorology

The closest National Climatic Data Center Cooperative Network weather station to the IIFP site with the longest length of service is the Hobbs weather station, at the Hobbs Regional Airport, approximately 13 km (8 mi) east of the proposed site, which has been in service since 1912. The most recent data available for the Hobbs weather station from the Western Regional Climate Center are from July 2010.

Table 3-3 presents a summary of temperatures from the Hobbs weather station from 1912 to 2010. July, on average, is the hottest month and January is the coldest month. The highest temperature measured over the period of record, 45.6°C) (114°F), occurred in June 1998. The lowest temperature measured, -21.7°C (-7°F), occurred in January 1962.

Table 3-3.   Monthly Temperature in Hobbs, New Mexico, 1912 to 2010

Month	Monthly Averages			Daily Extremes			
	Maximum	Minimum	Mean	High	Date	Low	Date
January	13.6°C (56.4°F)	-2.3°C (27.9°F)	5.7°C (42.2°F)	28.3°C (83°F)	1/11/1953	-21.7°C (-7°F)	1/11/1962
February	16.6°C (61.8°F)	0.0°C (32.0°F)	8.3°C (46.9°F)	30.6°C (87°F)	2/12/1962	-18.9°C (-2°F)	2/2/1985
March	20.6°C (69.1°F)	3.1°C (37.5°F)	11.8°C (53.3°F)	35.0°C (95°F)	3/27/1971	-17.2°C (1°F)	3/2/1922
April	25.4°C (77.8°F)	7.9°C (46.3°F)	16.7°C (62.1°F)	36.7°C (98°F)	4/30/1928	-7.8°C (18°F)	4/4/1920
May	29.8°C (85.6°F)	12.9°C (55.3°F)	21.3°C (70.4°F)	41.7°C (107°F)	5/30/1951	1.1°C (34°F)	5/2/1916
June	33.9°C (93.0°F)	17.5°C (63.5°F)	25.7°C (78.2°F)	45.6°C (114°F)	6/27/1998	4.4°C (40°F)	6/3/1919
July	34.4°C (93.9°F)	19.2°C (66.6°F)	26.8°C (80.2°F)	43.3°C (110°F)	7/15/1958	10.0°C (50°F)	7/1/1927
August	33.4°C (92.2°F)	18.7°C (65.6°F)	26.1°C (78.9°F)	41.7°C (107°F)	8/9/1952	8.3°C (47°F)	8/29/1916
September	29.9°C (85.8°F)	15.2°C (59.3°F)	22.5°C (72.5°F)	40.6°C (105°F)	9/5/1948	1.1°C (34°F)	9/23/1948
October	25.0°C (77.0°F)	9.1°C (48.4°F)	17.1°C (62.7°F)	36.7°C (98°F)	10/3/2000	-11.1°C (12°F)	10/29/1917
November	18.4°C (65.2°F)	2.7°C (36.8°F)	10.6°C (51.0°F)	31.1°C (88°F)	11/1/1952	-15.6°C (4°F)	11/29/1976
December	14.3°C (57.7°F)	-1.4°C (29.4°F)	6.5°C (43.7°F)	28.9°C (84°F)	12/9/1922	-18.3°C (-1°F)	12/24/1983

Source: WRCC, 2010

Table 3-4 summarizes precipitation at the Hobbs weather station from 1912 to 2010. September, on average, is the wettest month, while January and February receive the least precipitation. The one-day maximum rainfall of 19.05 cm (7.5 in) occurred in September 1995.

The NRC staff prepared an EIS for the National Enrichment Facility in Eunice, New Mexico (NRC, 2005). The NRC staff examined climatology data from four weather stations in the area: Eunice, New Mexico; Hobbs, New Mexico; Midland-Odessa, Texas; and Roswell, New Mexico. Table 3-5 describes these weather stations' locations relative to the proposed IIFP site, and the historic records available for each station.

The data presented in the National Enrichment Facility EIS indicate that the general wind patterns for Midland-Odessa, Hobbs, and Eunice were similar (NRC, 2005). Roswell data appeared to have a stronger northerly and westerly component.

Midland-Odessa and Hobbs had comparable climate data based on a comparative analysis of meteorological data at the four weather stations nearest the proposed IIFP site (Table 3-5).

Table 3-4.   Monthly Precipitation in Hobbs, New Mexico, from 1912 to 2010

Month	Rainfall						Snowfall		
	Mean	High	Year	Low	Year	1-Day Maximum	Mean	High	Year
January	1.14 cm (0.45 in)	7.52 cm (2.96 in)	1949	0.00	1924	3.07 cm (1.21 in) 1/11/49	3.30 cm (1.3 in)	31.75 cm (12.5 in)	1983
February	1.19 cm (0.47 in)	6.22 cm (2.45 in)	2010	0.00	1917	5.08 cm (2.0 in) 4/10	2.79 cm (1.1 in)	36.32 cm (14.3 in)	1973
March	1.42 cm (0.56 in)	7.57 cm (2.98 in)	2000	0.00	1918	5.08 cm (2.00 in) 3/20/02	1.27 cm (0.5 in)	25.40 cm (10.0 in)	1958
April	2.01 cm (0.79 in)	13.13 cm (5.17 in)	1922	0.00	1917	4.75 cm (1.87 in) 4/20/26	0.51 cm (0.2 in)	22.86 cm (9.0 in)	1983
May	5.05 cm (1.99 in)	35.13 cm (13.83 in)	1992	0.00	1938	13.21 cm (5.20 in) 5/22/92	0.0	0.0	1913
June	4.78 cm (1.88 in)	23.62 cm (9.30 in)	1921	0.00	1924	11.23 cm (4.42 in) 6/7/18	0.0	0.0	1913
July	5.38 cm (2.12 in)	23.90 cm (9.41 in)	1988	0.00	1954	11.35 cm (4.47 in) 7/19/88	0.0	0.0	1913
August	6.12 cm (2.41 in)	23.29 cm (9.17 in)	1920	0.10 cm (0.04 in)	1938	11.30 cm (4.45 in) 8/9/84	0.0	0.0	1913
September	6.58 cm (2.59 in)	32.99 cm (12.99 in)	1995	0.00	1939	19.05 cm (7.50 in) 9/15/95	0.0	0.0	1913
October	4.01 cm (1.58 in)	20.70 cm (8.15 in)	1985	0.00	1917	14.22 cm (5.60 in) 10/9/85	0.25 cm (0.1 in)	11.43 cm (4.5 in)	1976
November	1.42 cm (0.56 in)	11.00 cm (4.33 in)	1978	0.00	1915	9.65 cm (3.80 in) 11/4/78	1.52 cm (0.6 in)	41.91 cm (16.5 in)	1980
December	1.42 cm (0.56 in)	12.90 cm (5.08 in)	1986	0.00	1917	4.72 cm (1.86 in) 12/21/42	2.54 cm (1.0 in)	24.13 cm (9.5 in)	1986
Annual	40.54 cm (15.96 in)	81.76 cm (32.19 in)	1941	13.41 cm (5.28 in)	1917	19.05 cm (7.50 in) 9/15/95	12.7 cm (5.0 in)	68.83 cm (27.1 in)	1980

Source: WRCC, 2010

Table 3-5.   Weather Stations Located Near the Proposed IIFP Site

Station	Distances and Direction from Proposed Site	Length of Record (years)[1]	Station Elevation
Eunice, New Mexico	34 km (21 mi) south of site	1 (1993)	1,050 m (3,445 ft)
Hobbs, New Mexico	13 km (8 mi) east of site	16 (1982-1997)	1,115 m (3,658 ft)
Midland-Odessa, Texas	138 km (86 mi) southeast of site	16 (1982-1997)	872 m (2,861 ft)
Roswell, New Mexico	129 km (80 mi) northwest of site	16 (1982-1997)	1,118 m (3,668 ft)

Source: NRC, 2005 and WRCC, 2010
[1]Years of compiled data for climatology analysis.

Because Midland-Odessa was a first-order weather station with data completeness exceeding EPA requirements, NRC staff used the data from the Midland-Odessa weather station for its dispersion modeling for the National Enrichment Facility EIS. Hourly meteorological observations at Midland-Odessa were used to generate wind rose plots. Monthly wind speeds and prevailing wind directions at Midland-Odessa for the years 1987 to 1991 are presented in Figure 3-3. The annual mean wind speed was 17.7 km/hr (11 mph) and the prevailing wind direction was 180 degrees with respect to north. The maximum 5-second wind speed was 112.7 km/hr (70 mph) (NRC, 2005). At Hobbs, the average wind speed varied from 16.1 km/hr (10.0 mph) for the month of August to 21.6 km/hr (13.4 mph) for the month of April. The annual average wind speed recorded at Hobbs was 18.3 km/hr (11.4 mph). The prevailing wind direction was out of the north blowing to the south.

3.5.4   Air Quality

3.5.4.1   Regulatory Setting

3.5.4.1.1   Criteria Pollutants

Under the *Clean Air Act*, the EPA has established National Ambient Air Quality Standards (NAAQS) that specify the maximum concentrations for seven criteria air pollutants: CO, particulate matter with aerodynamic diameters of 10 microns or less (PM_{10}), particulate matter with aerodynamic diameters of 2.5 microns or less ($PM_{2.5}$), ozone, SO_2, lead, and NO_2. New Mexico also has ambient air quality standards in place (New Mexico Ambient Air Quality Standards [NMAAQS]), which are equal to or more stringent than the NAAQS. NMAAQS are enforced by New Mexico, and allowed under the Clean Air Act by EPA. Table 3-6 lists the Federal and New Mexico Ambient Air Quality Standards. Areas with air quality as good as or better than the standards are designated as "attainment areas." Areas with air quality that is worse than the standards are designated as "non-attainment areas." Areas that were designated non-attainment and subsequently re-designated as attainment due to meeting the standards are termed "maintenance areas." States with maintenance areas are required to develop an air quality maintenance plan as an element of the State Implementation Plan.

The EPA divided the nation into 247 Air Quality Control Regions (AQCRs) based on a number of factors that influence regional air quality including climate and meteorology, topography, vegetation, land use patterns, population characteristics, and growth projections. Lea County is in the Pecos-Permian Basin Intrastate AQCR (40 CFR 81.242) and is in attainment for all of the NAAQS, as is the rest of the Pecos-Permian Basin Intrastate AQCR (40 CFR 81.332). The

Figure 3-3. Wind Roses for Midland-Odessa, Roswell, Hobbs, and Eunice for 1993

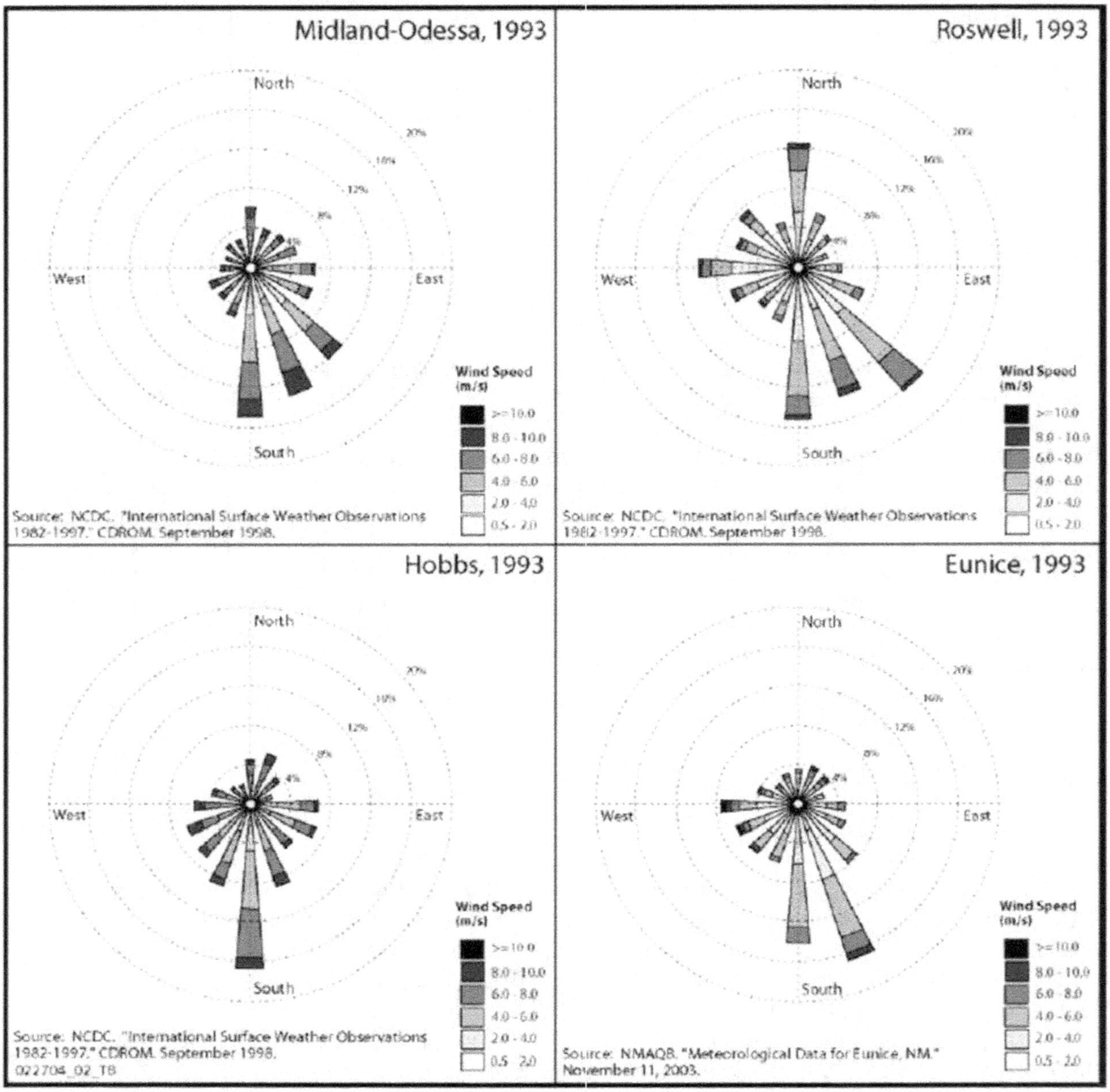

Source: NRC, 2005

closest non-attainment areas are in the El Paso-Las Cruces-Alamogordo Interstate AQCR (40 CFR 81.82), approximately 314 km (195 mi) southwest of the proposed IIFP facility. The Anthony area in Doña Ana County, New Mexico and the city of El Paso in El Paso County, Texas are designated as moderate non-attainment areas under the PM_{10} NAAQS (40 CFR 81.332 and 40 CFR 81.344).

3.5.4.1.2   Hazardous Air Pollutants

Provisions of the *Clean Air Act* as amended required EPA to establish technology-based standards for Hazardous Air Pollutants (HAP). Under Federal law, HAPs are those air pollutants listed in Section 112 of the Clean Air Act for which no NAAQS have been established.

Table 3-6.   Federal and New Mexico Ambient Air Quality Standards

Pollutant	Averaging Period	NAAQS	NMAAQS
CO	8-hour	9 ppm (10 mg/m^3)	8.7 ppm
	1-hour	35 ppm (40 mg/m^3)	13.1 ppm
NO$_2$	Annual	0.053 ppm	0.05 ppm
	24-hour	None[5]	0.10 ppm
	1-hour	0.100 ppm	None[5]
Ozone	8-hour	0.075 ppm	None
	1-hour[1]	0.12 ppm	None
SO$_2$	Annual	Revoked[6]	0.02 ppm
	24-hour	Revoked[6]	0.10 ppm
	3-hour	0.50 ppm	None
	1-hour	0.075 ppm	None
PM$_{2.5}$	Annual	15.0 µg/m^3	None
	24-hour	35 µg/m^3	None
PM$_{10}$	Annual	Revoked[7]	None
	24-hour	150 µg/m^3	None
Lead	Rolling 3-month	0.15 µg/m^3	None
Total Suspended Particulates	Annual Geometric Mean	Not an NAAQS Pollutant	60 µg/m^3
	30-day		90 µg/m^3
	7-day		110 µg/m^3
	24-hour		150 µg/m^3
Hydrogen Sulfide (H$_2$S)	1-hour[2]	Not an NAAQS Pollutant	0.010 ppm
	½-hour[3]		0.100 ppm
	½-hour[4]		0.030 ppm
Total Reduced Sulfur	½-hour[2]	Not an NAAQS Pollutant	0.003 ppm
	½-hour[3]		0.010 ppm
	½-hour[4]		0.003 ppm

Source: : 40 CFR 50; NMAC 20.2.3

[1]The 1-hour ozone NAAQS will not apply to an area one year after the effective date of the designation of that area for the 8-hour ozone NAAQS. The effective designation date for most areas is June 15, 2004 (40 CFR 50.9).

[2]For the state, except for the Pecos-Permian Basin Intrastate AQCR.

[3]For the Pecos-Permian Basin Intrastate AQCR.

[4]For within 5 miles of the corporate limits of municipalities within the Pecos-Permian Basin Intrastate AQCR.

[5]Regulatory agencies have not established standards.

[6]The 24-hour and annual SO$_2$ NAAQS was revoked by EPA on June 22, 2010.

[7]The annual PM$_{10}$ NAAQS was revoked by EPA on September 21, 2006.

µg/m^3 = micrograms per cubic meter
mg/m^3 = milligrams per cubic meter
ppm = parts per million

There are currently 188 hazardous air pollutants listed, including, but not limited to, the pollutants controlled by the National Emissions Standards for Hazardous Air Pollutants (NESHAP) program (40 CFR 61 and 63).

3.5.4.1.3   Prevention of Significant Deterioration

Under the *Clean Air Act*, the EPA established Prevention of Significant Deterioration (PSD) regulations, which apply to proposed new or modified sources in an attainment area that have the potential to emit NO_2, $PM_{2.5}$, PM_{10}, or SO_2 in excess of predetermined levels (40 CFR 52.21). Allowable deterioration to air quality can be expressed as the incremental increase in ambient concentrations of criteria pollutants, or PSD increment. Increments for criteria pollutants are based on the PSD classification of the area. Class I areas, which include certain national parks and wilderness areas, allow the lowest amount of permissible deterioration by precluding development near designated areas. All other areas of the United States are Class II areas where moderate, well-controlled industrial growth is allowed. The allowable PSD increments for Class I and Class II areas are identified in Table 3-7.

The proposed IIFP facility is in a PSD Class II area. There are no PSD Class I areas within 100 km (62 mi) of the proposed IIFP facility (40 CFR 81, Subpart D). The nearest PSD Class I areas to the proposed IIFP facility are Carlsbad Caverns National Park and the Guadalupe Mountains National Park, located about 114 km and 154 km (71 mi and 96 mi), respectively, southwest of the proposed site. Therefore, due to the distances involved to the closest PSD Class I areas; there is no reason to expect deterioration of air quality from the volumes of IIFP facility-generated NO_2, particulate matter, and SO_2.

Table 3-7.   Allowable Prevention of Significant Deterioration Increments

Pollutant	Averaging Period	Class I PSD Increment ($\mu g/m^3$)	Class II PSD Increment ($\mu g/m^3$)
NO_2	Annual	2.5	25
$PM_{2.5}$	Annual	1	4
	24-hour	2	9
PM_{10}	Annual	4	17
	24-hour	8	30
SO_2	Annual	2	20
	24-hour	5	91
	3-hour	25	512

Source: 40 CFR 52.21

3.5.4.1.4   Regional Haze

Regional haze is a visibility impairment caused by cumulative air pollutant emissions from numerous sources over a wide geographic area. The primary cause of regional haze in many parts of the country is light scattering from fine particles ($PM_{2.5}$) in the atmosphere. Course particles (PM_{10}) can also contribute to light extinction. Section 169 of the *Clean Air Act* established a national goal for visibility, defined as the "prevention of any future, and remedying of any existing, impairment of visibility in Class I areas...from manmade air pollution." Under the regional haze rule, States are required to develop State Implementation Plans to address visibility at designated mandatory PSD Class I areas, including designated national parks, wilderness areas, and wildlife refuges (40 CFR 51.309). A visibility analysis is required for each PSD Class I area located within 100 km (62 mi) of any new or modified major stationary sources whose emissions exceed PSD modeling thresholds. As discussed above, there are no PSD Class I areas within 100 km (62 mi) of the proposed IIFP so no visibility analysis is required.

3.5.4.1.5   General Conformity for Federal Actions

According to Section 176 of the *Clean Air Act* (40 CFR 51.853), a Federal agency must make a conformity determination in the approval of a project with air emissions that exceed specified thresholds in nonattainment and/or maintenance areas. This General Conformity Rule ensures that the actions taken by Federal agencies in non-attainment and maintenance areas meet national standards for air quality and do not cause further degradation to air quality which would be inconsistent with the attainment and maintenance of ambient air quality standards. The proposed project is not in a non-attainment or maintenance area; therefore, no general conformity analysis is required.

3.5.4.2   Existing Conditions

Air quality in Lea County, New Mexico is considered unimpaired. Farming, ranching, oil and gas development, a few industrial facilities, and vehicular traffic are the primary activities that would affect ambient air quality.

The closest air quality monitoring station to the proposed IIFP site is in Hobbs, New Mexico. The Hobbs station monitors NO_2, ozone, $PM_{2.5}$, and PM_{10}. The nearest air quality monitoring stations for CO, SO_2, and lead are in Rio Rancho, New Mexico, Artesia, New Mexico, and El Paso, Texas, respectively. The monitored criteria pollutant concentrations for the years 2006 through 2008 are summarized in Table 3-8.

Table 3-8.   Ambient Levels of Criteria Pollutants in Nearby Counties

Pollutant	Averaging Time	NAAQS	2006	2007	2008	Monitor Location
CO (ppm)	8-hour	9	1.4	0.6	NA	Rio Rancho, NM[1]
	1-hour	35	1.6	1.1	NA	
NO_2 (ppm)	Annual	0.053	0.008	0.006	0.006	Hobbs, NM
	1-hour	0.100	0.054	0.053	0.052	
Ozone (ppm)	8-hour	0.075	0.075	0.064	0.067	Hobbs, NM
SO_2 (ppm)	Annual	0.03	0.001	0.001	0.001	Artesia, NM
	24-hour	0.14	0.004	0.001	0.001	
	3-hour	0.50	0.017	0.005	0.001	
	1-hour	0.075	0.066	0.011	0.002	
$PM_{2.5}$ ($\mu g/m^3$)	Annual	15.0	6.82	7.26	6.85	Hobbs, NM
	24-hour	35	12.5	14.8	14.6	
PM_{10} ($\mu g/m^3$)	24-hour	150	60	55	39	Hobbs, NM
Lead ($\mu g/m^3$)	Quarterly	0.15	0.04	0.05	0.02	El Paso, TX

Source: EPA, 2009a
[1] The CO monitor in Rio Rancho did not operate in 2008.

3.6   Geology, Minerals, and Soil

3.6.1   Regional and Site Near-surface Geology

The proposed IIFP site is in the Llana Estacado section of the Southern High Plains physiographic region. The Llana Estacado is an isolated mesa that slopes gently to the east-southeast and covers a large part of eastern New Mexico and western Texas. The Mescalero Ridge escarpment, which defines the southwestern limit of the Llano Estacado (Figure 3-4) crosses the western and central portions of Lea County as a nearly perpendicular cliff (Nicholson and Clebsch, 1961).

The site is underlain by (in descending order) Quaternary-age alluvium, Triassic- and Cretaceous-age rocks, and Permian-age rocks that fill the Permian Basin. The Permian Basin underlies an area approximately 402 km (250 mi) wide and 483 km (300 mi) long and is a major oil and natural gas producing area (UTPB, 2010). Beginning in 1921, more than 40,000 exploration wells and 200,000 development wells have been drilled in the Permian Basin region (Scholle, 2000). The Basin produces 17 percent of the nation's crude oil (UTPB, 2010). The Basin is also a major source of potassium salts (potash) (UTPB, 2010). Oil, gas, and potash production in Lea County are summarized in Section 3.6.3.

According to the EPA, there were 95 point sources of criteria pollutants in Lea County, New Mexico for emissions year 2002 (EPA, 2009b). Emission data for 2002 are the most recent data available from EPA. Motor vehicles and various area sources also contributed to the criteria pollutant emissions in Lea County. Table 3-9 presents a summary of the 2002 annual Criteria Air Pollutants emissions for Lea County.

The Ogallala Formation consists of valley-fill deposits of clay, silt, fine- to coarse-grained sand, gravel and caliche (hardened calcium carbonate), the distributions of which vary both vertically and horizontally. The formation ranges in thickness from 0 to as much as 107 m (350 ft) (Fahlquist, 2003; Tillery, 2008). Locally, the top of the Ogallala Formation consists of a resistant layer of caliche as thick as 18 m (60 ft) (Nicholson and Clebsch, 1961).

Table 3-9.   Lea County Criteria Pollutant Emissions in 2002

Pollutant	Point metric tons per year (tons per year)	Mobile metric tons per year (tons per year)	Area metric tons per year (tons per year)	Total metric tons per year (tons per year)
CO	7,250 (7,992)	13,376 (14,744)	618 (681)	21,244 (23,417)
NO_2	25,605 (28,225)	1,386 (1,528)	128 (141)	27,119 (29,894)
SO_2	7,197 (7,933)	67.8 (74.7)	68.5 (75.5)	7,334 (8,084)
$PM_{2.5}$	214 (236)	47.9 (52.8)	2,630 (2,899)	2,892 (3,188)
PM_{10}	244 (269)	57.2 (63.1)	24,747 (27,279)	25,048 (27,611)
VOC	1,996 (2,200)	1,067 (1,176)	1,373 (1,513)	4,436 (4,890)

Source: EPA, 2009b

Quaternary-age alluvial deposits underlying the site consist of sand ranging up to 1 m (3 ft) thick (Hunt, 1977) that mantles the underlying late Tertiary-age Ogallala Formation (NMBGMR, 2003).

3.6.2   Seismicity and Volcanism

The proposed IIFP site is in a seismically quiet region, with local earthquakes of relatively small magnitude (moment magnitude of less than 2 on the Modified Mercalli-Revised 1931 scale [MM]). No Quaternary faults or folds, thought to be associated with most earthquakes of moment magnitude 6 or greater over the last 1.6 million years, exist in the southeast New Mexico/west Texas region (Crone and Wheeler, 2000; Machette et al., 1988; Yarger, 2009). The nearest faulting is more than 161 km (100 mi) west of the site and is associated with the Rio Grande Rift.

Seismic activity in southeastern New Mexico is typically of small magnitude and generally caused by oil field injection activities. However, the largest recent major earthquake (5.0 MM) in New Mexico occurred south of Eunce in January, 1992 (Sanford et al., 2002; Sanford et al., 2006; Yarger, 2009). A seismic event of 5.0 MM would be felt outside only and observed inside by swinging doors or swaying wall pictures.

The New Mexico Institute of Mining and Technology using instrumental data has estimated probabilistic seismic hazards for New Mexico of duration magnitude 2.0 MM or greater for the time period 1962 through 1998 (Sanford et al., 2002; Sanford et al., 2006). Figure 3-5 shows the probabilistic seismic hazard map in the format of peak horizontal ground acceleration (PGA) at 10 percent probability of exceeding 6 MM in a 50-year period (or, approximately once every 500 years). PGA is a measure of earthquake force at ground surface and is an index of hazard for structures. The units for PGA are in percent gravity (g). As shown in Figure 3-5, the highest predicted PGA, approximately 0.18 g, is approximately 40 km (25 mi) north of Socorro. The IIFP site area has a predicted PGA of approximately 0.02 g. A PGA of 0.02 g is considered the acceleration level at which considerable damage can begin to occur to poorly designed or weakly-built structures of masonry, adobe, or stone (Sanford et al., 2002; Sanford et al., 2006).

New Mexico has experienced almost 700 volcanic events over the past 5 million years, ranging from small basalt flows to large eruptions. The volcanic events are roughly aligned with two zones of structural weakness that cross New Mexico: the Colorado Plateau Transition Zone and the Rio Grande Rift. The most recent volcanic activities were the eruptions of two relatively large basalt flows associated with the tectonic activity along Rio Grande Rift: the Carrizozo and McCarty's basalts (Limburg, 2009). The Carrizozo basalt covers approximately 329 km^2 (127 mi^2) near the town of Carrizozo, approximately 258 km (160 mi) west of the site. Studies indicate that the age of the basalt flow is between 4,800 and 5,200 years. The McCarty basalt flow covers approximately 344 km^2 (133 mi^2) near the town of Grants, approximately 547 km (340 mi) northwest of the site. Isotope studies indicate that the Grants flow is approximately 3,000 years old (Zimbelman and Johnston, 2001).

3.6.3   Minerals

Mineral resources in Lea County include industrial minerals such as fluorite and gypsum; construction materials such as potash, caliche, sand, and gravel; and energy sources such as coal, oil, and gas (Figure 3-6).

Although there are no designated mining districts in Lea County (McLemore et al., 2007), industrial and construction materials including potash, salt, sulfur, sand, gravel and caliche are mined at the eight active commercial mines/pits/mills in Lea County listed on Table 3-10.

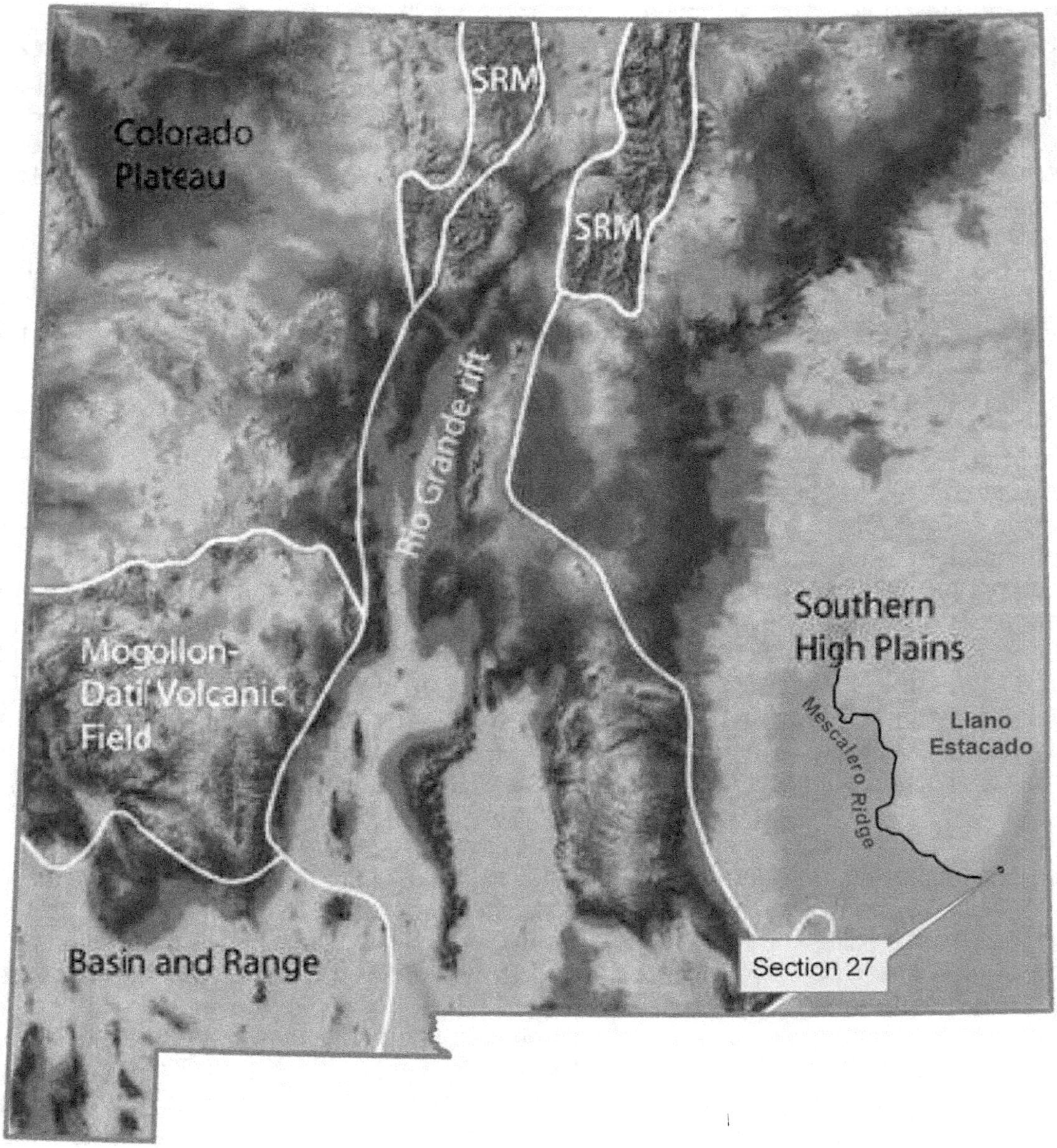

Colorado
Plateau

SRM

SRM

Rio Grande rift

Mogollon-
Datil Volcanic
Field

Southern
High Plains

Llano
Estacado

Mescalero Ridge

Basin and Range

Section 27

New Mexico

N
W   E
S

FEP/DUP
Environmental Impact Statement
Figure 3-4 Physiography of New Mexico

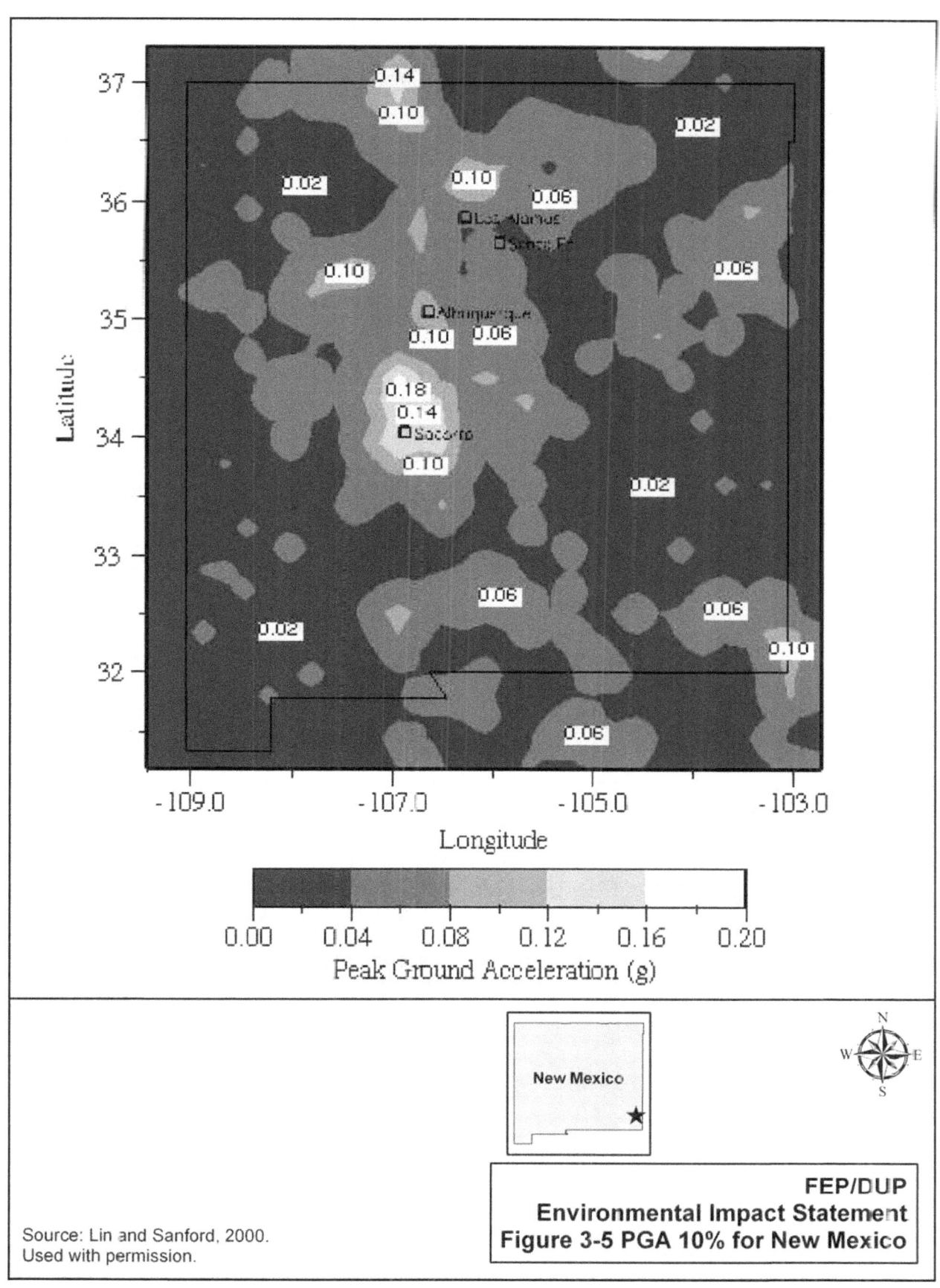

Source: Lin and Sanford, 2000.
Used with permission.

FEP/DUP
Environmental Impact Statement
Figure 3-5 PGA 10% for New Mexico

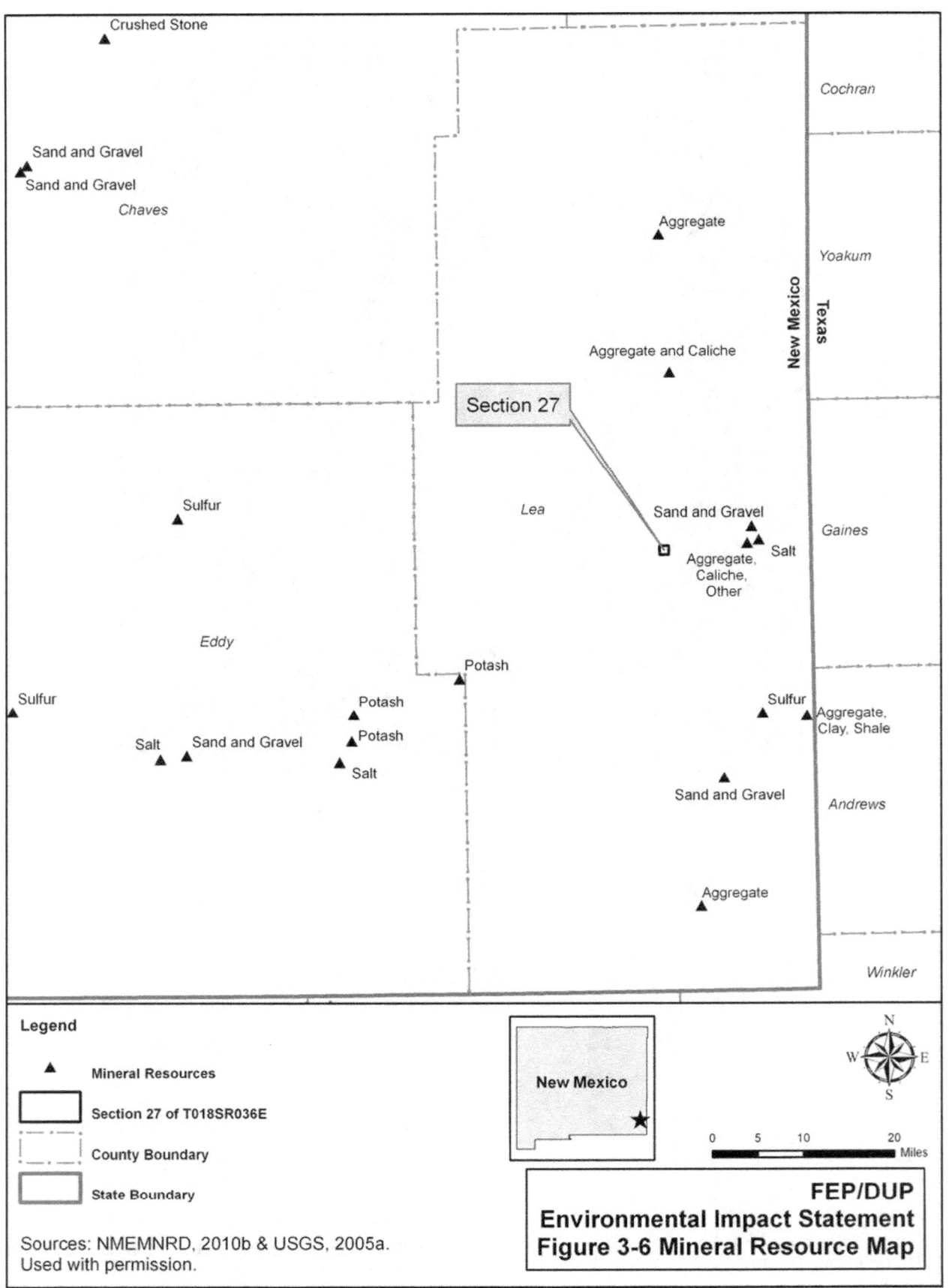

Legend

▲ Mineral Resources

☐ Section 27 of T018SR036E

County Boundary

State Boundary

Sources: NMEMNRD, 2010b & USGS, 2005a.
Used with permission.

New Mexico

N
W E
S

0   5   10   20
              Miles

FEP/DUP
Environmental Impact Statement
Figure 3-6 Mineral Resource Map

Table 3-10.   Summary of Active Mines, Mills, Pits and Quarries in Lea County

Name	Commodity	Operation
Lea County Pit	Base course, crushed rock, caliche, top soil, sand	Pit/Mill
Hawthorne Pit	Caliche	Pit
Constructors	Aggregate, caliche, other	Pit
Eunice Pit	Sand and gravel	Pit
Old Baldy Pit	Aggregate	Pit
Intrepid Potash	Potash	Mill
Rowland Salt	Salt	Mine
Eunice Plant Sulfur	Sulfur	Pit

Sources: USGS, 2005a; Pfeil, et al., 2001; NMEMNRD, 2010a

Caliche caps the Llano Estacado to a maximum thickness of 18 m (60 ft). It is mined throughout southeastern New Mexico, including Lea County, for construction and cement uses. Lea County is riddled with hundreds of small, abandoned caliche pits that were used by the New Mexico Department of Transportation for road construction material. There is a small caliche pit in the southeastern corner of Section 27. Caliche is currently mined at the Hawthorne Pit north of Lovington in Lea County.

Coal is not mined in Lea County (McLemore et al., 2007).

The New Mexico portion of the Permian Basin contains 1,112 designated, discovered oil reservoirs and 672 designated, discovered gas reservoirs. Large active oil and gas fields have existed in Lea County for more than 50 years (Leedshill-Herkenhoff et al., 2000).

According to the New Mexico Oil & Gas Wells database (NMEMNRD, 2010b), 715 oil or gas wells are within a 10-km (6-mi) radius of the proposed IIFP site. Seven of these wells are within a 1.6-km (1-mi) radius. The seven wells were drilled between 1987 and 1999, but subsequently abandoned. One abandoned well is in the extreme southwestern corner of Section 27, but no oil or gas wells are located within the proposed facility boundary. The locations of the seven wells are shown in Figure 3-7, and well details are summarized in Table 3-11. The proposed IIFP site has been explored for oil and gas and caliche. The site has very limited leasable, locatable, or marketable mineral resources (NMEMNRD, 2010a; NMEMNRD,2010b; NMT, 2010; Pfeil, et al., 2001; USGS, 2005a).

3.6.4   Soil

Soils occupying the southern High Plains in Lea County generally comprise shallow to deep gravelly and loamy soils, or deep sandy soils formed from windblown and water-deposited materials in the Quaternary and late Tertiary periods. Soft or hard caliche is generally found below the soils in most of the southern High Plains.

Soils underlying the proposed IIFP site include those of Kimbrough, Lea, Portales, Stegall and Slaughter soil associations. The distribution of these soil associations are shown in Figure 3-8, and the soil characteristics are summarized in Table 3-12.

In October 2010, a study was conducted by GL Environmental, Inc. for IIFP (GL Environmental, 2010a). They collected two soil samples from the site, which were analyzed to characterize

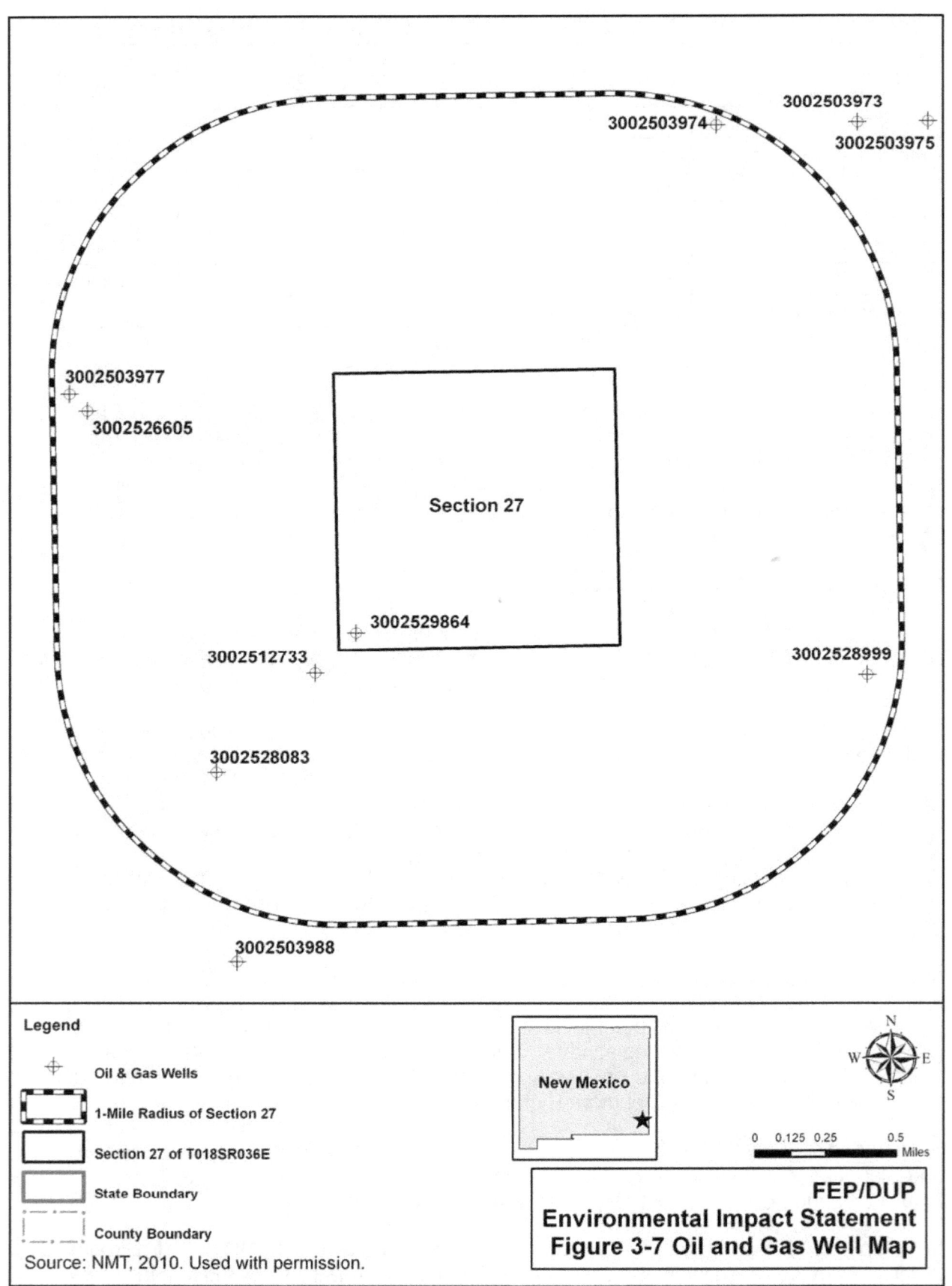

3002503973
3002503974
3002503975

3002503977
3002526605

Section 27

3002529864

3002512733

3002528999

3002528083

3002503988

Legend

✛ Oil & Gas Wells

1-Mile Radius of Section 27

Section 27 of T018SR036E

State Boundary

County Boundary

New Mexico

N
W E
S

0   0.125 0.25   0.5
                        Miles

FEP/DUP
Environmental Impact Statement
Figure 3-7 Oil and Gas Well Map

Source: NMT, 2010. Used with permission.

Table 3-11. Oil and Gas Wells within 1.6 km (1 mi) of the Site

Grid Name	American Petroleum Institute (API) Number	Spud Date[1]	Depth Meters (ft)	Comments
Yates Petroleum Corp	3002529864	1987	3,360 (11,025)	Plugged and abandoned
Chevron USA Inc	3002528999	1999	3,475 (11,400)	Abandoned
Basin Alliance LLC	3002528083	1982	3,708 (12,164)	Abandoned
Westbrook Oil Corp	3002526605	1999	1,728 (5,670)	Abandoned
Shell Oil Company	3002512733	1999	1,612 (5,289)	Abandoned
Getty Oil Company	3002503974	1999	1,433 (4,700)	Abandoned
Texas Pacific & Pure	3002503977	1999	3,733 (12,245)	Abandoned

Source: NMT, 2010
[1]Spud date is the date when the drill bit first hits the ground.
API = American Petroleum Institute

Table 3-12. Site Soil Characteristics

Soil Association	Soil Map Symbol	Section 27 Hectares (Acreage)	Description
Kimbrough gravelly loam	KO	17.7 (43.7)	Gravelly loam from zero to 15.2 cm (6 in). Cemented material from 15.2 to 40.6 cm (6 to 16 in). Well-drained with a very low capacity to transmit water.
Kimbrough-Lea Complex	KU	227.3 (561.7)	Loam from zero to 66 cm (26 in). Cemented material from 25.4 to 66 cm (10 to 26 in). Well-drained with a very low capacity to transmit water.
Portales loam	PC	18 (44.5)	Farmland of Statewide Importance. Loam from zero to 20.3 cm (8 in). Clay loam from 20.3 to 152.4 cm (3 to 60 in). Well-drained with a high capacity to transmit water.
Portales-Stegall loam	PS	0.4 (0.9)	Farmland of Statewide Importance. Loam from zero to 22.9 cm (9 in). Clay loam from 22.9 to 71.1 cm (9 to 28 in). Cemented material from 71.1 to 96.5 cm (28 to 38 in). Variable from 96.5 to 152.4 cm (38 to 60 in). Well-drained with a very low to moderate capacity to transmit water.
Stegall and Slaughter soils	SS	2.0 (5.0)	Farmland of Statewide Importance. Loam from zero to 5.1 cm (2 in). Clay from 5.1 to 38.1 cm (2 to 15 in). Cemented material from 38.1 to 63.5 cm (15 to 25 in). Variable from 63.5 to 152.4 cm (25 to 60 in). Well-drained with a very low to moderately high capacity to transmit water.

Source: NRCS, 2010

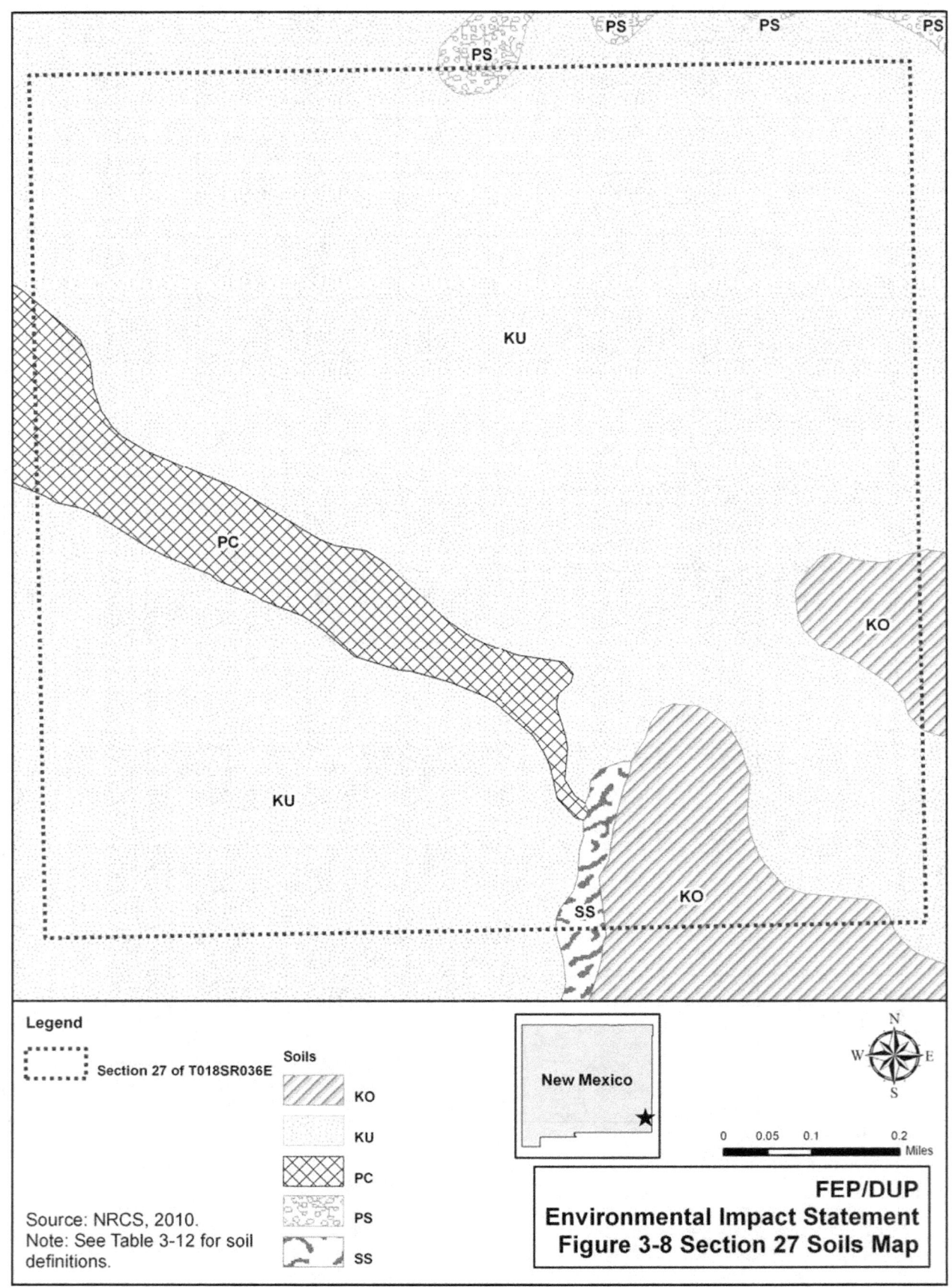

Legend

..... Section 27 of T018SR036E

Soils

KO

KU

PC

PS

SS

Source: NRCS, 2010.
Note: See Table 3-12 for soil
definitions.

New Mexico

N
W   E
S

0   0.05   0.1   0.2
                        Miles

FEP/DUP
Environmental Impact Statement
Figure 3-8 Section 27 Soils Map

baseline soil conditions. The soil samples were collected from a depth of 6 in and analyzed for radiological parameters, RCRA metals, volatile organic compounds (VOCs) and semi-volatile organic compounds (SVOCs).

U-234 was reported in the two soil samples at concentrations from 4.42×10^{-7} to 5.95×10^{-7} microcuries per gram (μCi/g). U-235/-236 was reported in concentrations from 5.58×10^{-9} to 2.60×10^{-8} μCi/g. U-238 results were from 5.86×10^{-7} to 5.95×10^{-7} μCi/g. All isotope concentrations are consistent with background levels in the site area.

Detected RCRA metals included barium with concentrations from 88.5 to 109 milligrams per kilograms (mg/kg), cadmium from 0.27 to 0.42 mg/kg, chromium from 10.0 to 12.2 mg/kg, and lead from 11.7 to 14.7 mg/kg. All other RCRA metals were at less than laboratory method detection limits. These elements are not uncommon in soils, but levels may have been elevated due to past petrochemical-related operations in the area. No VOCs or SVOCs were detected (GL Environmental, 2010a).

New Mexico has farmland protection programs to help slow the conversion of farmland to developed uses. Farmland is usually divided into three distinct categories: prime farmland, unique farmland, and farmland of statewide or local importance. Prime farmland is land of exceptional physical and chemical soil characteristics that can be used in agriculture with minimum input of nutrients, labor, etc. Prime farmland cannot be committed to urban development or water storage. Unique farmland is of lower quality than prime farmland but is still able to produce high-value food or grain products. Farmland of statewide or local importance does not meet the criteria for prime or unique farmland, but the soil is still considered important for the production of food and fiber.

The proposed IIFP site has approximately 20.4 ha (50.5 ac) of soils classified as farmland of statewide importance. The soils on the proposed site do not include tracts of land that have been designated for agriculture by state law (Carter, 2010).

3.7   Water Resources

These sections consider the groundwater and surface water use, and groundwater and surface water quality that could affect water use or quality at the site, or be affected by the construction or operation of the proposed IIFP facility.

3.7.1   Groundwater

Regional and site-specific data on the physical and hydrologic characteristics of the groundwater resources at, and in the vicinity of, the site are summarized in this section to provide basic data for an evaluation of impacts on the aquifers of the area.

3.7.1.1   Regional Groundwater

The High Plains aquifer, also known as the Ogallala aquifer, is a regional aquifer system that underlies 450,660 km^2 (174,000 mi^2) in parts of eight States: Colorado, Kansas, Nebraska, New Mexico, Oklahoma, South Dakota, Texas, and Wyoming (USGS, 2010a; McGuire et al., 2003). Because of its large size, the High Plains aquifer has been geographically subdivided into three aquifer regions: the southern High Plains, central High Plains, and the northern High Plains. About 27 percent of the irrigated land in the United States overlies this aquifer system, which yields about 30 percent of the nation's groundwater for irrigation. In addition, the aquifer system provides drinking water to 82 percent of the population within the aquifer boundary (USGS, 2010a).

The proposed IIFP site and surrounding region are underlain by the southern High Plains aquifer (Hart and McAda, 1985). The southern High Plains aquifer is an unconfined aquifer and is composed primarily of Quaternary-age alluvial sediments and the Tertiary-age Ogallala Formation. The Ogallala Formation, which underlies about 80 percent of the High Plains, is the principal geologic unit forming the aquifer (USGS, 2010a). The Ogallala aquifer is typically underlain by impermeable clays and shale, although in some places the underlying Cretaceous-age formations are hydraulically connected to the aquifer. Beneath the Ogallala aquifer in the Lea County underground water basin (UWB) are the Triassic-age Lower Dockum Group Santa Rosa aquifer and the deeper Permian-age rocks, which include the Rustler Formation, Capitan aquifer, and San Andres aquifer (Figure 3-9) (NMOSE, 2009; McCoy and Perry, 2004).

3.7.1.2   Local Groundwater

Groundwater resources in Lea County include hydrogeologic strata within five UWBs (Figure 310). The UWBs are areas of underground water with reasonably ascertainable boundaries declared by the New Mexico Office of the State Engineer (NM OSE). The four primary basins, from north to south, are the Lea County UWB, the Capitan UWB, the Carlsbad UWB, and the Jal UWB. A small area (approximately 142 km^2 [55 mi^2]) of a fifth UWB, the Roswell UWB, lies beneath west-central and northeast Lea County. The UWBs are designated based on their distinct hydrogeologic configurations, which do not typically end at county or State boundaries. The four primary UWBs include the following primary aquifers: Lea County UWB (Ogallala aquifer), Capitan UWB (Capitan aquifer), Carlsbad UWB (Santa Rosa aquifer), and the Jal UWB (Alluvial aquifer) (Leedshill-Herkenhoff et al., 2000).

The Lea County UWB is discussed in detail below because it underlies the proposed IIFP site. Following that discussion, site and vicinity groundwater are more specifically characterized.

3.7.1.2.1   Lea County Underground Water Basin

The proposed IIFP site is above the Lea County UWB, which encompasses 5,646 km^2 (2,180 mi^2) and covers most of northern Lea County and small portions of Chaves and Eddy Counties in southeast New Mexico (Stephens & Assoc., 2009). The basin boundaries are shown in Figures 3-9, 3-10, and 3-11.

The Ogallala aquifer is the primary water source in the Lea County UWB, which extends the width of Lea County to the east and west. To the south, the Lea County UWB is bounded by the Mescalero Ridge and associated escarpment (Figures 3-11 and 3-4), which indicates the southern extent of the High Plains aquifer. The maximum saturated thickness of the Ogallala aquifer within the UWB is about 76 m (250 ft) (Leedshill-Herkenhoff et al., 2000). The depth to groundwater in the Ogallala Formation is approximately 9.1 m (30 ft) in the site area (Figure 3-12).

Generally, the Ogallala Formation has an upward fining of sediments, which may have a significant effect on the distribution of porosity and permeability in the aquifer, controlling both the amount of water that can be stored and its movement through the aquifer (Stephens & Assoc., 2009).

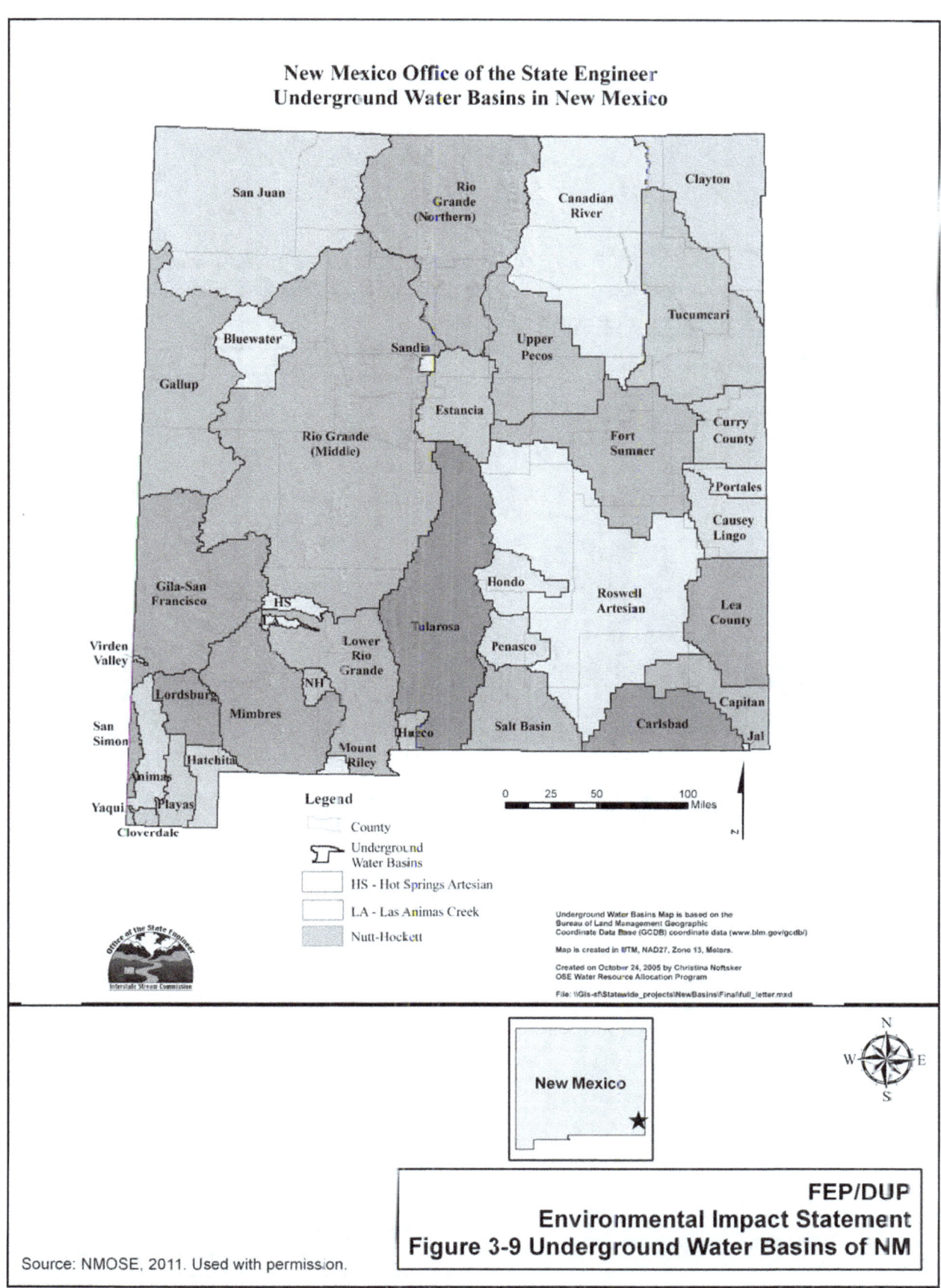

New Mexico Office of the State Engineer
Underground Water Basins in New Mexico

San Juan

Rio Grande (Northern)

Canadian River

Clayton

Bluewater

Gallup

Sandia

Upper Pecos

Tucumcari

Estancia

Rio Grande (Middle)

Fort Sumner

Curry County

Portales

Causey Lingo

Gila-San Francisco

Hondo

Roswell Artesian

Lea County

HS

LA

Virden Valley

Lower Rio Grande

Tularosa

Penasco

Lordsburg

NH

Mimbres

San Simon

Hatchita

Hueco

Mount Riley

Salt Basin

Carlsbad

Capitan

Jal

Animas

Yaqui

Playas

Cloverdale

Legend

County

Underground Water Basins

HS - Hot Springs Artesian

LA - Las Animas Creek

Nutt-Hockett

0 25 50 100 Miles

Underground Water Basins Map is based on the
Bureau of Land Management Geographic
Coordinate Data Base (GCDB) coordinate data (www.blm.gov/gcdb/)

Map is created in UTM, NAD27, Zone 13, Meters.

Created on October 24, 2005 by Christina Noftsker
OSE Water Resource Allocation Program

File: \\Gis-sf\Statewide_projects\NewBasins\Finalfull_letter.mxd

New Mexico

N
W E
S

FEP/DUP
Environmental Impact Statement
Figure 3-9 Underground Water Basins of NM

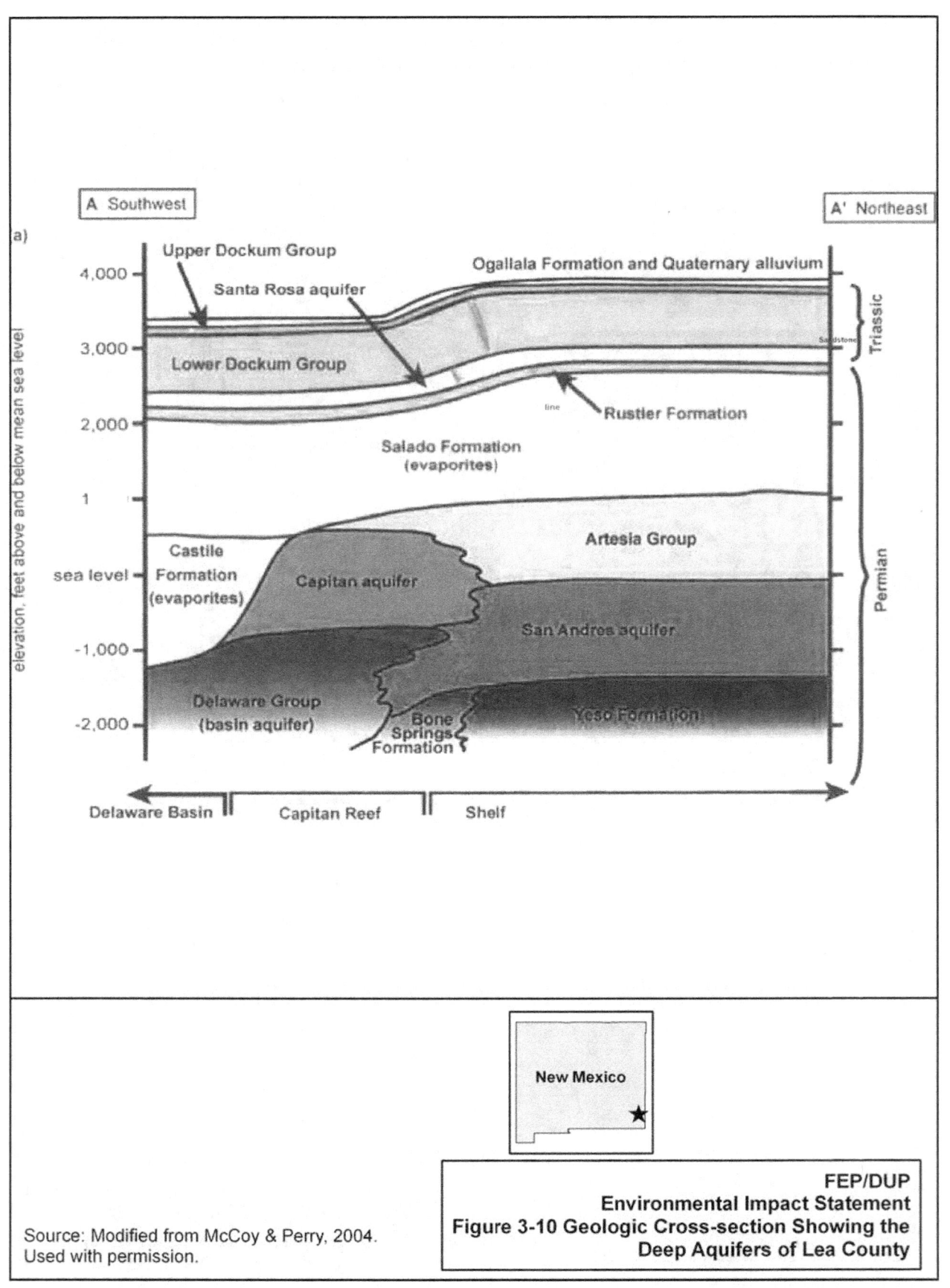

A Southwest

A' Northeast

a)

Upper Dockum Group

Ogallala Formation and Quaternary alluvium

Santa Rosa aquifer

4,000

3,000

Lower Dockum Group

Triassic

Sandstone

Rustler Formation

line

2,000

Salado Formation
(evaporites)

1

Artesia Group

Castile
Formation
(evaporites)

Capitan aquifer

sea level

San Andres aquifer

Permian

-1,000

Delaware Group
(basin aquifer)

Yeso Formation

-2,000

Bone
Springs
Formation

elevation, feet above and below mean sea level

Delaware Basin

Capitan Reef

Shelf

New Mexico

Source: Modified from McCoy & Perry, 2004.
Used with permission.

FEP/DUP
Environmental Impact Statement
Figure 3-10 Geologic Cross-section Showing the
Deep Aquifers of Lea County

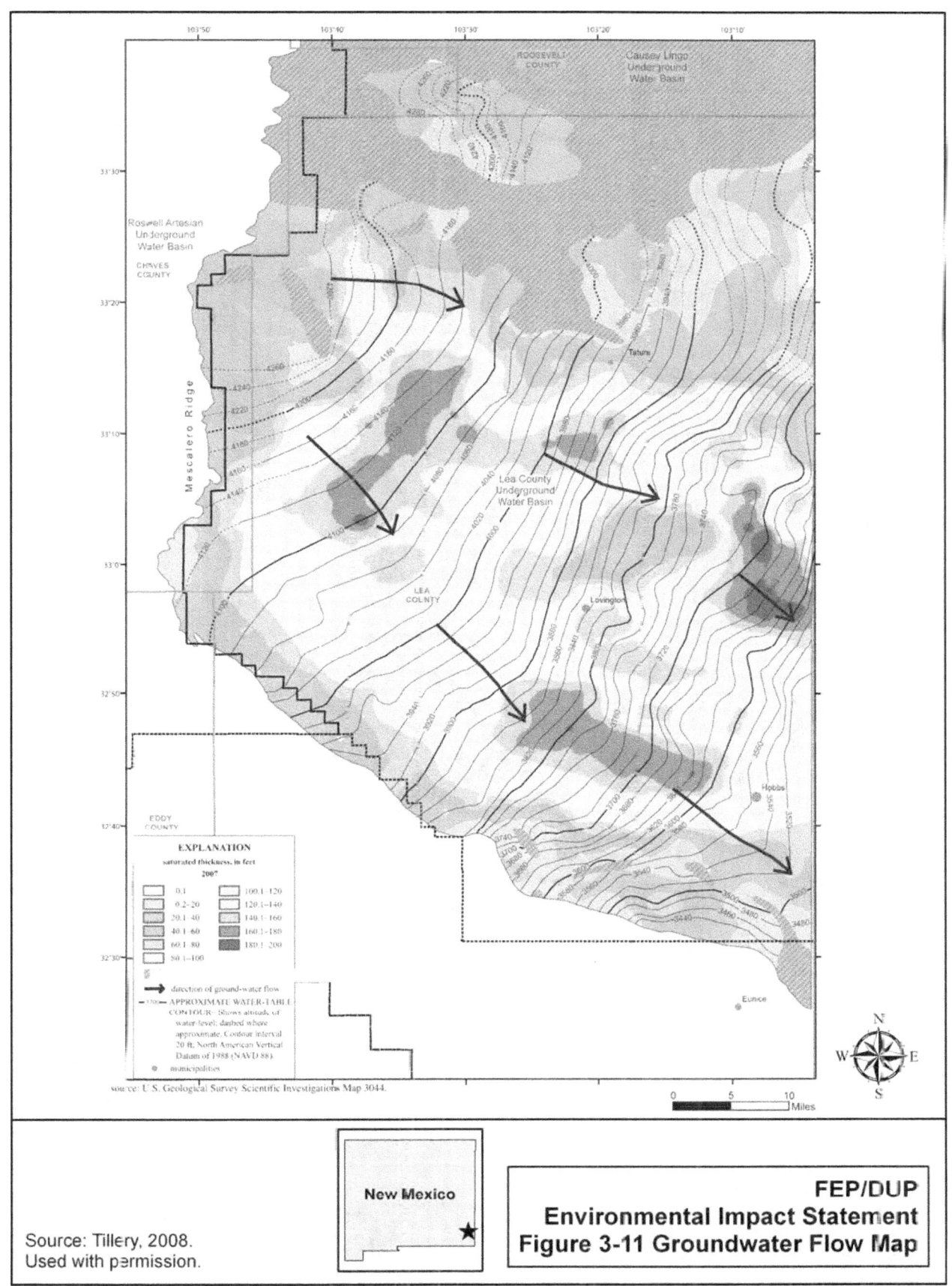

New Mexico

FEP/DUP
Environmental Impact Statement
Figure 3-11 Groundwater Flow Map

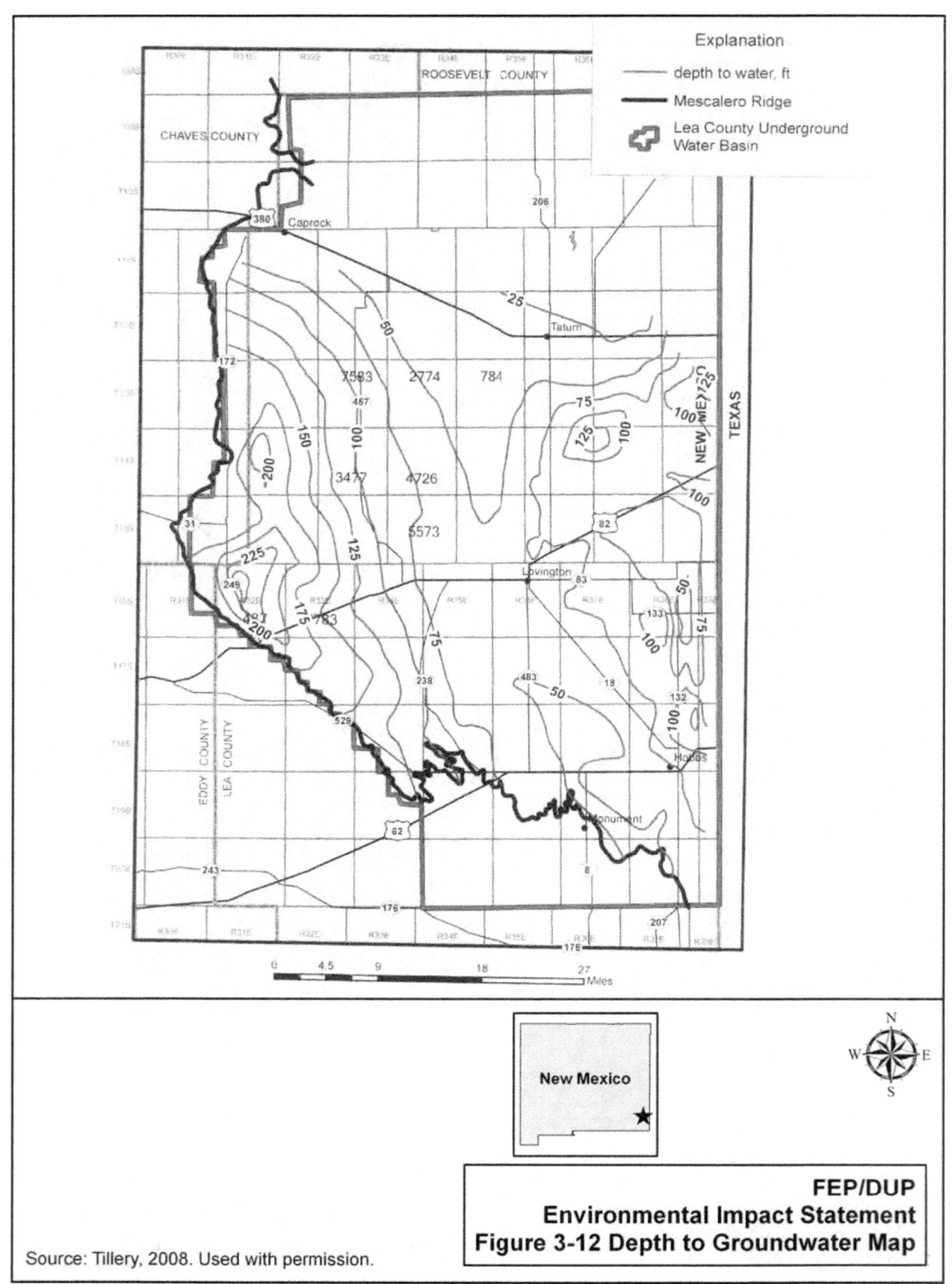

Explanation

— depth to water, ft

— Mescalero Ridge

Lea County Underground Water Basin

0   4.5   9   18   27
Miles

New Mexico

FEP/DUP
Environmental Impact Statement
Figure 3-12 Depth to Groundwater Map

Source: Tillery, 2008. Used with permission.

The hydraulic conductivity of the Ogallala aquifer in the Lea County UWB, as reported by a number of different studies (McGuire et al., 2003; Musharrafieh and Chudnoff, 1999; Stephens & Assoc., 2009), ranges from 0.9 to 80 m per day (3 to 262 ft per day), with higher hydraulic conductivities near Hobbs (i.e., near the proposed IIFP site) and eastward toward the Texas border. Irrigation well yields in the aquifer range from 757 to nearly 7,571 liters per minute (L/min) (200 to nearly 2,000 gallons per minute [gpm]) (Leedshill-Herkenhoff et al., 2000).

Discharge from the Ogallala aquifer in Lea County occurs through groundwater pumping and subsurface flow. The largest amount of natural groundwater discharge is the subsurface flow into Texas (Figure 3-11). A small amount of groundwater discharges through the Quaternary alluvium to southern Lea County (McAda, 1984).

The principal source of recharge to the aquifer occurs from precipitation infiltrating into the subsurface, primarily in areas covered by dune sand or playa lakes. Annual average recharge is estimated to range from 1.3 to 2.5 cm (0.5 to 1 in) (Tillery, 2008). It is estimated that approximately 3,840,000 ha-m (31,100,000 ac-ft) of groundwater is presently in storage in the UWB, of which only 45 percent (approximately 1,730,000 ha-m [14,000,000 acre-ft]) can actually be recovered because the saturated thickness of much of the aquifer is too shallow for water recovery to be feasible (Musharrafieh and Chudnoff, 1999; Leedshill-Herkenhoff et al., 2000).

Under pre-pumping conditions, recharge of the Ogallala aquifer was in equilibrium with natural discharge. Because current pumping for municipal, industrial, irrigation and other uses exceed the Ogallala's recharge rate, the aquifer has experienced significant drawdown. The water level in the Ogallala aquifer has declined as much as 30 m (97 ft) in the Lea County UWB from 1914 to 2007. The area of maximum saturated thickness is generally near or coincident with the area of maximum water-level decline, which is north of Hobbs and near the Texas state line (Tillery, 2008). Groundwater in the Ogallala aquifer flows east-southeast towards Texas (Figure 3-11). Depths to groundwater in the Lea County UWB range from 6.1 m (20 ft) in the Monument area to 76 m (250 ft) near the exposed caprock of the Mescalero Ridge (Figure 3-12) (Tilllery, 2008; Musharrafieh and Chudnoff, 1999).

Modeling by Musharrafieh and Chudnoff (1999) and observed water level declines indicate portions of the aquifer may become unsaturated by the year 2045. Other portions of the aquifer are also predicted to have a saturated thickness inadequate to sustain existing water rights (NMOSE, 2009).

Due to the limited groundwater supply within the southern High Plains aquifer, the NM OSE issued an order on March 10, 2009 closing the southern High Plains aquifer to the filing of applications under NMSA Section 72-12-3, which is the statute that regulates wells for new appropriations other than those applications filed under Section 72-12-1. The order does not apply to applications filed under NMSA Section 72-12-1.1 (wells required for relatively small amounts of water for single or multiple households, or for drinking or sanitary uses in conjunction with a commercial operation); Section 72-12-1.2 (livestock wells); or Section 72-12-1.3 (wells used for a period not to exceed one year for specifically listed purposes) (NMOSE, 2009).

Applications filed under NMSA Section 72-12-3 to appropriate groundwater from the units listed on Table 3-13 are considered on a case-by-case basis (NMOSE, 2009).

Table 3-13. Summary of Potential Deep-Aquifer Groundwater Sources for Lea County

Aquifer	Geologic Age	Typical Depth to Top of Aquifer m (ft) bgs	Typical Thickness m (ft)	Estimate Yields L/min (gpm)
Santa Rosa aquifer	Triassic	150 to 335 (500 to 1,100)	60 to 76 (200 to 250) [730 (2,400) max]	23 to 379 (6 to 100)
Rustler Formation	Upper Permian	210 to 410 (700 to 1,350)	24 to 43 (80 to140) [110 (360) max]	38 to 379 (10 to 100)
Capitan aquifer	Permian	850 to 1,400 (2,800 to 4,600)	270 to 430 (900 to1,400) [670 (2,200) max]	189 to 4,921 (50 to 1,300)
San Andres aquifer	Permian	910 to 1,500 (3,000 to 5,000)	210 to 610 (700 to 2,000)	833 (220)

Source: McCoy and Perry, 2004
bgs = below ground surface
gpm = gallons per minute
L/min = litters per minute

The NM OSE guidelines for new groundwater withdrawal applications for all UWBs include block administration and local assessment methods to limit aquifer drawdown. The block administration consists of 1.6 km^2 (1 mi^2) blocks that correspond to model cells in the groundwater-flow model (Musharrafieh and Chudnoff, 1999). A 40-year planning period ending in 2045 has been selected for block administration (NMOSE, 2009).

Model cells predicted to become unsaturated or with an inadequate saturated thickness for continued well operation require a higher level of restriction. Areas requiring such restriction are designated Critical Management Areas (CMAs). The CMAs include those model cells predicted to have a saturated thickness of 17 m (55 ft) or less by the year 2045 (Figure 3-13).

Key aspects of the NM OSE guidelines for new groundwater applications include the following: (1) water rights can be moved from one block to another throughout the basin, (2) the administrative groundwater flow model will be used to determine regional drawdowns resulting from an application, (3) applications to move water rights cannot create more drawdown than 0.0076 m/yr (0.025 ft/yr) in a CMA, or 0.061 m/yr (0.20 ft/yr) in a non-CMA, and (4) local area impacts from the proposed water-rights application will be performed and will include evaluation of impacts to the saturated thickness and reductions in water columns of existing wells (NMOSE 2009). On March 10, 2009, the NM OSE issued an order closing the Lea County UWB to the filing of groundwater applications (NMOSE, 2009). In 1999, in order to meet the projected groundwater demands of Lea County, 138 applications were filed by the Lea County Water Users Association to appropriate 6,389 ha-m (51,797 ac-ft) per year of groundwater, which was essentially all the unappropriated groundwater in the Lea County UWB (Leadshill-Herkenhoff et al., 2000).

A portion of the applications to appropriate groundwater, totaling 4,215.2 ha-m (34,173 ac-ft) per year, have been assigned to Lea County during (at least) the 40-year planning period to allow the County to hold the subject water rights unused until the rights can be put to beneficial use at projects currently under construction, select future projects, and homes and businesses in unincorporated areas. The proposed IIFP site has been included in Lea County's 40-Year Water Development Plan as an area of proposed development where up to 21.6 ha-m (175 ac-ft) per year of the Lea County's water rights would be put to beneficial use. The estimated

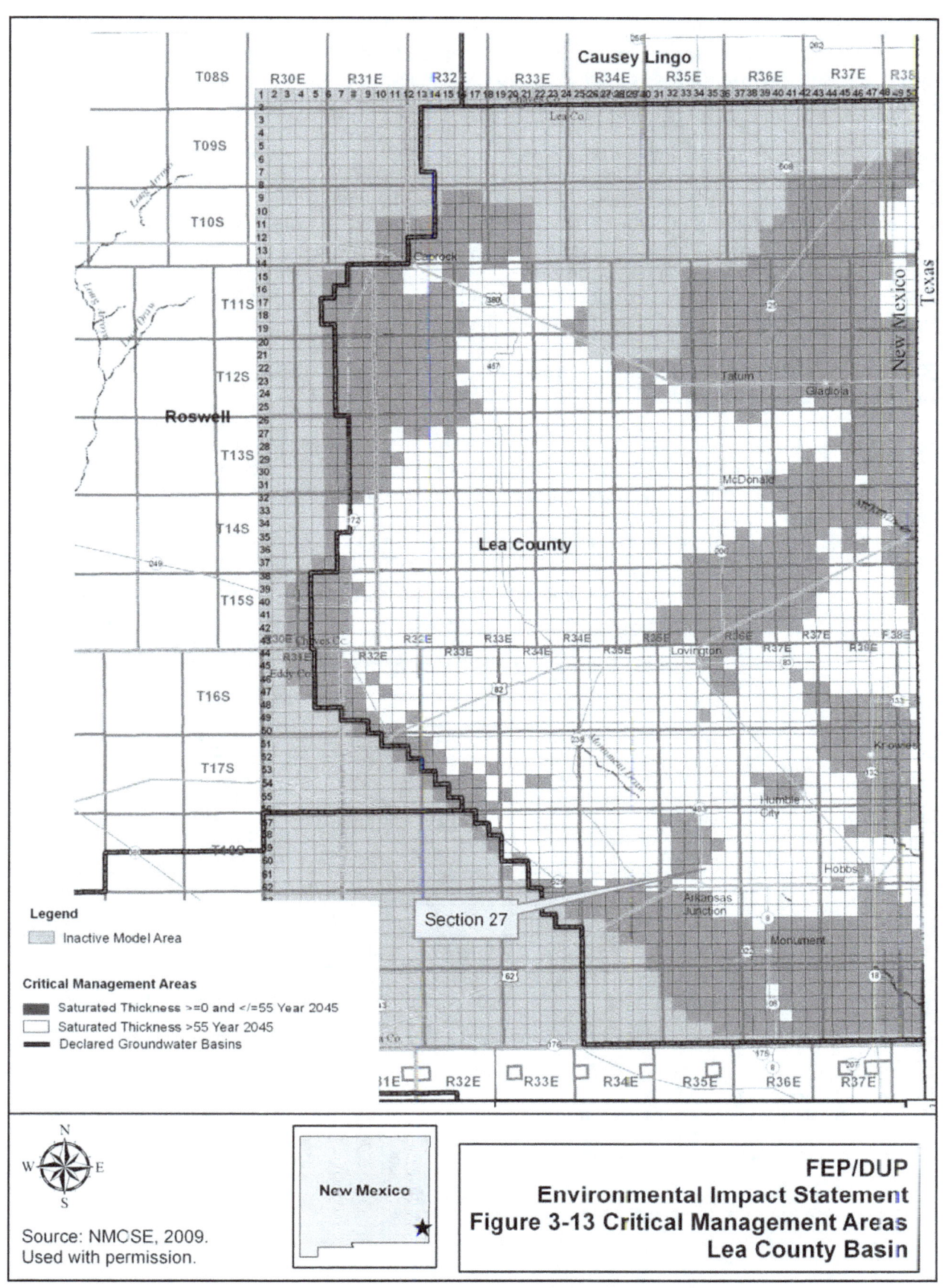

Legend

Inactive Model Area

Critical Management Areas

Saturated Thickness >=0 and </=55 Year 2045
Saturated Thickness >55 Year 2045
Declared Groundwater Basins

Source: NMCSE, 2009.
Used with permission.

New Mexico

**FEP/DUP
Environmental Impact Statement
Figure 3-13 Critical Management Areas
Lea County Basin**

quantity of groundwater that Lea County would need by the end of the 40-year planning period for all projects that currently exist, are being constructed, or have a high likelihood of being constructed in the near future is 1,173.5 ha-m (9,514 ac-ft) per year (Leadshill-Herkenhoff et al., 2000).

3.7.1.2.2   Site and Vicinity Groundwater

A data query of the NM OSE Statewide well database on water supply wells within 1.6 km (1 mi) of the site is summarized in Table 3-14, and the locations of the wells are shown in Figure 3-14. The depth to groundwater in the 10 water supply wells within 1.6 km (1 mi) of the site ranges from 8.5 to 21 m (28 to 70 ft) below ground surface (bgs). One well (L 04011) has insufficient data and was most likely not completed. The wells are installed in the Ogallala aquifer at depths ranging from 35 to 62.5 m (115 to 205 ft) bgs.

Xcel Energy's Cunningham Station, located just west of Section 27 has four groundwater monitoring wells (M-3, CU-6, -8, and -9) in Section 27 (Figure 3-14).

According to the NM OSE Statewide well database (2010), 261 groundwater wells are within a 10-km (6-mi) radius of the site (Figure 3-15). Most of the wells are categorized as prospecting wells. There is a domestic well approximately 2.6 km (1.6 mi) northwest of the site.

Table 3-14.   Summary of Supply Wells within 1.6 km (1 mi) of the Site

Owner	POD Number	Well Use	Drill Date	Total Depth m (ft)	Depth to Water m (ft)	Well Yield L/min (gpm)
Abbott, Murrell	L 03757	PRO	1957	38.1 (125)	13.7 (45)	NA
Abbott, Murrell	L 03928	PRO	1958	35 (115)	18.3 (60)	NA
NA	L 04011	PUB	NA	NA	NA	NA
Abbott Bros	L 05176 X	IND	1965	60.4 (198)	16.8 (55)	200 (53)
Abbott Bros	L 05176 X-2 (M-3)	IND	1965	50 (164)	16.8 (55)	151 (40)
Abbott Bros	L 05176 X-3	IND	1965	58 (190)	21.3 (70)	159 (42)
Abbott Bros	L 05176 X-4	IND	1965	54 (177)	21.3 (70)	144 (38)
Abbott Bros	L 05176 X-5	IND	1965	62.5 (205)	15.2 (50)	151 (40)
Abbott Bros	L 05176 X-6	IND	1965	61.9 (203)	8.5 (28)	151 (40)
NA	L 07469	DOM	1976	48.8 (160)	21.3 (70)	NA
Keith, Ronny	L 12341 POD1	SAN	2009	58 (190)	21.3 (70)	NA

Source: NMOSE, 2010
DOM = Domestic
IND = Industrial
NA = Not available
POD = Point of diversion
PRO = Prospecting/development of natural resources
PUB = Construction of public works
SAN = Sanitary in conjunction with a commercial use

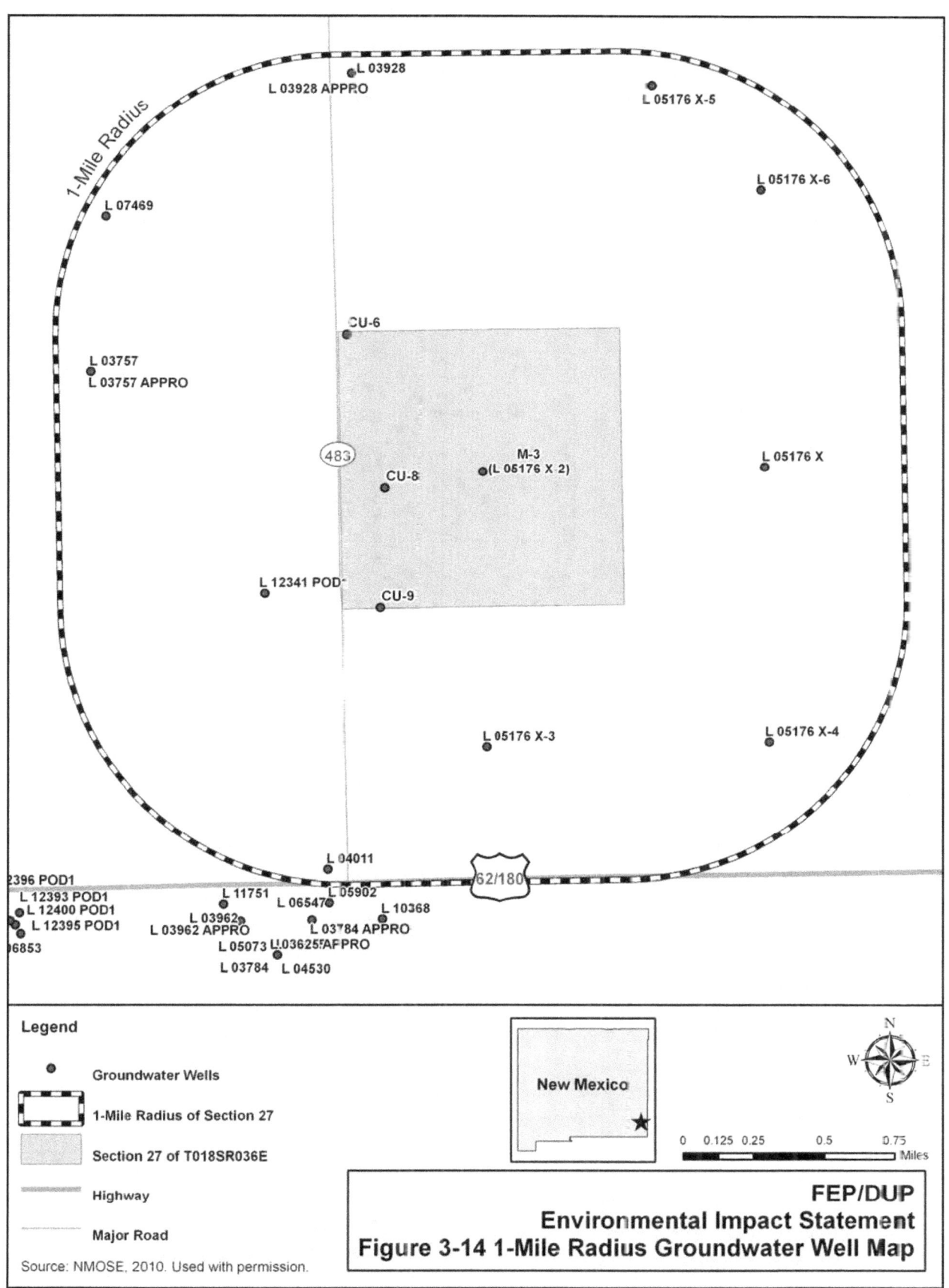

L 03928
L 03928 APPRO
L 05176 X-5
L 05176 X-6
1-Mile Radius
L 07469
CU-6
L 03757
L 03757 APPRO
483
M-3
(L 05176 X-2)
L 05176 X
CU-8
L 12341 POD
CU-9
L 05176 X-3
L 05176 X-4
L 04011
62/180
2396 POD1
L 12393 POD1
L 11751
L 05902
L 12400 POD1
L 06547
L 10368
L 12395 POD1
L 03962
L 03784 APPRO
L 03962 APPRO
06853
L 05073
L 03625 APPRO
L 03784
L 04530

Legend

● Groundwater Wells

▦ 1-Mile Radius of Section 27

▢ Section 27 of T018SR036E

Highway

Major Road

Source: NMOSE, 2010. Used with permission.

New Mexico

N
W   E
S

0   0.125 0.25   0.5   0.75
Miles

FEP/DUP
Environmental Impact Statement
Figure 3-14 1-Mile Radius Groundwater Well Map

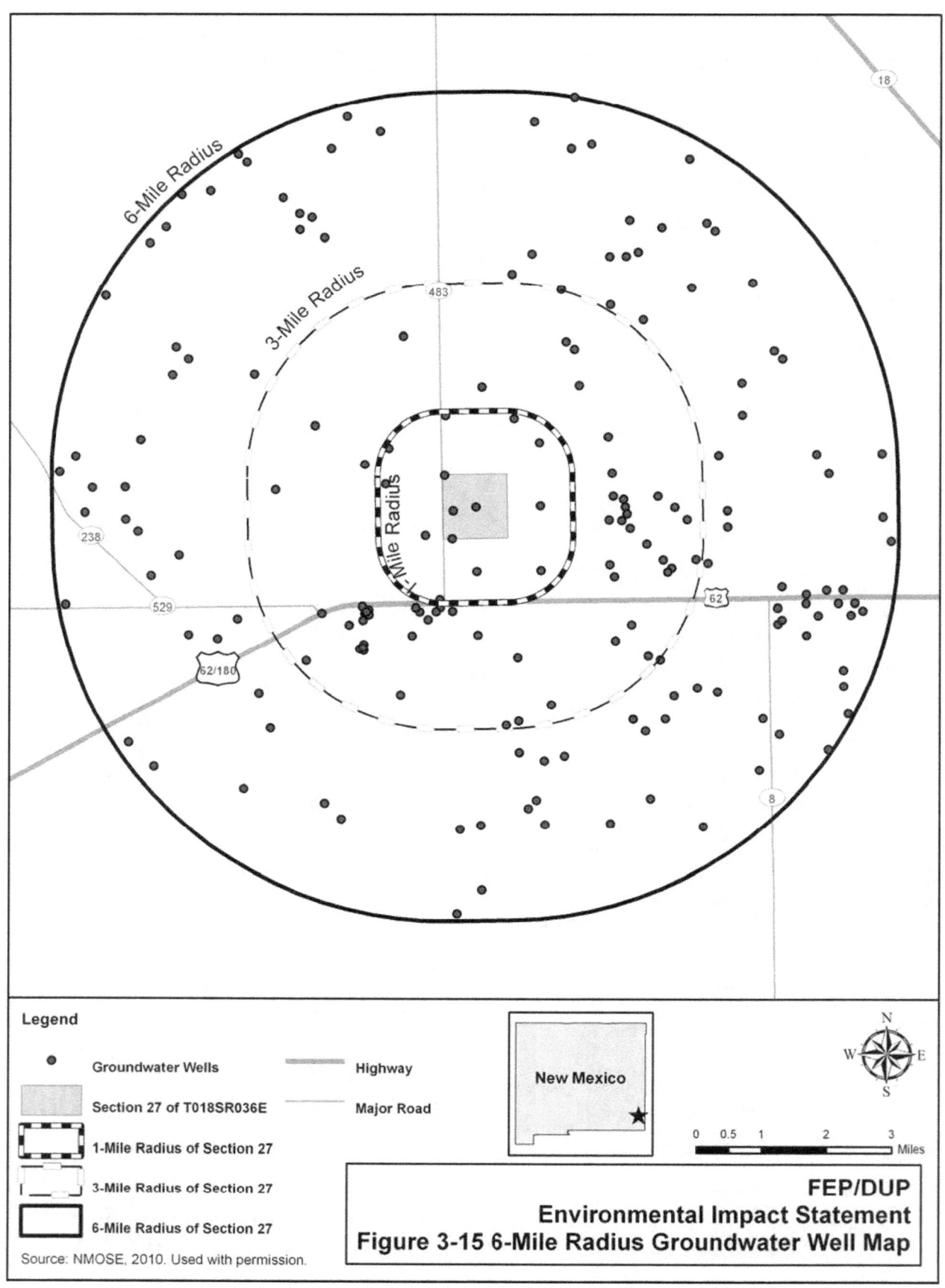

6-Mile Radius

3-Mile Radius

1-Mile Radius

483

238

529

62/180

62

18

8

Legend

- Groundwater Wells
- Section 27 of T018SR036E
- 1-Mile Radius of Section 27
- 3-Mile Radius of Section 27
- 6-Mile Radius of Section 27

Highway

Major Road

New Mexico

N
W E
S

0 0.5 1   2   3
Miles

FEP/DUP
Environmental Impact Statement
Figure 3-15 6-Mile Radius Groundwater Well Map

The nearest municipal water system is the City of Hobbs municipal system, approximately 16 km (10 mi) east northeast of the site. The system comprises 29 active groundwater supply wells, which are grouped into five well fields or systems. The wells range in depth from 54 to 81.7 m (177 to 268 ft) bgs, and the depth to water ranges from 23 to 51 m (75 to 167 ft) bgs. Yields for the wells range from 927 to 3,407 L/min (245 to 900 gpm). The combined yield from the five systems is estimated at 59,620 L/min (15,750 gpm) when the pumps are running 24 hours a day (Stephens & Assoc., 2009).

3.7.1.3 Groundwater Use

Groundwater use for Lea County, as reported by the NM OSE, is summarized in Table 3-15. Irrigation systems are the largest users (72.8 percent) of groundwater in the county, followed by mining (9.9 percent) and public water supply systems (7.2 percent). Smaller amounts of groundwater are used by industry, electric power generators, livestock, and domestic/commercial users.

Table 3-15.   Summary of Groundwater Use in Lea County, 2005

Groundwater Use Category	Groundwater Use ha-m (ac-ft)
Commercial (self-supplied)	403 (3,264)
Domestic (self-supplied)	175 (1,419)
Industrial (self-supplied)	751 (6,088)
Irrigated Agriculture	16,698 (135,371)
Livestock (self-supplied)	453 (3,670)
Mining (self-supplied)	2,265 (18,365)
Power (self-supplied)	545 (4,415)
Public Water Supply	1,648 (13,360)
Total	22,937 (185,952)

Source: NMOSE, 2008
ha-m = hectare meters
ac-ft – acre feet

3.7.1.4 Groundwater Quality

This section considers the groundwater quality of aquifers that could affect water use at the proposed site, or be affected by the construction or operation of the proposed IIFP facility.

3.7.1.4.1   Regional Groundwater Quality

In 2001, the U.S. Geological Survey (USGS) National Water-Quality Assessment Program collected groundwater samples from 48 wells screened in the Southern High Plains aquifer, primarily from domestic wells in eastern New Mexico and western Texas. Depths of wells sampled ranged from 30 to 152 m (100 to 500 ft), with a median depth of 61.3 m (201 ft). Depths to water ranged from 10 to 136 m (34 to 445 ft) bgs, with a median depth of 40.8 m (134 ft).

Of 240 parameters analyzed in the 48 wells, EPA public drinking-water standards or guidelines were exceeded in one or more wells for arsenic, boron, chloride, dissolved solids, fluoride, manganese, nitrate, radon, strontium, or sulfate. Pesticides were detected at very low concentrations (<1 mg/L) in fewer than 20 percent of the samples (Fahlquist, 2003).

Groundwater quality data from the City of Hobbs municipal drinking water system between 2005 and 2009 are summarized in Table 3-16. Each of the 29 wells comprising the system is screened in the Ogallala aquifer. The data indicate that groundwater quality in the aquifer near Hobbs is good and water quality standard exceedances are rare.

3.7.1.4.2 Site Groundwater Quality

The three monitoring wells along the western Section boundary (Figure 3-13) were installed by Xcel Energy to monitor for the presence of contaminants in groundwater that could originate from an unlined cooling tower pond at the Cunningham Station, and runoff from agricultural fields. Groundwater analytical data collected from monitoring wells CU-6, -8, and -9 indicate that of seven constituents sampled, sulfate, total dissolved solids, and chloride exceeded the New Mexico Water Quality Bureau Control Commission Standards for Groundwater (NMAC 20.6.2); however, boron, chlorite, pH and nitrates (including nitrate nitrogen) are below or within the standards. Groundwater quality data were not available for monitoring well M-3 (GL Environmental, 2010b).

3.7.2 Surface Water

The Southern High Plains Basin in New Mexico encompasses 14,211 km^2 (5,487 mi^2) in Curry, Roosevelt, Chavez, and Lea Counties (NMWQCC, 2002). The Mescalero Ridge (see Figure 3-4) separates the Southern High Plains (and the associated Texas Gulf Basin watershed) from the Pecos River Basin watershed. The proposed IIFP site lies just east of the Mescalero Ridge, in the Southern High Plains Basin and the Texas Gulf Basin watershed (see Figure 3-4).

No perennial streams traverse the Southern High Plains although there are some ephemeral streams that occasionally have large flows after rain storms. Surface water in Lea County is limited to intermittent streams, lakes, and numerous small playa lakes that result from heavy rainfall during the summer (Leedshill-Herkenhoff, Inc. et al., 2000). Surface drainage to playa lakes captures 80 to 90 percent of rainfall (Sublett and Peery, 2009). There is no true drainage system off the High Plains within Lea County.

Several depressions that hold water after rainfalls dot Section 27. Two dry stream beds bisect the southern portion of the Section from the northwest to the southeast, outside the boundaries of the proposed facility (Figure 3-16). The US ACE has determined that these ephemeral surface waters are not jurisdictional wetlands (USACE, 2011).

The nearest surface waters flow through Monument Draw, which is an ephemeral stream that can have large flows (Leedshill-Herkenhoff, Inc. et al., 2000). The headwaters of Monument Draw's nearest tributary is approximately 4 km (2.5 mi) from the nearest Section boundary and the main reach is approximately 10 km (6 mi) from the nearest Section boundary. The nearest permanent lake is approximately 32 km (20 mi) southwest of the proposed IIFP site (Figure 3-17). The site is in an area classified as an undetermined risk for flooding (Zone D) by the Federal Emergency Management Agency (FEMA) (Figure 3-18) (eHow, undated). Properties in Zone D lie outside areas that are known floodplains, but may still flood.

Table 3-16. Summary of City of Hobbs Municipal Water System Groundwater Water Quality, 2005 – 2009

Parameter	MCL[a] (µg/L[b])	Number of Detections	Detected Concentrations (µg/L[b])		
			Minimum	Maximum	Average
1,2-Dichloroethane	5	3	0.47	0.61	0.52
Antimony, Total	6	4	0.09	0.13	0.11
Arsenic	10	13	5.5	8.1	7.29
Barium	2,000	13	43.51	89	69.77
Benzene	5	2	0.58	0.81	0.70
Beryllium, Total	4	1	0.25	0.25	0.25
Bromodichloromethane	100	3	0.08	0.3	0.16
Bromoform	100	4	0.092	12	6.43
Chloroform	100	4	0.057	0.24	0.13
Chromium	100	13	2.9	18.8	7.09
Combined Uranium	30	6	0.00321	0.00927	0.01
Dibromochloromethane	100	3	0.055	0.37	0.23
Dichloromethane	5	5	4.35	5.62	4.89
Ethylbenzene	700	1	0.50	0.50	0.50
Fluoride (mg/L)	4	13	0.719	1.13	0.91
Gross Beta Particle Activity (pCi/L)	4	6	2.869	7.305	4.33
Iron (mg/L)	0.3[c]	1	0.0134	0.0134	0.0134
Nickel	100	13	0.3	3.51	1.46
Nitrate (as N) (mg/L)	10	12	2.69	5.82	4.01
Nitrate plus Nitrite (as N) (mg/L)	10	30	2.7	6.97	4.24
pH (standard units)	6.5 to 8.5[c]	1	7.24	7.24	7.24
Radium-226 (pCi/L)	5	2	0.175	0.382	0.23
Radium-228 (pCi/L)	5	1	1.082	1.082	1.082
Selenium	50	13	0.00589	18	5.24
Total Haloacetic Acids)	60	11	1	105.3	14.01
Total Trihalomethanes	80	20	0.602	13.95	6.85
Thallium, Total	2	2	0.05	0.05	0.05
Xylenes, Total	10,000	6	0.7	2.05	1.37

Note: Includes water quality data for five ground storage reservoirs and Well 5, which pumps directly into the distribution system. Samples were collected after the raw water had been chlorinated.
Source: Stephens & Associates, 2009.
[a]Maximum contaminant level specified in EPA National Primary Drinking Water Regulations (40 CFR 141 (2010))
[b]Unless otherwise noted
[c]EPA National Secondary Drinking Water Regulations (40 CFR 143 (2010))
µg/L = micrograms per liter
MCL = maximum contaminant level
mg/L = milligrams per liter
pCi/L = picocuries per liter

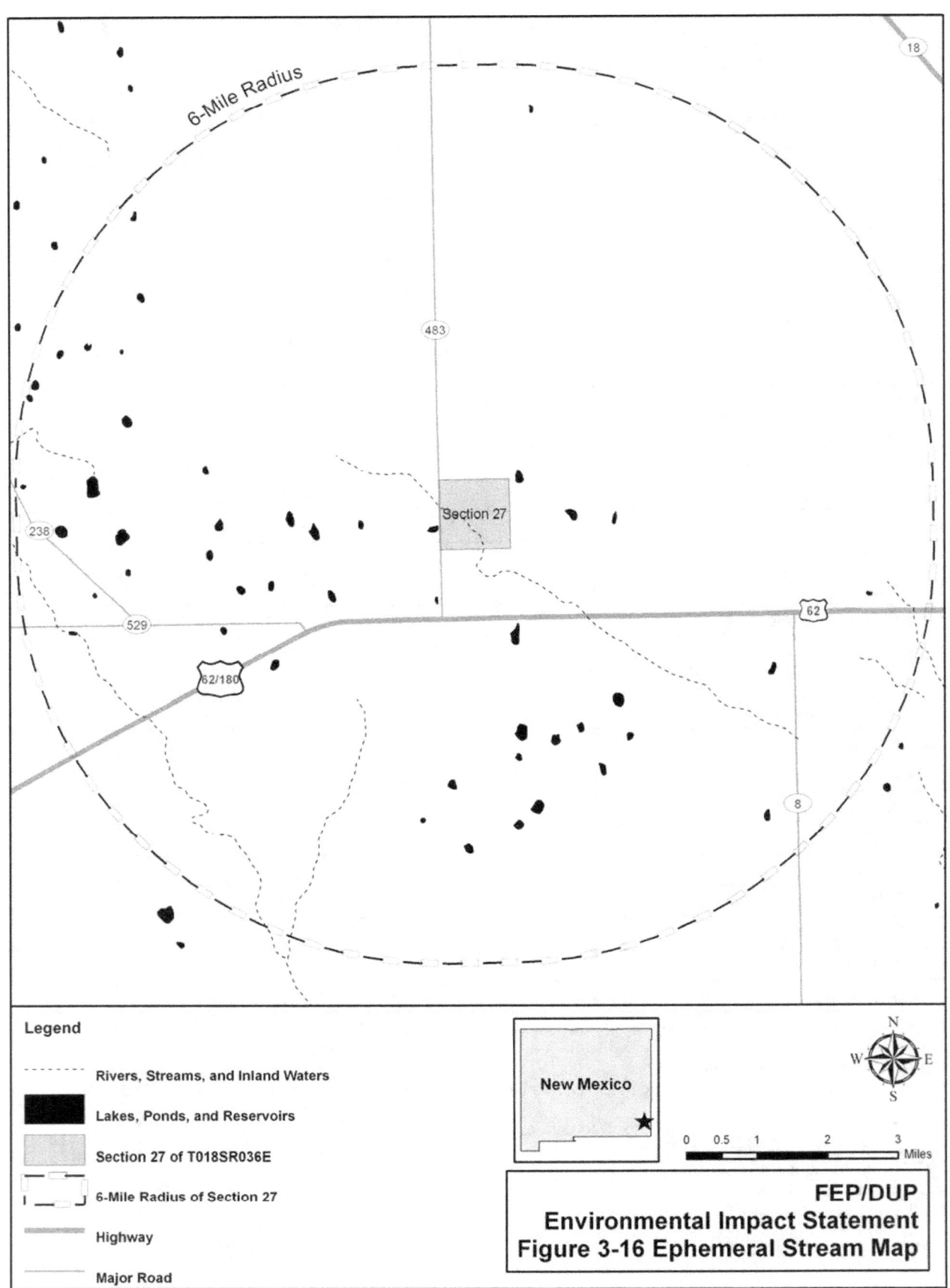

Legend

------ Rivers, Streams, and Inland Waters

■ Lakes, Ponds, and Reservoirs

▨ Section 27 of T018SR036E

⬚ 6-Mile Radius of Section 27

━━ Highway

── Major Road

New Mexico

N
W E
S

0 0.5 1 2 3 Miles

FEP/DUP
Environmental Impact Statement
Figure 3-16 Ephemeral Stream Map

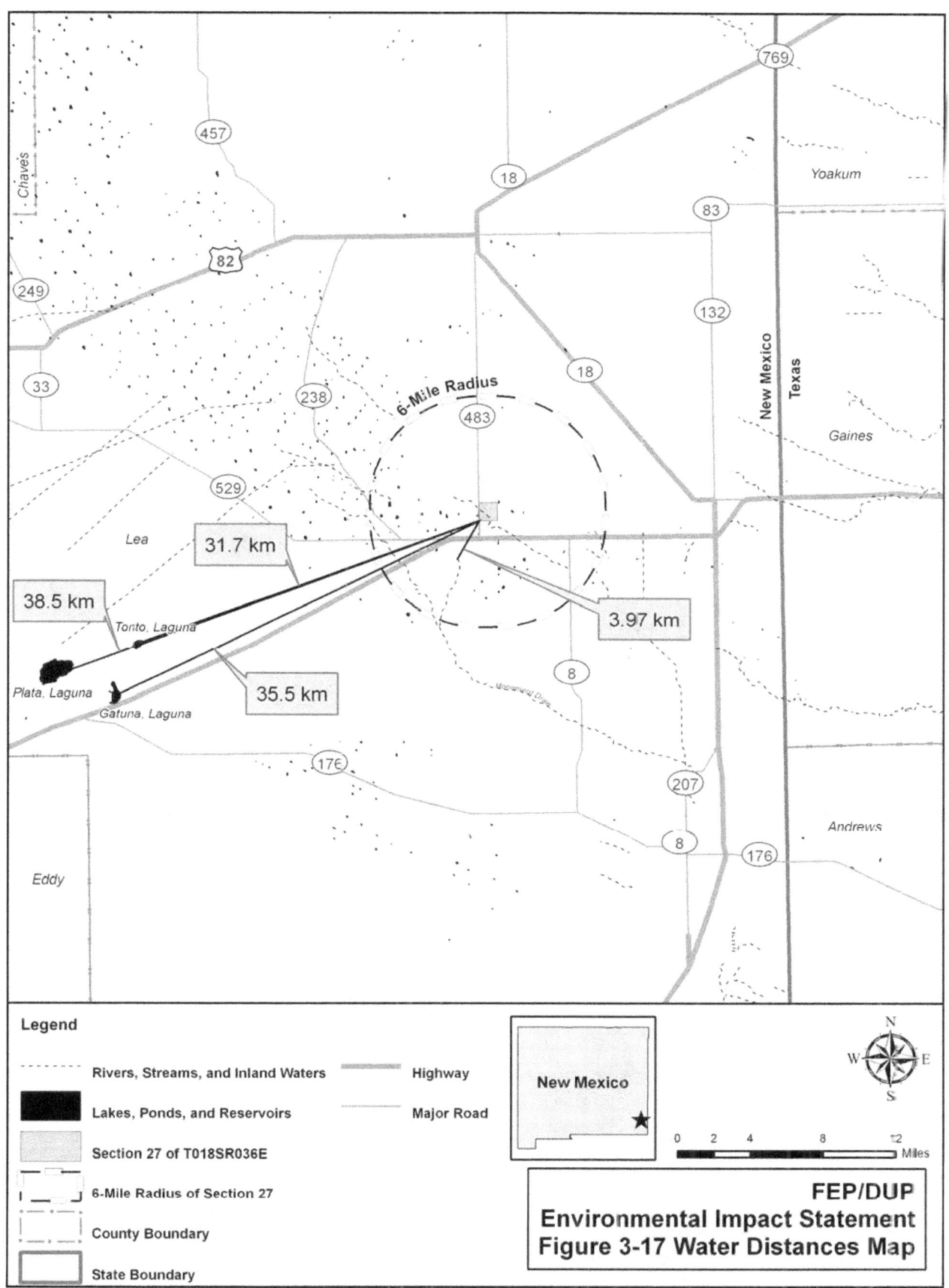

Legend

- ------ Rivers, Streams, and Inland Waters
- ■ Lakes, Ponds, and Reservoirs
- Section 27 of T018SR036E
- 6-Mile Radius of Section 27
- County Boundary
- State Boundary

— Highway
— Major Road

New Mexico

0 2 4 8 12
Miles

FEP/DUP
Environmental Impact Statement
Figure 3-17 Water Distances Map

Labels on map: 769, 457, 18, 83, 132, 18, 83, 249, 82, 33, 238, 483, 529, 457, 17€, 207, 8, 176, 8

31.7 km
38.5 km
35.5 km
3.97 km

6-Mile Radius

Chaves
Yoakum
New Mexico
Texas
Gaines
Lea
Tonto, Laguna
Plata, Laguna
Gatuna, Laguna
Eddy
Andrews

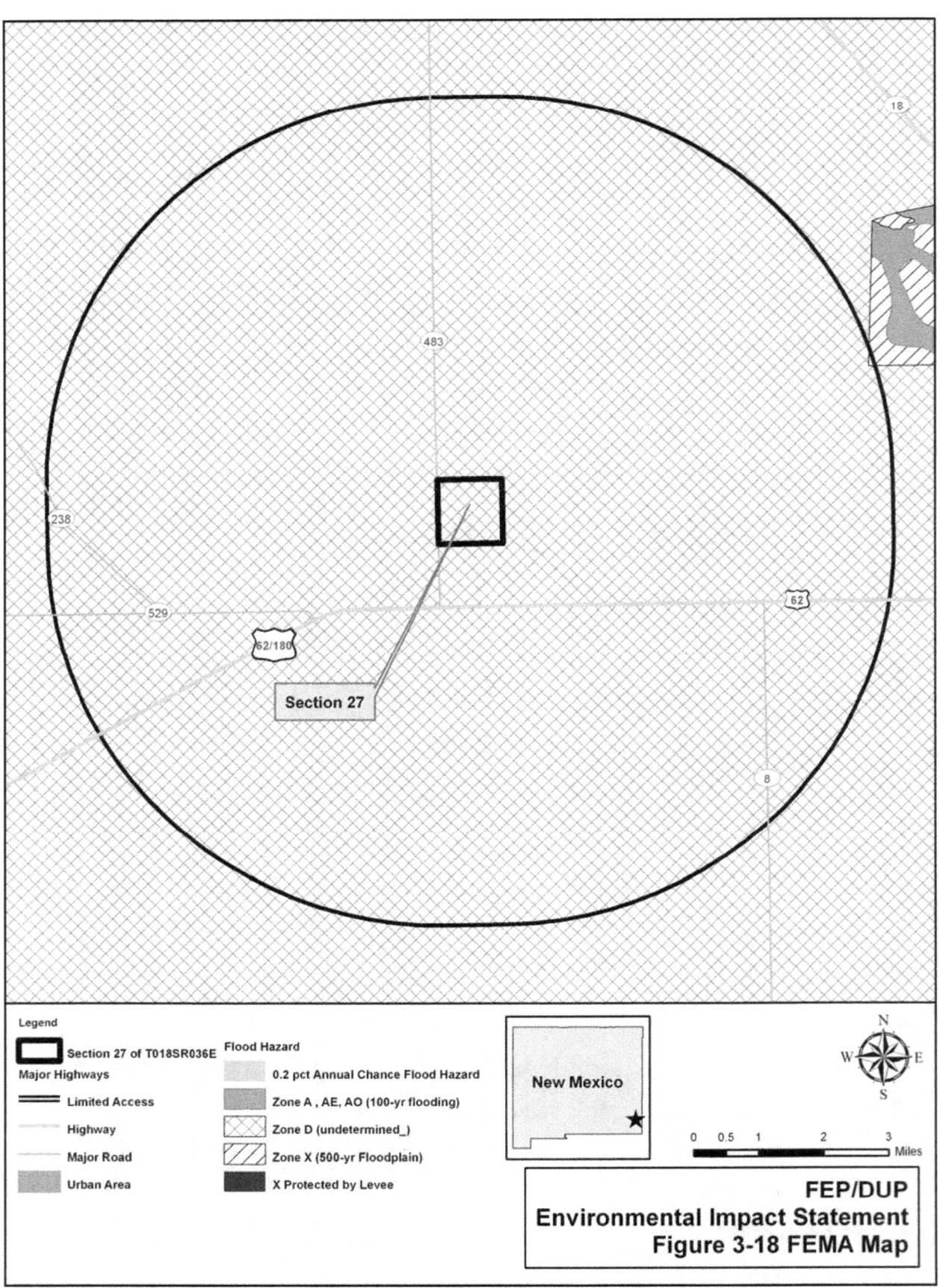

Legend

Section 27 of T018SR036E

Major Highways

Limited Access

Highway

Major Road

Urban Area

Flood Hazard

0.2 pct Annual Chance Flood Hazard

Zone A , AE, AO (100-yr flooding)

Zone D (undetermined_)

Zone X (500-yr Floodplain)

X Protected by Levee

Section 27

New Mexico

N
W   E
S

0   0.5   1   2   3
                                    Miles

FEP/DUP
Environmental Impact Statement
Figure 3-18 FEMA Map

3.8   Ecological Resources

This section describes the ecological communities on the 259-ha (640-ac) proposed IIFP site, which includes the 16-ha (40-ac) proposed facility, and in the vicinity of the proposed site. It also discusses important species that occur or have the potential to occur on the site or in the vicinity, and habitats in the vicinity that are important to those species.

Surveys were conducted for IIFP for vegetation and wildlife, including the dunes sagebrush lizard (*Sceloporus arenicolus*) and the lesser prairie chicken (*Tympanuchus pallidicintus*), two Federal candidate species (GL Environmental, 2010c; GL Environmental, 2010d; SORA, 2011). The dunes sagebrush lizard's range encompasses the IIFP site and a BLM resource management plan has been proposed to preserve habitat for the lesser prairie-chicken and the dunes sagebrush lizard, roughly 11 km (7 mi) from the IIFP site, and create a travel corridor for the lesser prairie-chicken roughly 52 km (32 mi) from the site, as discussed in Section 3.8.6.3.

3.8.1   Ecosystems in the Proposed Facility

As described in Section 3.6.1, the site is on the Llano Estacado of the Southern High Plains physiographic region. The western portion of the Llano Estacado supports shortgrass prairie habitat, and the southern portion is transitional to the more arid Chihuahuan Desert. The site lies in a transitional zone between two distinct ecoregions: Western Great Plains Shortgrass Prairie and Apacherian-Chihuahuan Mesquite Upland Scrub (Figure 3-19) (USGS, 2010b).

3.8.1.1   Western Great Plains Shortgrass Prairie

Western Great Plains Shortgrass Prairie habitat covers approximately 55 percent of the proposed site. The short grasses that dominate the system are extremely drought- and grazing-tolerant. This ecosystem is characterized by blue grama grass (*Bouteloua gracilis*). Scattered shrub and dwarf-shrub species such as sand sagebrush (*Artemisia filifolia*), prairie sagewort (*Artemisia frigida*), little sagebrush (*Artemisia tridentate*), fourwing saltbush (Atriplex canescens), crispleaf buckwheat (*Eriogonum effusum*), broom snakeweed (*Gutierrezia sarothrae*), and pale desert-thorn (*Lycium pallidum*) may also be present. Climate, fire, and grazing maintain this system (NatureServe, 2009; USGS, 2010b).

3.8.1.2   Apacherian-Chihuahuan Mesquite Upland Scrub

Apacherian-Chihuahuan Mesquite Upland Scrub habitat covers approximately 45 percent of the proposed site. This ecosystem often occurs as invasive upland shrublands such as those that are concentrated in the foothills and piedmont of the Chihuahuan Desert (NatureServe, 2009). Vegetation is dominated typically by honey mesquite (*Prosopis glandulosa*) or velvet mesquite (*Prosopis velutina*) and succulents. Grass cover is typically low and composed of desert grasses such as low woollygrass (*Dasyochloa pulchella*), bush muhly (*Muhlenbergia porteri*), curlyleaf muhly (*Muhlenbergia setifolia*), and tobosagrass (*Pleuraphis mutica*) (NatureServe, 2009). During the last century, the area occupied by this ecosystem has increased through conversion of desert grasslands as a result of drought, overgrazing by livestock, and decreases in fire frequency (NatureServe, 2009; USGS, 2010b).

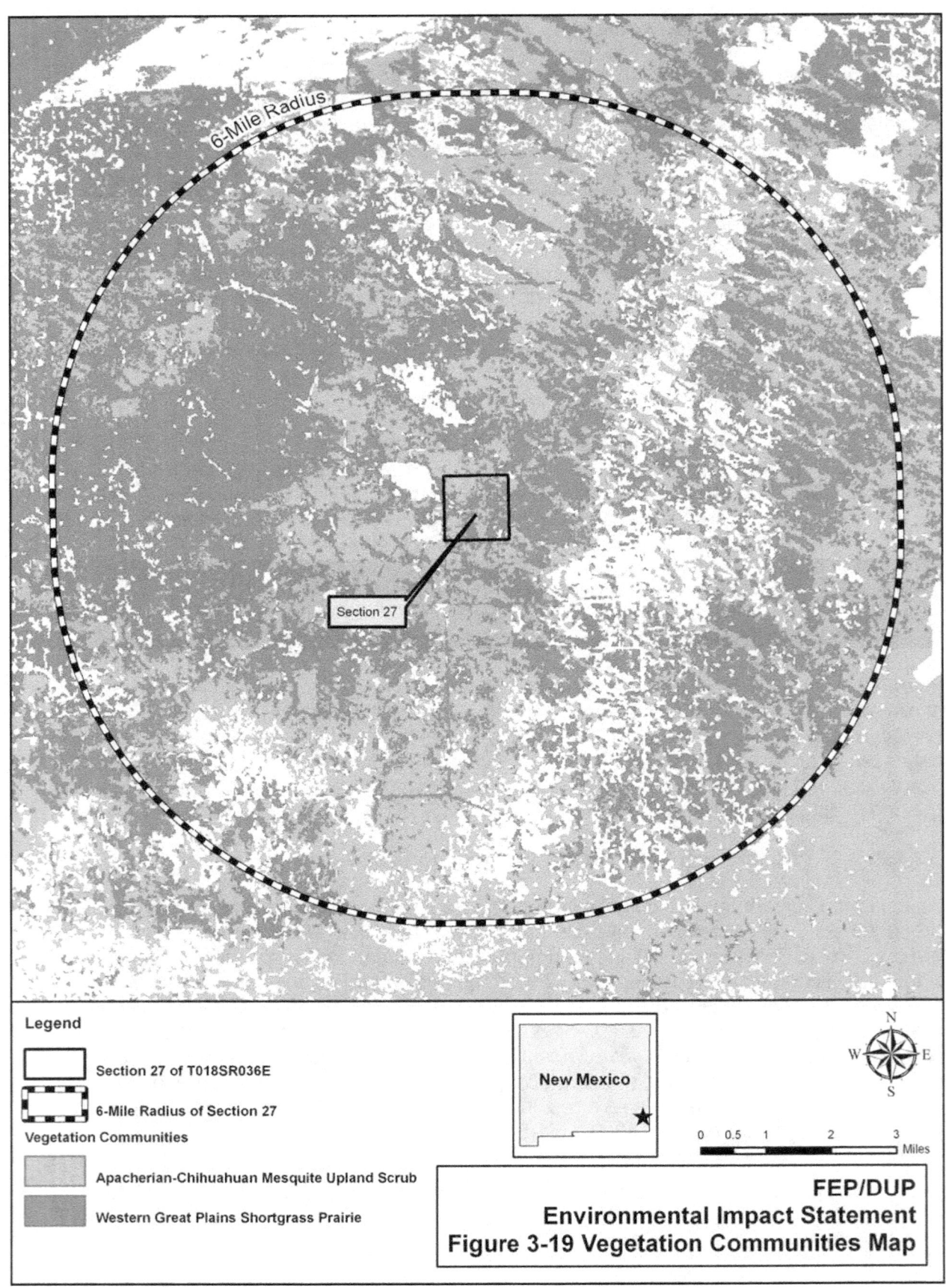

Legend

☐ Section 27 of T018SR036E

▣ 6-Mile Radius of Section 27

Vegetation Communities

▢ Apacherian-Chihuahuan Mesquite Upland Scrub

▣ Western Great Plains Shortgrass Prairie

New Mexico

0 0.5 1 2 3 Miles

FEP/DUP
Environmental Impact Statement
Figure 3-19 Vegetation Communities Map

3.8.1.3   Wetland and Riparian Habitat

As described in Section 3.7.2, there are no wetlands or stream systems within the footprint of the proposed facility. Depressions that hold ephemeral water, which are located throughout Section 27, are important breeding and nursery sites for amphibians, and can be important stopovers for migrating waterfowl and shorebirds. Vegetated arroyos, such as the one running across the southern part of Section 27 (and outside the 16-ha [40-ac] facility footprint), serve as excellent wildlife corridors, and nesting habitat for birds (New Mexico Department of Game and Fish [NMGF] correspondence in Appendix 3).

3.8.2   Vegetation of the Proposed Facility

Most of the plant species on the proposed facility are typical of Plains-Mesa Grassland and Desert Grassland communities (Dick-Peddie, 1993). These communities are characterized by significant amounts of grasses and less than 10 percent of total cover being forbs and shrubs (Dick-Peddie, 1993).

A vegetation survey was conducted by GL Environmental on behalf of IIFP at the proposed 16-ha (40-ac) facility on October 16, 2010 to determine total vegetative cover and relative cover (GL Environmental, 2010c). Total vegetative cover represents the percentage of ground that has living vegetation on it compared to bare ground or litter. Relative cover represents the fraction of total vegetative cover that is composed of a certain species or category of plants (e.g., perennial plants). Total vegetative cover is approximately 45 percent with 98 percent of this cover consisting of perennial grasses, including blue grama, burrograss (*Scleropogon bevifolius*), black grama (*Bouteloua eripoda*), and James Galleta grass (*Pleuraphis jamesii*). Shrubs included honey mesquite and hedgehog cactus (*Echinocereus* sp.). Forbs on the site included Texas croton (*Croton texensis*), Texas blueweed, (*Helianthus ciliaris*), and curly cup gumweed (*Grindelia squarrosa*) (GL Environmental, 2010c).

3.8.3   Wildlife that Could Occur on the Site and Proposed Facility

Wildlife that could occur on the IIFP site include species typical of arid grassland and desert habitats. Table 3-17 lists mammals, birds, amphibians, and reptiles that could be present on the site, based on habitat requirements, and presents information regarding their preferred habitats. The table was compiled from the lesser prairie-chicken survey (SORA, 2011) conducted by SORA on behalf of IIFP for the proposed site in April 2011 and surveys conducted in 2004 for the Louisiana Enrichment Services (LES) National Enrichment Facility, located approximately 33 km (20 mi) southeast of the proposed IIFP (NRC, 2005). Comparison with the LES site is appropriate because both facilities are in the transition zone from the shortgrass prairie to the Chihuahuan desert (USDA, 2004).

A diverse assemblage of animals, including several commercially and recreationally important game species are typical of this habitat. Pronghorn "antelope" (*Antilocapra americana*) and mule deer (*Odocoileus hemionus*) are plentiful in eastern New Mexico (NMGF, 2009). NMGF has assigned sections of the State to specific Antelope Management Units in order to better manage pronghorn antelope populations. The proposed IIFP site is within Antelope Management Unit 26, one of several management units in southeastern New Mexico. An estimated 44 pronghorn were harvested from this Management Unit in 2007-2008 (NMGF, 2008a). The site is also a part of New Mexico Game Management Unit 31. Approximately 500 mule deer were harvested from Game Management Unit 31 in 2009 (NMGF, 2010a).

Table 3-17. Mammal, Bird, Amphibian, and Reptile Species Likely to be Present at the Proposed Site and Vicinity

Common Name	Scientific Name	
Mammals		**Preferred Habitat**
Black-tailed jackrabbit	*Lepus californicus*	Grasslands and open areas
Black-tailed prairie dog	*Cynomys ludovicianus*	Shortgrass prairie
Cactus mouse	*Peromyscus eremicus*	Grasslands, prairies, and mixed vegetation
Collared peccary	*Dicotyles tajacu*	Brushy, semi-desert, chaparral, mesquite, and oaks
Coyote	*Canis latrans*	Open space, grasslands, and brush country
Deer mouse	*Peromyscus maniculatus*	Grasslands, prairies, and mixed vegetation
Desert cottontail	*Sylvilagus audubonii*	Arid lowlands, brushy cover, and valleys
Mule deer	*Odocoileus hemionus hemionus*	Desert shrubs, chaparral, and rocky uplands
Ord's kangaroo rat	*Dipodomys ordii*	Hard desert soils
Plain's pocket gopher	*Geomys bursarius*	Deep soils of the plains
Pronghorn	*Antilocapra americana*	Sagebrush flats, plains, and deserts
Raccoon	*Procyon lotor*	Brushy, semi-desert, chaparral, and mesquite
Southern Plains woodrat	*Neotoma micropus*	Grasslands, prairies, and mixed vegetation
Spotted ground squirrel	*Spermophilus spilosoma*	Brushy, semi-desert, chaparral, mesquite, and oaks
Striped skunk	*Mephitis mephitis*	All land habitats
Swift fox	*Vulpes velox*	Rangeland with short grasses and low shrub density
White-throated woodrat	*Neotoma albigula*	Grasslands, prairies, and mixed vegetation
Yellow-faced pocket gopher	*Pappogeomys castanops*	Deep soils of the plains
Birds		**Seasonal Preference**
American kestrel	*Falco sparverius*	Year round
Ash-throated flycatcher	*Myiarchus cinerascens*	Spring
Bewick's wren	*Thyromanes bewickii*	Spring and summer
Black-chinned hummingbird	*Archilochus alexandri*	Spring
Blue grosbeak	*Guiraca caerulea*	Summer
Bullock's oriole	*Icterus bullockii*	Summer
Cassin's sparrow	*Aimophila cassinii*	Spring and Fall
Cactus wren	*Campylorhynchus brunneicapillus*	Year round

Table 3-17. Mammal, Bird, and Amphibian, and Reptile Species Likely to be Present at the Proposed Site and Vicinity (Continued)

Common Name	Scientific Name	
Birds (Continued)		**Seasonal Preference**
Chihuahuan raven	*Corvus cryptoleucus*	Year round
Common raven	*Corvus corax*	Summer and winter
Crissal thrasher	*Toxostoma dorsale*	Migrant
Eastern meadowlark	*Sturnella magna*	Year round
European starling	*Sturnus vulgaris*	Year round
Gambel's quail	*Lophortyx gambelii*	Rare
Great-tailed grackle	*Quiscalus mexicanus*	Year round
Green-tailed towhee	*Pipilo chlorurus*	Migrant
Horned lark	*Eremophila alpestris*	Spring
House finch	*Carpodacus mexicanus*	Spring and summer
Killdeer	*Charadrius vociferus*	Spring and summer
Lark bunting	*Calamospiza melanocorys*	Winter
Lark sparrow	*Chondestes grammacus*	Spring and summer
Loggerhead shrike	*Lanius ludovicianus*	Summer
Long-eared owl	*Asio otus*	Summer and winter
Mallard	*Anas platrhynchos*	Spring
Mourning dove	*Zenaida macroura*	Spring and summer
Nighthawk	*Chordeiles minor*	Spring
Northern mockingbird	*Mimus polyglottos*	Summer
Northern bobwhite	*Colinus virginianus*	Spring
Pyrrhuloxia	*Cardinalis sinuatus*	Winter
Red-tailed hawk	*Buteo jamaicensis*	Winter
Red-winged blackbird	*Agelaius phoeniceus*	Year round
Roadrunner	*Geococcyx californianus*	Uncommon
Sage sparrow	*Amphispiza belli*	Uncommon
Say's phoebe	*Sayornis saya*	Spring
Scaled quail	*Callipepla squamata*	Spring and summer
Scissor-tailed flycatcher	*Tyrannus forficatus*	Migrant
Scott's oriole	*Icterus parisorum*	Summer and winter
Swainson's hawk	*Buteo swainsoni*	Summer
Turkey vulture	*Cathartes aura*	Summer
Vermillion flycatcher	*Pyrocephalus rubinus*	Winter and migrant
Vesper sparrow	*Pooecetes gramineus*	Spring
Western burrowing owl	*Athene cunicularia hypugaea*	Spring
Western kingbird	*Tyrannus verticalis*	Summer

Table 3-17. Mammal, Bird, and Amphibian, and Reptile Species Likely to be Present at the Proposed Site and Vicinity (Continued)

Common Name	Scientific Name	
Amphibians/Reptiles		Preferred Habitat
Coachwhip	*Masticophis flagellum*	Mixed grass prairie and desert grasslands
Collared lizard	*Crotaphytus collaris*	Desert grasslands
Eastern fence lizard	*Sceloporus undulates*	Mixed grass prairie and desert grasslands
Garter snake	*Thamnophis* sp.	Desert grasslands
Ground snake	*Sonora semiannulata*	Desert grasslands
Longnose leopard lizard	*Gambelia wislizenii*	Mixed grass prairie and desert grasslands
Lesser earless lizard	*Holbrookia maculata*	Mixed grass prairie and desert grasslands
Longnosed snake	*Rhinocheilus lecontei*	Desert grasslands
Ornate box turtle	*Terrapene ornate ornata*	Desert grasslands and shortgrass prairie
Pine-gopher snake	*Pituophis melanoleucus*	Shortgrass prairie and desert grasslands
Plains blackhead snake	*Tantilla nigriceps*	Shortgrass prairie and desert grasslands
Plains spadefoot toad	*Spea bombifrons*	Shallow to standing pools of water
Rattlesnakes	*Crotalus* sp.	Shortgrass prairie and desert grasslands
Six-lined racerunner	*Cnemidophorus sexlineatus*	Mixed grass prairie and desert grasslands
Tiger salamander	*Ambystoma tigrinum*	Tall-grass and mixed prairie
Texas horned lizard	*Phrynosoma cornutum*	Desert grasslands
Western whiptail lizard	*Cnemidophorus tigris*	Mixed grass prairie and desert grasslands

Source: NRC, 2005; USDA, 2004; eBird, 2011; SORA, 2011

Lea County also provides opportunities to hunt small birds, most notably scaled quail (*Callipepla squamata*) and Northern bobwhite (*Colinus virginianus*). Scaled quail occur primarily in semi-arid rangelands with mixed scrub communities (shrubs, grasses, and bare ground) (NMGF, 2008b). The vegetation on the proposed IIFP site provides habitat for scaled quail. Northern bobwhites also occur in southeastern New Mexico, including Lea County (NMGF, 2008b). Northern bobwhite habitat in New Mexico is characterized by large expanses of native warm-weather grasses mixed with annual weeds and legumes, with dense, brushy areas for escape cover and roosting (NMGF, 2008b). The near absence of dense thickets on the proposed IIFP site suggests that it offers only marginal Northern bobwhite habitat.

3.8.4   Wildlife Travel Corridors for Resident and Migratory Species

3.8.4.1   Migratory Species

Southeastern New Mexico, including Lea County, is within the Central Flyway, one of the four major North American bird migration corridors between nesting and wintering grounds (CFC, undated; TPWD, 2007). Birds of prey associated with the Central Flyway include the American kestrel (*Falco sparverius*), ferruginous hawk (*Buteo regalis*), Swainson's hawk (*Buteo swainsoni*) and others. Waterfowl that use the Central Flyway to move between breeding areas in Canada and wintering areas in Texas and Mexico include the mallard (*Anas platrhynchos*), American widgeon (*Anas americana*), green-winged teal (*Anas crecca*) and others. Songbirds that migrate along the Central Flyway include the American goldfinch (*Spinus tristis*), Western kingbird (*Tyrannus verticalis*), lark bunting (*Calamospiza melanocorys*), vesper sparrow (*Pooecetes gramineus*) and others. Common shorebirds associated with the Central Flyway include the killdeer (*Charadrius vociferus*), greater yellowlegs (*Tringa melanoleuca*), spotted sandpiper (*Actitis macularia*), least sandpiper (*Calidris minutilla*) and others (Stokes and Stokes, 1996). Depending on the availability of food and water that may be temporarily present in the depressions that dot Section 27 during seasonal migrations, migratory birds such as these could occasionally be present on or in the vicinity of the site.

3.8.4.2   Resident Species

Wildlife corridors are typically linear habitats that link larger habitats. They can serve a region (e.g., a river followed by migratory waterfowl), a landscape (e.g., a transmission corridor right-of-way that connects two natural areas), or a local site (e.g., a gully or strip of trees that deer use to move between bedding and feeding areas). There are no terrain features at the proposed IIFP site that would serve as wildlife corridors.

3.8.5   Critical Habitats

Under the *Endangered Species Act* "critical habitat" is defined as: (1) specific areas within the geographical area occupied by the [listed] species at the time of listing, if they contain physical or biological features essential to conservation, and those features may require special management considerations or protection, and (2) specific areas outside the geographical area occupied by the [listed] species if the agency determines that the area itself is essential for conservation.

The nearest critical habitat is the Pecos River, approximately 146 km (91 mi) northwest of the site, which supports the Pecos bluntnose shiner (*Notropis simus pecosensis*). This fish is listed as Federally threatened (USFWS, 2010a).

3.8.6   Wildlife Sanctuaries, Wildlife Management Areas, Refuges, and Preserves

Wildlife sanctuaries, management areas, refuges, and preserves are areas designated by the NM GF as open to wildlife-associated recreation activities beyond the traditional uses of hunting and fishing.

3.8.6.1   Green Meadow Lake

Green Meadow Lake is a New Mexico-designated Wildlife Area approximately 16 km (10 mi) northeast from the proposed site (NMGF, Undated 1). Migratory waterfowl using the Central Flyway may rest at the lake during migrations.

3.8.6.2   Prairie Chicken Wildlife Area

The New Mexico-designated Prairie Chicken Wildlife Area comprises parcels throughout southeastern New Mexico that provide habitat for the preservation and restoration of the lesser prairie-chicken (NMGF, Undated 2). The closest Prairie Chicken Wildlife Area is more than 80 km (50 mi) from the proposed site.

3.8.6.3   Lesser Prairie-Chicken Habitat Preservation Area of Critical Environmental Concern and Proposed Lesser Prairie-Chicken Expansion Corridor

In 2008, the BLM issued a Record of Decision (ROD) to implement a resource management plan (RMP) for all resources on approximately 343,983 surface ha (850,000 surface ac) of public land in parts of Chaves, Eddy, Lea, and Roosevelt Counties in southeastern New Mexico. To meet some of the objectives of this RMP, the BLM will establish a 23,472-ha (58,000-ac) Lesser Prairie-Chicken Habitat Preservation Area of Critical Environmental Concern (ACEC) to maintain and enhance habitat for the lesser prairie-chicken and the dunes sagebrush lizard. The entire RMP area lies west and south of the proposed facility location; the nearest part of the RMP area is approximately 11 km (7 mi) due south of the site (Figure 3-20) (BLM, 2008).

Additionally, the BLM has proposed a Lesser Prairie-Chicken Expansion Corridor in southeastern New Mexico in order to maintain a north-south travel way for lesser prairie-chickens. No final decision has been made about the corridor (BLM, 2010) which is 51.5 km (32 mi) from the proposed site at its nearest boundary (Figure 3-20).

The IIFP proposed site does not provide optimal habitat for the lesser prairie-chicken and is not included in the ACEC nor in the proposed corridor.

3.8.7   Special-Status Species

The *Endangered Species Act* defines an endangered species as any species which is in danger of extinction throughout all or a significant portion of its range, and a threatened species as any species which is likely to become an endangered species within the foreseeable future throughout all or a significant portion of its range.

According to the New Mexico Rare Plant Technical Council there are no special-status plant species in Lea County (NMRPTC, 1999).

The U.S. Fish and Wildlife Service (USFWS) maintains lists of endangered and threatened species, candidate species, and species of concern for Lea County (USFWS, 2010b; USFWS, 2010c). The Northern aplomado falcon (*Falco femoralis septentrionalis*), the black-footed ferret (*Mustela nigripes*), and the least tern (*Sterna antillarum athalassos*) are listed as Federally endangered species occurring in Lea County (USFWS, 2010b; NMGF, 2010b).

Candidate species are those that the USFWS has sufficient information to propose that they be added to the Federal list of threatened and endangered species, but the listing action has been precluded by other higher priority listing activities. Two candidate species are listed as potentially occurring in Lea County: the lesser prairie-chicken and the dunes sagebrush lizard. On December 14, 2010, the USFWS issued a proposal to list the dunes sagebrush lizard as a Federally endangered species (USFWS, 2010d). The USFWS also maintains a list of species of concern, however, these species are not protected by law. There are eight Federal species

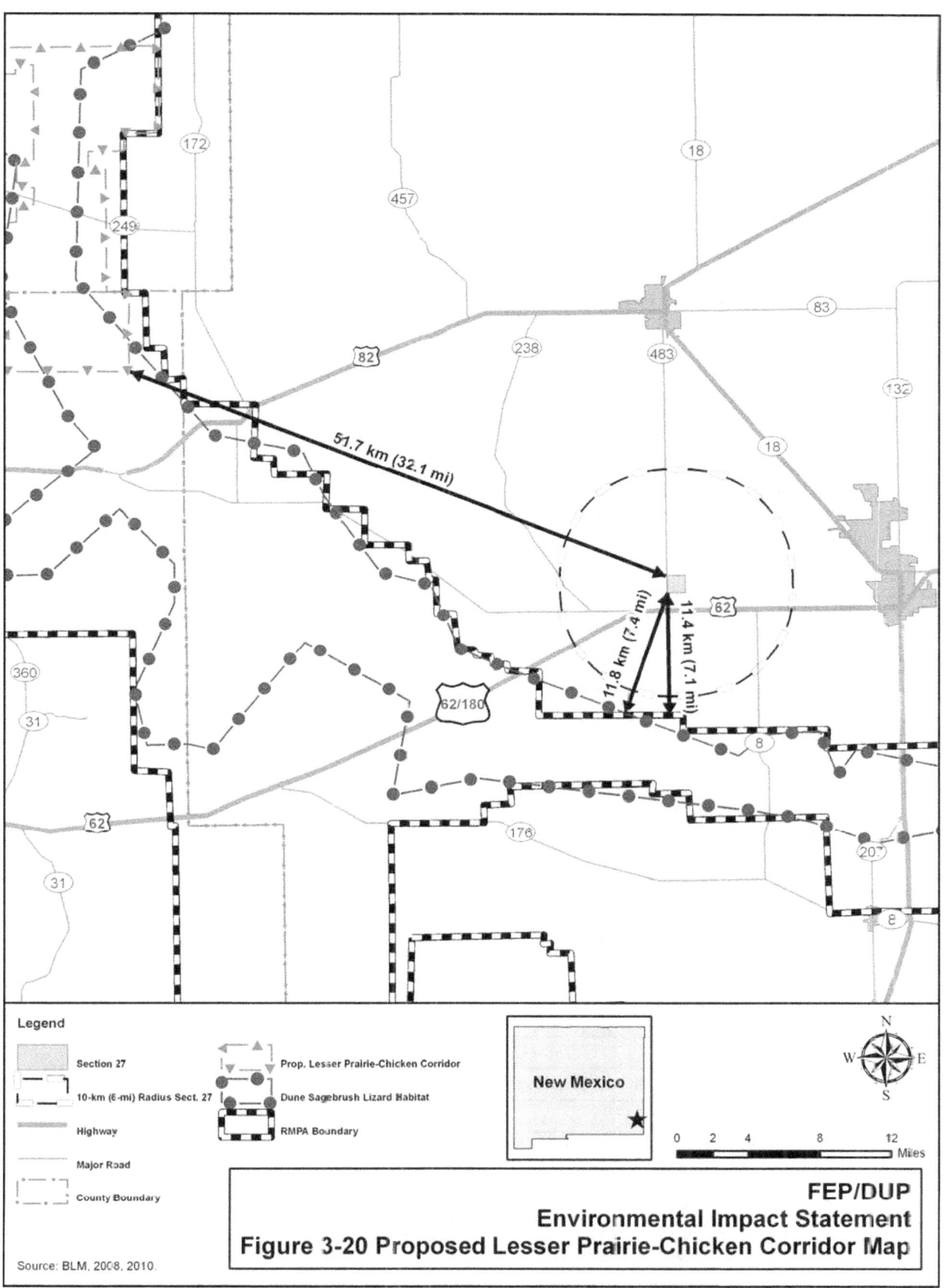

51.7 km (32.1 mi)

11.8 km (7.4 mi)

11.4 km (7.1 mi)

62/180

Legend

Section 27

10-km (6-mi) Radius Sect. 27

Highway

Major Road

County Boundary

Prop. Lesser Prairie-Chicken Corridor

Dune Sagebrush Lizard Habitat

RMPA Boundary

New Mexico

N
W E
S

0   2   4   8   12
Miles

FEP/DUP
Environmental Impact Statement
Figure 3-20 Proposed Lesser Prairie-Chicken Corridor Map

Source: BLM, 2008, 2010.

of concern in Lea County: black-tailed prairie dog (*Cynomys ludovicianus*), swift fox (*Vulpes velox*), American peregrine falcon (*Falco peregrinus anatum*), arctic peregrine falcon (*Falco peregrinus tundrius*), Baird's sparrow (*Ammodramus bairdii*), Bell's vireo (*Vireo bellii*), western burrowing owl (*Athene cunicularia hypugaea*), and the yellow-billed cuckoo (*Coccyzus americanus*) (USFWS, 2010b).

Based on the best available information, the swift fox and western burrowing owl could occur on or visit the proposed site. It is unlikely that the dunes sagebrush lizard (GL Environmental, 2010d) or the lesser prairie-chicken (SORA, 2011) would occur at the site. The black-tailed prairie dog has not been reported as occurring within Lea County; and the American peregrine falcon, arctic peregrine falcon, and Baird's sparrow have been reported only rarely or very rarely in Lea County. No preferred habitat for the Northern aplomado falcon, least tern, black-footed ferret, Bell's vireo, or yellow-billed cuckoo occurs on the proposed site.

Endangered, threatened, candidate species, and species of concern listed by the USFWS and the State of New Mexico for Lea County are described in the following sections and presented in Table 3-18.

Table 3-18.   Rare, Threatened or Endangered Species Listed for Lea County, New Mexico

Common Name	Scientific Name	Federal Status[a]	State Status[a]
Mammals			
Black-footed ferret	*Mustela nigripes*	E[2]	-
Black-tailed prairie dog	*Cynomys ludovicianus*	S[2]	-
Swift fox	*Vulpes velox*	S[1]	-
Birds			
American peregrine falcon	*Falco peregrinus anatum*	S[2]	T[1]
Arctic peregrine falcon	*Falco peregrinus tundrius*	S[2]	T[1]
Baird's sparrow	*Ammodramus bairdii*	S[2]	T[1]
Mammals			
Bald eagle	*Haliaeetus leucocephalus*	-	T[1]
Bell's vireo	*Vireo bellii*	S[2]	T[1]
Broad-billed hummingbird	*Cynanthus latirostris magicus*	-	T[1]
Least tern[b]	*Sterna antillarum athalassos*	E[1]	E[1]
Lesser prairie-chicken	*Tympanuchus pallidicinctus*	C[2]	-
Northern aplomado falcon	*Falco femoralis septentrionalis*	E[2]	E[1]
Western burrowing owl	*Athene cunicularia hypugaea*	S[2]	-
Yellow-billed cuckoo	*Coccyzus americanus*	S[2]	-
Amphibians/Reptiles			
Dunes sagebrush lizard	*Sceloporus arenicolus*	PE[3]	E[1]

Sources: [1] NMGF, 2010b; [2] USFWS, 2010b; [3] USFWS, 2010d
[a] C = Candidate, E = Endangered, T = Threatened, S = Species of Concern, PE = Proposed Endangered, - = Not listed.
[b] The least tern is not listed by the USFWS as occurring in Lea County, however, it is listed by the New Mexico Department of Game and Fish as occurring in Lea County.

3.8.7.1   Federally Endangered Species

Northern Aplomado Falcon

The Northern aplomado falcon is listed as both a Federal and State endangered species. The preferred habitat in New Mexico for this species consists of open yucca desert land from the Rio Grande westward and north to Deming and Separ. The few nests known to occur in New Mexico were in areas of yucca grassland (USGS, 2005b; NMGF, 2010c). This habitat does not occur on the proposed site.

Black-Footed Ferret

The black-footed ferret is a Federally listed endangered species, but is not listed by the State of New Mexico. The historic range of the black-footed ferret included all of New Mexico; however, it was extirpated from most of its range, including New Mexico, by the 1960s. Black-footed ferrets are being reintroduced to their historic range, and the State of New Mexico has pursued reintroduction efforts (USFWS, 2008). The black-footed ferret is limited to open habitat, the same habitat used by prairie dogs: grasslands, steppe, and shrub steppe. The black-footed ferret has co-evolved with the prairie dog; their ranges and habitat closely overlap; however, the prairie dog has fewer protective regulations than the ferret (USFWS, 2008; USGS, 2005b). The preferred habitat does not occur on the proposed site.

Least Tern

The least tern is Federally listed as endangered and is also listed as endangered by the State of New Mexico. Its historic distribution was coincident with the major river systems of the Midwest as its habitat includes barren shorelines of lakes, rivers, and reservoirs (USGS, 2005b). The least tern has not been documented in Lea County, but has been reported as a migrant in Eddy County, just west of the proposed site, and has been documented breeding at Bitter Lake National Wildlife Refuge in Chaves County, approximately 161 km (100 mi) northwest of the proposed site (NMGF, 2010c; USGS, 2005b). No rivers, lakes, or reservoirs occur on the site; therefore, no habitat for this species is present on the proposed site.

3.8.7.2   Federally Proposed Endangered Species

Dunes Sagebrush Lizard

On December 14, 2010, the USFWS issued a proposal to modify the listing of the dunes sagebrush lizard from its current status as a Federal candidate species to that of endangered; this species is already listed as endangered by the State of New Mexico (USFWS, 2010c).

The range of the dunes sagebrush lizard within New Mexico appears to be confined to areas of active sand dunes vegetated by shinnery oak (*Quercus havardii*) in the extreme southeastern portion of the state and adjoining areas of Texas; although adjacent open habitats may be used in some places (NMGF, 2010c). The range stretches from eastern Chaves County, southernmost Roosevelt, and northernmost Lea Counties, southward and eastward into northeastern Eddy and south/central Lea counties. The closest part of the range lies 11.9 km (7.4 mi) south of the boundary of the facility site (Figure 3-20) (BLM, 2008; Center for Biological Diversity, 2002). Shinnery oak and sand dunes do not occur on the proposed IIFP site and; therefore, it is unlikely for the dunes sagebrush lizard to occur at the proposed site (GL Environmental, 2010d).

3.8.7.3   Federal Candidate Species

Lesser Prairie-Chicken

The lesser prairie-chicken is a candidate species for Federal protection due to habitat loss (USGS, 2005b). See Section 3.8.6.3 for a discussion on the BLM Lesser Prairie-Chicken ACEC and proposed Lesser Prairie-Chicken Expansion Corridor.

Lesser prairie-chickens are most common in dwarf shrub-mixed grass vegetation, interspersed with short-grass or mixed-grass habitats. They are also found in shinnery oak and bunch sumac and squaw bush (USGS, 2005b). Lea County is historically known to have habitat for lesser prairie-chickens, but the proposed site is at the southern periphery of their range (BLM, 2010; SORA, 2011). The IIFP site could provide suitable habitat for the lesser prairie-chicken, though there are limited water sources on the site (SORA, 2011).

3.8.7.4   Federal Species of Concern

American Peregrine Falcon

The American peregrine falcon is Federal species of concern and is listed as threatened by the State of New Mexico. It breeds in mountain areas and migrates essentially statewide; however, this species has only been reported rarely in Lea County (NMGF, 2010b).

Arctic Peregrine Falcon

The arctic peregrine falcon is a Federal species of concern and is listed as threatened by the State of New Mexico. This species is migratory and is found in a variety of habitats including forests, grasslands, and the Chihuahuan Desert Scrub (NMGF, 2010c). It is a very rare migrant through the State of New Mexico, but was reported in Lea County in 2007 (NMGF, 2010b).

Baird's Sparrow

The Baird's sparrow is a Federal species of concern and is listed as threatened by the State of New Mexico. Found in a variety of habitats ranging from desert grasslands to prairies and mountain meadows, the Baird's sparrow is a transient species in eastern and southern New Mexico and is considered rare to uncommon in Lea County (NMGF, 2010b).

Bell's Vireo

Bell's vireo is a Federal species of concern and is listed as threatened by the State of New Mexico. It winters south of the Mexican border and is a rare summer resident in Lea County (NMGF, 2010b). In New Mexico, this species occurs in riparian and wooded lowland habitats (NMGF, 2010c), none of which occur on the proposed site.

Black-Tailed Prairie Dog

The black-tailed prairie dog is a Federal species of concern, but it is not listed by the State of New Mexico. The black-tailed prairie dog commonly occurs in shortgrass prairie habitats (USGS, 2005b; NMGF, 2010c). However, it has not been reported in Lea County (NMGF, 2010b).

Swift Fox

The swift fox is a Federal species of concern, but it is not listed by the State of New Mexico. It is a year-round resident throughout the State, inhabiting shortgrass, midgrass and mixed prairies and adapting to overgrazed pastures, plowed fields, and fencerows (NMGF, 2010c).

Western Burrowing Owl

The Western burrowing owl is a Federal species of concern, but it is not listed by the State of New Mexico. Habitats include well-drained grasslands, prairies, steppes, deserts, and agricultural lands. The owls normally migrate south in late fall, but may not migrate if there are abandoned mammal burrows, which the owl uses for nests (NMGF, 2010c). This species has been reported only in the summer in Lea County; however, it has been documented as a year round resident in southern New Mexico (USGS, 2005b; NMGF, 2010c).

Yellow-Billed Cuckoo

The yellow-billed cuckoo population in eastern New Mexico is listed as a Federal species of concern, while the population in western New Mexico is a Federal candidate species; but it is not listed by the State of New Mexico. Yellow-billed cuckoos are often associated with riparian forests and deciduous woodlands (USGS, 2005b; NMGF, 2010c). They have been reported to occur in Lea County during the fall (USGS, 2005b; NMGF, 2010b).

3.8.7.5   New Mexico Threatened Species

Bald Eagle

The bald eagle was removed from the Federal endangered and threatened species list in 2007; however, it remains listed as threatened by the State of New Mexico and it still receives protection under the *Bald and Golden Eagle Protection Act, Lacey Act,* and *Migratory Bird Treaty Act.* It is a rare visitor to Lea County (NMGF, 2010b).

Broad-billed Hummingbird

The broad-billed hummingbird is listed as a threatened species by the State of New Mexico. It is rare in Eddy County (adjacent to Lea County) and is not known to occur in Lea County. It is usually associated with riparian woodlands (NMGF, 2010c), none of which occur on the proposed site.

3.9   Socioeconomics and Environmental Justice

This section describes the socioeconomic resources that have the potential to be affected by the construction and operation, and decommissioning of the IIFP facility at a rural site near Hobbs, New Mexico. The section is divided into six major subsections: (1) demography, including minority and low-income populations (environmental justice); (2) employment and income; (3) taxes; (4) housing; (5) public utilities; and (6) public services. These subsections include discussions of spatial (e.g., regional, vicinity, and site) and temporal (e.g., 10-year increments of population growth) considerations, where appropriate. Supporting analyses are provided in Appendix D – Socioeconomics.

NRC staff collected and analyzed regional socioeconomic data, including the commuting points of origin and destination of all workers among Lea County and its neighboring counties, to determine the appropriate socioeconomic ROI.

The NRC staff considered counties with their land area mostly within the 80-km (50-mi) radius of the site, or with a small portion of the area within the 80-km (50-mi) radius, but with a large population center within the 80-km (50-mi) radius, which was assumed to be a reasonable commuting distance. Two counties in New Mexico and three counties in Texas have these characteristics: Lea County and Eddy Counties, in New Mexico, and Andrews, Gaines, and Yoakum Counties in Texas.

Commuting patterns of working residents in Lea County demonstrate a preference for a work site in Lea County, and residents of the surrounding counties have demonstrated a reluctance to drive to a worksite in Lea County. However, Carlsbad and the Waste Isolation Pilot Plant are in Eddy County, approximately 80 km (50 mi) from the proposed site, and some residents with the appropriate skill set for the IIFP facility may commute to the proposed IIFP facility. Despite the limited employment opportunities in Andrews, Gaines, and Yoakum Counties, few residents of those counties work in Lea County, even with its larger employment base. Therefore, it is reasonable to assume that most of the IIFP workforce will come from Lea or Eddy Counties.

Changes in population are the key driver of impacts to socioeconomics. Therefore, the proposed action has the potential to impact socioeconomics (employment, population, income, housing, infrastructure, and community services) within Lea and Eddy Counties, because those are the counties most likely to incur population increases due to the proposed action, and it is unlikely to affect socioeconomic variables in the Texas counties.

Based on this analysis, NRC staff assumes that the socioeconomic ROI for this project is Lea and Eddy Counties, New Mexico. The majority of the socioeconomic impacts would be expected to occur in Lea County because the proposed IIFP site is in Lea County, and because of Lea County's population characteristics, commuting patterns, and amenities. See Figure 3-21 for the counties and major populated areas within the ROI.

3.9.1 Demography

3.9.1.1 Populations within the Socioeconomic ROI

The socioeconomic ROI comprises Lea and Eddy Counties, New Mexico. The proposed IIFP site would be in unincorporated Lea County, New Mexico. The nearest population center, Hobbs, is approximately 22.5 km (14 mi) east of the proposed site. The nearest residence is approximately 2.6 km (1.6 mi) northwest of the proposed site (Figure 3-2).

Table 3-19 lists selected population characteristics of the counties in the ROI, and for comparison, New Mexico. Population characteristics, including race, ethnicity, and population density of the counties in the ROI broadly reflect those same characteristics in New Mexico. The ROI 2009 estimated population of 112,938 is about 5.6 percent of the 2009 estimated New Mexico population (Table 3-19). The racial and Hispanic demographics of the ROI residents generally reflect the racial and Hispanic demographics of residents in New Mexico as a whole. However, the ROI has a noticeably greater percentage than the state of persons who identified themselves as of the white race and a markedly smaller percentage of persons who identified themselves as "American Indian and Alaskan Native." Both ROI counties are sparsely populated, as is New Mexico. New Mexico's average density is 15 persons per square mile,

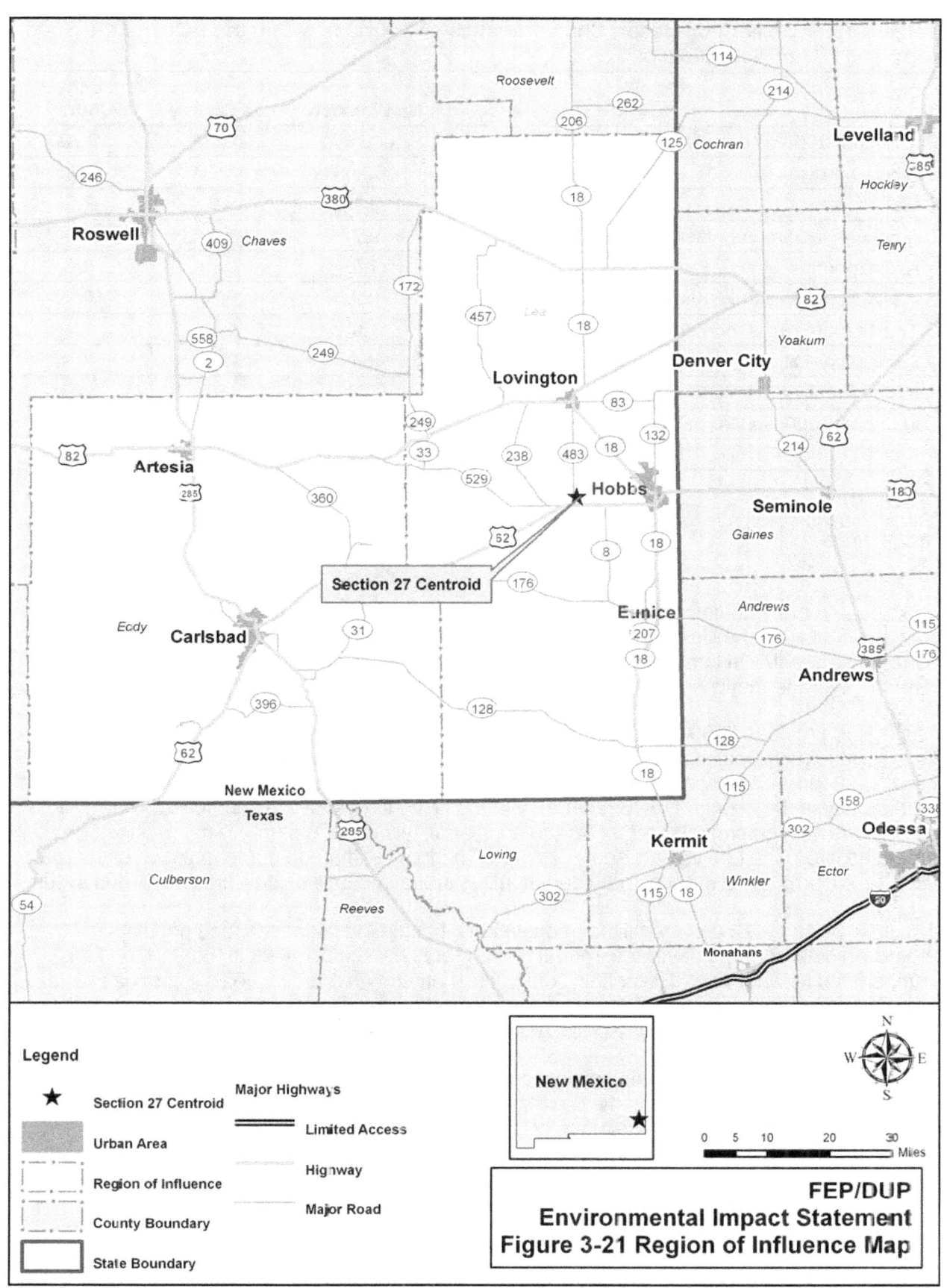

Section 27 Centroid

Legend

★ Section 27 Centroid
Urban Area
Region of Influence
County Boundary
State Boundary

Major Highways
Limited Access
Highway
Major Road

New Mexico

N
W E
S

0 5 10 20 30 Miles

FEP/DUP
Environmental Impact Statement
Figure 3-21 Region of Influence Map

Table 3-19.   Select Population Characteristics of Counties within the ROI and the State of New Mexico

	New Mexico	Lea County	Eddy County
Population, 2009 estimate[a]	2,009,671	60,232	52,706
White, percent	83.6	90.5	93.5
Black, percent	3.1	5.8	2.5
American Indian and Alaskan Native, percent	9.7	1.4	1.6
Asian, percent	1.5	0.7	0.8
Native Hawaiian and other Pacific Islander, percent	0.2	0.1	0.2
Two or more races, percent	1.9	1.5	1.4
Hispanic or Latino Origin, percent[b]	45.6	49.6	43.4
Average Family Size, 2008[c]	3.23	2.93	3.04
Land Area, 2000, square mile[a]	121,356	4,393	4,182
Persons per square mile, 2000[a]	15.0	12.6	12.4

Source:
[a]USCB, 2010a
[b]Hispanics may be of any race, so are also included in applicable race categories
[c]USCB, 2010b

and Lea and Eddy counties' densities are between 12 and 13 persons per square mile. The average density in the United States is about 80 persons per square mile (USCB, 2010a). The average family size in Lea County (2.93 people) and in Eddy County (3.04 people) is smaller than the average family size in New Mexico (3.23 people) (USCB, 2010b).

Table 3-20 provides 2009 estimated population information for Lea and Eddy Counties and their incorporated municipalities. In 2009, the population of Lea County was estimated to be 60,232 (USCB, 2010c). Slightly more than half of the county's population resides in Hobbs, the largest municipality in the county (USCB, 2010d). Hobbs is the largest city in southeastern New Mexico and serves as a commercial center for the population within the 80-km (50-mi) radius of the proposed site. The Lea County seat, Lovington, had an estimated 2009 population of 10,108 (USCB, 2010d). Other incorporated communities in the county include Eunice, Jal, and Tatum.

Tables 3-21 and 3-22 provide historic populations, population estimates, and population projection data, including average annual growth rates, for the counties in the ROI, and for comparison, New Mexico. Historically, the population growth rates in the ROI counties have generally lagged the population growth rate of New Mexico. The projected population growth rates for the counties also lag the projected growth rates for the state.

In 2009, the Eddy County population was estimated to be 52,706 (USCB, 2010a). Carlsbad, the county seat, is the largest city in the county with approximately half of the county population (USCB, 2010d). Northeastern sections of Carlsbad are within 80 km (50-mi) of the proposed site (Figure 3-1) but most of the town is just outside the 80-km (50-mi) radius. Other incorporated communities in Eddy County include Artesia, Hope, and Loving.

Table 3-20. Population Estimates of ROI Counties and Incorporated Municipalities, 2009

Political Jurisdiction	2009 Estimated Population	Percent of County Population
Lea County[a]	60,232	--
Eunice[b]	2,809	4.7
Hobbs[b]	30,838	51.2
Jal[b]	2,074	3.4
Lovington[b]	10,108	16.8
Tatum[b]	767	1.3
Eddy County[a]	52,706	--
Artesia[b]	11,338	21.5
Carlsbad[b]	26,259	49.8
Hope[b]	109	0.2
Loving[b]	1,366	2.6

Source:
[a]USCB, 2010c
[b]USCB, 2010d

Table 3-21. Historic Population in the ROI, 1990 to 2009

Political Jurisdiction	1990[a]	2000[b]	2009[b]
New Mexico	1,515,069	1,819,046	2,009,671
Lea County	55,765	55,511	60,232
Eddy County	48,605	51,653	52,706
Average Annual Growth Rate			
Political Jurisdiction		1990 to 2000[a,b]	2000 to 2009[b]
New Mexico		1.85%	1.11%
Lea County		-0.05%	0.91%
Eddy County		0.61%	0.22%

Source:
[a]USCB, 2000b
[b]USCB, 2010c

Table 3-22. Projected Population in the ROI, 2005 to 2035

Political Jurisdiction	2005[a]	2010[a]	2020[a]	2030[a]	2035[a]
New Mexico	1,969,292	2,162,331	2,540,145	2,864,796	3,018,289
Lea County	57,006	60,896	67,479	72,928	75,716
Eddy County	52,167	54,145	58,294	60,764	61,605
Average Annual Growth Rate					
Area		2005 to 2010	2010 to 2020	2020 to 2030	2030 to 2035
New Mexico		1.89%	1.62%	1.21%	1.05%
Lea County		1.33%	1.03%	0.78%	0.75%
Eddy County		0.75%	0.74%	0.42%	0.28%

Source: BBER, 2008
[a]Population projections are built on slightly different base year numbers than those presented in Table 3-22.

3.9.1.2 Environmental Justice: Minority and Low Income Populations

3.9.1.2.1 Methodology

On February 11, 1994, the President signed Executive Order 12898, "Federal Actions to Address Environmental Justice in Minority Populations and Low-Income Populations," which directs all Federal agencies to develop strategies that consider environmental justice in their programs, policies, and activities. Environmental justice is described in the Executive Order as "identifying and addressing, as appropriate, disproportionately high and adverse human health or environmental effects of its programs, policies, and activities on minority populations and low-income populations." On December 10, 1997, the Council on Environmental Quality (CEQ) issued Environmental Justice Guidance under the National Environmental Policy Act (CEQ, 1997). The NRC has provided general guidelines on the evaluation of environmental analyses in Environmental Review Guidance for Licensing Actions Associated with NMSS [Nuclear Material Safety and Safeguards] Programs (NUREG-1748) (NRC, 2003), and issued a final policy statement on the Treatment of Environmental Justice Matters in NRC Regulatory and Licensing Actions (69 FR 52040) and environmental justice procedures to be followed in NEPA documents prepared by the NRC's Office of Nuclear Material Safety and Safeguards (NMSS).

NRC's NMSS environmental justice guidance, as found in Appendix C to NUREG-1748 (NRC, 2003), recommends that the area for assessment for a facility in a rural area be a circle with a radius of approximately 6.4 km (4 mi) whose centroid is the facility being considered. However, the guidance also states that the scale should be commensurate with the potential impact area. Therefore to ensure consistency with the accident analysis, which considers airborne impacts to populations within an 80-km (50-mi) radius, the NRC staff concludes that an environmental justice assessment area with an 80-km (50-mi) radius would be appropriate. As such, New Mexico and Texas and each county with some land area within the 80-km (50-mi) radius of the proposed IIFP site (i.e., centroid of Section 27) are appropriate areas for comparative analysis.

A minority or low-income community may be considered as either a population of individuals living in geographic proximity to one another or a dispersed/transient population of individuals (e.g., migrant workers) where either type of group experiences common conditions of environmental exposure (NRC, 2003). NUREG-1748 defines minority categories as: American Indian or Alaskan Native, Asian, Native Hawaiian or other Pacific Islander, African American (not of Hispanic or Latino origin), some other race, and Hispanic or Latino ethnicity (of any race) (NRC, 2003). The 2000 Census introduced a multiracial category. Anyone who identifies themselves as white and a minority is counted as that minority group. Individuals that identify themselves as more than one minority are counted in a "two or more races" group (NRC, 2003). Low-income is defined as being below the poverty level as defined by the U.S. Census Bureau (NRC, 2003).

The NRC-recommended area for evaluating census data is the census block group, which is delineated by the United States Census Bureau and is the smallest area unit for which race and poverty data are available (NRC, 2003). The NRC staff used ESRI ArcGIS® 9.3 software which accessed the 2000 decennial census, to identify block groups with low-income or minority populations within 80 km (50 mi) of the proposed IIFP site. NRC staff included a block group if any part of its fell within 80 km (50 mi) of the proposed site; 96 block groups were identified as being within, or partially within the 80-km (50-mi) radius.

NRC guidance indicates that a significant minority or low-income population exists if at least one of these conditions exists:

The minority or low-income population of the block group is more than 50 percent of the entire block group population.

The minority or low-income population percentage of the block group is significantly greater (typically at least 20 percentage points) than the minority or low-income population percentage in the geographic areas chosen for comparative analysis.

3.9.1.2.2   Minority Populations

Using the U.S. Census Bureau (USCB) 2000 census data, NRC staff calculated (1) the percentage of each block group's population represented by each minority category for each of the 96 block groups within the 80 km (50-mi) radius, (2) the percentage that each minority category represented of the entire populations of New Mexico and Texas, and (3) the percentage that each minority category represented of each of the counties that has some land within the 80-km (50-mi) radius of the proposed site. If the percentage of any minority in any block group exceeded 50 percent of the block group's total population or exceeded the minority's corresponding county or state percentages by more than 20 percent, then that block group was identified as having a significant minority population.

Table 3-23 identifies the number of block groups that met the 50 percent criterion or the more-than-20-percent criterion (some block groups may meet both criteria) for their corresponding state and/or county. If a block group met one or both of the criterion for either the state or the county, it was not double-counted. Of the 96 census block groups within the 80-km (50-mi) radius, 16 have a significant percentage of minority residents. Thirty-two block groups have a significant percentage of Hispanic ethnicity residents. Figures 3-22 through 3-24 provide graphical representations of the data presented in Table 3-23.

Seasonal agricultural (migrant) workers may make up a portion of the minority population within the 80-km (50-mi) radius. Although migrant worker population counts are not available from the USCB, the U.S. Department of Agriculture has collected information on farms that employ migrant labor. The number of farms that employ migrant laborers in each county which falls wholly or partially within the 80-km (50-mi) radius are: in New Mexico Lea County (9), Eddy County (12), and Chaves County (19) and in Texas Loving County (1), Winkler County (2), Andrews County (2), Gaines County (27), Yoakum County (9), Terry County (26), and Cochran County (15) (USDA, 2007). The number of these farms which fall wholly or partially within the 80-km (50-mi) radius is not known.

There are no Federally recognized Native American reservations within the 80-km (50-mi) radius of the proposed IIFP site (NPS, Undated).

3.9.1.2.3   Low-Income Populations

The NRC guidance defines low-income households based on statistical poverty thresholds (NRC, 2003).

Using the USCB 2000 census data, NRC staff calculated the percentage of each block group's population represented by low-income households for each of the 96 block groups within the 80 km (50-mi) radius, and the percentage of low-income households in New Mexico and Texas and in each of the counties that had some land within the 80-km (50-mi) radius of the site f the percentage of any low-income block group exceeded 50 percent of the block group's total population or exceeded the corresponding county or State low-income percentages by more

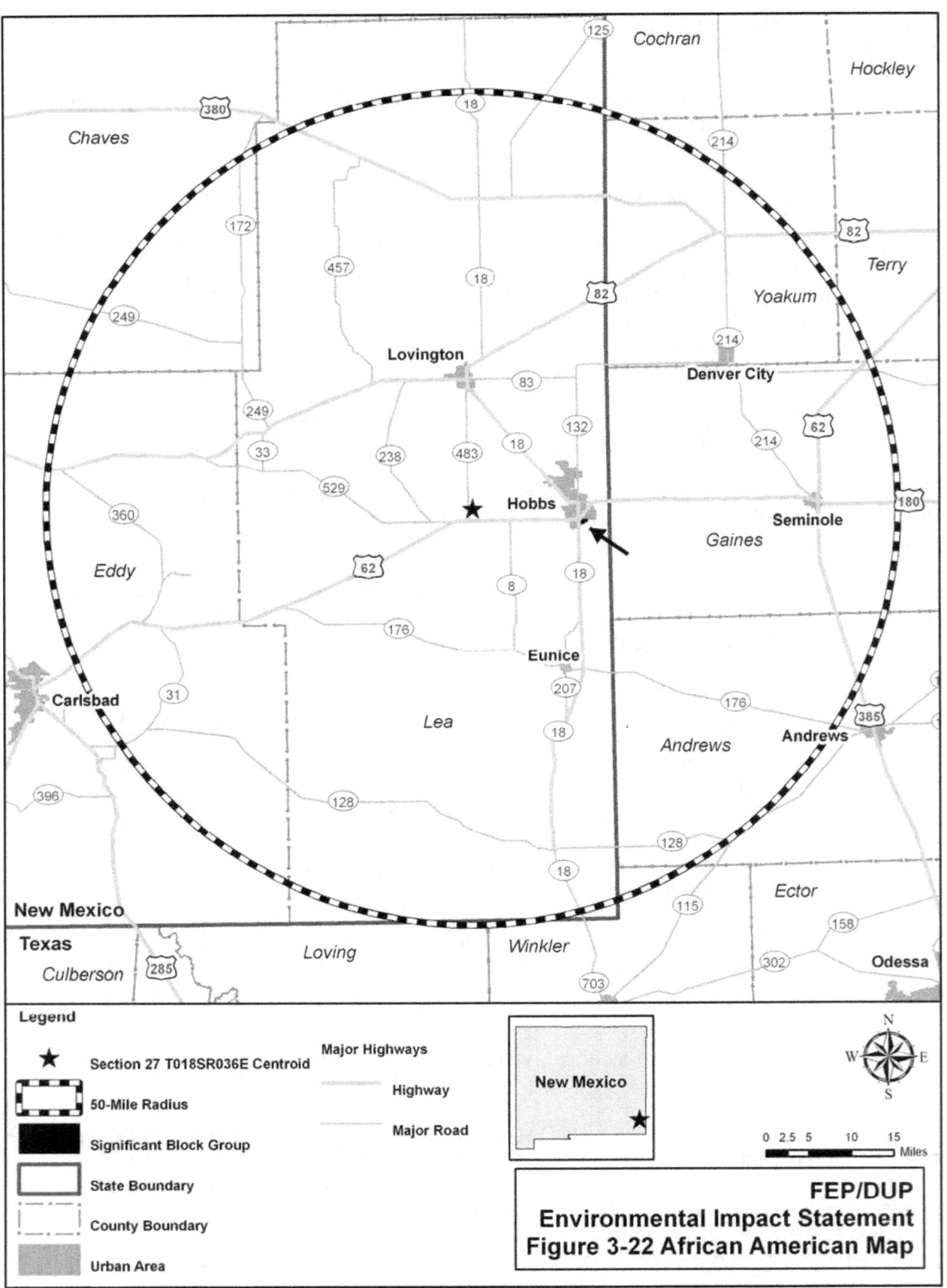

Legend

★ Section 27 T018SR036E Centroid

▨ 50-Mile Radius

■ Significant Block Group

□ State Boundary

┆ County Boundary

▨ Urban Area

Major Highways

Highway

Major Road

New Mexico

★

0 2.5 5   10   15
Miles

FEP/DUP
Environmental Impact Statement
Figure 3-22 African American Map

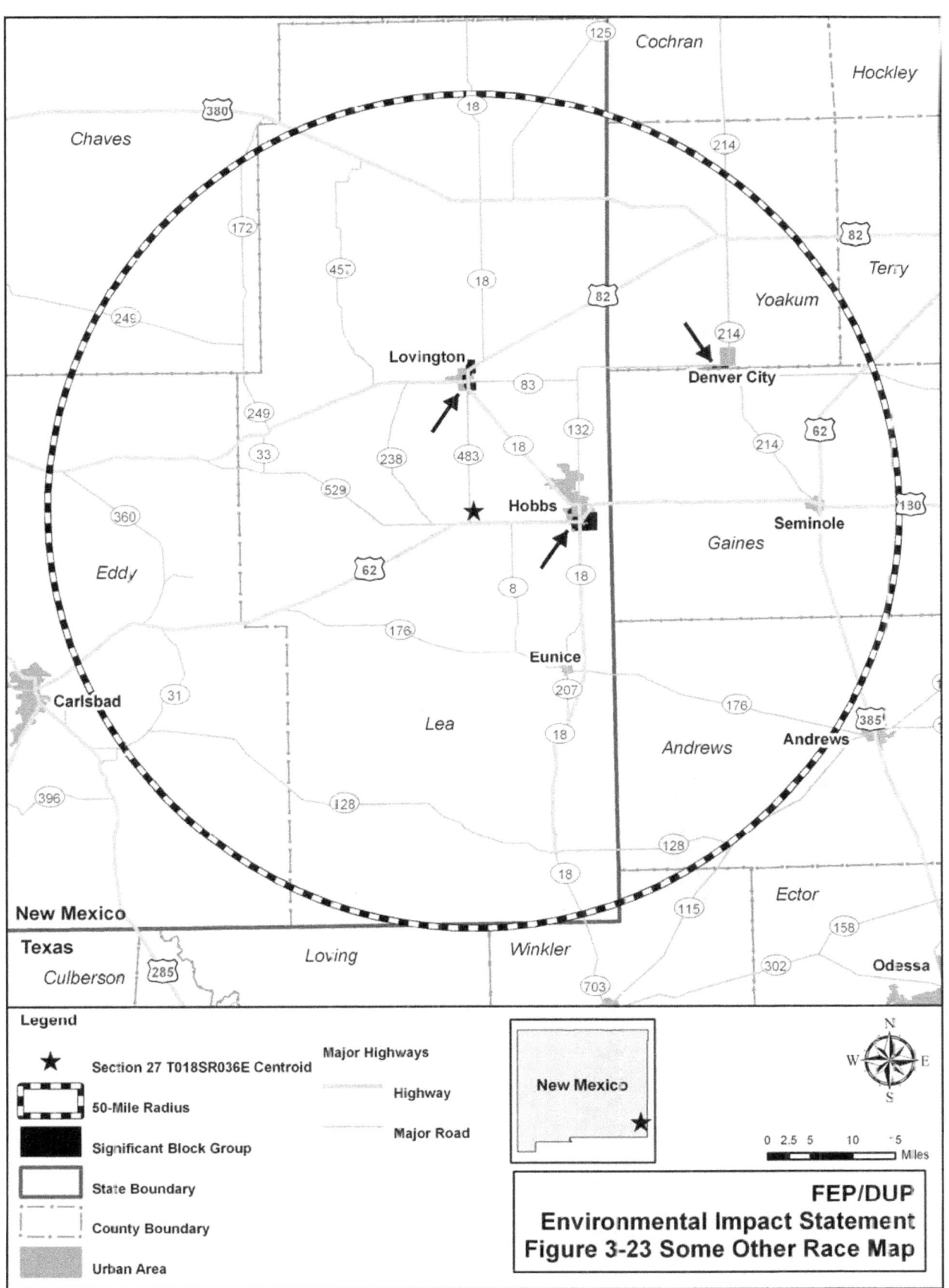

Legend

★ Section 27 T018SR036E Centroid

⬛ 50-Mile Radius

⬛ Significant Block Group

▭ State Boundary

County Boundary

Urban Area

Major Highways

Highway

Major Road

New Mexico

FEP/DUP
Environmental Impact Statement
Figure 3-23 Some Other Race Map

0   2.5 5   10   15 Miles

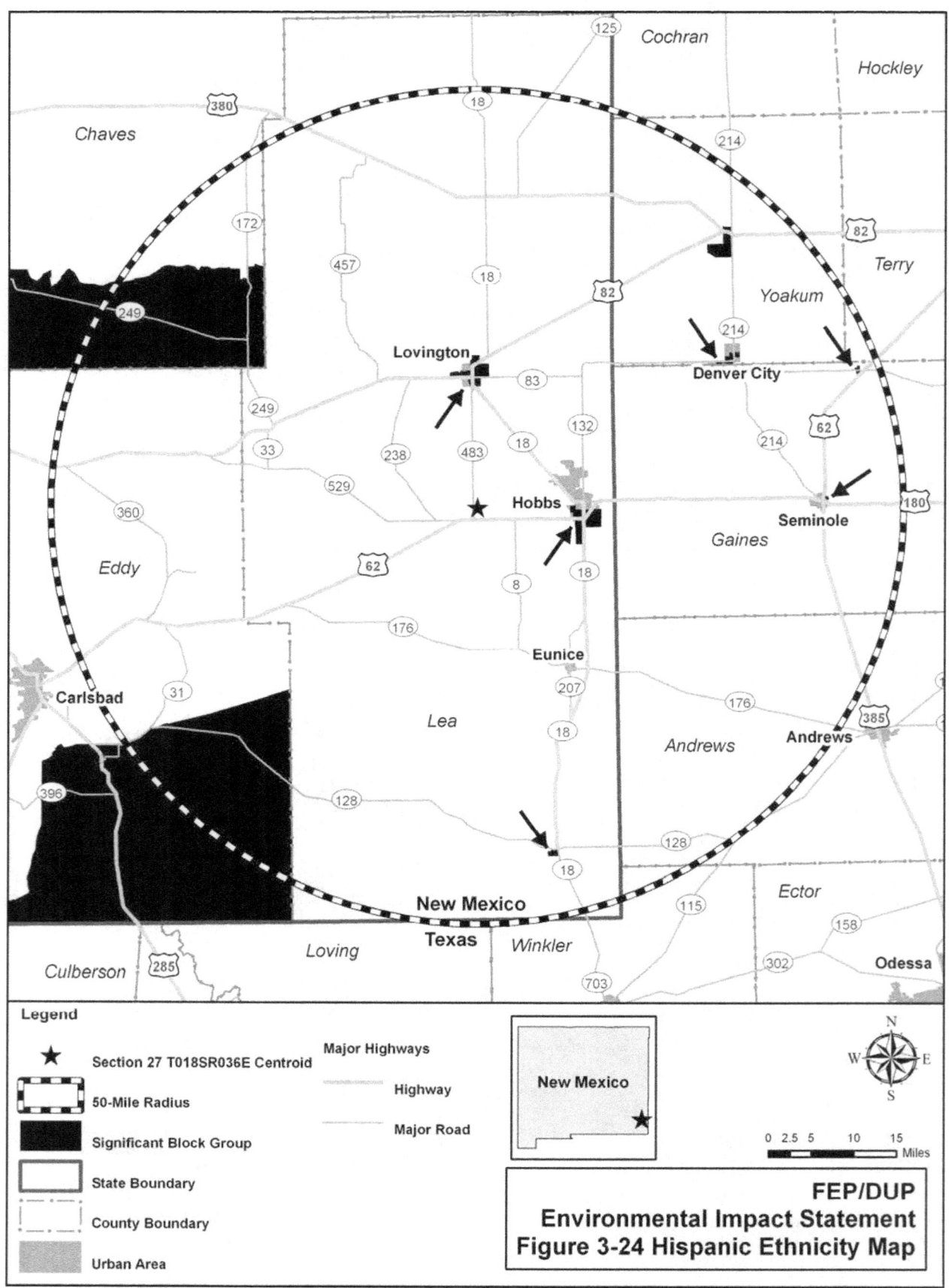

Table 3-23. Block Groups within 80 km (50 mi) of the Proposed IIFP Site with Significant Minority or Low-Income Populations (Meeting 50 Percent Criteria or Exceeding Respective County or State Percentages by 20 Percent)

County Name	Number of Block Groups	African American	American Indian or Alaskan Native	Asian	Native Hawaiian or Other Pacific Islander	Some Other Race	Two or More Races	Hispanic Ethnicity	Low-Income Households
Chaves	2	0	0	0	0	0	0	1	0
Eddy	3	0	0	0	0	0	0	1	0
Lea	64	1	0	0	0	14	0	24	10
Andrews	3	0	0	0	0	0	0	0	0
Cochran	1	0	0	0	0	0	0	0	0
Gaines	13	0	0	0	0	0	0	3	0
Loving	1	0	0	0	0	0	0	0	0
Terry	1	0	0	0	0	0	0	0	0
Winkler	1	0	0	0	0	0	0	0	0
Yoakum	7	0	0	0	0	1	0	3	0
Total	96	1	0	0	0	15	0	32	10

State/County	African American (%)	American Indian or Alaskan Native (%)	Asian (%)	Native Hawaiian or Other Pacific Islander (%)	Some Other Race (%)	Two or More Races (%)	Hispanic Ethnicity (%)	Low-Income Households (%)
New Mexico (State)	1.89	9.54	1.06	0.08	17.04	3.65	42.08	16.78
Chaves County	1.97	1.13	0.53	0.06	21.25	3.12	43.83	19.12
Eddy County	1.56	1.25	0.45	0.09	17.67	2.64	38.76	16.72
Lea County	4.37	0.99	0.39	0.04	23.81	3.27	39.65	19.90
Texas (State)	11.53	0.57	2.70	0.07	11.69	2.47	31.99	13.98
Andrews County	1.65	0.88	0.71	0.02	16.79	2.87	40.00	16.74
Cochran County	4.53	0.83	0.21	0.05	27.35	2.55	44.13	21.67
Gaines County	2.28	0.76	0.15	0.01	14.17	2.35	35.77	19.03
Loving County	0.00	0.00	0.00	0.00	8.96	1.49	10.45	0.00
Terry County	5.00	0.53	0.22	0.02	14.28	3.40	44.09	20.53
Winkler County	1.85	0.45	0.20	0.00	20.35	2.34	44.00	18.58
Yoakum County	1.39	0.71	0.12	0.01	25.48	1.65	45.93	18.20

Source: USCB, 2000a; USCB, 2000b; USCB, 2000c; USCB, 2000d; USCB, 2000e; USCB, 2000f;

than 20 percent, then that block group was identified as having a significant low-income population. Again, if the block group met one or both criteria, for either the state or county, it was not double-counted.

Table 3-23 lists the number of block groups in each county within the 80-km (50-mi) radius that meets the 50 percent criterion or the more than 20 percent criterion for its corresponding State or county. Ten census block groups within the 80-km (50-mi) radius have a significant percentage of low-income households. Figure 3-25 locates the low-income block groups.

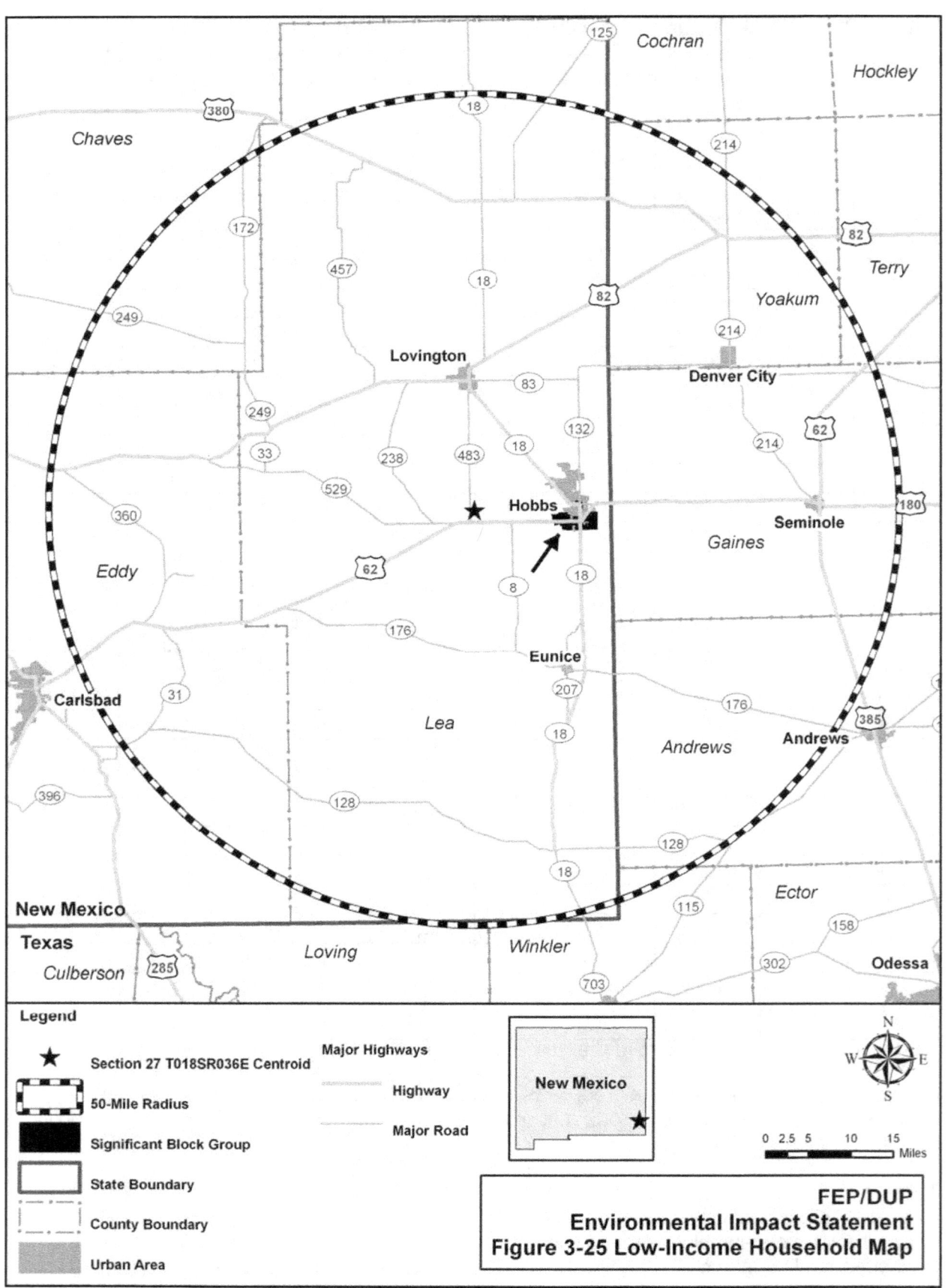

Legend

★ Section 27 T018SR036E Centroid

⬛ 50-Mile Radius

⬛ Significant Block Group

☐ State Boundary

⬚ County Boundary

▨ Urban Area

Major Highways

Highway

Major Road

New Mexico

0 2.5 5   10   15
Miles

FEP/DUP
Environmental Impact Statement
Figure 3-25 Low-Income Household Map

3.9.2 Employment and Income

3.9.2.1 Employment

Table 3-24 summarizes employment trends in the ROI from 2001 to 2008. From 2001 to 2008, growth in employment in the ROI was greater than population growth. The number of jobs in the ROI grew from 54,649 (29,463 for Lea County plus 25,186 for Eddy County) in 2001 (BEA, 2011a; BEA, 2011b) to 68,314 in 2008 (BEA, 2010), an increase of 25 percent. The ROI population increased from 105,562 to 110,903 (59,129 for Lea County plus 51,774 for Eddy County) (USCB, 2010c), an increase of 5 percent. Within the ROI, 2008 employment was dominated by jobs in mining (20.6 percent), government and government enterprises (10.4 percent), retail trade (10.1 percent), and construction (8.3 percent). With the exception of employment in farming, a sector that represents less than 0.5 percent of all jobs in the ROI, the number of jobs in all industrial sectors grew from 2001 to 2008 (BEA, 2010). A major employer in the ROI is the DOE's Waste Isolation Pilot Plant, in Eddy County. Nearly 600 individuals are employed at the facility (ECP, 2008). From 2001 to 2008, the ROI unemployment rate decreased from 4.6 to 3.0 percent. However, from 2008 to June of 2010, the rate increased from 3.0 to 7.0 percent.

Table 3-25 presents information about labor statistics in the ROI, and for comparison, New Mexico. The size of the ROI labor force grew from 47,199 to 57,708 (22.3 percent) between 2001 and 2008, but shrank in 2009 to 57,590 and declined again in 2010, to 56,945 (BLS, 2010a). The unemployment rate in the ROI has consistently remained below the unemployment rate in the State. As a point of comparison, the unemployment rate in the United States in June 2010 was 9.6 percent (BLS, 2010b).

3.9.2.2 Income

Table 3-26 presents income statistics for the ROI counties, their major population centers, and New Mexico. In 2008, various measures of income in Eddy County were higher and the rates of poverty in Eddy County lower than in New Mexico. With the exception of median household income, the various measures of income in Lea County were lower and the rates of poverty higher in Lea County than in New Mexico. In 2008, the poverty threshold ranged from $10,326 to $47,915, depending on family characteristics. Families and individuals residing in Lea County were more likely to be living below the poverty level than those living in Eddy County.

3.9.3 Taxes

3.9.3.1 Income Taxes

Corporate Income Taxes

New Mexico imposes a corporate income tax on the total net income (including New Mexico and non-New Mexico income) of every domestic and foreign corporation doing business in or from the State, or which has income from property or employment within the State. The percentage of New Mexico income is then applied to the gross tax. For corporations with a total net income exceeding $1,000,000 annually, corporate income tax is $56,000 plus 7.6 percent of net income over $1,000,000 (NMTRD, 2010a).

New Mexico also levies a corporate franchise tax of $50 per year (NMTRD, 2010a).

Table 3-24.   Employment in ROI, by Industry, 2001 to 2008

	Lea County 2001[a]	Lea County 2008[b]	Lea County, Percent Change, 2001 to 2008	Eddy County 2001[c]	Eddy County 2008[b]	Eddy County, Percent Change, 2001 to 2008	ROI Jobs 2008	ROI, Percent Change in Jobs, 2001 to 2008
Total Employment	29,463	37,622	27.7%	25,186	30,692	21.9%	68,314	25.0%
Farm Employment	1,026	725	-29.3%	975	798	-18.2%	1,523	-23.9%
Nonfarm Employment	28,437	36,897	29.7%	24,211	29,894	23.5%	66,791	26.9%
Private Employment	24,622	33,176	34.7%	20,604	26,007	26.2%	59,183	30.9%
Forestry, fishing, and related activities	108	101	-6.5%	163	177	8.6%	278	2.6%
Mining	5,484	8,339	52.1%	3,042	4,585	50.7%	12,924	51.6%
Utilities	247	489	98.0%	134	145	8.2%	634	66.4%
Construction	2,131	3,460	62.4%	1,506	2,597	72.4%	6,057	66.5%
Manufacturing	428	805	88.1%	897	922	2.8%	1,727	30.3%
Wholesale Trade	1,199	1,255	4.7%	555	669	20.5%	1,924	9.7%
Retail Trade	3,371	3,393	0.7%	2,972	3,126	5.2%	6,519	2.8%
Transportation and Warehousing	996	1,588	59.4%	972	1,097	12.9%	2,685	36.4%
Information	251	382	52.2%	345	309	-10.4%	691	15.9%
Finance and Insurance	722	879	21.7%	701	909	29.7%	1,788	25.7%
Real Estate and Rental and Leasing	779	1,130	45.1%	645	890	38.0%	2,020	41.9%
Professional, Scientific, and Technical Services	603	1,019	69.0%	715	1,315	83.9%	2,334	77.1%
Management of Companies and Enterprises	121	137	13.2%	54	219	305.6%	356	103.4%
Administrative and Waste Services	1,446	2,039	41.0%	1,563	1,924	23.1%	3,963	31.7%
Educational Services	(D)	(D)	NA	97	118	21.6%	NA	NA
Health Care and Social Assistance	(D)	(D)	NA	2,450	2,835	15.7%	NA	NA

Table 3-24. Employment in ROI, by Industry, 2001 to 2008 (Continued)

	Lea County 2001[a]	Lea County 2008[b]	Lea County, Percent Change, 2001 to 2008	Eddy County 2001[c]	Eddy County 2008[b]	Eddy County, Percent Change, 2001 to 2008	ROI Jobs 2008	ROI, Percent Change in Jobs, 2001 to 2008
Arts, Entertainment, and Recreation	219	605	176.3%	208	207	-0.5%	812	90.2%
Accommodation and Food Services	1,685	2,200	30.6%	1,898	2,175	14.6%	4,375	22.1%
Other Services, except Public Administration	1,652	1,921	16.3%	1,687	1,788	6.0%	3,709	11.1%
Government and Government Enterprises	3,815	3,721	-2.5%	3,607	3,887	7.8%	7,608	2.5%
Federal, Civilian	119	105	-11.8%	488	764	56.6%	860	43.2%
Military	178	160	-10.1%	165	139	-15.8%	299	-12.8%
State and Local	3,518	3,456	-1.8%	2,054	2,984	1.0%	6,440	-0.5%
State Government	286	289	1.0%	741	750	1.2%	1,039	1.2%
Local Government	3,232	3,167	-2.0%	2,213	2,234	0.9%	5,401	-0.8%

Source:
[a]BEA, 2010
[b]BEA, 2011b
[c]BEA, 2011a
NA Not Applicable
(D) Information not shown (by BEA) to avoid disclosure of confidential information, but estimates for this item are included in the totals

3-75

Table 3-25. Labor Statistics, ROI and New Mexico, 2001 to 2010

	Lea County	Eddy County	ROI	New Mexico
2001 (annualized) Labor Force	23,702	23,497	47,199	863,682
2008 (annualized) Labor Force	29,895	27,813	57,708	961,259
2009 (annualized) Labor Force	28,890	28,700	57,590	955,904
2010 (June) Labor Force	28,103	28,842	56,945	962,423
Percent Change 2001 to 2008	26.1%	18.4%	22.3%	11.3%
Percent Change, 2001 to 2010	18.6%	22.7%	20.6%	11.4%
2001 Unemployment Rate	4.3%	5.0%	4.6%	4.9%
2008 Unemployment Rate	2.9%	3.1%	3.0%	4.5%
2009 Unemployment Rate	7.6%	5.5%	6.6%	7.2%
June, 2010 Unemployment Rate	8.0%	6.1%	7.0%	8.5%

Source: BLS, 2010a

Table 3-26. Income Statistics, ROI Counties and Population Centers, and New Mexico, 2008[a]

	Lea County	Eddy County	New Mexico
Median Household Income	$43,638	$45,858	$43,202
Median Family Income	$47,853	$57,658	$51,724
Per Capita Income	$20,319	$25,151	$22,781
Families below Poverty level, percent	15.7	10.2	13.7
Individuals below Poverty Level, percent	18.9	14.4	17.9

Source: USCB, 2010b
[a]All dollar values are expressed in 2008 inflation-adjusted dollars

Individual Income Taxes

New Mexico imposes an individual income tax on the net income of every resident and nonresident employed or engaged in business in or from the State or deriving any income from any property or employment within the State. The rates vary depending upon filing status and income.

The top tax bracket is 4.9 percent (NMTRD, 2010b).

3.9.3.2   Sales Tax/Gross Receipts Tax

New Mexico is one of a minority of states that has a gross receipts tax structure instead of a sales tax structure. Gross receipts are the total amount of money or value of other considerations received from (NMTRD, 2011):

- Selling property in New Mexico

- Leasing or licensing property used in New Mexico

- Granting a right to use a franchise used in New Mexico

- Performing services in New Mexico

- Selling research and development services performed outside New Mexico, the product of which is initially used in New Mexico

Although the gross receipts tax is imposed on businesses, it is common for a business to pass the gross receipts tax on to the purchaser either by separately stating it on the invoice or by combining the tax with the selling price (NMTRD, 2011).

The gross receipts tax rate varies throughout the state from 5.125 percent to 8.6875 percent, depending on the location of the business. It varies because the total rate combines rates imposed by the state, counties, and, if applicable, municipalities where the businesses are located. The business pays the total gross receipts tax to the state, which then distributes the counties' and municipalities' portions to them (NMTRD, 2011).

The current gross receipts taxes in Lea and Eddy Counties are presented in Table 3-27.

Table 3-27.   Gross Receipts Tax Rates in the ROI, as of July, 2010

Lea County	Rates[a]	Eddy County	Rates
Eunice	6.8125%	Artesia	7.1875%
Hobbs	6.8125%	Carlsbad	7.4375%
Jal	6.8125%	Hope	6.6250%
Lovington	6.8750%	Loving	6.8125%
Lovington Industrial Park	5.5%	Remainder of County	5.75%
Tatum	6.8125%	--	--
Remainder of County	5.5%	--	--

Source: NMTRD, 2011
[a] Rates include State, county, and municipal gross receipts taxes, combined

3.9.3.3   Property Taxes

Four governmental entities in New Mexico are authorized to tax: the state, counties, municipalities, and school districts (NRC, 2005). Property assessment rates are 33.3 percent of appraised values (NRC, 2005; Eddy County, 2007). The tax applied to the assessed property value is a combination of state, county, municipal, and school district levies (NRC, 2005). The Lea County tax rate for nonresidential property outside the city limits of Hobbs is $24.949 per $1,000 of net taxable value of a property (Lea County, 2009). Rates for nonresidential properties are higher within the city limits of Hobbs. Residential property tax rates are lower for properties outside of Hobbs, and higher for those within Hobbs.

New Mexico and its local governments offer industrial revenue bonds (IRBs) as a way to encourage company relocations and expansions that provide jobs and economic opportunities for residents and communities. IRBs allow projects to qualify for certain tax incentives, including a property tax exemption on most real and personal property constituting a project's property, and possible exemptions from gross receipts tax and use tax related to the acquisition of equipment and other personal property for use in the business to be conducted at the project (City of Albuquerque, 2011). International Isotopes, the parent corporation of IIFP, has an IRB agreement with Lea County, New Mexico (IIFP, 2011). As a result, IIFP is generally exempt from property taxes. However, the school district and the New Mexico Junior College are not part of this IRB agreement. Table 3-28 contains annual property tax revenue data for the local entities that would have the authority to levy a property tax on the proposed IIFP facilities.

Table 3-28.   IIFP Annual Property Tax Information

Property Taxing Entity	Total Annual Tax Revenues	Estimated IIFP Facility in Lieu of Property Tax Payment	IIFP Facility Estimated in Lieu of Property Tax Payment as Percent of Total Annual Tax Revenues
Hobbs Municipal School District	$71,126,000[a]	$78,300 - $123,300	<1%
New Mexico Junior College	$37,201,924[b]	$139,200 - $219,200	<1%

Sources:
[a]NCES, 2008
[b]NCES, 2009

3.9.4   Housing

Table 3-29 summarizes housing data for Lea County, Eddy County and the largest city within each county, Hobbs and Carlsbad, respectively. A variety of types, prices, and settings comprise the housing inventory in the socioeconomic ROI. In 2008 there were 46,971 housing units in Lea and Eddy Counties (USCB, 2010b). The two largest population centers in the counties, Hobbs and Carlsbad, had approximately half of the total ROI housing inventory (USCB, 2010b). Within the ROI, approximately 12.4 percent (5,823 units) of the units were vacant (USCB, 2010b). Of the 41,148 occupied units, 29,021 were owner-occupied (70.5 percent) and 12,127 (29.5 percent) were renter-occupied (USCB, 2010c). The median value of an owner-occupied unit was $82,200 in Lea County and $85,600 in Eddy County (USCB, 2010b). For comparison, the median value of an owner-occupied house in New Mexico was $154,900 in 2008 (USCB, 2010b).

In 2008, the median monthly rent was $661 in New Mexico and slightly less in both Lea and Eddy Counties (USCB, 2010b). Mobile homes accounted for 16.9 percent of the housing in Lea County and 14.0 percent of the housing in Eddy County (USCB, 2010e). Mobile homes made up 16.7 percent of the housing inventory in New Mexico (USCB, 2010e). The housing inventory in the ROI grew by 1.1 percent from 2008 to 2009, while the growth in the ROI's population was 0.2 percent (USCB, 2010a; USCB, 2010d).

Table 3-29.   Housing Characteristics in ROI Counties and Population Centers, 2008

County/ Population Center	Housing Units, 2008[a]	Occupied Units, 2008[a]	Owner-Occupied Units, 2008[a]	Renter-occupied, 2008[a]	Vacant Housing Units, 2008[a]	Percent Units Vacant of All Units, 2008[a]	Mobile Homes, 2008[b]
Lea County	24,495	21,653	14,912	6,741	2,842	11.6%	4,134
Hobbs	12,299	10,854	6,998	3,856	1,445	11.7%	1,201
Eddy County	22,476	19,495	14,109	5,386	2,981	13.3%	3,142
Carlsbad	11,565	10,073	6,954	3,119	1,492	12.9%	556
ROI Total	46,971	41,148	29,021	12,127	5,823	12.4%	7,276

Source:
[a]USCB, 2010b
[b]USCB, 2010f

3.9.5   Public Utilities

3.9.5.1   Major Public Water Suppliers

EPA lists two major public water suppliers in Lea County and three major public water suppliers in Eddy County (EPA, 2010b). Major public water systems are those that serve more than 3,300 people. Table 3-30 presents water production and use statistics for these suppliers. Most of the major water suppliers in the ROI have excess capacity.

Table 3-30.   Major Public Water Suppliers in ROI, 2007 – 2009

Water System Name[a]	County Served[a]	Population Served[a]	Primary Water Source Type[a]	Average Daily Use (MGD)[b]	Maximum Capacity (MGD)[b]
Hobbs Municipal Water Supply	Lea	33,000	Groundwater	7.0	N/A
Lovington Municipal Water Supply	Lea	9,643	Groundwater	2.5	6
Artesia Municipal Water System	Eddy	14,000	Groundwater	2.3	8.64
Carlsbad Municipal Water System	Eddy	27,000	Groundwater	3.8	28.8
Otis Mutual Domestic Water Consumer's Association	Eddy	5,000	Groundwater	1.0	NA

Source:
[a]EPA, 2010b
[b]NMED, 2010b
MGD = million gallons per day

3.9.5.2   Major Public Wastewater Treatment Facilities

The New Mexico Environment Department (NMED) lists four major public wastewater treatment facilities in Lea and Eddy Counties (NMED, 2010c). Major wastewater treatment facilities are those that serve more than 3,000 people. Table 3-31 presents wastewater treatment production and capacity statistics for these facilities. All of the major wastewater treatment facilities in the ROI have excess capacity.

Table 3-31.   Major Wastewater Treatment Facilities in ROI[a]

Facility Name[b]	2009 Population Served[b]	Average Daily Production (MGD)	Maximum Permitted Capacity (MGD)[b]
City of Artesia	11,208	1.3[c]	3.0
City of Carlsbad	26,352	2.5[d]	8.5
City of Hobbs	31,151	3.4[e]	7.2
City of Lovington	10,206	0.8[f]	2.7

Source:
[a]Includes permitted, municipal wastewater treatment plants serving at least 3,000 persons.
[b]NMED, 2010c
[c]Artesia, 2010
[d]Carlsbad, 2010
[e]Hobbs, 2010
[f]ovington, 2010

3.9.6 Community Services

3.9.6.1 Education

Table 3-32 summarizes information about public school districts and schools in the ROI. Lea County has 5 public school districts, with 36 schools for early childhood education (Age 3) through Grade 12. The total enrollment in the county public schools was 12,588 students in 2008 (NCES, 2010a). There are also three private schools in the county with a total 2008 enrollment of 111 students (NCES, 2010b). In addition, there are two colleges in the county, both in Hobbs. New Mexico Junior College had a 2009 enrollment of 2,300 and University of the Southwest had an enrollment of 528, with an undergraduate enrollment of 317 (NCES, 2010c).

Table 3-32. Public School Districts in the ROI, 2008

County	Schools	Number of Schools in District	Number of Students
Lea County	Eunice Municipal Schools	3	589
	Hobbs Municipal Schools	17	8,038
	Jal Public Schools	3	405
	Lovington Public Schools	10	3,247
	Tatum Municipal Schools	3	309
Eddy County	Carlsbad Municipal Schools	15	3,581
	Artesia Public Schools	10	6,058
	Loving Municipal Schools	3	620

Source: NCES, 2010a

Eddy County has 28 schools in 3 public school districts with a 2008 enrollment of 10,259 students (NCES, 2010a). There is also a private school in Carlsbad with a 2008 enrollment of 68 students (NCES, 2010b). New Mexico State University has a campus in Carlsbad with an enrollment of 2,050 (NCES, 2010c).

3.9.6.2 Fire Protection

In 2010, there were 468 active career and volunteer firefighters in the ROI (USFA, 2010). Twenty fire departments, operating out of 37 fire stations, are in the ROI (USFA, 2010). The proposed IIFP site would be within the jurisdiction of the City of Hobbs Fire Department (HFD, 2010), which is staffed with 70 career firefighters (USFA, 2010). Lea County and Eddy County have mutual aid agreements among all the municipal and independent fire departments to assist with additional response services (HFD, 2010; LPD, 2010). Table 3-33 provides information about fire protection in the ROI.

Table 3-33. Fire Protection in the ROI, 2010

County	2009 County Population[a]	Active Firefighters 2010[b]	Ratio of Residents to Active Firefighters, 2010	Number of Fire Stations, 2010[b]
Lea	60,232	176	342	9
Eddy	52,706	292	181	28
ROI Total	112,938	468	241	37

Sources:
[a]USCB, 2010a; [b]USFA, 2010

3.9.6.3 Law Enforcement

In 2009, there were 89 county and 196 municipal law enforcement officers serving the two ROI counties (FBI, 2009a; FBI, 2009b; LPD, 2010). Law enforcement services in the ROI are provided by the Lea County and Eddy County Sheriff Departments and the Artesia, Carlsbad, Eunice, Hobbs, Jal, Lovington, and Tatum municipal police departments (FBI, 2009a; FBI, 2009b). Hope and Loving also maintain police departments (LPD, 2010). The Lea County Sheriff's Department has jurisdiction in the unincorporated portion of Lea County, including the proposed site (HFD, 2010; LPD, 2010). The New Mexico State Police could provide a second level of response to any sheriff or police department via existing mutual aid agreements (HFD, 2010; LPD, 2010). Table 3-34 provides information about law enforcement in the ROI.

3.9.6.4 Hospitals and Physicians

Lea County has two general medical and surgical hospitals. The Lea Regional Medical Center in Hobbs is the closest hospital to the proposed site. It has 214 staffed beds (AHA, 2007). Nor-Lea General Hospital, in Lovington, has 12 staffed beds (AHA, 2007). Eddy County also has two general medical and surgical hospitals. The Carlsbad Medical Center has 127 staffed beds and the Artesia General Hospital, in Artesia, has 20 staffed beds (AHA, 2007). The ROI has 146 practicing physicians; 69 physicians in Lea County representing 24 specialties and 77 physicians representing 25 specialties in Eddy County (AMA, 2010). Both Lea County and Eddy County are considered to be medically underserved areas (HRSA, Undated).

Table 3-34. Law Enforcement in ROI Counties and Incorporated Places, 2009

County/City[a,e]	Population 2009[a,e]	Law Enforcement Officers 2009[b,c]	Ratio of residents-to-Law Enforcement Officers, 2009
Lea County	--	43	--
Eunice	2,809	7	--
Hobbs	30,833	70	--
Jal	2,074	5	--
Lovington	10,108	23	--
Tatum	767	3	--
Lea County Totals	**60,232**	**151**	**399**
Eddy County[b]	--	46	--
Artesia	11,338	33	--
Carlsbad	26,259	50	--
Hope[d]	109	1[d]	--
Loving[d]	1,366	4[d]	--
Eddy County Totals	**52,706**	**134**	**393**

Sources:
[a]USCB, 2010a
[b]FBI, 2009a
[c]FBI, 2009b
[d]LPD, 2010
[e]USCB, 2010d

3.10   Traffic and Transportation

3.10.1   Roadways

Figure 3-26 shows the major highways near the proposed IIFP site. The site is approximately 1.6 km (1 mi) north of US 62/180 and immediately east of NM 483, in close proximity to the intersection of the two roadways.

From the east, US 62/180 crosses into New Mexico from Texas approximately 29 km (18 mi) from the IIFP site, runs through the City of Hobbs, intersects NM 483 at the IIFP site, and continues to Carlsbad, NM. Near the proposed IIFP site, US 62/180 is called Carlsbad Highway and is a 4-lane divided highway that provides access to the proposed site from the east and west.

NM 483 is a 2-lane north-south highway that connects Lovington, New Mexico to US 62/180. East of the proposed site north-south roadways NM 8 and NM 18 provide access to US 62/180 from Eunice, New Mexico and several unincorporated areas. NM 132 provides access to US 62/180 in Hobbs from points in Texas. NM 529 is an east-west roadway that intersects US62/180 just west of the proposed IIFP site (Figure 3-26).

Table 3-35 provides annual average daily traffic (AADT) data for the roadways in the vicinity of the proposed IIFP site (Figure 3-26). These roadways are used as trucking routes. The numbers in the left column correspond to the numbered locations on Figure 3-26.

Table 3-35.   AADT Volumes for Roadways that Access the Proposed IIFP Site

	Roadway	Location	AADT	Year
1	US 62/180	NM 8 Junction	7,868[a]	2008
2	NM 483	US 62/180 Junction	955[b]	2008
3	NM 132	US 18/NM 218 Junction	4,604[b]	2008
4	NM 8	0.256-mi South of US 62/180	1,302[b]	2008
5	NM 18	US 62/180 Junction	12,407[a]	2007
6	NM 176	NM 8 Junction	2,124[a]	2008
7	NM 529	West of US 62/180 Junction	2,393[b]	2008

Source: NMDOT, 2009
[a]The AADT was derived from recent coverage counts.
[b]The AADT was derived using an Annual Growth Factor, generalized from coverage counts within the traffic segment and updated with loop and growth factors.

3.10.2   Railroads

The Texas-New Mexico Railroad is an active rail line through Hobbs, New Mexico, approximately 16.2 km (10 mi) east of the proposed IIFP site. It is a shortline railroad operating between Monahans, Texas and Lovington, New Mexico. The rail line is predominantly used for freight transport associated with the oil and gas industry, and typical freight includes chemicals, minerals, construction aggregate, industrial waste, and scrap. There is a freight dock and warehouse in the Hobbs area. Train frequency is daily six days per week (IPH, 2010).

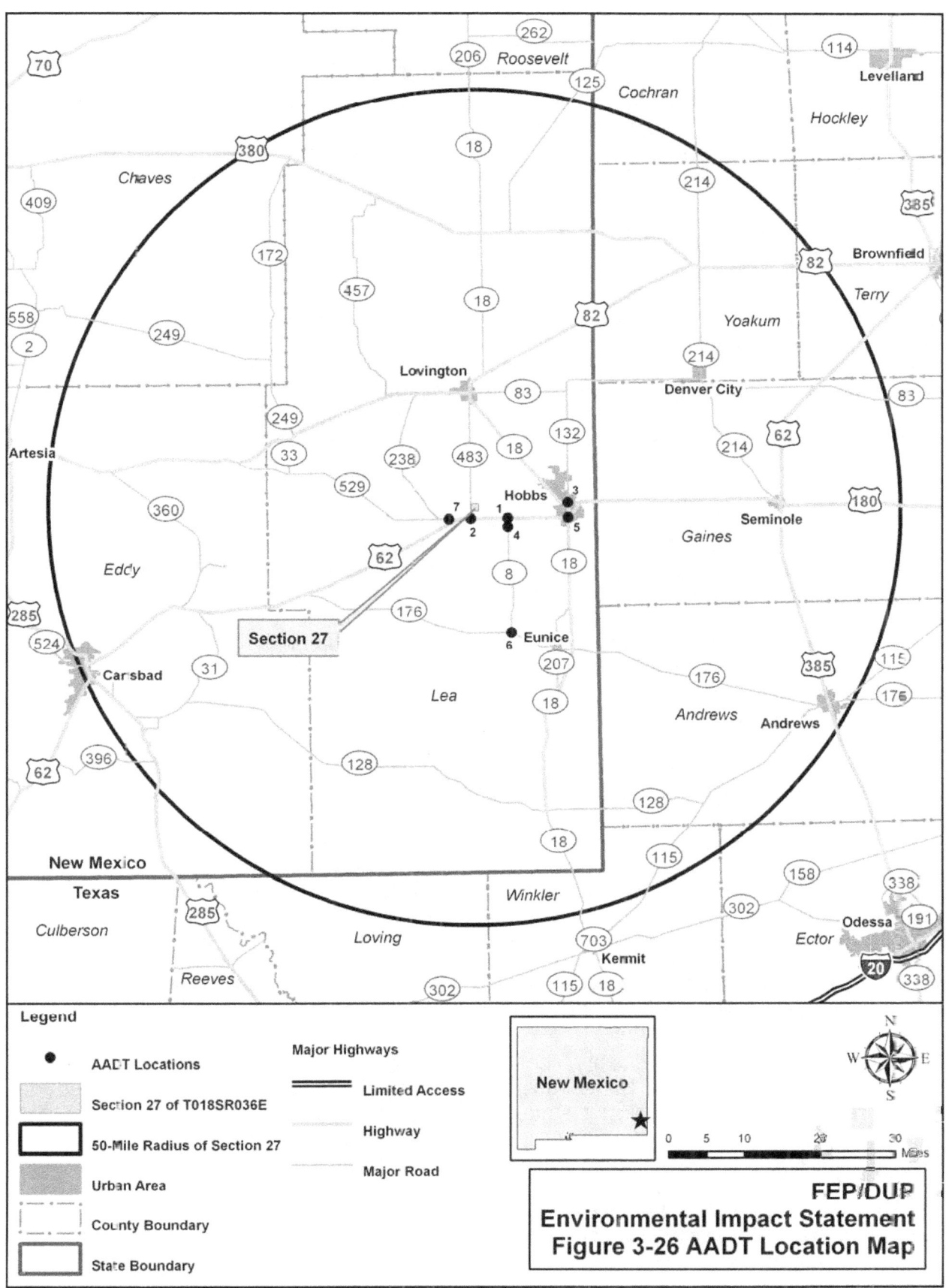

Environmental Impact Statement
Figure 3-26 AADT Location Map

FEP/DUP

Legend

● AADT Locations

Section 27 of T018SR036E

50-Mile Radius of Section 27

Urban Area

County Boundary

State Boundary

Major Highways

Limited Access

Highway

Major Road

New Mexico

Section 27

3.10.3   Airports

The Lea County Regional (Hobbs) Airport is approximately 13 km (8 mi) from the proposed IIFP site just northwest of the Hobbs city limits. The airport currently supports only general aviation, but may support domestic flights in the future (AIRNAV, 2010). Two additional airports are in Lea County: The Lea County-Zip Franklin Memorial Airport is 5 km (3 mi) west of Lovington (approximately 26 km [16 mi] from the IIFP site) and the Lea County-Jal Airport is 6 km (3.7 mi) northeast of Jal (approximately 64 km [40 mi] from the IIFP site). These airports support general aviation operations (AIRNAV, 2010).

3.11   Noise

The definition of noise is "unwanted or disturbing sound." Sound measurements are described in terms of frequencies and intensities. The decibel (dB) is used to describe the sound pressure level. The A-scale on a sound level meter best approximates the audible frequency response of the human ear, and is commonly used in noise measurements. Sound pressure levels measured on the A-scale of a sound meter are abbreviated dB(A). In noise measurements, sound pressure levels are typically averaged over a given length of time, because instantaneous levels can vary widely.

The intensity of sound decreases with increasing distance from the source. Typically, sound levels for a point source will decrease by 6 dB(A) for each doubling of distance. This may vary depending on the terrain, topographical features, and frequency of the noise source.

Generally, sound level changes of 3 dB(A) are barely perceptible, while a change of 5 dB(A) is readily noticeable by most people. A 10 dB(A) increase is usually perceived as a doubling of loudness, and conversely, noise is perceived to be reduced by one-half when a sound level is reduced by 10 dB(A).

Sound levels can vary for indoor and outdoor noise sources. For example, a jet flying overhead at 1,000' will produce a sound level of 100 dB(A), the same as an inside subway train. A typical outdoor commercial area is equivalent to a normal speech conversation indoors, at 65 dB(A), and a quiet rural nighttime environment will mimic an empty concert hall, at 25 dB(A).

3.11.1   Noise Level Standards

Noise level standards are established by Federal agencies including the U.S. Department of Housing and Urban Development (HUD) (24 CFR 51), the Environmental Protection Agency (EPA, 1974), Federal Highway Administration (23 CFR 772), and the Occupational Safety and Health Administration (OSHA) (29 CFR 1910).

Neither the city of Hobbs, Lea County, nor New Mexico have ordinances or regulations governing noise. There are no Native American Tribes within 10 km (6 mi) of the proposed site. Therefore, the facility is not subject to state, tribal, or local noise ordinances.

The EPA has defined a goal of 55 dB(A) for average day-night sound levels in outdoor spaces (EPA 1974). OSHA standards prescribe the maximum noise levels that employees can be exposed to within a facility. For an 8-hour work period, sound levels must remain below 90dB(A) or noise abatement measures must be taken, in order to comply with OSHA [29 CFR

1910.95(b)(2)]. HUD guidelines are that noise levels at 65 dB(A) or below are acceptable in a residential setting in normal situations.

3.11.2 Noise Receptors in the Vicinity of the Proposed Facility

The determination of noise impacts is based on the relationship between the ambient noise levels and the established noise abatement criteria for the project area. Noise sensitive areas are created to represent common noise environments within the same activity category, and are represented by receptors, which represent a discrete or representative location within the noise sensitive area. Activity categories include land uses such as residences, hotels, motels, active sport areas, schools, places of worship, hospitals, parks and others. No noise sensitive areas are within 10 km (6 mi) of the site, based on a review of aerial photographs.

The nearest commercial facilities are Xcel Energy's Cunningham Generating Station, approximately 1.6 km (1 mi) west of the proposed site, Xcel Energy's Maddox Generating Station approximately 3.7 km (2.3 mi) east of the proposed site, and the Colorado Energy Station approximately 2.4 km (1.5 mi) northeast of the proposed site. The nearest residence is approximately 2.6 km (1.6 mi) northwest of the site. No recreational facilities are within 10 km (6 mi) of the proposed site.

3.11.3 Noise in the Vicinity of the Proposed Facility

Noise sources in the vicinity of the site include ambient noise from the natural setting, highway noise from NM 483, and occasional noise associated with the overhead and underground utilities. The noise from the proposed facility would be associated with construction and operation activities and associated employee traffic. During operations, intermittent noise could be expected from delivery/disposal of the depleted uranium cylinders and other materials, commuting workers' vehicles, and operating equipment such as forklifts. Noise levels near the closed-loop cooling towers could be relatively high, but otherwise, most noise sources would be within buildings and would not be audible outside. Baseline ambient noise levels are the basis for comparison with predicted noise levels from construction and operation. It is typical in this type of topography and setting for background ambient noise levels to be between 50 and 60 dB(A).

Noise levels at uranium deconversion facilities in Paducah, KY and Piketon, OH would be comparable. DOE reported estimated or actual noise levels for those facilities (during operation) to range from 40 to 46 dB(A) at the closest receptors, which are residences approximately 1.6 km (1 m) from the proposed facility; similar to the distance to sensitive receptors at the proposed site (DOE, 2004a; DOE 2004b).

3.12 Public and Occupational Health

This section describes the natural and manmade sources of radiation and chemicals and the levels of exposure that may occur in the vicinity of the proposed IIFP facility.

3.12.1 Background Radiological Exposure

Figure 3-27 depicts the major sources of background radiation in the United States. As shown on Figure 3-27, humans are exposed to ionizing radiation from both natural and manmade sources. In the United States, each source contributes on average approximately one-half of an individual's total annual radiation dose. The total annual exposure to individuals from both natural and manmade sources of radioactivity is approximately 6.2 millisieverts (620 millirem) (NCRP, 2009).

A major proportion of natural radiation comes from naturally occurring airborne sources such as radon and thoron (an isotope of radon). The proposed IIFP site is in an area characterized by radon concentrations of 2 to 4 picocuries per liter (pCi/L) and is defined as of moderate radon potential. Moderate radon potential indicates that one-third to one-half of the structures have more than 0.148 becquerel per liter (Bq/L) or 4 pCi/L of indoor radon. In May 2004, direct background radiation was measured by the NMED Radiation Control Bureau to be 8 to 10 microrad per hour, which corresponds to 0.70 to 0.88 millisieverts (mSv) (70 to 88 millirem [mrem]) per year. The measured range falls within the average annual direct background radiation for the United States (NRC, 2005). Additionally cosmic radiation, which primarily consists of positively charged ions from protons to larger nuclei from sources outside our solar system, is continuously penetrating the earth's atmosphere, adding to the overall amount of natural background exposure each individual receives. As shown on Figure 3-27, the total contribution from natural background radiation to each individual is approximately 3.1 mSv (310 mrem) (NCRP, 2009).

Manmade sources include x-rays for medical purposes and consumer products such as smoke detectors. The National Council on Radiation Protection and Measurements (NCRP) released a report, Ionizing Radiation Exposure of the Population of the United States, (NCRP Report No. 160; NCRP 2009) in 2009 that updated the findings of the previously issued NCRP Report No. 93 (NCRP, 1987). The 2009 NCRP report found significant increases in radiation exposures related to medical procedures and treatments. The approximate doses to individuals for medical procedures such as x-rays comprise approximately 48 percent of the total dose received by individuals living in the United States (NCRP, 2009). Natural sources include cosmic sources, radionuclides within a person's body, radionuclides in soils, and radon and thoron inhalation.

DOE established radiological monitoring programs in southeastern New Mexico prior to the Waste Isolation Pilot Plant project to determine the level of background radiation. DOE estimated an annual dose in southeastern New Mexico of approximately 0.65 mSv (65 millirem) from atmospheric particulate matter, ambient radiation, soil, surface water and sediment, groundwater, and biota (NRC, 2005). These doses are within expected ranges and do not indicate any unexpected environmental concentrations in the area. Based on natural and manmade sources, residents living near the proposed IIFP facility could be expected to receive, on average, an annual dose of approximately 6.2 mSv (620 millirem).

3.12.2   Background Chemical Characteristics

The 16-ha (40-ac) area that would contain the proposed facility is undeveloped land. There is no known past activity on this land that would make its background chemical characteristics different than other undeveloped land. No site-specific chemical data are available.

Figure 3-27. Major Sources and Levels In The Vicinity of the Proposed FEP/DUP Background Radiation Exposure

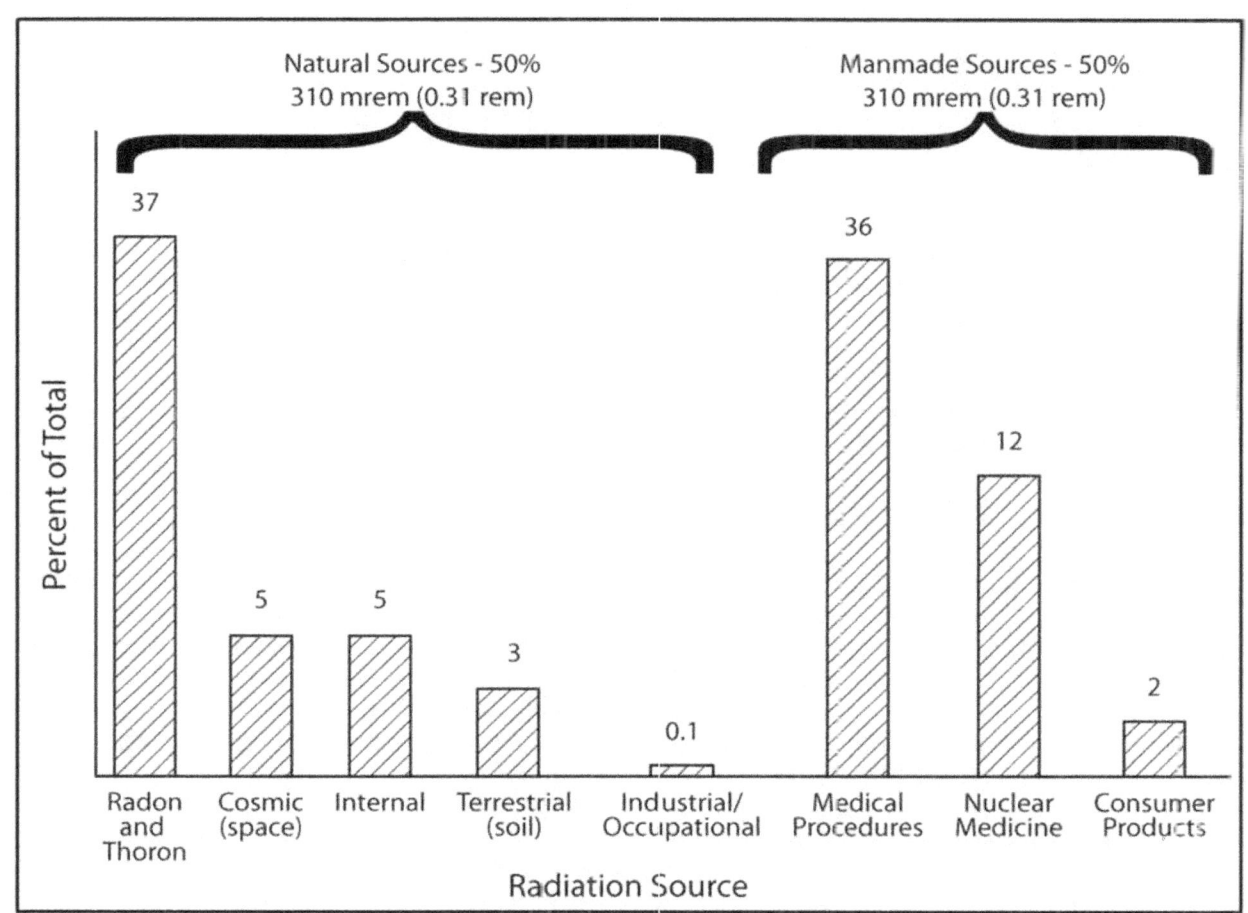

3.13 References

(AHA, 2007) American Hospital Association. 2007. AHA Guide, America's Directory of Hospitals and Health Care System.

(AIRNAV, 2010) AirNav, LLC. 2010. Lea County Airports. Available at http://www.airnav.com/airport. Accessed on September 21, 2010.

(AMA, 2010) American Medical Association. 2010. Physician-Related Data Resource, Lea County New Mexico and Eddy County New Mexico. Available at http://www.ama-assn.org/. Accessed August 26, 2010.

(Anderson, Undated) Anderson, H. A. "Hat Ranch," Handbook of Texas Online. Available at http://www.tshaonline.org/handbook/online/articles/aph04. Accessed November 20, 2010. ADAMS Accession No. ML 112710277.

(Artesia, 2010) City of Artesia Water and Wastewater Department. 2010. Personal Communication between M. Stroud, (City of Artesia Water and Wastewater Department) and P. Baxter (TtNUS). September 28, 2010. ADAMS Accession No. ML112840402.

(BBER, 2008) University of New Mexico, Bureau of Business and Economic Research. 2008. Population Projections for New Mexico and Counties. Released August 2008. Available at http://bber.unm.edu/demo/table1.htm. Accessed September 16, 2010. ADAMS Accession No. ML103430027.

(BEA, 2010) U.S. Department of Commerce, Bureau of Economic Analysis. 2010. Total Employment, "Table CA25N Total Employment by Industry, New Mexico, Lea County New Mexico, and Eddy County New Mexico, 2008," Updated August 9, 2010. Available at http://bea.gov/regional/reis/action.cfm. Accessed September 21, 2010. ADAMS Accession No. ML103430029.

(BEA, 2011a) U.S. Dept of Commerce, Bureau of Economic Analysis. 2011. "Table CA25N Total Full-Time and Part-Time Employment by NAICS Industry, Lea County and Eddy County, New Mexico, 2001" Updated April 21, 2011. Available at http://bea/gov/regional/reid/action/cfm. Accessed May 31, 2011. ADAMS Accession No. ML112710284.

(BEA, 2011b) U.S. Dept of Commerce, Bureau of Economic Analysis. 2011. "Table CA25N Total Full-Time and Part-Time Employment by NAICS Industry, Eddy County New Mexico, 2001" Updated April 21, 2011. Available at http://bea/gov/regional/reid/action/cfm. Accessed May 31, 2011. ADAMS Accession No. ML112710284.

(BLM, 2007) U.S. Bureau of Land Management. 2007. Visual Resource Inventory. April 30, 2007. ADAMS Accession No. ML 112710288.

(BLM, 2008) U.S. Bureau of Land Management. 2008. Record of Decision and Approved Resource Management Plan Amendment, BLM NM/PL-08-05-1610. Pecos District Office, Roswell, New Mexico. April. Available at http://www.blm.gov/pgdata/etc/medialib/blm/nm/field_offices/roswell/rfo_planning/special_status_species.Par.34868.File.dat/pdf_sss_rod_r mpa_May_2008.pdf. Accessed August 11, 2010. ADAMS Accession No. ML103090063.

(BLM, 2010) U.S. Bureau of Land Management. 2010. Lesser Prairie Chicken Corridor. Roswell Field Office. Available at http://www.blm.gov/nm/st/en/fo/Roswell_Field_Office/lpc_corridor.html. Accessed August 16, 2010. ADAMS Accession No. ML103090065.

(BLS, 2010a) U. S. Department of Labor, Bureau of Labor Statistics. 2010. Local Area Unemployment Statistics (LAUS), 2001 to 2010, New Mexico, Lea County and Eddy County New Mexico. Available at http://data.bls.gov/. Accessed September 21, 2010. ADAMS Accession No. ML103430031.

(BLS, 2010b) U. S. Department of Labor, Bureau of Labor Statistics. 2010. Labor Force Statistics from the Current Population Survey, USA. Available at http://data.bls.gov/. Accessed September 21, 2010. ADAMS Accession No. ML103430033.

(Carlsbad, 2010) City of Carlsbad Waste Water Treatment Plant. 2010. Personal Communication between A. Sena (City of Carlsbad Waste Water Treatment Superintendent) and P. Baxter (TtNUS). November 17, 2010. ADAMS Accession No. ML112840422

(Carter, 2010) Carter, P.S. 2010. Criteria for disturbing soils on the site classified at Farmland of Statewide Importance. Personal Communication between P. Carter (District Conservationist, NRCS, NM) and Krista Dearing (TtNUS) on August 16, 2010. ADAMS Accession No. ML112840429.

(CCS, 2006) Center for Climate Studies. 2006. Appendix D: New Mexico Greenhouse Gas Inventory and Reference Case Projections, 1990 – 2020. Prepared for the New Mexico Environment Department. November, 2006. ADAMS Accession No. ML112710300.

(Center for Biological Diversity, 2002) Center for Biological Diversity. 2002. Petition to List the Sand Dune Lizard, *Sceloporus arenicolus,* as a Threatened or Endangered Species Under the U. S. Endangered Species Act. Available at http://www.biologica diversity.org/species/reptiles/ dunes_sagebrush_lizard/pdfs/petition.pdf. Accessed October 24, 2011. ADAMS Accession No. ML112710308.

(CEQ, 1997) Council on Environmental Quality. 1997. Environmental Justice Guidance Under the National Environmental Policy Act. December 1997. Executve Office of the President. ADAMS Accession No. ML103430030.

(CFC, Undated) Central Flyway Council. Undated. "The Central Flyway." Available at http://central.flyways.us/. Accessed August 17, 2010. ADAMS Accession No. ML103090070.

(City of Albuquerque, 2011) City of Albuquerque. 2011. "Industrial Revenue Bonds Explained." Available at http://www.cabq.gov/econdev/irbs.html. Accessed March 21, 2011.

(Clampitt, 2008) Clampitt, M. A. 2008. Hobbs and Lea County (Images of America: New Mexico). Arcadia Publishing.

(Crone and Wheeler, 2000) Crone, A.J. and R. L. Wheeler, R.L. 2000. Data for Quaternary Faults, Liquefaction Features, and Possible Tectonic Features in the Central and Eastern United States, East of the Rocky Mountain Front. U.S. Geological Survey Open File Report 00-0260. Available at http://pubs.usgs.gov/of/2000/ofr-00-0260/ofr-00-0260.pdf. ADAMS Accession No. ML103080519.

(CRS, 2007) Congressional Research Service. 2007. CRS Report for Congress State Greenhouse Gas Emissions: Comparison and Analysis. Jonathan Ramseur. Prepared for Members and Committees of Congress. December 5, 2007. Available at http://opencrs.com/document/RL34272/. Accessed March 22, 2011. ADAMS Accession No. ML112710313.

(Daras, 2009) Daras, S. 2009. Cultural Resource Survey of 640 acres for the Arkansas Junction Site, Lea County, New Mexico. 2009 Report No. 1224. Prepared for New Mexico State Land Office by Lone Mountain Archeological Services, Inc. 2009. ADAMS Accession No. ML113120204.

(Dick-Peddie, 1993) Dick-Peddie. W. 1993. New Mexico Vegetation Past, Present, and Future. University of New Mexico Press. 1993.

(DOE 2004a) U.S. Department of Energy. 2004. Final Environmental Impact Statement for Construction and Operation of a Depleted Uranium Hexafluorice Conversion Facility at the

Paducah, Kentucky Site, DOE/EIS-0359. Washington, D,C., June 2004. ADAMS Accession No. ML050380331.

(DOE 2004b) U.S. Department of Energy. 2004. Final Environmental Impact Statement for Construction and Operation of a Depleted Uranium Hexafluoride Conversion Facility at the Portsmouth Ohio Site, DOE/EIS-0360. Washington, D.C. June 2004. ADAMS Accession No. ML050380334.

(DOE, 2006) U.S. Department of Energy, 2006. Waste Isolation Pilot Plant Contact Handled (CH) Waste Documented Safety Analysis, DOE/WIPP-95-2065 Rev. 10, Washington, D.C., November, 2006. Available at http://www.energy.ca.gov/nuclear/yucca/documents/AG-155-2007-005754.pdf. Accessed August 21, 2010. ADAMS Accession No. ML103080100.

(eBird, 2011) Audubon and Cornell Lab of Ornithology. 2011. Bird Observations for Lea County, New Mexico. Available at http://ebird.org/ebird/GuideMe?step=saveChoices& getLocations=counties&parentState=US-NM&bMonth=01&bYear=1900&eMonth=12&eYear= 2010&reportType=location&counties=US-NM-025&continue.x=36&continue.y=4&continue =Continue. Accessed January 20, 2011.

(ECP, 2008) Eddy County Comprehensive Plan, 2008. Eddy County Comprehensive Plan, October. Available at http://www.co.eddy.nm.us/, http://www.co.eddy.nm.us/EddyCty-FinalRPT10-08.pdf. Accessed August 31, 2010. ADAMS Accession No. ML103430034.

(Eddy County, 2007) Eddy County, New Mexico. 2007. Eddy County, New Mexico. County Treasurer. 2007. Available at http://www.co.eddy.nm.us/treasurer.html. Accessed August 31, 2010.

(eHOW) eHOW. Undated. "FEMA Flood Zone Definitions. Available at http://www.ehow.com/ about_5043095_fema-flood-zone-definitions.html/ Accessed October 29, 2010.

(EPA, 1974) U. S. Environmental Protection Agency. 1974. Information on Levels of Environmental Noise Requisite to Protect Public Health and Welfare with an Adequate Margin of Safety," EPA/ONAC 550/9-74-004. March 1974. Available at http://www.nonoise.org/ library/levels74/levels74.htm. Accessed on December 13, 2010. ADAMS Accession No. ML110110692.

(EPA, 2009a) U.S. Environmental Protection Agency. 2009. Lea County, New Mexico Emissions by Category Report - Criteria Air Pollutants, Texas. Washington, D.C., January 10, 2009. Available at http://www.epa.gov/oar/data/geosel.html. Accessed on May 31, 2011. ADAMS Accession No. ML112710378.

(EPA, 2009b) U.S. Environmental Protection Agency. 2009. County Air Quality Report – Criteria Air Pollutants for Eddy, Lea, and Sandoval Counties, New Mexico, and El Paso County, Texas. Washington, D.C. January, 10, 2009. Available at http://www.epa.gov/oar/data/geosel.html. Accessed on May 31, 2011.

(EPA, 2010a) U.S. Environmental Protection Agency. 2010. Inventory of U.S. Greenhouse Gas Emissions and Sinks: 1990–2008. Executive Summary. April 15, 2010. Available at http://www.epa.gov/climatechange/emissions/usgginv_archive.html. Accessed March 10, 2011. ADAMS Accession No. ML112710366.

(EPA, 2010b) U.S. Environmental Protection Agency. 2010. "Safe Drinking Water Information System (SDWIS)." Available at http://oaspub.epa.gov/enviro/sdw_form_v2.create_page?state_abbr=NM. Accessed August 26, 2010. ADAMS Accession No. ML 103430026.

(Fahlquist, 2003) Fahlquist, L. 2003. Ground-Water Quality of the Southern High Plains Aquifer, Texas and New Mexico. U.S. Geological Survey Open File Report 03-345. Available at http://co.water.usgs.gov/nawqa/hpgw/reports/ofr03_345.pdf. Accessed July 28, 2010. ADAMS Accession No. ML103080521.

(FBI, 2009a) Federal Bureau of Investigation. 2009. Uniform Crime Report, "Table 80: Full-time Law Enforcement Employees, by State, by Metropolitan and Nonmetropolitan Counties, 2009." Available at http://www.fbi.gov/ucr/. Accessed September 21, 2010. ADAMS Accession No. ML103430037.

(FBI, 2009b) Federal Bureau of Investigation. 2009. Uniform Crime Report, "Table 78: Full-time Law Enforcement Employees, by State, by Cities, 2008." Available at http://www.fbi.gov/ucr/. Accessed September 21, 2010. ADAMS Accession No. ML103430041.

(Gerald, 1974) Gerald, R.E. 1974. Aboriginal Use and Occupation by Tigua, Manso, and Suma Indians. Garland Publishing. 1974.

(GL Environmental, 2010a). GL Environmental, Inc. 2010. 2010 Soil and Vegetation Characterization Report. Prepared for International Isotopes Fluorine Products. ADAMS Accession No. ML112140543.

(GL Environmental, 2010b) GL Environmental, Inc. 2010. Existing Groundwater Condition in Section 27, Range 18 South, Township 36 East. Prepared for International Isotopes Fluorine Products. ADAMS Accession No. ML112720073.

(GL Environmental,. 2010c) GL Environmental, Inc. 2010. 2010 Soil and Vegetation Survey Report. Prepared for International Isotopes Fluorine Products. November 29, 2010. ADAMS Accession No. ML112720310.

(GL Environmental, 2010d) GL Environmental, Inc. 2010. Status and Habitat of the Dune Sagebrush Lizard at the Proposed Site for the International Isotopes Fluorine Products Facility in Lead County, New Mexico. Prepared for International Isotopes Fluorine Products. November 11, 2010. ADAMS Accession No. ML112720311.

(Hart and McAda, 1985) Hart, D.L. and D. P. McAda, 1985. Geohydrology of the High Plains Aquifer in Southeastern New Mexico. U.S. Geological Survey Hydrologic Atlas 679. Available at http://pubs.er.usgs.gov/#search:basic/query=HA-679/page=1/page_size=25:0. Accessed August 9, 2010. ADAMS Accession No. ML103080964.

(Hat Ranch, 2010) Hat Ranch. 2010. "Ranch History." Hat Ranch Quarter Horses, http://www.hatranchquarterhorses.com/history.html .Accessed December 12, 2010. ADAMS Accession No. ML112710379.

(HFD, 2010) City of Hobbs Fire Department. 2010. Fire Protection and Law Enforcement in Hobbs and New Mexico. Personal Communication between C. Allen (Hobbs Fire Department) and P. Baxter (TtNUS), September 1, 2010. ADAMS Accession No. ML112840437.

(Hickerson, 1994) Hickerson, N.P. 1994. The Jumanos: Hunters and Traders of the South Plains. University of Texas Press. 1994.

(Hobbs, 2010) City of Hobbs. 2010. Personal communication between L. Wilson (City of Hobbs City Hall) and P. Baxter (TtNUS). September 28 - 30, 2010. ADAMS Accession No. ML112840425.

(Hobbs Country Club, 2010) Hobbs Country Club. 2010. Available at http://www.hobbscountryclub.com/golf/proto/hobbscountryclub/ directions/directions.htm. Accessed November 11, 2010.

(Hobbs Motorsports Park, 2010) Hobbs Motorsports Park. 2010. Available at http://www.hobbsmotorsportspark.com/general/contact.htm. Accessed November 11, 2010.

(HRSA, Undated) U. S. Department of Health and Human Resources, Health Resources and Services Administration. Undated. Medically Underserved Areas by State and County. Available at http://muafind.hrsa.gov/. Accessed August 27, 2010. ADAMS Accession No. ML103430043.

(Hunt, 1977) Hunt, C.B. 1977. Surficial Geology of Southeast New Mexico. New Mexico Bureau of Mines & Mineral Resources, Geologic Map 41. ADAMS Accession No. ML103080523.

(ICC, 1979) U.S. Indian Claims Commission. 1979. "U.S. Indian Claims Commission, August 13, 1946-September 30, 1978: Final Report and Map (Indian Land Areas Judicially Established, 1978)."U.S. Government Printing Office. 1979. Available at http://www.nps.gov/history/nagpra/ DOCUMENTS/ClaimsMAP.HTM and http://www.nps.gov/history/nagpra/DOCUMENTS/ JUDICIAL.PDF. Accessed August 15, 2010. ADAMS Accession No. ML103080097.

(IIFP, 2009) International Isotopes Fluorine Products. Inc. 2009. Fluorine Extraction Process and Depleted Uranium De-conversion Plant (FEP/DUP) Environmental Report, Revision A, ER-IFP-001. December 27, 2009. ADAMS Accession No. ML100120758.

(IIFP, 2011) International Isotopes Fluorine Products, Inc. 2011. Fluorine Extraction Process and Depleted Uranium De-conversion Plant (FEP/DUP) Official Responses to Environmental Report RAIs, Revision A. March 31, 2011. ADAMS Accession No. ML110970481.

(INIS, 2011) International Isotopes, Inc. 2011. "International Isotopes Inc. Provides an Update Summary on the Progress of the Uranium Deconversion and Fluorine Extraction Processing Facility" Press Release. Idaho Falls, ID, February 7, 2011. Available at http://www.intisoid.com/ wp-content/uploads/2010/08/INISAnnouncesContinuedProgress TowardstheLicensingand-Construction.pdf. Accessed March 28, 2011. ADAMS Accession No. ML112710381.

(IPCC, 2007) Intergovernmental Panel on Climate Change. 2007. Summary for Policy Makers in: Climate Change 2007: The Physical Science Basis. Contribution of Working Group I to the Fourth Assessment Report of the Intergovernmental Panel on Climate Change. S. Solomon, D. Qin, M. Manning, Z. Chen. M. Marquis, K.B. Averyt, M. Tignor and H.L. Miller (eds.). Cambridge University Press, Cambridge, United Kingdom and New York, NY, USA. Available at http://www.ipcc.ch/publications_and_data/ar4/wg1/en/spm.html. Accessed October 24, 2011. ADAMS Accession No. ML112710385.

(IPH, 2010) Iowa Pacific Holdings, LLC 2010. Texas-New Mexico Railroad. Available at http://www.iowapacific.net/texas-new-mexico-railroad.html. Accessed on September 21, 2010. ADAMS Accession No. ML 103430159.

(Kelley, 1986) Kelley, J.C. 1986. Jumano and Patarabueye: Relations at La Junta de los Rios. University of Michigan, Museum of Anthropology. Anthropological Papers No. 77.

(Lea County, 2009) Lea County, New Mexico. 2009. "2009 Tax Levies, Lea County, New Mexico." Available at http://www.leacounty.net/. Accessed August 31, 2010. ADAMS Accession No. ML103430051.

(Leedshill-Herkenhoff et al., 2000) Leedshill-Herkenhoff, Inc., John Shomaker & Associates, Inc., and Montgomery and Andrews. 2000. Region 16: Lea County Regional Water Plan. Prepared for the Lea County Water User's Association. Available at http://www.ose.state. nm.us/isc_regional_plans16.html. Accessed July 25, 2010.

(Limburg, 2009) Limburg, E. 2009. Recent and Potential Volcanism of New Mexico. April 16, 2009. Available at http://dspace.nmt.edu/dspace/handle/10136/296. Accessed August 16, 2010.

(Lin and Sanford, 2000) Lin, K., and A.R., Sanford. 2000. Some Characteristics of a Probabilistic Seismic Hazard Map for New Mexico. July 2000. New Mexico Institute of Mining and Technology Geophysics Oper-File Report 92. Available at http://www.ees.nmt.edu/Geop/ nmquakes/R92/R92.HTM. Accessed August 15, 2010

(Lovington, 2010) City of Lovington Wastewater Treatment Plant. 2010. Personal Communication between M. De La Cruz (City of Lovington Wastewater Treatment Plant) and P. Baxter (TtNUS). September 30, 2010. ADAMS Accession No. ML112840439.

(LPD, 2010) Lovington Police Department. 2010. "Law Enforcement in Hope, Lovington, New Mexico and Mutual Aid Agreements." Personal Communication between Chief M.Bird (Loving Police Department) and P Baxter (TtNUS). September 1, 2010. ADAMS Accession No. ML112840414.

(Machette, et al., 1998) Machette, M.N., Personius, S.F., Kelson, K.I., Haller, K.M., and R.L. Dart. 1998. Map of Quaternary Faults and Folds in New Mexico and Adjacent Areas. U.S. Geological Survey Open File Report 98-521, (Revised 2000). ADAMS Accession No. ML103080536.

(Main, 1992) Main, R. 1992. Limited Testing at Maljamar Lea County, New Mexico. Archaeological Notes 78. Prepared for Museum of New Mexico, Office of Archaeological Studies. Santa Fe, NM. 1992.

(McAda, 1984) McAda, D.P. 1984. Projected Water-Level Declines in the Ogallala Aquifer in Lea County, New Mexico. U.S. Geological Survey Water-Resources Investigations Report 84-4062. Available at http://pubs.er.usgs.gov/#search:basic/query=water%20resources %20investigation%20report%2084-4062/page=1/page_size=25:0. Accessed July 28, 2010. ADAMS Accession No. ML103080967.

(McCoy and Perry, 2004) McCoy, A.M., and R.L. Perry. 2004. Second Draft, Lea County Deep Aquifer Study. Prepared for Lea County Water Users Association. John Shomaker & Associates, Inc. ADAMS Accession No. ML103080447.

(McGuire et al., 2003) McGuire, V.L., M. R. Johnson, R. L. Schieffer, J. S. Stanton, S. K. Sebree, and I. M. Verstraeten. 2003. Water Storage and Approaches to Groundwater Management, High Plains Aquifer, 2000 Circular 1243. U.S. Geological Survey. Available at http://pubs.er.usgs.gov/#search:basic/query=circular%201243/page=1/page_size=25:0. Accessed July 29, 2010. ADAMS Accession No. ML103080969.

(McLemore, et. al., 2007) McLemore, V., T., Hoffman, G.K., Mansell, M., Jones, G. R., Krueger, C.B., and M. Wilks. 2007. Mining Districts in New Mexico. July 2007. New Mexico Bureau of Geology and Mineral Resources Open File Report 494. Available at http://geoinfo.nmt.edu/publications/openfile/downloads/OFR400-499/476-499/494/494_CDROM/NMMD_10July07_fin.pdf. Accessed July 25, 2010. ADAMS Accession No. ML103080540.

(Merlan, 2010) Merlan, Thomas. 2010. Historic Homesteads and Ranches in New Mexico: A Historic Context. State of New Mexico, Historic Preservation Division. ADAMS Accession No. ML112710390.

(Musharrafieh and Chudnoff, 1999) Musharrafieh, G., and M. Chudnoff. 1999. Numerical Simulation of Groundwater Flow for Water Rights Administration in the Lea County Underground Water Basin New Mexico, Technical Report 99-1. Prepared for the New Mexico Office of the State Engineer. January 1999. Available at http://www.ose.state.nm.us/publications_library_hydrologyrpts.html. Accessed July 27, 2010. ADAMS Accession No. ML103080980.

(NatureServe, 2009) NatureServe. 2009. International Ecological Classification Standard: Terrestrial Ecological Classifications. NatureServe Central Databases. Arlington, Va. Available at http://www.natureserve.org/getData/vegData/nsDescriptions.pdf. Accessed August 17, 2010.

(NCDC, 2010) National Climatic Data Center. 2010. Storm Events – Lea County, New Mexico. Available at http://www4.ncdc.noaa.gov/cgi-win/wwcgi.dll?wwevent~storms. Accessed August 24 2010. ADAMS Accession No. ML103080177.

(NCES, 2008) National Center for Education Statistics. 2008. Search for Public School Districts - District Name: Hobbs Municipal Schools. Available at http://nces.ed.gov/ccd/districtsearch/. Accessed March 10, 2011. ADAMS Accession No. ML112710441.

(NCES, 2009) National Center for Education Statistics. 2009. Finance 2008-2009. Institution: New Mexico Junior College (187903). Available at http://nces.ed.gov/ipeds/datacenter/Default.aspx. Accessed March 10, 2011. ADAMS Accession No. ML112710447.

(NCES, 2010a) National Center for Education Statistics. 2010. Search for Public School Districts, Lea and Eddy Counties, New Mexico. Available at http://nces.ed.gov/ccd/districtsearch/. Accessed August 30, 2010. ADAMS Accession No. ML103430054.

(NCES, 2010b) National Center for Education Statistics. 2010. Search for Private Schools, Lea and Eddy Counties, New Mexico. Available at http://nces.ed.gov/ccd/districtsearch/. Accessed August 31, 2010. ADAMS Accession No. ML103430061.

(NCES, 2010c) National Center for Education Statistics. 2010. College Navigator. Available at http://nces.ed gov/collegenavigator/. Accessed August 30, 2010 – September 1, 2010. ADAMS Accession No. ML103430065.

(NCRP, 1987) National Council on Radiation Protection and Measurements. 1987. Ionizing Radiation Exposure of the Population of the United States. NCRP Report No. 93. September 1, 1987.

(NCRP, 2009) National Council on Radiation Protection and Measurements. 2009. Ionizing Radiation Exposure of the Population of the United States. NCRP Report No. 160, Bethesda, MD. March 3, 2009.

(Nicholson and Clebsch, 1961) Nicholson, A. and A. Clebsch. 1961. Geology and Ground-Water Conditions in Southern Lea County, New Mexico. New Mexico Bureau of Mines and Mineral Resources Ground-Water Report 6. ADAMS Accession No. ML103080877.

(NMBGMR, 2003) New Mexico Bureau of Geology and Mineral Resources. 2003. Geologic Map of New Mexico. ADAMS Accession No. ML103080881.

(NMBGMR, 2010) New Mexico Bureau of Geology and Mineral Resources, 2010. Physiographic Map of New Mexico. Available at http://geoinfo.nmt.edu/tour/. Accessed March 7, 2011. ADAMS Accession No. ML112710476.

(NMDOT, 2009) New Mexico Department of Transportation. Consolidated Highway Database (CHDB). Website no longer accessible. Accessed April 10, 2009. ADAMS Accession No. ML112860287.

(NMED, 2010a) New Mexico Environment Department. 2010. Inventory of New Mexico Greenhouse Gas Emissions: 2000 – 2007. Prepared by the New Mexico Environment Department. March 15, 2010. Available at http://www.nmenv.state.nm.us/cc/documents/ GHGInventoryUpdate3_15_10.pdf. Accessed March 11, 2011. ADAMS Accession No. ML112710489.

(NMED, 2010b) New Mexico Environment Department. 2010. "RE: FOIA Request." Personal Communication between M. Huber (New Mexico Environment Department) and N. Hill (TtNUS). September 22, 2010. ADAMS Accession No. ML112850086.

(NMED, 2010c) New Mexico Environment Department, Ground Water Quality Bureau. 2010. Ground Water Regulated Facilities, List of Current Discharge Permits, updated 07/29/2010. Available at http://www.nmenv.state.nm.us/gwb/NMED-GWQB-Permits.htm. Accessed August 31, 2010. ADAMS Accession No. ML112710484.

(NMEMNRD, 2010a) New Mexico Energy, Minerals, and Natural Resources Department. 2009 Annual Report. Available at http://www.emnrd.state.nm.us/main/documents/EMNRD2009 AnnualReportWeb.pdf. Accessed July 28, 2010. ADAMS Accession No. ML103080893.

(NMEMNRD, 2010b) New Mexico Energy, Minerals, and Natural Resources Department. 2010. New Mexico Oil & Gas Wells. Available at http://mapserve.nmt.edu/Website/ NMOG/viewer.htm. Accessed August 12, 2010.

(NMGF, 2008a) New Mexico Department of Game and Fish. 2008. "2007-2008 Pronghorn Antelope Hunter Harvest Report." New Mexico Dept. of Game and Fish, Santa Fe, NM.

Available at http://www.wildlife.state.nm.us/recreation/hunting/harvest/documents/07-08PronghornAntelopeHarvestReportv2.pdf. Accessed August 17, 2010. ADAMS Accession No. ML103090074.

(NMGF, 2008b) New Mexico Department of Game and Fish. 2008. "New Mexico's Quail: Biology, Distribution, and Management Recommendations." New Mexico Dept. of Game and Fish, Santa Fe, NM. Available on line at http://www.wildlife.state.nm.us/publications/documents/ QuailHabitat.pdf. Accessed August 17, 2010. ADAMS Accession No. ML103090080.

(NMGF, 2009) New Mexico Department of Game and Fish. 2009. "New Mexico Pronghorn Management." Presentation to the New Mexico State Game Commission. Available at http://www.wildlife.state.nm.us/commission/presentations/documents/PronghornManagement.p df. Accessed August 17, 2010. ADAMS Accession No. ML103090060

(NMGF, 2010a) New Mexico Department of Game and Fish. 2010. New Mexico Deer Harvest Survey Report for the 2009-2010 Season. Available at http://www.wildlife.state.nm.us/ recreation/hunting/harvest/documents/2009-2010_ DeerHarvestReport.pdf. Accessed October 14, 2010. ADAMS Accession No. ML103090088.

(NMGF, 2010b) New Mexico Department of Game and Fish. 2010. Biota Information System of New Mexico Database Query for Lea County. Available at http://www.bison-m.org/reports.aspx?rtype=13&category=&county='025',&gapveg=&habitat=&landuse=&otherdist=&statenm=&status='101','102','105','106','107','112','201','202','215'. Accessed August 17, 2010. ADAMS Accession No. ML103090102.

(NMGF, 2010c) New Mexico Department of Game and Fish. 2010. "Biota Information System of New Mexico." Available at http://www.bison-m.org/speciesbooklet.aspx. Accessed October 14, 2010. ADAMS Accession No. ML103090099.

(NMGF, Undated 1) New Mexico Department of Game and Fish. Undated. Green Meadows Lake NM State Wildlife Area. Available at http://www.wildlife.state.nm.us/conservation/ wildlife_management_areas/documents/GreenMeadowLk.pdf. Accessed August 9, 2010. ADAMS Accession No. ML103090353.

(NMGF, Undated 2) New Mexico Department of Game and Fish. Undated. Prairie Chicken Wildlife Areas. Available at http://www.wildlife.state.nm.us/recreation/gain/documents/ maps/PCAs.pdf. Accessed August 13, 2010. ADAMS Accession No. ML103090360.

(NMHPD, 2001) New Mexico Historic Preservation Division. 2001. Office of Cultural Affairs. New Mexico Historic Preservation: A Plan for 2002-2006. Santa Fe, NM. Available at http://www.nmhistoricpreservation.org/plan/2002-2006plan.pdf. Accessed July 25, 2010. ADAMS Accession No. ML103080080.

(NMHPD 2005) New Mexico Historic Preservation Division, Department of Cultural Affairs. August 15, 2005. New Mexico Register, Volume XVI, Number 15. Title 4, Cultural Resources; Chapter 10, Cultural Properties and Historic Preservation; Part 15, Standards for Survey and Inventory. Santa Fe, NM. ADAMS Accession No. ML112710497.

(NMHPD, 2011) New Mexico Historic Preservation Division. New Mexico's Rich Cultural Heritage: Listed State and National Register Properties. Santa Fe, NM. 2011. Available at http://www.nmhistoricpreservation.org/documents/PropertiesByCounty.pdf. Accessed April 26, 2011. ADAMS Accession No. ML112710469.

(NMMA, Undated) New Mexico Museum of Art. Undated. "Growing New Mexico: History: Ranching." New Mexico Art Tells New Mexico History. Available at http://www.nmartmuseum.org/online/nmhistory/growing-new-mexico/ranching/history-ranching.html. Accessed July 8, 2011.

(NMOSE, 2008) New Mexico Office of the State Engineer. 2008. New Mexico Water Use by Categories, 2005, Technical Report 52. Available at http://www.ose.state.nm.us/PDF/ Publications/Library/TechnicalReports/TechReport-052.pdf. Accessed July 25, 2010. ADAMS Accession No. ML103080984.

(NMOSE, 2009) New Mexico Office of the State Engineer. 2009. Proposed Lea County Underground Water Basin Guidelines for Review of Water Right Applications. Adopted May 25, 2009. Accessed July 29, 2010. ADAMS Accession No. ML103080986.

(NMOSE, 2010) New Mexico Office of State Engineer. 2010. Lea County Wells with Well Logs. Available at http://nmwrrs.ose.state.nm.us/nmwrrs/index.html. Accessed July 29, 2010.

(NMOSE, 2011) New Mexico Office of the State Engineer. Underground Water Basin Map in New Mexico. 2011. Available at ttp://www.ose.state.nm.us/PDF/Maps/underground_water.pdf. Accessed June 5, 2011. ADAMS Accession No. ML112710537.

(NMRPTC, 1999) New Mexico Rare Plant Technical Council. 1999. New Mexico Rare Plants. Albuquerque, New Mexico: New Mexico Rare Plants Home Page. Available at http://nmrareplants.unm.edu (Latest update: July 22, 2010). Accessed on July 26, 2010.

(NMT, 2010) New Mexico Tech Institute of Mining and Technology. 2010. New Mexico Oil and Gas. Available at http:// mapserve.nmt.edu/Website/NMOG/viewer.htm. Accessed August 12, 2010. ADAMS Accession No. ML112710538.

(NMTRD, 2010a) New Mexico Taxation and Revenue Department. 2010. FYI-350. For Your Information. Corporate Income Tax and Corporate Franchise Tax. Rev. 7/10. Available at http://www.tax.newmexico.gov/. Accessed August 31, 2010. ADAMS Accession No. ML103430069.

(NMTRD, 2010b) New Mexico Taxation and Revenue Department. 2010. Personal Income Tax. Available at http://www.tax.newmexico.gov/All-Taxes/Pages/Personal-Income-Tax.aspx. Accessed August 31, 2010.

(NMTRD, 2011) New Mexico Taxation and Revenue Department. 2011. "Gross Receipts Tax." Available at http://www.tax.newmexico.gov/All-Taxes/Pages/Gross-Receipts-Tax.aspx. Accessed March 31, 2011.

(NMWQCC, 2002) New Mexico Water Quality Control Commission. 2002. "Water Quality and Water Pollution in New Mexico." A report prepared for submission to the Congress of the United States by the State of New Mexico pursuant to Section 305(b) of the Federal Clean Water Act. Santa Fe. NMED/SWQ-02/1. Available at http://www.nmenv.state.nm.us/ swqb/305b/2002/2002_305b_Report-Introduction.pdf. Accessed October 11, 2011.

(NPS, Undated) National Park Service, U. S. Department of the Interior. Undated. "Indian Reservations in the Continental United States," Available at http://www.nps.gov/. Accessed September 2, 2010. ADAMS Accession No. ML103430075.

(NRC, 2003) U.S. Nuclear Regulatory Commission. 2003. Environmental Review Guidance for Licensing Actions Associated with NMSS Programs. Final Report. NUREG-1748. August 2003. Division of Waste Management, Washington, D.C. ADAMS Accession No. ML ML032450279.

(NRC, 2005) U.S. Nuclear Regulatory Commission. 2005. Environmental Impact Statement for the Proposed National Enrichment Facility in Lea County, New Mexico. Final Report. NUREG 1790. June 2005. ADAMS Accession No's. ML051730238 (v1) and ML051730292 (v2).

(NRCS, 2010) U.S. Department of Agriculture, Natural Resource Conservation Service. 2010. Custom Soil Resource Report for Lea County, New Mexico. August 16, 2010. ADAMS Accession No. ML103080906.

(Ocotillo Park Golf Course, 2010) Ocotillo Park Golf Course. 2010. Ocotillo Park website, Directions. Available at http://home.valornet.com/lyle/location.htm. Accessed November 11, 2010. ADAMS Accession No. ML112710544.

(Pfeil, et al., 2001) Pfeil, J.P., A.J. Leavitt, M. E. Wilks, S. Azevedo, L. Hemenway, K. Glesener, and J.M. Baker. 2001. Mines, Mills and Quarries in New Mexico. Mining and Minerals Division, New Mexico Energy, Minerals, and Natural Resources Department and New Mexico Bureau of Geology and Mineral Resources New Mexico Institute of Mining and Technology. ADAMS Accession No. ML103080933.

(Rothman and Holder, 1998) Rothman, H. K. and D. Holder. 1998. Promise Beheld and the Limits of Place: A Historic Resource Study of Carlsbad Caverns and Guadalupe Mountains National Parks and the Surrounding Areas. Department of the Interior, National Park Service, Washington, DC. 1998. ADAMS Accession No. ML103080082.

(Sanford, et. al., 2002) Sanford, A.R., K. Lin, I. Tsai, I., and L.H. Jaksha. 2002. Earthquake Catalogs for New Mexico and Bordering Areas: 1969-1998. New Mexico Bureau of Geology and Mineral Resources, Circular 210.

(Sanford, et. al., 2006) Sanford, A., T. Mayeau, J. Schlue, R. Aster, and L. Jaksha. 2006. Earthquake Catalogs for New Mexico and Bordering Area II: 1999-2004. November 2006. In New Mexico Geology, Volume 28, Number 4.

(Scholle, 2000) Scholle, P. 2000. An Introduction and Virtual Geologic Field Trip to the Permian Reef Complex, Guadalupe and Delaware Mountains, New Mexico-West Texas. Available at http://geoinfo.nmt.edu/staff/scholle/guadalupe.html. Accessed July 27, 2010.

(SORA, 2011) SORA. 2011. Lesser Prairie-Chicken Survey on the International Isotope Flourine Products Project Site-2011. Prepared for International Isotopes, Inc., May 24, 2011. ADAMS Accession No. ML112720317.

(Stephens & Assoc., 2009) Stephens & Associates, Inc. 2009. Preliminary 40-Year Water Plan, City of Hobbs, New Mexico. October 30, 2009. Prepared for City of Hobbs, New Mexico. ADAMS Accession No. ML103080960.

(Stokes and Stokes, 1996) Stokes, D.W. and L.Q. Stokes. 1996. Stokes Field Guide to Birds. Little, Brown, and Company. Boston.

(Sublett and Peery, 2009) L. Sublett and R. Peery. 2009. 40-Year Development Plan, Lea County, New Mexico. John Shomaker & Associate, Inc. July 2, 2009. ADAMS Accession No. ML112710546.

(Tillery, 2008) Tillery, A., 2008. Current (2004 – 2007) Conditions and Changes in Groundwater Levels from Predevelopment to 2007, Southern High Plain Aquifer, Southeast New Mexico-Lea County Underground Water Basin. U.S. Geological Survey Scientific Investigations Map 3044. Available at http://pubs.usgs.gov/sim/3044. Accessed July 28, 2010. ADAMS Accession No. ML103080935.

(TPWD, 2007) Texas Parks and Wildlife Department. 2007. Migratory Flyways of North America: Central Flyway. Available at http://www.tpwd.state.tx.us/huntwild/wild/birding/migration/flyways/central/. Accessed August 16, 2010.

(USACE, 2011) U.S. Army Corps of Engineers. 2011. Approved Jurisdictional Determination Form. SPA-2011-00030-LCO. 2011. ADAMS Accession No. ML112710549.

(USCB, 2000a) U. S. Census Bureau. 2000. *Summary File 1 for New Mexico: Census 2000.* Available at http://www2.census.gov/census_2000/datasets/Summary_File_1/New_Mexico/. Accessed July 23, 2010. ADAMS Accession No. ML112710551.

(USCB, 2000b) U. S. Census Bureau. 2000. *Summary File 1 for Texas: Census 2000.* Available at http://www2.census.gov/census_2000/datasets/Summary_File_1/Texas/. Accessed July 23, 2010. ADAMS Accession No. ML112710551

(USCB, 2000c) U. S. Census Bureau. 2000. *Summary File 3 for New Mexico: Census 2000.* Available at http://www2.census.gov/census_2000/datasets/Summary_File_3/New_Mexico/. Accessed July 23, 2010. ADAMS Accession No. ML112710554.

(USCB, 2000d) U. S. Census Bureau. 2000. *Summary File 3 for Texas: Census 2000.* Available at http://www2.census.gov/census_2000/datasets/Summary_File_1/Texas/. Accessed July 23, 2010. ADAMS Accession No. ML112710554.

(USCB, 2000e) U. S. Census Bureau. 2000. P92. Poverty Status in 1999 of Households by Household Type by Age of Householder. Detailed Tables. Census 2000 Summary File 3 (SF 3) Sample Data. Available at http://factfinder.census.gov/. Accessed July 26, 2010.

(USCB, 2000f) U. S. Census Bureau. 2000. P3. Race and P4. Hispanic or Latino, and Not Hispanic or Latino by Race (Total Population). Detailed Tables. Census 2000 Summary File 1 (SF 1) 100-Percent Data. Available at http://factfinder.census.gov/. Accessed September 7, 2010.

(USCB, 2010a) U. S. Census Bureau. 2010. State and County Quick Facts, "Lea County and Eddy County New Mexico." Revised August 16. Available at http://quickfacts.census.gov/. Accessed September 21, 2010. ADAMS Accession No. ML103430121.

(USCB, 2010b) U. S. Census Bureau. 2010. American Fact Finder, "Fact Sheet, Lea County New Mexico, Eddy County New Mexico, Hobbs New Mexico, and Carlsbad New Mexico,"

Revised August 16. Available at http://factfinder.census.gov. Accessed September 21, 2010. ADAMS Accession No. ML103430104.

(USCB, 2010c) U. S. Census Bureau. 2010. Population Estimates, Vintage 2009, "Table 1. Annual Estimates of the Resident Population for Counties of New Mexico: April 1, 2000 to July 1, 2009." Available at http://www.census.gov/. Accessed August 26, 2010. ADAMS Accession No. ML 103430109.

(USCB, 2010d) U. S. Census Bureau. 2010. Population Estimates, Vintage 2009, "Table 4. Annual Estimates of the Resident Population for Incorporated Places of New Mexico: April 1, 2000 to July 1, 2009." Available at http://www.census.gov/. Accessed August 26, 2010. ADAMS Accession No. ML 103430116.

(USCB, 2010e) U. S. Census Bureau. 2010. American Fact Finder, "Fact Sheet: 2006 – 2008. American Community Survey 3-Yar Estimates." Available at http://factfinder.census.gov/servlet/ADPTable?_bm=y&-geo_id=05000US35025&-qr_no. Accessed August 29, 2010. ADAMS Accession No. ML103430131.

(USCB, 2010f) U. S. Census Bureau. 2010. State and County Quick Facts, "New Mexico and Lea County, New Mexico," Revised August 16. Available at http://quickfacts.census.gov/. Accessed September 21, 2010. ADAMS Accession No. ML103430126.

(USDA, 2004) U. S. Department of Agriculture. 2004. Assessment of Grassland Ecosystem Conditions in the Southwestern United States. Volume 1. Gen. Tech. Rep. RMRS-GTR-135-vol. 1. Fort Collins, CO. Available at http://www.fs.fed.us/rm/pubs/rmrs_gtr135_1.pdf. Accessed on July 27, 2010. ADAMS Accession No. ML103090363.

(USDA, 2007) U.S. Department of Agriculture. 2007. 2007 Census of Agriculture. Table 7: Hired Farm Labor. Available at http://www.agcensus.usda.gov/. Accessed August 30, 2010. ADAMS Accession No. ML103430132.

(USFA, 2010) U. S. Fire Administration. 2010. National Fire Department Census, New Mexico. Available at http://www.usfa.dhs.gov/. Accessed September 21, 2010. ADAMS Accession No. ML103430147.

(USFWS, 2008) United States Fish and Wildlife Service. 2008. Black-footed Ferret (Mustela nigripes) 5-year Status Review: Summary and Evaluation. South Dakota Field Office, Pierre, South Dakota. Available at http://www.fws.gov/ecos/ajax/docs/five_year_review/doc2364.pdf. Accessed October 24, 2011. ADAMS Accession No. ML112710563.

(USFWS, 2010a) United States Fish and Wildlife Service. 2010. FWS Critical Habitat for Threatened and Endangered Species Mapper. Available at http://criticalhabitat.fws.gov/crithab/. Accessed January 19, 2011. ADAMS Accession No. ML112710570.

(USFWS, 2010b) United States Fish and Wildlife Service. 2010. New Mexico Ecological Services Field Office, "Listed and Sensitive Species in Lea County." New Mexico Ecological Service Field Office. Available at http://www.fws.gov/southwest/es/NewMexico/SBC_view.cfm?spcnty=Lea. Accessed on August 9, 2010. ADAMS Accession No. ML103090370.

(USFWS, 201Cc) United States Fish and Wildlife Service. 2010. Consultation Response Letter regarding Federal Endangered, Threatened. Proposed, and Candidate Species and Species of Concern in New Mexico. New Mexico Ecological Services Field Office, Albuquerque, New Mexico. August, 10, 2010. ADAMS Accession No. ML102370576.

(USFWS, 2010d) United States Fish and Wildlife Service. 2010. Endangered and Threatened Wildlife and Plants; Endangered Status for Dunes Sagebrush Lizard, 75 FR 239, pages 77801 – 77817. December 14, 2010. ADAMS Accession No. ML112710578.

(USGS, 2005a) U.S. Geological Survey. 2005. USGS Mineral Resources On-Line Spatial Database. Available at http://tin.er.usgs.gov/mrds/find-mrds.php. Accessed August 11, 2010. ADAMS Accession No. ML103080936.

(USGS, 2005b) U.S. Geological Survey. USGS National Gap Analysis Program. Digital Animal-Habitat Models for the Southwestern United States. Version 1.0, 200. Center for Applied Spatial Ecology, New Mexico Cooperative Fish and Wildlife Research Unit, New Mexico State University. Available at http://fws-nmcfwru.nmsu.edu/swregap/habitatreview/ Review.asp. Accessed on October 15, 2010. ADAMS Accession No. ML103090472.

(USGS, 2010a) United States Geological Survey. National Water-Quality Assessment (NAWQA) Program. High Plains Regional Ground Water (HPGW) Study. Available at http://co.water.usgs.gov/nawqa/hpgw/HPGW_home.html. Accessed August 15, 2010. ADAMS Accession No. ML103080991.

(USGS, 2010b) U.S. Geological Survey. National Biological Information Infrastructure, Gap Analysis Program (GAP). February 2010. National Land Cover, Version 1. 2010. Available at http://www.nbii.gov/portal/server.pt/community/maps_and_data/1850. Accessed on October 15, 2010. ADAMS Accession No. ML103090471.

(UTPB, 2010) University of Texas of the Permian Basin. 2010. Graph and Data on the Permian Basin Oil Production and Reserves. Available at http://ceed.utpb.edu/energy-resources/petroleum-library/permian-basin-statistics/graphs-and-data-on-permian-basin-oil-production-and-reserves/. Accessed August 17, 2010. ADAMS Accession No. ML103080937.

(WRCC, 2010) Western Regional Climate Center. 2005. Climate Summary of New Mexico, Hobbs New Mexico". Available at http://wrcc.dri.edu/narratives/NEWMEXICO.htm. Accessed August 18 2010. ADAMS Accession No. ML103080182.

(Yarger, 2009) Yarger, F. 2009. Seismic Probability in Lea County, NM: A Brief Analysis, New Mexico Institute of Mining and Technology, New Mexico Center for Energy Policy. ADAMS Accession No. ML103080942.

(Zimbelman and Johnston, 2001) Zimbelman, J.R. and A.K. Johnston. 2001. Improved Topography of the Carrizozo Lava Flow: Implications for Emplacement Conditions. In Volcanology in New Mexico. New Mexico Museum of Natural History and Science. Bulletin 18. ADAMS Accession No. ML103080947.

4.0   ENVIRONMENTAL IMPACTS

This chapter describes the potential environmental impacts associated with the construction, operation, and decommissioning of the proposed IIFP facility. Section 4.1 addresses both construction and operations impacts from the proposed action. Plant decommissioning at the termination of the license is included as part of the proposed action. Cumulative impacts of the proposed action and past, present, and reasonably foreseeable future actions are presented in Section 4.2. The no-action alternative is discussed in Section 4.3.

4.1   Proposed Action

As defined in Chapter 2 of this EIS, the proposed action is to construct, operate, and decommission a chemical plant for the deconversion of commercially generated DUF_6 inventories into depleted uranium oxide and other deconversion products.

The impacts discussions are organized by the subject areas described in Chapter 3, "Description of the Affected Environment." NRC staff significance criteria, SMALL, MODERATE, and LARGE, are used throughout the analyses. These are defined as follows (NRC, 2003):

SMALL: The environmental effects would not be detectable or are so minor that they would neither destabilize, nor noticeably alter any, important attribute of the resource.

MODERATE: The environmental effects would be sufficient to noticeably alter, but not destabilize, important attributes of the resource.

LARGE: The environmental effects would be clearly noticeable and are sufficient to destabilize important attributes of the resource.

Section 4.1.1 addresses environmental impacts of construction and Section 4.1.2 addresses environmental impacts of operations.

4.1.1   Environmental Impacts of Construction

The impacts of construction on each of the major resource areas described in Chapter 3 are presented in this Section. Impacts of preconstruction activities (defined and identified in Chapter 2) are evaluated as cumulative impacts in Section 4.2.

4.1.1.1   Land Use

Impacts on land use result from commitment of the land for the proposed use and therefore its potential exclusion from other possible uses. Land use impacts occur when the presence of a project would limit possible future land uses near the proposed project. For example, land use impacts could occur if the project restricts future access to mineral resources.

The current land uses on the proposed 259-ha (640-ac) site are cattle grazing, and access to and maintenance of utility rights-of-way and monitoring wells. The proposed site is not subject to local or county zoning, land use planning, or associated review process requirements, and there are no potential conflicts of land use plans, policies, or controls (Appendix A). The conversion of 16 ha (40 ac) of the 259-ha (640-ac) site from its current land use to the proposed facility without disturbing the remaining 240 ha (600 ac) would not conflict with any existing Federal, State, local, or Tribal land uses or restrict current or planned mineral resources exploitation (Appendix A).

Construction of the proposed IIFP facility would modify the current land use by restricting cattle grazing on the entire 259-ha (640-ac) site. Currently, approximately 93 percent of land in Lea County is used as range land for grazing (approximately 1.0-million ha [2.6-million ac]). Restricting grazing on the 259-ha (640-ac) site would result in a loss of 0.02 percent of the land available for grazing. The proposed facility footprint was selected by IIFP to avoid, to the extent possible, existing utility rights-of-way so that the change in land use would not limit access to or maintenance of the rights-of-way. In addition, it is not likely that any of Xcel Energy's Cunningham Station's monitoring wells (CU-6, CU-8, and CU-9) in Section 27 (the proposed IIFP site) would be affected by the construction of the facility or the associated infrastructure such as the access road. NRC staff expects the remainder of the proposed IIFP site (Section 27) to be left in its present condition, largely undeveloped for the duration of the facility's operation and decommissioning.

Consequently, the NRC staff finds that the impacts to land use from the construction of the facility within the 259-ha (640-ac) site would be SMALL, due to the abundance of other nearby grazing land and the ability to continue to access utility rights-of-way.

4.1.1.2   Historic and Cultural Resources

This section describes the potential environmental impacts on historic and cultural resources resulting from construction of the proposed IIFP facility. As Chapter 3 states, historic and cultural resources include archaeological sites, historic structures, and places of cultural importance to groups for maintaining their heritage. Cultural resources are nonrenewable; that is, once altered, the information contained in cultural resources cannot be recovered. NRC staff identified separate Areas of Potential Effect (APEs) for archaeology and architecture; and for construction versus operations. For construction impacts, the APE for archaeology is considered to include the 16-ha (40-ac) IIFP facility, approach road, and parking lot where disturbance would occur. For construction impacts, the APE for historic architectural resources, would include the IIFP facility (limits of disturbance), and any areas where noise or construction activity would be heard or visible.

NRC staff identified no historic properties, districts, resources or significant culturally important or archaeological sites within the cultural resources APE during files research at the New Mexico State Historic Preservation Office (NM SHPO), and tribal consultation. The field survey conducted by IIFP's archaeological consultant also identified no significant cultural resources although it found three isolated artifact occurrences – a brown chert San Jose projectile point fragment, a gray quartzite hammerstone, and three glass vessel fragments. The artifacts were recorded with the SHPO, but are not National Register-eligible as determined by the archaeological consultant for the State Land Office.

As discussed in Section 3.3.4, on October 14, 2010, the New Mexico Trust Land Archaeologist in the office of the State of New Mexico Commissioner of Public Lands recommended a finding of no effect to historic properties, districts, resources or significant historic/precontact archaeological sites because the cultural resource survey found no significant cultural resources in the proposed IIFP site, and no historic properties have been identified within the APE. The NM SHPO concurred with this determination on October 25, 2010 (Appendix B).

NRC staff also consulted with Native American Tribes which were identified from a list maintained by the NM SHPO, consistent with the NHPA Section 106. The NRC staff received three responses (Appendix B). On July 13, 2010, the Ysleta del Sur Pueblo stated that the Pueblo believes the project will not adversely affect traditional, religious, or culturally significant sites, but requested consultation should human remains or artifacts regulated by the *Native American Graves Protection and Repatriation Act* be discovered. On June 15, 2011, the Tribal Historic Preservation Officer for the Comanche Tribe noted that he had no comments on the project. On July 13, 2011, the Shawnee Tribe's Tribal Historic Preservation Department concurred that no known historic properties would be impacted by the project, but also requested consultation if archaeological materials are encountered during construction or operation of the facility.

Based on the history of the region, its lack of permanent surface water, and the results of previous investigations, no significant archaeological, cultural, or historic resources are present on the site. Therefore, the NRC staff finds that impacts to historic properties, districts, resources or significant historic/precontact archaeological sites during construction would be SMALL.

4.1.1.3   Visual Resources

As discussed in Section 3.4, the scenic value of the proposed IIFP site is low ("least valued") and is further diminished by the presence, within 5 km (3 mi), of four industrial facilities, three of which are visible from the proposed site. Consequently, the NRC staff concludes that the impacts to visual resources during construction would be SMALL.

4.1.1.4   Climatology, Meteorology, and Air Quality

4.1.1.4.1   Greenhouse Gases

This section presents an assessment of the effect of construction of the proposed IIFP facility on the concentration of CO_2 and other greenhouse gases in the atmosphere.

During construction, air emissions would come from (1) construction vehicles and equipment, (2) personal vehicles of the construction workforce, (3) delivery vehicles bringing materials and equipment to the proposed site, and (4) vehicles transporting construction-related wastes from the proposed site to area landfills and treatment/disposal facilities.

NRC staff used NONROAD (EPA, 2005) and MOVES (EPA, 2009a) computer models to calculate the estimated emissions for construction vehicles and equipment at 1,320 metric tons (1,455 tons) of CO_2 equivalent emissions (Table 4-1). During construction, an estimated 140 workers would commute to and from the proposed site with an assumed average (round-trip) distance of 64 km (40 mi) daily for 250 days per year (IIFP, 2011b). This is based on the assumption that the workforce would live in the nearby population centers of Hobbs, Lovington, and Carlsbad, New Mexico. A weighted average of the distance to each population center and the 2000 population was used to determine average daily trip distance (IIFP, 2011a). Over the 12-month construction period, the workforce commuting distance would be 2,253,082 km (1,400,000 mi). It was also estimated that over the course of the construction period, there would be 20 truck deliveries each day each traveling 64 km (40 mi). The EPA MOVES (EPA, 2009a) was used to calculate estimated CO_2 equivalents from all anticipated construction traffic as 790 metric tons (871 tons) (Table 4-2).

Finally, onsite storage and dispensing of fuels during construction would result in minor greenhouse gas emissions; however, because neither the specific volume nor the chemical speciation's of these evaporative losses are known, resulting greenhouse gas emissions have not been estimated.

Total CO_2 emissions expected during the 12-month construction period are 2,110 metric tons (2,326 tons); 1,320 metric tons (1,455 tons) from construction equipment and 790 metric tons (871 tons) from workforce commuting and deliveries. Using calendar year 2000 as a reference point (the latest year for which New Mexico greenhouse gas emission data are available), and as shown in Table 3-1, total net CO_2 emissions for New Mexico for the year 2000 were 62 million metric tons (68 million tons) of CO_2 equivalents. For the United States for that same year, total net CO_2 emissions were 5,977 million metric tons (6,588 million tons) (EPA, 2010a). By comparison, during the 12 months of construction, the proposed IIFP facility CO_2 emissions are projected to be 2,110 metric tons (2,326 tons) or 0.003 percent of New Mexico's statewide output and 0.00004 percent of the projected nationwide CO_2 emissions for the same period. Therefore, the NRC staff concludes that impacts from the construction of the proposed IIFP facility from the emissions of CO_2 and other greenhouse gases would be SMALL.

Table 4-1. CO_2 Emissions from Construction Equipment

Activity	Annual CO_2 Emissions	
	(ton)	(MT)
Construction Equipment	1,455	1,320

MT = metric ton

Table 4-2.   Emissions from Workforce Commuting and Delivery Activities During Construction

Activity	Total Distances for 12 months		Annual CO_2 Emissions	
	(mi)	(km)	(ton)	(MT)
Commuting traffic (workforce)	1,400,000	2,300,000	660	599
Delivery Truck traffic	200,000	320,000	211	191
Total for workforce commuting and deliveries	1,600,000	2,600,000	871	790

MT = metric ton

4.1.1.4.2   Air Quality

Air quality impacts from the operation of construction equipment during facility construction were evaluated based on the construction schedules and parameters provided by IIFP.

Impacts to ambient air resources would occur from the construction of the proposed IIFP facility. As discussed in more detail below, the impacts would not be significant. Because the proposed IIFP site is located in an air quality control region that is designated as attainment with Federal and state ambient air quality standards, a General Conformity evaluation is not required.

Construction of the proposed project would produce criteria pollutants (i.e., CO, NO_2, PM_{10}, $PM_{2.5}$, SO_2, lead, and volatile organic compounds (VOCs), an ozone precursor) and hazardous air pollutants (HAPs) emissions from construction equipment, delivery vehicles, commuter vehicles, and onsite refueling activities. Particulate emissions in the form of fugitive dust from soil transfers, land grading, and vehicle and equipment travel on unpaved roads would also be generated.

The NRC staff used emission factors to estimate annual emissions from the construction of the proposed IIFP facility. Emission factors for highway vehicles (i.e., worker commute vehicles and delivery vehicles) were determined using the EPA MOVES Model (EPA, 2009a). Emission factors for non-road vehicles (including construction engines and equipment) were determined using the EPA NONROAD model (EPA, 2005). Emission factors for all other sources associated with construction were obtained from the EPA document AP-42, "Compilation of Air Pollutant Emission Factors" (EPA, 1995a).

Emissions from each type of non-road equipment are a function of equipment-specific factors, including engine horsepower, load factor, and hours of operation. IIFP estimated that the construction project would be completed in approximately 12 months based on the assumption that construction would occur 10-hrs per day, 5 days per week, and would involve conventional construction equipment (e.g., dozers, graders, excavators, dump trucks, lifts). IIFP compiled a list of construction equipment that identifies for each month of construction, the quantity and average monthly hours of operation for each piece of equipment (See Table 4-3).

Emissions from highway vehicles (i.e., worker commute vehicles and delivery vehicles) are a function of the vehicle-specific factors, including type of vehicle and age, fuel type, and vehicle miles traveled. The NRC staff used Lea County, New Mexico-specific default values for vehicle type, age, and fuel type to determine highway vehicle emission factors. IIFP estimated that over the 12-month construction period, 140 workers would commute to and from the proposed site an average daily trip distance of 64 km (40 mi) for 250 days each year; and delivery trucks would make 20 delivery trips per day (at an average round trip distance of 64 km [40 mi]) to transport materials and equipment and remove wastes (IIFP, 2011a).

Fugitive dust generated during land clearing and soil transfer operations is dependent on a number of factors including silt and moisture content of the soil, wind speed, and disturbed area. To estimate fugitive dust emissions, IIFP used the EPA emission factor of 1.2 tons/acre/month of activity (EPA, 1995a). This emission factor represents total suspended particulates (i.e., particles less than 30 microns in diameter). IIFP assumed that the emission factor would drop to 0.3 tons/acre/month after the first month of construction, when the majority of earth moving activities would be complete. Multiplication factors of 0.15 and 0.075 were used to adjust the emission factor for PM_{10} and $PM_{2.5}$, respectively. IIFP also assumed that water would be applied to disturbed areas. This would reduce fugitive dust emissions by about 50 percent (IIFP, 2011a).

Small quantities of VOCs and HAPs emissions would be released from the refueling and onsite maintenance of construction equipment. Diesel fuel would be stored on site during construction and would be hand pumped into construction equipment and support vehicles. Annual VOCs and HAPs fugitive emissions are a function of diesel fuel consumption.

Table 4-3. Construction Equipment and Hours of Operation by Type and Month

Equipment	Max HP	Month 1 No.	Month 1 Hr	Month 2 No.	Month 2 hr	Month 3 No.	Month 3 hr	Month 4 No.	Month 4 hr	Month 5 No.	Month 5 hr	Month 6 No.	Month 6 hr	Month 7 No.	Month 7 hr	Month 8 No.	Month 8 hr	Month 9 No.	Month 9 hr	Month 10 No.	Month 10 hr	Month 11 No.	Month 11 hr	Month 12 No.	Month 12 hr
Tractor/backhoe	150	2	160	1	80	1	80	1	80	1	80	1	80	1	80	1	80	1	80	1	80	1	80	1	80
Grader	400	1	160	0	0	0	0	0	0	0	0	0	0	0	0	0	0	0	0	0	0	0	0	0	0
Excavator	500	1	160	0	0	0	0	0	0	0	0	0	0	0	0	0	0	0	0	0	0	0	0	0	0
Dump Truck	300	2	160	2	160	2	160	1	100	1	100	1	100	1	100	1	100	1	100	1	100	1	100	1	100
Dozer	400	1	160	0	0	0	0	0	0	0	0	1	160	1	160	0	0	0	0	0	0	0	0	0	0
Air Compressor	325	1	80	2	160	2	160	2	160	2	160	2	160	2	160	2	160	2	160	2	160	2	160	2	160
Concrete Pump	125	0	0	1	160	1	160	1	160	1	160	1	160	1	160	1	160	1	160	1	160	1	160	1	160
Crane	175	0	0	1	160	1	160	1	160	1	160	1	160	1	160	1	160	1	160	1	160	1	160	1	160
Fuel Truck	250	1	50	1	50	1	50	1	50	1	50	1	50	1	50	1	50	1	50	1	50	1	50	1	50
Water Truck	250	1	50	1	50	1	50	1	50	1	50	1	50	1	50	1	50	1	50	1	50	1	50	1	50
Forklifts	200	1	50	2	160	2	160	2	160	2	160	2	160	2	160	2	160	2	160	2	160	2	160	2	160
Flatbed, 2-ton	200	1	80	1	80	1	80	2	160	2	160	2	160	2	160	2	160	2	160	2	160	2	160	2	160
Generator	33	0	0	1	80	1	80	2	160	2	160	2	160	2	160	2	160	2	160	2	160	2	160	2	160
Welder	50	1	30	2	60	2	60	5	160	5	160	5	160	5	160	3	160	3	160	3	160	3	160	3	160

Source: IIFP, 2011a
No. = number

The estimated maximum annual pollutant emissions from construction activities are presented in Table 4-4. Actual construction emissions are expected to be less, because conservative assumptions used in the modeling of construction activities tend to overestimate impacts. The applicant's environmental report described Best Management Practices (BMPs) they would use to reduce impacts to various resources at the proposed IIFP site, including those to minimize the impacts of construction activities on air quality. These BMPs are described in Chapter 5, Mitigation Measures and Commitments.

Table 4-4.   Estimated Maximum Annual Emissions from Construction of the IIFP Facility

Source	CO metric tons (tons)	NO$_2$ metric tons (tons)	PM$_{2.5}$ metric tons (tons)	PM$_{10}$ metric tons (tons)	SO$_2$ metric tons (tons)	VOC metric tons (tons)	HAP metric tons (tons)
Construction Equipment	4.00	8.55	0.639	0.660	0.268	0.778	-----
	(4.41)	(9.43)	(0.705)	(0.727)	(0.295)	(0.858)	-----
Delivery Vehicles	2.05	0.926	0.0394	0.0476	0.00224	0.174	0.174
	(2.26)	(1.02)	(0.0434)	(0.0524)	(0.00247)	(0.192)	(0.192)
Personal Vehicles	10.9	1.46	0.0266	0.0494	0.0116	1.03	1.03
	(12.1)	(1.61)	(0.0293	(0.0545)	(0.0128)	(1.14)	(1.14)
Fugitive Emissions	-----	-----	3.67	7.35	-----	0.00161	0.0016
	-----	-----	(4.05)	(8.10)	-----	(0.00177)	(0.00177)
Total	17.7	10.9	4.38	8.11	0.282	1.99	1.2
	18.7	(12.1)	(4.83)	(8.93)	0.31	(2.19)	(1.33)

Source: See Appendix C
CO = carbon monoxide
NO$_2$ = nitrogen dioxide
PM$_{2.5}$ = particulate matter less than 2.5 microns in diameter
PM$_{10}$ = particulate matter less than 10 microns in diameter
SO$_2$ – sulfur dioxide
VOC = volatile organic compounds
HAP = hazardous air pollutants

To estimate the impact to local air quality, the NRC staff compared the anticipated criteria pollutant and VOCs emissions from the proposed IIFP construction activities to baseline emissions from Lea County, New Mexico. As shown in Table 4-5, emissions from the proposed construction activities at the IIFP site would represent a very small portion of the annual criteria pollutant emissions in Lea County.

NRC staff used EPA's SCREEN3 model (EPA 1995b) to estimate the maximum concentrations of pollutants at the proposed IIFP site property line that would be associated with construction activities. As shown in Table 4-6, the estimated incremental increases in ambient background concentrations due to the proposed construction activities would be above the National NAAQS for NO$_2$, PM$_{2.5}$, and PM$_{10}$ emissions. HAPs and VOCs are not included in Table 4-6 because there are no regulatory metrics for comparison (HAPs and VOCs emissions are regulated by source controls and permit requirements).

Table 4-5.   Comparison of Maximum Annual Emissions from Construction of the IIFP Facility to Lea County Baseline Conditions

Source	CO metric tons (tons)	NO$_2$ metric tons (tons)	PM$_{2.5}$ metric tons (tons)	PM$_{10}$ metric tons (tons)	SO$_2$ metric tons (tons)	VOC metric tons (tons)
Site Construction	17.0 (18.70)	10.90 (12.10)	4.38 (4.83)	8.11 (8.93)	0.28 (0.31)	1.99 (2.19)
Lea County Baseline (a)	21,244 (23,417)	27,119 (29,894)	2,892 (3,188)	25,048 (27,611)	7,334 (8,084)	4,436 (4,890)
Net Increase over Baseline	0.080%	0.040%	0.151%	0.032%	0.004%	0.045%

Source: EPA, 2009b
Source: See Appendix C
CO = carbon monoxide
NO$_2$ = nitrogen dioxide
PM$_{2.5}$ = particulate matter less than 2.5 microns in diameter
PM$_{10}$ = particulate matter less than 10 microns in diameter
SO$_2$ – sulfur dioxide
VOC = volatile organic compounds
HAP = hazardous air pollutants

Table 4-6.   Comparison of Predicted Maximum Downwind Concentrations Due to Construction Activities to NAAQS

Pollutant	Averaging Time	NAAQS µg/m^3	Incremental Concentration Increase µg/m^3	Incremental Concentration Increase as Percentage of NAAQS
CO	1-hr	10,000	116	1.2%
	8-hr	40,000	81.3	0.20%
NO$_2$	1-hr	100	269	269%
	Annual	188	26.9	14%
PM$_{2.5}$	24-hr	35	142	406%
	Annual	15	35.5	237%
PM$_{10}$	24-hr	150	277	185%
SO$_2$	1-hr	200	8.5	4.3%
	3-hr	1,300	7.7	0.6%

Source: See Appendix C

As discussed above, the estimated emissions during construction of the proposed IIFP facility represent a very small fraction of the current emissions in Lea County. Because conservative assumptions that tend to overestimate impacts were used to produce these estimates, actual emissions from the construction activities are expected to be lower. Pollutant emissions from construction activities have the potential to change the existing ambient air quality in the vicinity of the proposed IIFP facility. Overall, the construction impacts would be localized and short-term. Therefore, the NRC staff concludes that the air quality impacts resulting from the construction of the proposed IIFP facility would be MODERATE for NO$_2$, PM$_{2.5}$, and PM$_{10}$ emissions and SMALL for CO, and SO$_2$ emissions. BMPs identified by the applicant to use during construction to reduce impacts to air quality are described in Chapter 5, Mitigation

Measures and Commitments. The NRC staff finds that the BMPs committed to by IIFP for the proposed facility would be sufficient to maintain impacts as MODERATE to SMALL.

4.1.1.5 Geology, Minerals, and Soil

Section 3.6 describes the geology, minerals, and soils at the proposed IIFP site. Alluvium and/or caliche would be removed during site preparation. If the materials are of the appropriate quality they would be used for roads or facility foundations at the site. The existing caliche pit described in Section 3.6.3 could be a source of caliche for site roads. Impact to topography and bedrock would be limited to clearing and excavating areas for facility and road construction. These impacts would be largely limited to the 16-ha (40-ac) facility, with the exception of the roadway construction that would extend beyond the facility, through the 259-ha (640-ac) site, to NM 483.

The proposed IIFP site has been explored for oil and gas and caliche. The site has very limited leasable, locatable, or marketable mineral resources. Therefore, the proposed IIFP construction activities would not result in loss of mineral resources.

As discussed in Section 3.6.2, the site is in an area of limited seismic and volcanic activity (Crone and Wheeler, 2000; Machette et al. 2000; Yarger, 2009), and the statistical probability of risk of earthquake damage is very low. Because excavation depth is limited to near-surface geology, construction activities are not expected to cause seismic or fault-related impacts.

Soil erosion due to stormwater runoff and wind erosion must be managed during construction. As described in Section 3.6.4, the proposed site does not contain any prime farmland; therefore, prime farmland would not be impacted. During construction, BMPs would be employed to limit soil loss and mitigate any impacts. These would include:

- Soil stabilization (e.g. temporary and permanent seeding),

- Structural controls (e.g. hay bales and sediment fences), and

- Management practices (e.g. construction sequencing, materials delivery sequencing, and physical delineation of disturbed areas).

Construction excavation would be limited to near-surface geology at the 16-ha (40-ac) facility, and would result in minimal loss of mineral resources. BMPs identified in the applicant's environmental report would be used to limit soil loss. Therefore, the NRC staff finds that the BMPs committed to by IIFP would be sufficient to ensure that construction impacts of the proposed IIFP facility to geology, minerals, and soil would be SMALL.

4.1.1.6 Water Resources

This section discusses the potential impacts of construction of the proposed IIFP facility on water resources.

4.1.1.6.1 Groundwater

As discussed in Section 3.7.1.2, the site is within the Lea County underground water basin (UWB). The Ogallala aquifer is the primary water supply source in the Lea County UWB, and is proposed as the water supply source for the proposed IIFP project. The aquifer is encountered at a depth of approximately 9.1 m (30 ft) beneath the proposed site (GL Environmental, 2010).

Excavation through the dense caliche at the site could be required for sewer systems, roads, pads, and building foundations. Excavation depths are not expected to exceed the depth to groundwater (GL Environmental, 2010; IIFP, 2009a).

4.1.1.6.1.1   Groundwater Use

As discussed in Section 3.7.1.2, all the groundwater wells in the site vicinity obtain water from the Ogallala aquifer. The closest domestic groundwater supply well is (L-07469) approximately 1.6 km (1 mi) northwest of the proposed site (IIFP, 2011a). The well depth is 48.8 m (160 ft) and the depth to water is 21.3 m (70 ft) (NMOSE, 2010).

No municipal water line runs near the proposed site. The construction activities for the facility would require relatively low volumes of water. Up to two groundwater production wells would be installed in the Ogallala aquifer to supply all the water for the proposed facility. One of these production wells would be installed on the site by Lea County prior to the property transfer. This well would be designed to meet the production needs for both proposed Phase 1 and 2 operations. This well would be drilled to a depth estimated to be between 61 to 76.2 m (200 to 250 ft) below ground surface with an estimated pumping capacity of 1,325 L/min (350 gpm). A second production well may be needed during IIFP operations for emergency preparedness purposes. This well, if required, would be installed by IIFP during construction. If the second production well is installed, it would be located to minimize interference with the first production well (INIS, 2011). All new monitoring and production wells installed at the site would be installed in accordance with the New Mexico Office of the State Engineer and Lea County Water Users Association regulations.

IIFP states that during construction, water use is projected to equal 3,600 L/day (960 gpd) on average, and 12,500 L/day (3,300 gpd) maximum (IIFP, 2011a). Groundwater may be used during construction for personal use, construction activities, and dust suppression. Groundwater pumped from the site well would be a consumptive loss because the groundwater would either be consumed or evaporated. For more information on groundwater appropriations and use restrictions refer to Section 4.1.2.6.1.

As discussed in Section 3.7.1.2.1, because current pumping is in excess of the Ogallala's recharge rate, the aquifer has experienced significant drawdown in the last several decades. In 2009, the New Mexico Office of the State Engineer (NM OSE) prepared guidelines on the procedures for processing water rights applications filed with the Lea County UWB. The NM OSE developed administrative guidelines in order to promote the orderly development of water resources in the Lea County UWB, while meeting statutory obligations regarding existing water rights, availability of unappropriated water, conservation of water with the State, and public welfare of the State through a 40-year planning period beginning on January 1, 2005 and ending on January 1, 2045.

On March 10, 2009, the NM OSE issued an order closing the Lea County UWB to the filing of groundwater applications (NMOSE, 2009a). In 1999, in order to meet the projected groundwater demands of Lea County, 138 applications were filed by the Lea County Water Users Association to appropriate 6,389 ha-m (51,797 ac-ft) per year of groundwater, which was essentially all the unappropriated groundwater in the Lea County UWB (Leadshill-Herkenhoff et al., 2000).

A portion of the applications to appropriate groundwater, totaling 4,215.2 ha-m (34,173 ac-ft) per year, have been assigned to Lea County during (at least) the 40-year planning period to allow the County to hold the subject water rights unused until the rights can be put to beneficial use at projects currently under construction, select future projects, and homes and businesses in unincorporated areas. The proposed IIFP site has been included in Lea County's 40-Year Water Development Plan as an area of proposed development where up to 21.6 ha-m (175 ac-ft) per year of the Lea County's water rights would be put to beneficial use. The estimated quantity of additional groundwater that Lea County would need by the end of the 40-yr planning period for all projects that currently exist, are being constructed, or have a high likelihood of being constructed in the near future is 1,173.5 ha-m (9,514 ac-ft) per year (Leadshill-Herkenhoff et al., 2000).

Groundwater rights for the facility would be obtained from the Lea County unallocated water rights under State Engineer Office Water Right File No. L-04719-A for industrial water use. The proposed facility would obtain a joint Lea County and New Mexico Office of State Engineer (NM OSE) Water Rights Agreement for 6.2 ha-m (50 ac-ft) per year (far less than the 21.6 ha-m [175 ac-ft] per year estimation in the Lea County Water Development Plan). The site would use approximately 0.5 percent of the estimated annual 40-year planning period groundwater demand of 9,514 ac-ft per year for Lea County, and only 0.15 percent of the 4,215.2 ha-m (34,173 ac-ft) per year of unappropriated water rights that have been assigned to Lea County.

The proposed site production well(s) could be the sole source of water during construction; or the applicant could use water from tanker trucks. Groundwater from onsite wells, or tankers filled from the Hobbs city water system, would be required during Phase 1 construction, mostly for dust suppression, fill compaction, and concrete formation. Unlike preconstruction, where water is assumed to be brought in on tanker trucks, IIFP may have operating wells at some point during construction. This water is not anticipated to be recycled and reused like plant process water during operations (discussed in Section 4.1.2.6.1), however the volumes required during construction would be less than a third of the volumes required during operations. Regardless if water comes from onsite wells or from the Hobbs municipal system via tanker trucks, the water use is accounted for in, and consistent with, Lea County's water use plan. Therefore, the NRC staff finds that groundwater use impacts during construction would be SMALL.

4.1.1.6.1.2 Groundwater Quality

The Ogallala aquifer beneath the site is unconfined, and is recharged by natural precipitation that percolates to the groundwater table. As a consequence, any contaminants (e.g., diesel fuel, hydraulic fluid, antifreeze, or lubricants) spilled during construction and not controlled by spill control measures could affect the unconfined aquifer, although migration of contaminants would be slowed by the approximately 9.1-m (30-ft) thick indurated caliche layer that overlies the aquifer at the site (GL Environmental, 2010).

Any spills of diesel fuel, hydraulic fluid, antifreeze, or lubricants during construction would be cleaned up quickly in accordance with the proposed facility's Spill Prevention Control and Countermeasure Plan, to prevent the contaminant from entering the groundwater.

All construction activities would comply with the site's National Pollutant Discharge Elimination System (NPDES) General Permit to discharge stormwater associated with construction and the associated stormwater pollution prevention plan. Surface water flow from rain events would be directed to the site's proposed catch basins.

During construction activities, portable sanitary facilities would be used until a permanent sanitary waste treatment facility is functional. The waste collected from these temporary facilities would be disposed of offsite by a licensed sanitary waste disposal contractor. No process waste effluent will be generated during this construction phase.

Therefore, the NRC staff finds that impacts to groundwater quality due to construction-related activities would be SMALL.

4.1.1.6.2   Surface Water

As discussed in Section 3.7.2, no permanent surface water, including jurisdictional waters, is present on the proposed IIFP site. Therefore, the NRC staff finds that impacts to surface water quality or quantity due to construction-related activities would be SMALL.

4.1.1.7   Ecological Resources

This section discusses the potential impacts of construction of the proposed IIFP facility on ecological resources. Most of the potential ecological disturbances due to habitat loss from land clearing of the proposed IIFP site would occur during preconstruction, and thus is evaluated as a cumulative impact. Other potential ecological disturbances could include: noise and vibrations from heavy equipment and traffic, fugitive dust, and the presence of construction personnel. Approximately 16 ha (40 ac) of land would be affected by construction, which represents less than 10 percent of the total 259 ha (640 ac) site. Leaving a majority of the site undisturbed would allow mobile resident wildlife within the disturbed areas an opportunity to relocate to the undisturbed areas. These undisturbed areas are expected to be left undisturbed for the life of the proposed facility. Some wildlife may suffer stresses or mortality during construction. Human encounters with some wildlife could increase due to loss of habitat. No construction facilities or equipment would be 61 m (200 ft) high or taller, so no aviation safety lights will be required (FAA, 1992). Security lighting will be directed downward, and construction will not require night shifts so night-migrating and nocturnal animals would not be affected.

As described in Section 3.7.2, there are no wetlands or permanent stream systems, and therefore no riparian habitat within the facility footprint.

As described in Section 3.8, no Federal or state-listed threatened or endangered species are likely to be found on the site as their preferred habitats are not found on the site. No unique or critical habitats occur on the site. No threatened or endangered species are known from the site or in the vicinity. No commercially or recreationally important species use the habitat at the site exclusively.

Maintenance practices such as the use of chemical herbicides, roadway maintenance, and clearing practices could be employed during construction. Land clearing would destroy the Western Great Plains Shortgrass Prairie and Apacherian-Chihuahuan Mesquite Upland Scrub vegetation communities within the 16 ha (40 ac) facility footprint. However, neither of these vegetation communities provides unique habitat in the area and the impacted area of 16 ha (40 ac) constitutes a small fraction of these vegetation communities in the vicinity of the proposed IIFP site (see Figure 3-19). Therefore, the NRC staff finds that the loss of 16 ha (40 acres) of either habitat type would have a SMALL impact on native vegetation in the vicinity of proposed action.

During construction, the presence of humans, the presence of construction equipment and associated noise and vibrations, and general construction activities could result in animals currently using the property, such as birds, foxes and other small mammals, to avoid the construction area. Many other species, such as rodents, and some reptiles, are small, have limited mobility, occur in habitats that provide concealment, or spend at least a portion of their lives underground. During site clearing and grading activities, it is likely that some individuals of these species will not survive the construction activities. Rodents and larger mammals and reptiles may be killed along access roads by vehicles moving to and from the site. There are many square miles of undeveloped land surrounding the IIFP site (i.e., Section 27) which have native vegetation and habitats suitable for native species. As discussed in Section 3.8.3 the species of wildlife present or that could be present are typical of those found in the habitat in the surrounding area. Because the area surrounding the proposed IIFP site is largely undeveloped (see Section 3.2), there is sufficient suitable habitat in the vicinity of the project to support displaced animals. Therefore the NRC staff finds that impacts to ecological resources from construction of the proposed facility would be SMALL.

4.1.1.8   Socioeconomic Resources and Environmental Justice

This section analyzes the potential social and economic (socioeconomic) impacts associated with construction of the proposed IIFP facility. As discussed in Section 3.9, most socioeconomic impacts would occur within a two-county area (Lea and Eddy Counties, New Mexico), where the majority of the construction workforce would live and spend their wages. These two counties comprise the socioeconomic Region of Interest (ROI) in these analyses which are based on IIFP's estimate of a peak of 140 construction workers (IIFP, 2011a). Construction would begin in 2012 and be completed in 2013. Wage and salary spending would have a small positive economic benefit in the ROI, and expenditures associated with materials, equipment, and supplies would produce local and State tax revenue. The migration of workers and their families into the area would slightly increase the population of the ROI and affect housing availability and community services such as education, fire protection, law enforcement, medical resources, and the availability of public utilities.

The major factor influencing socioeconomic impacts of construction is the number of construction workers who would relocate to the area with their families. An NRC staff study, Migration and Residential Location of Workers at Nuclear Power Plant Construction Sites (BMI, 1981) evaluated behaviors and characteristics of construction workers at nuclear power plant construction sites. It provides a methodology for estimating in-migrating workforce sizes and residential distribution patterns at nuclear power plant construction sites. There is no evidence that the fundamental nuclear construction workforce characteristics and behaviors have changed appreciably since the study's publication. The proposed IIFP facility is a nuclear fuel cycle facility, and, as such would require construction methods, quality control, and safety procedures similar to those used for constructing nuclear power plants. Therefore, the current analysis assumed that the construction behaviors and characteristics identified in the BMI study would be a fair representation of the expected IIFP construction workforce, and therefore the worker migration patterns and family characteristics described in the BMI study remain valid assumptions. The BMI study indicates the following construction worker characteristics:

- Between 15-35 percent of the construction workforce would migrate into the ROI.

- Approximately 70 percent of in-migrating construction workers are likely to bring families (this may be an overestimate for a construction job with a duration of one year, however, to be conservative, the NRC staff maintained this assumption).

- Average family size of a construction worker is 3.25 persons.

- Average number of school-aged children per construction worker family is 0.8.

- Approximately 50 percent of the in-migrating construction workforce will remain in the ROI following construction.

IIFP anticipates the peak number of construction workers would be 140. In 2008, construction and mining employment provided more than 28 percent of all non-farm employment in the ROI. They are two of the largest employment sectors in the ROI. Because of the presence of workers with construction experience, the NRC staff estimates that 80 percent of the total construction workforce would already reside within the ROI and 20 percent would migrate into the ROI. An estimate of 20 percent in-migrants is within the range of the in-migrating construction workforce identified in the BMI study when there is already an existing, viable construction workforce within the ROI. Table 4-7 depicts workforce in-migration, family, and workforce retention characteristics based on the BMI study and the maximum workforce for the construction of the proposed IIFP facility. These projections are used throughout this analysis.

4.1.1.8.1   Population

As presented in Section 3.9.1.2, the population within the ROI was 112,938 persons in 2009. Construction of the proposed IIFP facility would directly employ a maximum of 140 people, of which 80 percent would be ROI residents. The other 20 percent (28 workers) would migrate into the socioeconomic ROI (Table 4-7). Of the 28 employees that would migrate into the ROI, 70 percent (20 workers) would bring their families (Table 4-7). Eight construction workers would not bring families. Using an average family size of 3.25 from the BMI study, the total construction workforce in-migration would result in 72 new residents (Table 4-7) in the two-county ROI. An increase of 72 residents would result in a 0.06 percent increase in the ROI population.

4.1.1.8.2   Employment and Income

Workers already residing in the ROI would fill 80 percent of the construction jobs or 112 jobs (see Table 4-7). These workers represent 0.2 percent of the June 2010 labor force within the ROI. If all 112 of the jobs were filled by unemployed workers, the unemployment rate in the ROI would decrease by 0.2 percent. The remaining 28 jobs would be filled by workers migrating into the ROI (Table 4-7). The in-migrating workers would increase the labor force by 0.05 percent.

The U.S. Department of Commerce Bureau of Economic Analysis (BEA), Economic and Statistics Division uses an economic model called RIMS II, which incorporates buying and selling linkages among regional industries and uses a multiplier specific to an industry to estimate the economic impact within the region. The multiplier is the number of times the final increase in consumption exceeds the initial dollar spent. In this analysis, the NRC staff uses the multiplier for the construction industry in the ROI to estimate the number of indirect jobs that would result from the in-migration associated with the construction of the proposed IIFP facility. Indirect jobs are often non-technical and non-professional positions in the retail and service sectors. The 12 indirect jobs that would be created (Table 4-8) would likely be filled by ROI residents. If all 12 jobs were filled by unemployed workers, those workers would represent 0.3 percent of the unemployed labor force in June 2010.

Table 4-7. Assumptions for Workforce Characterization during Peak Construction Period

Workforce characterization	
Peak number of workers on site during construction [1]	140
Workforce migration	
Percent of construction workers migrating into ROI [2]	20%
Total number of construction workers migrating into ROI during construction peak [3]	28
Families	
Percent of construction workers who bring families [2]	70%
Percent of construction workers who do not bring families [3]	30%
Average construction worker family size (worker, spouse, children) [2]	3.25
Number of construction workers who would move into ROI and bring families	20
Post-construction workforce retention	
Number of construction workers who would move into ROI and would not bring families	8
Number of construction worker family members who would move into ROI	44
Total number of workers and family members migrating into ROI (new population in ROI)	72
School-age children	
Number of school-age children per family [2]	0.8
Number of school-age children in ROI	16
Percent of in-migrating workers that would leave ROI, post-construction [2]	50%
Number of in-migrating workers that would leave ROI, post-construction	14
Total number of in-migrating workers and family members that would leave ROI, post-construction	36
Number of school-age children of in-migrating workers that would leave ROI, post-construction	8

Source:
[1] IIFP, 2011a
[2] BMI, 1981
[3] See Appendix D
Note: there are slight variations in the calculations due to rounding,

Table 4-8. Direct and Indirect Employment

Construction workforce peak (Table 4-7)	140
Number of construction workers who migrate into ROI (20 percent of construction workforce peak) (Table 4-7)	28
Employment multiplier for construction in ROI (BEA, 2010)	0.4324
Indirect jobs resulting from in-migrating construction workers (i.e., 28x0.4324=12)	12

Expenditures for goods and services to support construction activities would occur both inside and outside the ROI. Approximately $70 million to $94 million in capital costs would be spent for construction (IIFP, 2009a). Also, construction workers would spend a portion of their earnings on goods and services within the ROI. Construction worker wages are estimated to average $32,700 annually (IIFP, 2011a).

The NRC staff finds that due to the size of the available workforce in the ROI, the effect of construction on employment and income within the ROI would be SMALL and beneficial.

4.1.1.8.3   Taxes

Construction-related activities, purchases, and workforce expenditures would generate several types of taxes, including individual income taxes, gross receipts taxes, and property taxes. Increased tax collections are viewed as a benefit to the State of New Mexico, Lea County, the Hobbs Municipal School District, the New Mexico Junior College, the communities in Lea County, and other locales where plant-related spending would occur.

Income Taxes

New Mexico imposes a tax on the net income of every resident and nonresident employed or engaged in business in or from the State or deriving any income from any property or employment within the State. The rates vary depending upon filing status and income (Section 3.9.3). Construction wages would be taxed as income and the NRC staff finds that those tax payments would have a SMALL, beneficial impact.

IIFP would not pay corporate income tax during construction.

Gross Receipts Taxes

New Mexico has a gross receipts tax structure instead of a sales tax structure. Like sales taxes, gross receipts taxes are generated by the purchase of goods and services. The gross receipts tax rate varies throughout the state from 5.125 percent to 8.6875 (Section 3.9.3).

IIFP estimates (in 2009 dollars) that construction capital costs would be between $70 million and $94 million. Some portion of those expenditures would occur within the ROI and adjacent counties. The expenditures would generate gross receipts tax revenues for both the counties and New Mexico (IIFP, 2009a). The NRC staff finds that these revenues would be SMALL and beneficial. Because IIFP would have an industrial revenue bond with Lea County, some expenditure would be exempt from gross receipts taxes.

Regional spending on goods and services by the construction workforce would generate gross receipts tax revenues for Lea County and Eddy County municipalities, the two counties, New Mexico, and other locales where spending occurs. The NRC staff finds that this increase in gross receipt taxes would create a SMALL, beneficial impact.

Property Taxes

Property taxes in Lea County are derived using property assessment values (33.3 percent of appraised values) and the tax rates of the taxing entities. The annual payment in lieu of tax (PILT) to the Hobbs Municipal School District is based on a tax rate of $7.60 per $1,000 of assessed value (IIFP, 2011a). The annual PILT to the New Mexico Junior College is based on a tax rate of $4.30 per $1,000 of assessed value (IIFP, 2011a). Based on the estimated assessed value of the IIFP land and attachments to the land and on the equipment and materials, the estimated PILTs during the construction would be $261,000 in 2012 and

$293,400 in 2013 (IIFP, 2011a). According to Table 3-28, the Hobbs Municipal School District's total 2007-2008 revenues were about $71 million and the New Mexico Junior College's total 2008-2009 revenues were about $37 million. Therefore, the NRC staff finds that the property tax impact of construction would be SMALL (less than 1 percent of the school revenues in all cases) but nonetheless, beneficial.

4.1.1.8.4   Housing

In 2008, 5,823 vacant housing units were within the ROI (Section 3.9.4). Construction would result in an influx of approximately 28 construction workers (Table 4-7), all of whom would need housing. Housing the 28 in-migrating construction workers would require 0.5 percent of the vacant housing units within the ROI. The in-migrating workers would not adversely affect the existing housing inventory. In addition, the ROI has temporary housing in hotel/motel rooms available for short-term leasing and areas available for trailers and recreational vehicles that some workers may elect to live in.

Because the existing vacant housing inventory would be sufficient to accommodate the expected population increase associated with the proposed IIFP construction project, the NRC staff finds that the impact of construction on housing would be SMALL.

4.1.1.8.5   Public Utilities

Public Water

All onsite potable, process, and fire water needed during the construction of the IIFP facility would be provided by one or two wells installed in the Ogallala aquifer (IIFP, 2009a). During construction, 72 people (Table 4-7) would relocate to the ROI and likely find housing within an area that is served by a public water utility. The major public water suppliers in the ROI serve 88,643 people (Table 3-30) and most have excess capacity. The in-migration during construction would result in a 0.08 percent increase in people who rely on the ROI's public water supply. The NRC staff finds that the impact of construction on public water supplies would be SMALL, because the excess capacity of water suppliers in the ROI is sufficient to support the in-migrating workforce. The construction site would not be connected to a public water supply.

Public Wastewater

There would be no onsite disposal of any solid or liquid waste during construction of the IIFP facility. The proposed IIFP site would not be connected to any public wastewater or sewage system (IIFP, 2011a). All wastes generated during construction, including sanitary wastes, would be shipped offsite for disposal. During construction, 72 people (Table 4-7) would relocate to the ROI and likely find housing within areas that are served by a public wastewater system. The major public wastewater treatment facilities in the ROI serve approximately 78,917 people (Table 3-31) and all have excess capacity. Construction in-migration would result in a 0.09 percent increase in people who rely on the ROI's public wastewater systems. Therefore, the NRC staff concludes that the impact of construction on public wastewater would be SMALL.

4.1.1.8.6   Community Services

Education

In 2008, there were 8 public school districts, containing 64 schools, educating 22,847 students in the ROI (Section 3.9.6.1). Construction would result in an influx of approximately 28 construction workers, 20 of whom would bring their families (Table 4-7). Each in-migrating family is estimated to have 0.8 school-aged children (Table 4-7); therefore, 16 children could require public school during construction (Table 4-7). The new student enrollment resulting from construction would represent an increase of 0.07 percent in the 2008 enrollment in the ROI. The increase in public school enrollment would be less than 1 percent of total enrollment and would be essentially undetectable. Therefore, the NRC staff concludes that the impact of construction on education would be SMALL.

Fire Protection

The in migrating workforce would increase the population in the ROI less than 0.1 percent (Section 4.1.1.8.1) and would result in filling 0.5 percent of the available housing. Therefore, there would not be a detectable increase in the demand for fire protection. Existing fire protection personnel, facilities, and equipment would be sufficient to support the population increase. Therefore, the NRC staff concludes that the impact of construction would be SMALL.

Law Enforcement

The in migrating workforce would increase the ROI population less than 0.1 percent (Section 4.1.1.8.1) and would not change the ability of existing law enforcement services to meet the needs of the population. Existing law enforcement personnel, facilities, and equipment would be sufficient to support the population increase; therefore, the NRC staff concludes that the impact of construction of the proposed IIFP facility would be SMALL.

Hospitals and Physicians

An ROI population increase of less than 0.1 percent (Section 4.1.1.8.1) would not measurably increase the demand for hospital and physician services. Therefore, the NRC staff concludes that the impact of construction on hospitals and physician services would be SMALL.

4.1.1.8.7   Environmental Justice

Environmental Justice refers to a Federal executive order that directs all Federal agencies, including the NRC, to identify and address disproportionately high and adverse human health and environmental effects on minority or low-income populations. Section 3.9.1.2 defines and identifies the minority and low-income populations within the 80-km (50-mi) radius of the proposed IIFP site. There are 96 block groups that fall completely or partially within the 80-km (50-mi) radius of the proposed site. Of the 96 block groups, 1 has a significant African American population, 15 have significant "some other race" populations, 32 have significant Hispanic populations, and 10 have significant low-income populations. The locations of these block groups are shown on Figures 3-22, 3-23, 3-24 and 3-25. The following discussion summarizes project impacts on the general population and addresses whether or not minority and low-income populations would experience disproportionately high and adverse impacts. The primary resource areas that could be affected by construction are soil, groundwater quality, groundwater quantity, air quality, ecology, and socioeconomics.

- Land Use – The primary land use on the proposed IIFP site is cattle grazing. Less than 10 percent of the 259-ha (640-ac) site area would be disturbed during construction, and cattle grazing would not be permitted on the site. Construction would not conflict with any existing Federal, State, local, or Indian Tribe land use plans, or planned development in the area. The NRC staff finds that the land use impacts resulting from construction and conversion from agricultural (grazing) land use to industrial use would be SMALL due to the abundance of other nearby grazing land (Section 4.1.1.1 Land Use).

- Soils – The largest potential for impacts on soils during construction would result from clearing and grading, which loosens soil and increases the potential for erosion by wind and water. BMPs would be implemented during construction to limit soil loss. The NRC staff finds that the construction impacts on soils would be SMALL and confined to the site (Section 4.1.1.5 Geology, Minerals, and Soils).

- Groundwater quality – Groundwater beneath the proposed IIFP site is unconfined and recharged by natural precipitation, therefore, uncontrolled spills during construction could temporarily and locally affect the aquifer. However, a site-specific Spill Prevention Control and Countermeasure Plan would be developed with procedures to manage spills. Therefore, the NRC staff finds that impacts on groundwater quality would be SMALL, localized, and temporary (Section 4.1.1.6 Water Resources).

- Groundwater quantity – No municipal water line runs near the proposed IIFP facility. Water brought in tanker trucks from Hobbs, or one or two wells onsite would supply all of the water for the construction activities. Average and peak site water requirements for construction are expected to be approximately 6.05 L/min (1.6 gpm) and 20.5 L/min (5.42 gpm), respectively. Because the IIFP site is not located in a critical management area (CMA, the legal drawdown limit for any wells on the property would be on average 0.06 m/yr (0.20 ft/yr). It is unlikely that the two wells would ever exceed the Lea County Underground Water Basin drawdown limit (Section 4.1.2.6 Water Resources). Therefore, the NRC staff finds that impacts to groundwater quantity would be SMALL.

- Air quality – Section 4.1.1.4 reports that site boundary concentrations of some criteria air pollutants would be higher than the NAAQS. Therefore, the NRC staff finds that Phase 1 construction impacts to air quality would be SMALL to MODERATE and localized.

- Ecology – Approximately 16 ha (40 ac) of land would be disturbed, which represents less than 10 percent of the total 259 ha (640 ac) site. Construction would destroy or displace local wildlife. No impacts to rare or unique habitats, threatened or endangered species, or commercially or recreationally valuable species would result from construction. The NRC staff finds that potential impacts to ecological resources during construction would be SMALL and localized (Section 4.1.1.7 Ecological Resources), based on the small area that would be impacted, compared to the available comparable habitat within the region.

- Socioeconomics – Construction would require a maximum of 140 workers, 28 of whom would migrate into the ROI, 20 of whom would bring families. The potential in-migrating population would increase the population within the ROI by 0.1 percent. The NRC staff concludes that this small increase in the population within the ROI would have a SMALL impact on employment, taxes, housing, community services, and public utilities (Section 4.1.1.8 Socioeconomics).

The NRC staff concludes that the impacts of construction on each of these resource areas would be SMALL (SMALL to MODERATE for air quality) and localized. Furthermore, the nearest minority or low-income population meeting the NRC definition is 22.5 km (14 mi) from the proposed site. Therefore, because potential impacts to all resource area impacts would be SMALL or MODERATE and localized, and the identified minority and low-income populations are not in close proximity to the proposed site, the NRC staff finds that impacts would not be disproportionately high and adverse for any populations in the region, including minority or low-income populations.

4.1.1.9 Traffic and Transportation

This section identifies the traffic and transportation impacts within the region during construction. The transportation mode for personnel, construction equipment, and materials deliveries would be exclusively by roadways to the proposed IIFP site. There are no plans to extend the railroad in Hobbs to transport goods or materials to the proposed site. Although routine rail freight and air freight is expected to be used for shipping materials or equipment to the region, this freight would be offloaded elsewhere and arrive at the site on trucks.

The principal highway routes that would handle this traffic include NM 483, which borders the site to the west, and US 62/180, which provides an east-west route to the nearest population centers. All traffic would access the site via NM 483, and most traffic would use US 62/180 to NM 483 to access the site. Some portion of the workforce may access the site from the north, using NM 83 to access NM 483 north of the site. At the junction of NM 483 and US 62/180, traffic would go east to Hobbs, Eunice and other Lea County municipalities or southwest to Eddy County. After the intersection of NM 483 and US 62/180 traffic associated with the site would be increasingly dispersed.

Peak construction would use 140 workers (IIFP, 2011b). Therefore, if each employee commutes alone, there would be an increase of 140 vehicles on NM 483. IIFP estimated 20 delivery or waste disposal trucks each day, for an additional 40 additional trips during one construction shift.

The Highway Capacity Manual 2000 (TRB 2000) indicates that the capacity of a two-lane highway is 1,700 passenger cars per hour for a single direction and 3,400 passenger cars per hour for both directions. The annual average daily traffic count (AADT) on NM 483, a two-lane highway, at the intersection of US 62/180 in 2008 was 955 vehicles per day (NMDOT, 2009). If all the vehicles on NM 483 in one day used the road in a single hour, and if the construction workforce used the road to access the site during that same hour, a maximum of 1,095 vehicles would be on the road. This is less than the design capacity of a two-lane highway. Traffic impacts on NM 483 due to 20 truck trips per shift would have a smaller impact than the scenario described here. The maximum construction traffic on US62/180, which is a four-lane highway, also would have a smaller impact that that of the scenario analyzed here. Therefore, the NRC staff concludes that impacts to traffic from construction would be SMALL.

The potential for traffic accidents increases with increased traffic. Assuming that the majority of the workers and trucks would travel from Hobbs (a distance of 23.3 km [14.5 mi]) and Lovington (24.9 km [15.5 mi]), and a small percentage would come from Carlsbad (80 km [50 mi]), the NRC staff estimates an average one-way commute distance of 32 km (20 mi). Delivery trucks would travel an average round trip distance of 80 km (50 mi). A 64-km (40-mi) daily commute by 140 commuting workers and an 80-km (50-mi) commute by 20 truck results in 10,600 km (6,600 mi) traveled each day for 250 work days per year during the peak construction period. In New Mexico in 2010, vehicle accidents resulted in 51.73 injuries and 1.73 fatal accidents per 160 million vehicle-km (100 million vehicle-mi) (UNM, 2010). Based on these rates, statistically, there would be one injury (risk of less than 0.85 injury crashes per year) and no fatalities (risk of less than 0.03 fatal crashes per year) as a result of the construction traffic.

Therefore, the NRC staff concludes that impacts to traffic due to construction would be temporary and SMALL.

4.1.1.10   Noise

As discussed in Section 3.11, noise from the construction of the proposed facility would be restricted to daylight hours, temporary, and attenuated with distance. Four industrial facilities are within 5 km (3.1 mi) of the site. The nearest residence is approximately 2.6 km (1.6 mi) northwest of the site and there are no recreational facilities within 8.0 km (5.0 mi) of the proposed site. Because the construction equipment noise will attenuate within a short distance of the proposed IIFP site, the nearest residence and other land uses would not be adversely affected by construction noise. The NRC staff finds that impacts due to noise would be SMALL, based on the distances to surrounding residences and recreational areas and the rate at which noise is attenuated with distance.

4.1.1.11   Public and Occupational Health Impacts

This section analyzes the potential impacts on public and occupational health from the proposed IIFP facility construction. The analysis is divided into two main sections: nonradiological impacts and radiological impacts.

The proposed action involves a major construction activity with the potential for industrial accidents, material-handling accidents, and construction accidents that could result in temporary injuries, long-term injuries and/or disabilities, and fatalities.

The number of potential fatal and nonfatal occupational injuries from construction of the proposed IIFP facility were estimated based on injury rate data from the U.S. Department of Labor's Bureau of Labor Statistics. As shown in Table 4-9, six nonfatal injuries and less than one fatality are expected during construction. Additionally, because of the commitment that IIFP is making to a safe design basis for facilities and programs, its safety culture, and adherence to the Integrated Safety Management System program and procedures, the occupational injury rates during construction of the proposed IIFP facility could be better than the industry average. Therefore, the NRC staff concludes that impacts to human health from occupational injuries during construction would be SMALL.

Table 4-9.   Nonfatal/Fatal Occupational Injuries Projected for IIFP Facility Construction

Category	Injury Rate	Expected Occurrences[1]
Nonfatal Injuries	4.3 injuries per 100 workers per year[2]	6.0
Fatal Injuries	9.7 fatalities per 100,000 workers per year[3]	less than 1 (1.4×10^{-2})

[1] Expected occurrences are based on an average of 140 workers during the construction of the facility for 12 months (IIFP, 2011a).
[2] The expected nonfatal injury rate (total recordable cases) is from BLS (2010a).
[3] The fatal injury rate is from BLS (2010b).

In addition to the potential occupational injuries that could result during construction, impacts to the public from air pollutants have been considered. Air pollutants would be generated by the internal combustion engines used in heavy equipment. As discussed in Section 4.1.1.4, the estimated air quality impacts from the air emissions during construction for the proposed IIFP facility would not measurably change the existing ambient air quality in the vicinity of the proposed IIFP facility. As a result, the NRC staff finds that the impacts to human health from air pollutants would be SMALL.

4.1.1.12   Waste Management

Construction of the proposed IIFP facility would generate waste materials that would be collected and transported offsite for recycling or disposal. Refuse and construction debris typical of industrial construction projects would be the predominant wastes generated during the construction phase (IIFP, 2011a). IIFP conservatively assumes that small quantities of low level radioactive wastes are also expected to be generated during the construction phase. This is because IIFP plans to install previously-used process vessels and standard unit operations equipment shipped from the decommissioned Sequoyah uranium conversion facility in Gore, Oklahoma, to the proposed IIFP facility (IIFP, 2011a; IIFP, 2011b). Because this equipment has been used for processing of radioactive materials, refuse and construction debris from its transport and installation could be disposed of as LLW as a precaution.

The anticipated construction wastes include paper, plastic, cardboard, packaging materials, wood scraps, metal scraps, roofing and insulation scraps, masonry and ceramic materials, and empty paint and coating containers. Small quantities of organic solvent-based residuals remaining from application of specialty paints, architectural coatings, sealants, and adhesives, and wastes from certain other materials that are used for construction may be required to be managed as hazardous waste. The specific compositions and quantities for these construction waste types would depend on the final facility design (IIFP, 2011a). Tables 4-10 through 4-12 provide the estimated annual quantities of solid, hazardous, and LLW currently anticipated by IIFP to be generated during construction, respectively.

The general design/build contractor selected for the proposed IIFP facility would have responsibility for the day-to-day supervision of onsite waste collection and storage and for arranging for removal of these wastes from the IIFP site. Good work practices would be used to collect and sort the wastes for recycling or disposal (e.g., using designated roll-off containers and collection areas for different types of wastes). Solid (nonhazardous, nonradioactive) wastes would be transported offsite to an approved local landfill. Hazardous waste generated throughout the construction phase would be temporarily stored onsite and then shipped to an offsite facility appropriate for handling the waste composition, in accordance with established recycling and hazardous waste management programs. Any radiological waste would be shipped offsite to licensed LLW disposal facilities (IIFP, 2011a). The management of stormwater at the proposed IIFP facility is discussed in Section 4.1.1.6.

The proposed IIFP facility would be located in the New Mexico Environment Department (NMED) district comprising Chaves, Eddy, Lincoln, and Lea Counties. At present disposal rates, the remaining disposal life of the three permitted solid waste landfills (Roswell Municipal Landfill, Sand Point Landfill, and Lea County Landfill) in that district ranges from 16 years to 63 years (NMED, 2009). This district also has an "industrial waste only" landfill, Lea Land, Inc. Industrial Landfill, with an anticipated remaining disposal life of more than 100 years (NMED, 2009). Nonhazardous wastes from the proposed IIFP facility would likely be transported to the Lea County landfill for disposal. The landfill accepts residential, commercial, private and public waste material from generators within a 161-km (100-mi) radius. The landfill is operated by the Solid Waste Authority of Lea County under NMED Permit # Stormwater Management

Table 4-10. Solid Waste Generation - Construction

Waste Type	Estimated Annual Amount
Air filters (vehicle)	23 – 45 kg (50 – 100 lbs)
Cardboard / packing	140 – 230 kg (300 – 500 lbs)
Fiber drums	140 – 230 kg (300 – 500 lbs)
Total	300 – 500 kg (650 – 1,100 lbs)

Source: IIFP, 2011a

Table 4-11. Hazardous Waste Generation - Construction

Waste Type	Estimated Annual Amount
Adhesives, resins, caulking residues	45 – 90 kg (100 – 200 lbs)
Lead (batteries)	45 –110 kg (100 – 250 lbs)
Oil filters	45 – 90 kg (100 – 200 lbs)
Paints, thinners, solvents, organic residues	45 – 230 kg (100 – 500 lbs)
Pesticides	45 – 68 kg (100 – 150 lbs)
Petroleum products, oils, lubricants residues	45 – 230 kg (100 – 500 lbs)
Total	270 – 820 kg (600 – 1,800 lbs)

Source: IIFP, 2011a

Table 4-12. Low Level Radioactive Waste Generation - Construction

Material	Estimated Annual Amount
Scrap metal	1,800 – 2,700 kg (4,000 – 6,000 lbs)
Spent blasting sand	45 kg (100 lbs)
Wood trash (pallets)	450 – 680 kg (1,000 – 1,500 lbs)
Total	2,300 – 3,400 kg (5,100 – 7,600 lbs)

Source: IIFP, 2011a

(SWM) -13030. The Lea County landfill receives approximately 74,800 metric tons (82,500 tons) annually (NMED, 2009). Nonhazardous waste generated from the proposed IIFP construction activities would result in a negligible increase (less than 0.5 metric ton or 0.0007 percent) in the waste that the Lea County landfill receives annually from all other sources. Therefore, the NRC staff finds that the solid waste management impacts resulting from construction of the IIFP facility would be SMALL.

Hazardous wastes generated during construction would be packaged and shipped offsite to licensed hazardous waste treatment and disposal facilities in accordance with Federal and State regulations (IIFP, 2011a). The projected annual hazardous waste generation would likely classify the proposed IIFP facility as small quantity generator (over 100 kg/mo [220 lb/mo] but less than 1,000 kg/mo [2,200 lb/mo]) during construction. Hazardous waste generators in New Mexico accounted for 978,000 metric tons (1,079,000 tons) of hazardous waste in 2009, with all but 3,700 metric tons (4,084 tons) originating at one facility operated by the Navajo Refining Company (EPA, 2010b). Less than 0.9 metric tons (1 ton) per year of hazardous wastes would be expected from construction of the proposed IIFP facility. The IIFP facility would, during construction, be one of the smaller hazardous waste generators in New Mexico and would contribute less than 0.00009 percent to the overall hazardous waste generated in the State. Therefore, the NRC staff concludes that the quantity of construction-generated hazardous waste material would result in SMALL impacts that could be managed effectively.

Radiological waste generated during construction, due to the use of previously-used radiological processing equipment, would be shipped offsite to licensed LLW disposal facilities (IIFP, 2011a). As shown in Table 4-12, up to 3.4 metric tons (3.8 tons) per year of LLW could be sent for disposal. That corresponds to approximately 22.5 drums per year. This LLW volume represents 0.008 percent of the annual commercial waste volume currently received at the EnergySolutions facility in Clive, Utah (NRC, 2010). All LLW generated will be Class A wastes as defined by 10 CFR 61.55 (IIFP, 2009a). The Clive facility accepts the majority of the United States' Class A LLW (as detailed in 10 C.F.R. § 61.55, Waste Classification, enforced by NRC) and is estimated to have capacity to accept this waste at current volume levels for more than 20 years (GAO, 2004). Thus, the NRC staff finds that the quantity of construction-generated LLW would result in SMALL impacts to LLW disposal capacity.

4.1.2 Environmental Impacts of Operation

The impacts of operations of the facility described in Chapter 2 on each of the major resource areas described in Chapter 3 (Affected Environment) are presented in this section. Impacts of the proposed Phase 2 facility activities identified in Chapter 2 are evaluated as cumulative impacts in Section 4.2.

4.1.2.1 Land Use

This section describes the potential impacts on land use during operation of the proposed IIFP facility.

During operations the primary current land use at the site, which is cattle grazing, would be eliminated by a fence surrounding the entire 259-ha (640-ac) site. Except for the 16-ha (40-ac) facility footprint, the remainder of the site (240 ha or 600 ac) would be remain undeveloped for the duration of the license. Operation of the proposed IIFP facility would be consistent with the industrial nature of land in the vicinity which supports four energy production facilities (Xcel Energy's Cunningham Station and Maddox Station, Colorado Energy Station, and the DCP Midstream Linam Ranch Plant). The proposed IIFP facility and retention of the remaining portion of the site as undeveloped land would not conflict with any existing Federal, State, local, or Native American tribal land use plans. The use of land for the facility would not interfere with any planned development in the area (Appendix A). The facility's location within Section 27 was selected to avoid, to the extent possible, utility rights-of-way, and operation of the facility will not prohibit access to the rights-of-way for maintenance. None of the Cunningham Station's monitoring wells in Section 27 would be affected by the operation of the facility

(GL Environmental, 2010). Because many square miles around the facility have similar habitat, and because the industrial nature of the facility would be consistent with local land use, the NRC staff concludes that land use impacts associated with operation of the proposed facility would be SMALL.

4.1.2.2   Historic and Cultural Resources

This section describes the potential environmental impacts on historic and cultural resources resulting from operation of the proposed IIFP facility. Impacts to historic or cultural resources would most likely occur during ground-disturbing activities associated with construction. Therefore, because operations would not require additional land disturbance, the NRC staff concludes that any impacts to historic properties, districts, resources or significant historic/precontact archaeological sites during facility operation would be SMALL.

4.1.2.3   Visual Resources

This section discusses the potential visual and scenic impacts that could result from operation of the proposed IIFP facility. Visual impacts could occur as a result of tall or massive structures being imposed on a landscape, or if plumes visible from a long distance were emitted from a facility.

The tallest proposed building would be 21.3 m (70 ft) high and emission stacks would be less than 30.5 m (100 ft) tall (IIFP, 2009a); well under the 61-m (200-ft) threshold that requires lights for aviation safety (FAA, 1992). The facility will not be visible from any recreational or historic facilities, and will not degrade the existing viewscape which includes four other industrial facilities. In addition, security lighting would be directed downward to minimize light pollution (IIFP, 2009a), therefore, the NRC staff concludes that impacts to visual resources would be SMALL.

4.1.2.4   Climate, Meteorology and Air Quality

4.1.2.4.1   Greenhouse Gases

This section presents an assessment of the effect operation of the proposed IIFP facility could have on the concentrations of CO_2 and other greenhouse gases in the atmosphere.

During operation, greenhouse gas emissions would result from workforce commuting, deliveries of feedstock and consumable materials to the proposed facility, return of empty feedstock containers to their points of origin, transfer of wastes to designated offsite disposal facilities and operation of a gas-fired boiler. An incidental amount of greenhouse gas emissions would result from the onsite storage and dispensing of fossil fuels to support operations, but is not evaluated here.

A workforce of 140 is assumed to commute a round-trip distance of 64 km (40 mi), assuming 250 round trips per year and taking no credit for carpooling or busing. Annually, the workforce would commute approximately 2,200,000 km (1,400,000 mi).

Deliveries and returns of DUF_6 cylinders and waste shipments are estimated at 2,650 round trips per year. Thus, an average of approximately 10 truck round trips would occur daily during a 5-day work week. The DUF_6 feed materials for the facility would be transported by 18-wheeled trucks via highway only and are expected to come from several facilities across the country: the URENCO USA facility approximately 53 km (33 mi) away; Global Laser Enrichment 2,600 km (1,616 mi) away; and the Eagle Rock Enrichment Facility approximately 1,796 km (1,116 mi) away. Waste from the IIFP facility would likely be transported to one of several disposal facilities. One low-level waste disposal facility is the Energy*Solutions* facility in Clive, Utah approximately 1,572 km (977 mi) from the IIFP site. Hazardous and mixed low-level radioactive wastes could be disposed at Waste Control Specialists which is approximately 61 km (38 mi) from the IIFP facility. Because it is difficult to anticipate the proportion of shipments among the DUF_6 feed materials suppliers and waste disposal sites, an average of the distances to the five facilities was used to establish a conservative scenario with respect to GHG emissions. An average roundtrip distance of 2,433 km (1,512 mi) was assumed. The resulting annual travel distance is 6,447,450 km (4,006,800 mi). Table 4-13 shows the estimated total transportation-related CO_2 emission associated with proposed IIFP facility operations. The total CO_2 emissions expected during IIFP facility operations from commuting of the operational workforce, deliveries of feedstock to the proposed facility, return of empty feedstock containers to their points of origin and delivery of operational wastes to designated offsite disposal facilities are 4,433 metric tons (4,886 tons) per year. NRC staff estimated these levels based on modeling summarized and presented in Appendix C (Air Emissions) of this EIS.

Table 4-13. **Annual Transportation-Related CO_2 Emissions During IIFP Facility Operations**

Activity	Total Workers	RT Distance (mi)	RT Distance (km)	Working Days/ Year	Total Distances per Year (mi)	Total Distances per Year (km)	Annual CO_2 Emissions (ton)	Annual CO_2 Emissions (MT)
Commuting traffic	140	40	64	250	1,400,000	2,300,000	660	599
Operational deliveries and waste removal shipments	N/A	1,512	2,433	N/A	4,000,000	6,400,000	4,226	3,833
Subtotal of CO_2 emissions as a result of transportation related impacts from IIFP facility operations	–	–	–	–	–	–	4,886	4,433

The proposed IIFP facility would also require a gas-fired boiler (the facility would install two boilers for redundancy, but only one at a time would operate). The estimated emissions of CO_2 equivalents from the boilers are 1,345 metric tons (1,483 tons) per year (IIFP, 2009a). Therefore, the total CO_2 emissions expected from facility operations are 5,778 metric tons (6,369 tons) per year.

Using calendar year 2000 as a reference point (the latest year for which New Mexico GHG emission data are available), and as shown in Table 3-1, total net CO_2 emissions for New Mexico for the year 2000 were 62 million metric tons (68 million tons) of CO_2 equivalents. For the United States for that same year, total net CO_2 emissions were 5,977 million metric tons (6,588 million tons) (EPA, 2010a). By comparison, during any typical year of IIFP facility operation, CO_2 emissions are projected to be 5,778 metric tons (6,369 tons), approximately

0.009 percent of the New Mexico statewide output or 0.0001 percent of the nationwide emissions for calendar year 2000. Therefore, the NRC staff concludes that impacts from the operation of the proposed IIFP facility from the emissions of CO_2 and other greenhouse gases would be SMALL.

4.1.2.4.2 Air Quality

Air quality would be affected during operation of the proposed uranium deconversion facility. As discussed in more detail below, the impact levels would not be significant.

4.1.2.4.2.1 Criteria Pollutant Emissions

Operation of the proposed project would produce criteria pollutant (i.e., CO, NO_2, PM_{10}, $PM_{2.5}$, SO_2, and VOC, an ozone precursor) emissions from natural-gas fired boilers, an emergency diesel generator, a fire-water pump, a hydrogen generator, and commuter/delivery vehicles.

IIFP used emission factors obtained from the EPA document AP-42, "Compilation of Air Pollutant Emission Factors" (EPA, 1995a) to estimate emissions from the natural gas boilers, emergency diesel generator, and fire water pump. Emissions from each equipment type are a function of equipment-specific factors, including engine horsepower, load factor, and hours of operation. During operations one boiler would operate continuously, providing 3000 pounds of steam per hour for the heating and autoclave feed systems. The diesel generator and fire water pump are assumed to be operated for emergency and testing purposes only (IIFP, 2011c). IIFP used equipment manufacturer data to estimate emissions from the hydrogen generator. Operation of the hydrogen generator would be on demand (IIFP, 2011a). Title V of the 1990 Clean Air Act Amendments requires facilities defined as "major stationary sources" to obtain a Title V operating permit. A major stationary source is any facility that has the potential to emit more than 100 tons of any criteria pollutant per year. As shown in Table 4-14, emissions resulting from operations at the proposed IIFP facility would be well below the 100 tons per year threshold. Therefore, the proposed project would not require a Title V operating permit.

NRC staff used emission factors to estimate annual criteria pollutant emissions from highway vehicles (i.e., worker commute vehicles and delivery vehicles). The emission factors were determined using the EPA MOVES Model (EPA, 2009a). Emissions from highway vehicles (i.e., worker commute vehicles and delivery and waste transport vehicles) are a function of the vehicle-specific factors, including type of vehicle and age, fuel type, and vehicle miles traveled. NRC staff used Lea County, New Mexico-specific default values for vehicle type, age, and fuel type to determine highway vehicle emission factors. IIFP estimates that during operations 140 workers would commute to and from the proposed site, an average daily trip distance of 64 km (40 mi), for 250 days each year. Delivery and waste transport trucks (presumed to be diesel-fueled, long-haul semi-trailer trucks averaging 10 mpg) would make on average, 10 trips per day (at an average round-trip distance of 2,433 km [1,512 mi]) to transport materials and remove wastes. Table 4-15 shows the estimated annual emissions as a result of a commuting workforce and material transport.

To estimate the impact to local air quality, NRC staff compared the total anticipated direct (facility) and indirect (highway vehicle) criteria pollutant and VOC emissions from the proposed IIFP facility to baseline emissions from Lea County, New Mexico. As shown in Table 4-16, emissions from the proposed project would represent a very small portion of the annual criteria

Table 4-14   Estimated Maximum Annual Criteria Pollutant Emissions from Phase 1 Operation of the IIFP Facility

Source	CO metric tons (tons)	NO$_2$ metric tons (tons)	PM$_{2.5}$ metric tons (tons)	PM$_{10}$ metric tons (tons)	SO$_2$ metric tons (tons)	VOC metric tons (tons)
Boilers	0.93 (1.03)	1.09 (1.20)	0.08 (0.09)	0.08 (0.09)	0.01 (0.01)	0.06 (0.07)
Generators	0.032 (0.035)	0.149 (0.164)	0.011 (0.012)	0.011 (0.012)	0.010 (0.011)	0.00308 (0.00340)
Firewater Pump	0.003 (0.003)	0.013 (0.014)	8.98×10^{-4} (9.9×10^{-4})	9.0×10^{-4} (9.9×10^{-4})	8.4×10^{-4} (9.3×10^{-4})	2.63×10^{-4} (2.90×10^{-4})
Hydrogen Generator	0.21 (0.23)	0.02 (0.02)	0 0	0 0	0 0	0 0
Total	1.18 (1.30)	1.27 (1.40)	0.093 (0.103)	0.093 (0.103)	0.020 (0.022)	0.067 (0.074)

Source: IIFP, 2011b

Table 4-15.   Estimated Maximum Annual Criteria Pollutant Emissions from Highway Vehicles during Phase 1 Operation of the IIFP Facility

Source	CO metric tons (tons)	NO$_2$ metric tons (tons)	PM$_{2.5}$ metric tons (tons)	PM$_{10}$ metric tons (tons)	SO$_2$ metric tons (tons)	VOC metric tons (tons)
Commuter Vehicles	10.9 (12.1)	1.46 (1.61)	0.0266 0.(0293)	0.0495 0.(0545)	0.0116 (0.0128)	1.03 (1.14)
Delivery Vehicles	41.0 (45.2)	18.6 (20.5)	0.789 (0.870)	0.953 (1.05)	0.0449 (0.0495)	3.50 (3.85)
Total	51.9 (57.3)	20.0 (22.1)	0.816 (0.899)	1.00 (1.10)	0.0565 (0.0623)	4.53 (4.99)

Source: IIFP, 2011b

Table 4-16.   Comparison of Maximum Annual Emissions from Phase 1 Operations to Lea County Baseline Conditions

Source	CO metric tons (tons)	NO$_2$ metric tons (tons)	PM$_{2.5}$ metric tons (tons)	PM$_{10}$ metric tons (tons)	SO$_2$ metric tons (tons)	VOC metric tons (tons)
Operations	53.1 (58.6)	21.3 (23.5)	0.908 (1.00)	1.09 (1.21)	0.0774 (0.0853)	4.60 (5.06)
Lea County Baseline[a]	21,244 (23,417)	27,119 (29,894)	2,892 (3,188)	25,048 (27,611)	7,334 (8,084)	4,436 (4,890)
Net Increase over Baseline	0.25%	0.079%	0.031%	0.0044%	0.0011%	0.10%

[a] Source: EPA, 2009b

pollutant emissions in Lea County. Because conservative assumptions that tend to overestimate impacts were used, actual emissions from operations would be less.

IIFP used EPA's SCREEN3 model (EPA, 1995b) to estimate the maximum concentrations of pollutants at the IIFP site property line that would be associated with operations. As shown in Table 4-17, the estimated incremental increases in ambient background concentrations due to the proposed operations would be below allowable PSD Class II increments and well below the National NAAQS. VOCs are not included in Table 4-17 because there are no regulatory metrics for comparison (VOC emissions are regulated by source controls and permit requirements).

Table 4-17. Predicted Property Boundary Concentrations Due to Phase 1 Operations, NAAQS and Allowable Class II PSD Increments

Pollutant	Averaging Time	NAAQS ($\mu g/m^3$)	Allowable Class II PSD Increment ($\mu g/m^3$)	Incremental Concentration Increase ($\mu g/m^3$)	Incremental Concentration Increase as Percentage of NAAQS
CO	1-hr	10,000	NA	5.8	0.06%
	8-hr	40,000	NA	4.1	0.010%
NO_2	Annual	100	25	0.059	0.6%
$PM_{2.5}$	24-hr	35	9	0.17	0.5%
	Annual	15	4	0.042	0.3%
PM_{10}	24-hr	150	8	0.17	0.11%
SO_2	1-hr	200	NA	0.096	0.05%
	3-hr	1,300	512	0.086	0.007%

Source: IIFP, 2011a; EPA, 2009a

4.1.2.4.2.2   Nonradioactive Process Effluents

Radioactive and nonradioactive gaseous effluents would be generated during operation of the proposed IIFP facility (IIFP, 2011a). Radioactive gaseous effluents are addressed in Section 4.1.2.11.

IIFP estimated annual nonradioactive process emissions for operation (IIFP, 2011a). Nonradioactive gaseous effluents would include HF, SiF_4, BF_3 CaF_2, calcium hydroxide [$Ca(OH_2)$], and B_2O_3. Gaseous effluents from the DU_6 to DU_4, SiF_4 and BF_3 processes (comprised mostly of nitrogen, air, some relatively low amounts of the product gases and other trace fluorides) would undergo treatment in the plant KOH scrubbing system to remove approximately 99.9 percent of the fluoride components before being released to the atmosphere via a monitored stack (IIFP, 2011a). The plant KOH scrubbing system is described in Section 2.1.6.4.1.

The nonradioactive process annual emissions are shown in Table 4-18. The combined estimated annual fluoride releases, including HF (52.6 kg [116 b]), SiF_4 (3.7 kg [8.2 lb]), BF_3 (64.1 kg [141 lb]), and CaF_2 (3.55 kg [7.82 lb]), are 124 kg (273 lbs). The annual total expressed as an hourly rate is 0.014 kg/hr (0.031 lb/hr). This pound per hour rate is well below the New Mexico threshold of 0.167 lb/hr for fluoride emissions.

Table 4-18. Annual Nonradioactive Gaseous Emissions from the Operation of the Proposed IIFP Facility

Emission	Estimated Releases Total Emissions							
	Plant KOH Scrubbing System Stack	DUF_4 Dust Collector Stack	FEP Dust Collector	DUF_4 Vacuum Transfer Dust Collector	CaF_2 Dust Collector	Lime Silo Dust Collector	B_2O_3 Silo	Total Emission
	kg/yr (lb/yr)							
HF	3.03 (6.69)	41.54 (91.59)	0.35 (0.77)	7.77 (17.13)	-	-	-	52.6 (116.18)
SiF_4	0.0 (0.01)	-	3.71 (8.19)	-	-	-	-	3.7 (8.20)
BF_3	1.28 (2.83)	-	62.48 (137.75)	-	-	-	-	64 (140.58)
U	-	0.03 (0.07)	0.03 (0.06)	0.25 (0.55)	-	-	-	0.31 (0.68)
CaF_2	-	-	-	-	3.55 (7.82)	-	-	3.55 (7.82)
$Ca(OH)_2$	-	-	-	-	-	60.78 (134.00)	-	60.78 (134.00)
B_2O_3	-	-	-	-	-	-	4.93 (10.87)	4.93 (10.87)
Totals	4.50 (9.93)	41.58 (91.66)	66.57 (146.77)	8.02 (17.68)	3.55 (7.82)	60.78 (134.00)	4.93 (10.87)	189.75 (418.33)

Source: IIFP, 2011a.

4.1.2.4.2.3   Summary

As discussed above, the estimated criteria pollutant emissions for Phase 1 operation of the proposed IIFP facility represent a very small fraction of the current emissions in Lea County. Because conservative assumptions that tend to overestimate impacts were used, actual emissions from operations would be less. In addition, pollutant emissions, including nonradioactive process effluents, would not change the existing ambient air quality in the vicinity of the IIFP facility. The NRC staff, therefore, concludes that air quality impacts during operation of the proposed IIFP facility would be SMALL to MODERATE.

4.1.2.5   Geology, Minerals, and Soil

No impact to the underlying bedrock, mineral resources, or soil is expected during facility operations. As discussed in Section 3.6.2, the site is in an area of limited seismic and volcanic activity, therefore, the statistical probability of fault rupture near the site is very low and the NRC staff finds that any associated impact due to seismic activity would be SMALL. Additionally, operation of the proposed IIFP facility is not expected to cause seismic or fault-related impacts.

Seismic risks to the facility would be mitigated by incorporation of seismic criteria in the facility design to prevent spills or releases to the environment (IIFP, 2009a).

Consequently, the NRC staff concludes that impacts of operation of the proposed IIFP facility to geology, minerals, seismicity, and soils are expected to be SMALL, and that impacts to the facility from any seismic activity would be SMALL.

4.1.2.6 Water Resources

All facility water systems – for potable, process and fire protection water -- would use groundwater. Groundwater pumped from the site well(s) would be a consumptive use because the groundwater would either be consumed or evaporated.

4.1.2.6.1 Groundwater Use

As discussed in Section 4.1.1.6.1.1, groundwater would be supplied from up to two onsite production wells. These wells would meet an estimated operations demand from 7.9 L/min (2.1 gpm) to 11.8 L/min (3.1 gpm) (normal) up to 26.3 L/min (6.95 gpm) (maximum). The operation of the facility would require relatively low volumes of water because it would recycle process water and re-circulate cooling water. The project is projected to use less than 38,000 L (10,000 gal) of groundwater per day (IIFP, 2011a).

As discussed in Section 3.7.1.2.1, because current pumping is in excess of the Ogallala's recharge rate, the aquifer has experienced significant drawdown in the last several decades. In 2009, the NM OSE prepared guidelines on the procedures for processing water rights applications filed with the Lea County UWB. The NM OSE developed administrative guidelines in order to promote the orderly development of water resources in the Lea County UWB, while meeting statutory obligations regarding existing water rights, availability of unappropriated water, conservation of water with the State, and public welfare of the State through a 40-year planning period beginning on January 1, 2005 and ending on January 1, 2045.

On March 10, 2009, the NM OSE issued an order closing the Lea County UWB to the filing of groundwater applications (NMOSE, 2009a). In 1999, in order to meet the projected groundwater demands of Lea County, 138 applications were filed by the Lea County Water Users Association to appropriate 6,389 ha-m (51,797 ac-ft) per year of groundwater, which was essentially all the unappropriated groundwater in the Lea County UWB (Leadshill-Herkenhoff et al., 2000).

A portion of the applications to appropriate groundwater, totaling 4,215.2 ha-m (34,173 ac-ft) per year, have been assigned to Lea County during (at least) the 40-year planning period to allow the County to hold the subject water rights unused until the rights can be put to beneficial use at projects currently under construction, select future projects, and homes and businesses in unincorporated areas. The proposed IIFP site has been included in Lea County's 40-Year Water Development Plan as an area of proposed development where up to 21.6 ha-m (175 ac-ft) per year of the Lea County's water rights would be put to beneficial use. The estimated quantity of groundwater that Lea County would need by the end of the 40-year planning period for all projects that currently exist, are being constructed, or have a high likelihood of being constructed in the near future is 1,173.5 ha-m (9,514 ac-ft) per year (Leadshill-Herkenhoff et al., 2000).

Groundwater rights for the facility would be obtained from the Lea County unallocated water rights under State Engineer Office Water Right File No. L-04719-A, for industrial water use. The proposed facility would obtain a joint Lea County and New Mexico Office of State Engineer Water Rights Agreement for 6.2 ha-m (50 ac-ft) per year (far less than the 21.6 ha-m [175 ac-ft] per year estimation in the Lea County Water Development Plan).

Operation of the proposed IIFP facility would use approximately 0.50 percent of the estimated additional annual 40-year planning period groundwater demand of 1,173.5 ha-m (9,514 ac-ft) per year for Lea County, and only 0.15 percent of the 4,215.2 ha-m (34,173 ac-ft) per year of unappropriated water rights that have been assigned to Lea County.

As part of NM OSE's guidelines on the procedures for processing water rights applications filed with the Lea County UWB, the basin was divided into blocks corresponding to township and range. No permits were granted to appropriate water in a block unless one-third or more of the original groundwater storage in the block would be available at the end of the 40-year period. Blocks with an estimated saturated thickness of 16.8 m (55 ft) or less by 2045 are designated a CMA. The proposed IIFP site is not located in a CMA.

For wells installed in a non-CMA, the Lea County Water Users Association recommends a drawdown limit of 2.4 m (7.9 ft) over 40 years, or 0.06 m/yr (0.20 ft/yr). Groundwater model simulations run by the NM OSE Hydrology Bureau to provide estimated drawdowns for a range of pumping scenarios in different hydraulic conductivity zones indicates that using a high hydraulic conductivity value for the Ogallala aquifer of 12.5 to 18.3 m (41 to 60 ft) per day for a single well pumping 6.2 ha-m (50 ac-ft) per year in an area of the aquifer with a saturated thickness of 61 m (200 ft) results in an estimated drawdown of 0.11 m (0.36 ft) in 40 years (NMOSE, 2009b). Considering that (1) the permeability of the Ogallala aquifer is quite variable (ranging as low as 0.61 m [2 ft/day]), and (2) the two site wells would never independently pump 6.2 ha-m (50 ac-ft) per year, a drawdown of 0.11 m (0.36 ft) in 40 years is very conservative. It is highly unlikely that the wells would ever exceed the Lea County Water Users Association drawdown limit of 2.4 m (8 ft) over 40 years.

The NRC staff finds that adverse impacts on groundwater quantity (availability) due to pumping from the site's potential of two production wells during operation would be SMALL based on the following findings:

- The proposed IIFP site has been included in Lea County's 40-Year Water Development Plan as an area of proposed development where up to 21.6 ha-m (175 ac-ft) of the Lea County's water rights would be put to beneficial use.

- Groundwater rights for the proposed facility would be obtained from the Lea County unallocated water rights under State Engineer Office Water Right File No. L-04719-A for industrial water use. The facility would obtain a joint Lea County and New Mexico Office of State Engineer Water Rights Agreement for 6.2 ha-m (50 ac-ft) per year (far less than the 21.6 ha-m [175 ac-ft] per year estimation in the Lea County Water Development Plan).

- Operation of the proposed facility would use approximately 5 percent of the estimated annual 40-year planning period groundwater demand of 1,173.5 ha-m (9,514 ac-ft) per year for Lea County, and only 0.15 percent of the 4,215.2 ha-m (34,173 ac-ft) per year of unappropriated water rights that have been assigned to Lea County.

- Based on comparing the site wells to the most conservative NM OSE groundwater model scenario, it is highly unlikely that the wells would ever exceed the Lea County Water Users Association drawdown limit of 2.4 m (8 ft) over 40 years.

- The site production wells would be installed in accordance with all NM OSE and Lea County Water Users Association well permit regulations.

4.1.2.6.2 Groundwater Quality

During IIFP operation, stormwater from the site would be collected in two runoff retention/evaporation basins. In addition, stormwater discharges during facility operation would be controlled by a Stormwater Pollution Prevention Plan. All fluids that would otherwise be process effluents are treated and recycled or reused within the process. Therefore, it is expected that no process effluent will be discharged to surface waters or groundwater. Water discharged from the site sanitary waste treatment system would meet required concentrations of all contaminants stipulated in any permit or license required for that activity, including 10 CFR 20 and a Groundwater Discharge Permit/Liquid Waste Permit (IIFP, 2011a). An application for the Groundwater Discharge Permit has been submitted by IIFP to the NMED Groundwater Quality Bureau, which has issued a conceptual groundwater monitoring plan that is subject to change as more information becomes available during the discharge permit application process. NMED will require that total dissolved solids, sulfate, chloride, nitrate as nitrogen, total Kjeldahl nitrogen, fluoride, and isotopic uranium be monitored quarterly (IIFP 2011a).

Treated process water from the sanitary waste treatment system would be used for irrigation at the facility. Because of high evaporation rates, and the presence of the 9.1-m (30-ft) indurate caliche unit that underlies the site, the irrigation water is not expected to migrate to groundwater. There would be no onsite disposal of solid, hazardous, radioactive, or mixed waste at the proposed IIFP site.

The existing groundwater monitoring program at the site would be supplemented with the installation of at least four additional monitoring wells. Three of these monitoring wells are proposed to be located hydraulically downgradient (south) from the DUF$_6$ Cylinder Storage Pad, the Cylinder Pad Stormwater Retention Basin, and the Stormwater Retention/Evaporation Basin. The fourth monitoring well is proposed hydraulically upgradient (north) from the primary production facility, just within the site's security fence (IIFP, 2011a). The wells would be installed per the requirements of the NMED and sampled quarterly.

Any spills of chemicals, diesel fuel, or other contaminants during operations would be cleaned up quickly in accordance with the proposed facility's Spill Prevention Control and Countermeasure Plan, to prevent a pathway for the contaminant to enter the groundwater, thereby mitigating impacts to groundwater such that any inadvertent releases would result in localized and temporary impacts. Due to limited liquid effluent discharge from the facility operations (which would be treated prior to discharge as necessary); the lack of groundwater n the caliche, sand and gravel layer above the Ogallala aquifer; the quarterly groundwater monitoring plan; permanent waste disposal off site; the proposed facility's Spill Prevention Control and Countermeasure Plan; and the 9.1-m (30-ft) depth to groundwater (GL Environmental, 2010) at the proposed IIFP site, the NRC staff concludes that the impacts to groundwater quality from operations would be SMALL.

4.1.2.6.3   Surface Water

As discussed in Section 3.7.2, no permanent surface water or jurisdictional waters are present on the proposed IIFP site and, therefore, the operation of the proposed facility would not affect surface water.

4.1.2.7   Ecological Resources

No additional land beyond the approximately 16 ha (40 ac) footprint would be disturbed during operations. The remaining portion of the IIFP site is expected to be left undeveloped for the duration of the license.

Maintenance practices such as the use of chemical herbicides and roadway maintenance would be implemented during plant operation (IIFP, 2009a).

The tallest proposed building would be 21.3 m (70 ft) high and emission stacks would be less than 30.5 m (100 ft) tall (IIFP, 2009a); well under the 61 m (200 ft) threshold that requires lights for aviation safety (FAA, 1992). Security lighting and equipment would be directed downward to help to minimize light pollution and reduce the potential for adverse impacts to wildlife (IIFP, 2009a). This minimization of lights, which attract nocturnal insects and their predator species, and the low height of the structures, reduce the potential for adverse impacts on night-migrating birds.

No unique or critical habitats or threatened or endangered species occur on the site or in the vicinity. Commercially and recreationally important species would not be adversely affected by plant operations. The NRC staff finds that adverse impacts during operations to ecological resources would be SMALL.

4.1.2.8   Socioeconomic Resources and Environmental Justice

This section provides analyses of the socioeconomic impacts associated with operation of the proposed IIFP facility. Phase 1 operation would begin during the fourth quarter of 2013; after Phase 2 is completed, Phase 1 and Phase 2 operate concurrently. Wage and salary spending and expenditures associated with materials, equipment, and supplies would produce income and employment and local and State tax revenue, while the in migration of workers and their families into the area would affect the availability of housing, public utilities and community services such as education, fire protection, law enforcement and medical resources. Socioeconomic impacts of the proposed IIFP would occur within the two-county ROI (Lea and Eddy Counties, New Mexico), where the operations workforce will likely live and spend most of their incomes.

These analyses are based on the peak number of workers (140) employed at the IIFP facility for Phase 1 operation (IIFP, 2011a). The location of the IIFP facility was selected in part because local colleges and universities have existing training programs in partnership with the nearby URENCO USA centrifuge facility. These institutions, particularly the New Mexico Junior College, have the capability and are committed to provide training to ensure a skilled nuclear workforce (IIFP, 2009a). The New Mexico Junior College Workforce Training Program is designed to offer training requested by area employers, including specialized training for the nuclear service industry. Enrollment in the Workforce Training Program has increased to over 4,251 total trainees through 2009 (NMJC, 2011). Therefore, this analysis assumes that 80 percent of the IIFP worker force would be filled by residents within the ROI. Table 4-19 depicts the workforce in-migration, based on the assumption that 80 percent of the operation employees would be current ROI residents and that each in-migrating operation employee would move his family into the ROI. These projections are used throughout this analysis.

Table 4-19.   Assumptions for Workforce Characterization During Phase 1 Operation

Workforce characterization	
Peak number of workers onsite during Phase 1 operation [1]	140
Workforce migration	
Percent of operation workforce migrating into ROI	20%
Number of workers migrating into ROI during peak operation [2]	28
Families	
Percent of operation workers who bring families [2]	100%
Average New Mexico family size (2009) [3]	3.23
Number of operation workers who would move into ROI and bring families [2]	28
Number of In-migrating workers' family members [2]	62
Number of operation workers and family members migrating into ROI (new population in ROI) [2]	90
School-age children	
Number of school-age children per family [4]	0.8
Number of school-age children migrating into ROI [2]	22

Source:
[1]   IIFP, 2011a
[2]   For supporting analyses, see Appendix D
[3]   USCB, 2010a
[4]   BMI, 1981. This study is an analysis of nuclear construction workforces, however, it included information about nuclear plant non-construction workers [i.e., managers, engineers, supervisors, clerical, security, and medical personnel who were on the site during construction].

4.1.2.8.1   Population

As shown in Section 3.9.1.2, the population within the ROI was 112,938 in 2009. The IIFP Phase 1 operation would employ 140 people, of which 80 percent would be current ROI residents. The other 20 percent of the operations workforce (28 workers) and their families would migrate into the ROI (see Table 4-19). Using the 2009 New Mexico average family size of 3.23, the in-migration would result in 90 new residents (Table 4-19). An increase of 90 residents would result in less than a 0.1 percent increase in the 2009 population of the ROI.

4.1.2.8.2   Employment and Income

Approximately 80 percent, or 112, of the IIFP Phase 1 operation positions (140 x 0.8 = 112 jobs) would be filled by people currently residing in the ROI (Table 4-19). Those 112 workers would represent 0.2 percent of the June 2010 ROI labor force. If all 112 of these jobs were filled by unemployed workers in the ROI, the unemployment rate would decrease by 0.2 percent. Approximately 20 percent of the IIFP Phase 1 operation positions (28 jobs) would be filled by people migrating into the ROI from outside the region (Table 4-19). The in-migrating workers would represent 0.2 percent of the June 2010 labor force.

The in-migration of 28 workers would create indirect jobs within the ROI because of the multiplier effect (described in Section 4.1.1.8.2). In this analysis, the NRC staff used the BEA direct effect employment multiplier for the "All Other Basic Inorganic Chemical Manufacturing" classification to estimate the number of indirect jobs that would be created as a result of the in-migration of the project-related workers. Table 4-20 provides information about direct and indirect employment for Phase 1 operation. Indirect jobs are often non-technical, non-professional positions in the retail and service sectors and would likely be filled by unemployed workers already residing in the ROI. The 51 indirect jobs represent 1.3 percent of the unemployed labor force in June 2010.

Table 4-20. Direct and Indirect Employment during IIFP Phase 1 Operation

Phase 1 operations workforce peak (Table 4-19)	140
Number of Phase 1 operations workers who migrate into ROI (20 percent of operation workforce peak) (Table 4-19)	28
Employment multiplier for Phase 1 operations workers (indirect portion only) (BEA, 2010)	1.8173
Indirect jobs resulting from in-migrating Phase 1 operations workers (See Appendix D)	51

The regional economy would benefit from the capital investment expenditures and recurring costs associated with the operation of the IIFP facility. IIFP has provided estimates for some of these costs. The payroll associated with Phase 1 would be between $7,900,000 and $9,100,000 annually (IIFP, 2009a). IIFP employees and indirect workers would spend earnings on goods and services with the ROI. Additional costs associated with operations include replacement capital; waste disposal; insurance premiums and taxes; utilities; and maintenance materials and supplies. These expenditures would range from $17,315,000 to $23,727,000 annually.

The NRC staff finds that due to the size of the available workforce in the ROI, the effect of IIFP Phase 1 operations on employment and income within the ROI would be SMALL and beneficial.

4.1.2.8.3   Taxes

Phase 1 operations-related wages and purchases would generate several types of taxes, including corporate income taxes, individual income taxes, gross receipts taxes, and property taxes. Increased tax collections are viewed as a benefit to the State of New Mexico, Lea and Eddy Counties, the Hobbs Municipal School District, the New Mexico Junior College, the communities in Lea and Eddy Counties, and other locales where plant-related spending would occur.

Income and Gross Receipts Taxes

IIFP has estimated the income and gross receipts tax impacts of Phase 1 operation in Table 4-21. The NRC staff finds that the increase in tax revenues to the State and county would be SMALL and beneficial.

Table 4-21. Estimated Gross Receipts and Income Tax Payments to New Mexico and Lea County for the Phase 1 Operation Period, 2009 Dollars[a]

	New Mexico	Lea County
Gross Receipts Tax		
High Estimate	$118,100,000	$8,800,000
Low Estimate	$87,100,000	$6,500,000
NM Corporate Income Tax[b]		
High Estimate	$77,200,000	None[c]
Low Estimate	$57,100,000	None[c]

Source: IIFP, 2011a
[a] Tax values based on 2009 tax rates
[b] Based on average annual earnings for the Phase 1 increment
[c] Allocation would be made to the State of New Mexico

In addition to IIFP's corporate income and gross receipts tax payments, plant employees would pay State individual income and State and county gross receipts taxes. The NRC staff finds that these tax payments would have a SMALL, beneficial impact on New Mexico's and the counties' income tax revenues. Regional spending on goods and services by IIFP employees would generate gross receipts tax revenues for Lea and Eddy County municipalities, Lea and Eddy Counties, New Mexico, and other locales. The NRC staff finds that these additional tax revenues would create a SMALL, beneficial impact

Property Taxes

As stated in Sections 3.9.3, International Isotopes, Incorporated, the parent corporation of IIFP, has an IRB agreement with Lea County and is generally exempt from property taxes. However, two taxing entities are not part of the IRB agreement. For Phase 1 operation, IIFP would pay an amount in lieu of property tax to the Hobbs Municipal School District and to the New Mexico Junior College. Table 3-28 presents total revenue data for the Hobbs Municipal School District and the New Mexico Junior College, IIFP's estimated average annual tax payments to those schools, and those payments as a percentage of the schools' revenues. As shown in Table 3-28, the Hobbs Municipal School District's total 2007-2008 revenues were about $71 million and the New Mexico Junior College's total 2008-2009 revenues were about $37 million. IIFP's payments would represent a very small percentage of the school district and college's revenues. The NRC staff finds that the impact of the payment in lieu of taxes to each jurisdiction would be SMALL, and beneficial.

Therefore, the NRC staff concludes that operation of the proposed IIFP facility would have a SMALL beneficial impact on tax revenues.

4.1.2.8.4   Housing

In 2008, about 46,971 housing units were in the ROI, and 5,823 of them were vacant (Section 3.9.4). The Phase 1 operation of the IIFP facility would result in an influx of approximately 28 workers (Table 4-19), all of whom would need housing. Housing the 28 in-migrating workers would require 0.5 percent of the vacant housing units within the ROI. The in-migrating workers would not exhaust the existing housing inventory.

Therefore, the NRC staff concludes that operation of the proposed IIFP facility on the existing housing inventory would be SMALL.

4.1.2.8.5   Public Utilities

Public Water

All onsite potable, process, and fire water needed for the operation of the IIFP Phase 1 facility would be provided by no more than two wells installed in the Ogallala aquifer. The facility will not use public water (IIFP, 2011a).

Phase 1 operation will result in 90 people migrating into the ROI (Section 4.1.2.8.1). These new residents would likely select housing within areas that rely on a public water supplier. The major public water suppliers serve approximately 88,643 people (Table 3-30) and most, if not all, have excess capacity. The 90 new residents would result in a 0.1 percent increase in customers who rely on the public water suppliers. The NRC staff finds that the impact of Phase 1 operations on public water supplies would be SMALL, because the excess capacity of water suppliers in the ROI is sufficient to support the in-migrating workforce.

Public Wastewater

The IIFP facility would not be connected to any public wastewater or sewage system (IIFP, 2011a). The project will result in 90 people migrating into the ROI. These new residents would likely elect to reside within areas that rely on a public wastewater system. The major public wastewater treatment facilities serve approximately 78,917 people (Table 3-31) and have excess capacity. The 90 new residents would result in a 0.1 percent increase in customers who rely on the public wastewater systems. Therefore, because the increase in households is a small percentage of the existing public wastewater users, and the public wastewater facilities have excess capacity, the NRC staff concludes that impact of the proposed IIFP operation on public wastewater treatment systems would be SMALL.

4.1.2.8.6   Community Services

Education

During the 2008 school year, there were 8 public school districts, containing 64 schools educating 22,847 students in the ROI (Section 3.9.6.1). The operation of the IIFP facility would result in an influx of approximately 28 employees and their families (Table 4-18). Each in-migrating family is estimated to have 0.8 school aged children (Table 4-18); therefore, 22 additional children would be eligible for public school as a result of Phase 1 operation (Table 4-18). The new student enrollment would represent an increase of 0.1 percent of the 2008 enrollment. Therefore, the NRC staff concludes that the impact of the proposed Phase 1 operation on education would be SMALL.

Fire Protection

As discussed in Section 4.1.2.8.1, the population increase in the ROI associated with the operation of the IIFP Phase 1 facility would be less than 0.1 percent and would result in filling 0.5 percent of the available housing. Therefore, there would not be a detectable increase in the demand for fire protection. Existing fire protection personnel, facilities, and equipment would be sufficient to support the population increase Therefore, the NRC staff concludes that impact of operation of the proposed facility on fire protection would be SMALL.

Law Enforcement

The in migrating workforce would increase the ROI population less than 0.1 percent (Section 4.2.1.8) and would not affect the ability of existing law enforcement services to meet the needs of the population. Existing law enforcement personnel, facilities, and equipment would be sufficient to support the population increase; therefore, the NRC staff finds that the impact of Phase 1 operation of would be SMALL.

Hospitals and Physicians

An ROI population increase of less than 0.1 percent (Section 4.2.1.8) would not measurably increase the demand for hospital and physician services. The NRC staff finds that the impact of operation on hospitals and physician services would be SMALL.

4.1.2.8.7   Environmental Justice

The primary environmental resources that could be affected by the operation of the proposed IIFP facility are soil, groundwater quality, groundwater quantity, air quality, ecology, socioeconomics and human health. Section 3.9.1.2 defines and identifies the minority and low-income populations within the 80 km (50-mi) radius of the proposed IIFP facility. There are 96 block groups that fall completely or part ally within the 80 km (50-mi) radius. Of the 96 block groups, one has a significant African American population, 15 have significant "Some Other Races" populations, 32 have significant Hispanic populations, and 10 have significant low-income populations. Figures 3-22, 3-23, 3-24 and 3-25 locate these block groups. The following is a summary of the impacts on the resources area and addresses whether minority or low-income populations would experience disproportionately high and adverse impacts from the IIFP Phase 1 operation:

- Land Use – The NRC staff finds that operation of the facility would not affect land use beyond those impacts attributed to construction. Accordingly, the NRC staff finds that impacts to land use would be SMALL.

- Soils – The NRC staff finds that impact to soils during IIFP operation would be SMALL.

- Groundwater quality – During IIFP operation, stormwater from the site would be collected in two runoff retention/evaporation basins. No wastes from facility process systems would be discharged to stormwater. Furthermore, any stormwater discharges would be controlled by a Stormwater Pollution Prevention Plan. Treated process water would be used for irrigation at the facility but is not expected to migrate to groundwater. The NRC staff finds that effects to groundwater quality would be SMALL, localized, and temporary.

- Groundwater quantity – No municipal water line runs near the proposed IIFP facility. No more than two groundwater wells would supply all of the water for the facility. The operation of the facility would require relatively low volumes of water because it would recycle process water and re-circulate cooling water; groundwater use is estimated to be less than 37,854 (10,000 gal) per day. The proposed IIFP site has been included in Lea County's 40 Year Water Development Plan and would use approximately 5 percent of the estimates 40-year planning period demand. The NRC staff finds that it is highly unlikely that the two wells would ever exceed the Lea County UWB drawdown limit. The NRC staff finds that impacts to groundwater quantity during operation would be SMALL.

- Air quality - The estimated criteria pollutant emissions from Phase 1 operation represent a very small fraction of the current emissions in Lea County. Pollutant emissions, including nonradioactive process effluents, would not change the existing ambient air quality in the vicinity of the IIFP facility. The NRC staff finds that the air quality impacts would be SMALL to MODERATE.

- Public and Occupational Health - Operation of the IIFP facility would require shipment of DUF_6 cylinders to and from the facility and hazardous, mixed and LLW to disposal facilities. The transportation risk associated with IIFP transportation operations is 0.03 additional latent cancer fatalities (LCF) per year. The NRC staff finds that impacts from the proposed action on Public and Occupational health would be SMALL.

- Ecology – More than 90 percent of the IIFP site would be undisturbed by operations, no threatened or endangered species or critical or unique habitats occur on the site, and the site does not provide extensive habitat for any commercial or recreational species. Therefore, the NRC staff finds that operation of the IIFP facility would have a SMALL effect on ecological resources.

- Socioeconomics – Phase 1 IIFP operations would employ 140 employees, 28 of whom would migrate into the ROI with their families. The in-migrating workers and their families would increase the population within the ROI by 0.1 percent. The NRC staff concludes that these workers and their families would have a SMALL effect on housing, community services, and public utilities and a SMALL and beneficial effect on employment and taxes.

The NRC staff finds that the impacts of IIFP operation on the resources evaluated would be SMALL for most resources and SMALL to MODERATE for air quality and in some cases, beneficial. Furthermore, the nearest minority or low-income population is 22.5 km (14 mi) from the proposed facility. Therefore, because all resource area impacts are SMALL and the identified minority and low income populations are not in close proximity to the proposed site, the NRC staff finds that impacts would not be considered disproportionately high and adverse impacts to any population, including low-income or minority populations.

4.1.2.9   Traffic and Transportation

4.1.2.9.1   Traffic

Operations impacts would occur from commuting personnel and the transport of nonradiological and radiological materials to and from the proposed IIFP site. The impacts from each are discussed below.

The principal highway routes that would handle this traffic include NM 483, which borders the site to the west, and US 62/180, which provides an east-west route to the nearest population centers. All traffic would access the site via NM 483, and most traffic would use US 62/180 to NM 483 to access the site. Some portion of the workforce may access the site from the north, using NM 83 to access NM 483 north of the site. At the junction of NM 483 and US 62/180, traffic would go east to Hobbs, Eunice and other Lea County municipalities or southwest to Eddy County. After the intersection of NM 483 and US 62/180 traffic associated with the site would be increasingly dispersed.

IIFP operations would use 140 workers working three shifts per day (IIFP, 2011a) or 47 people per shift. Therefore, if each employee commutes alone, there would be an increase of 94 vehicles (47 ending a shift plus 47 starting a shift) on NM 483 for each shift change. Additionally, IIFP estimated 10 delivery or waste disposal trucks each day.

The Highway Capacity Manual 2000 (TRB 2000) indicates that the capacity of a two-lane highway is 1,700 passenger cars per hour for a single direction and 3,400 passenger cars per hour for both directions. The AADT on NM 483, a two-lane highway, at Arkansas Junction in 2008 was 955 vehicles per day (NMDOT, 2009). If all the vehicles on NM 483 in one day used the road in a single hour, including the anticipated 10 truck trips per day (IIFP, 2011a), and if two operations workforce shifts used the road to the site during that same hour, a maximum of 1,059 vehicles would be on the road. This is less than the design capacity of a two-lane highway. The impact of traffic increases due to facility operations on US 62/180, which is a four-lane highway, would be smaller than the impact on NM 483. Therefore, the NRC staff concludes that impacts to traffic from operations would be SMALL.

Using the same assumptions for operations as for construction, 140 operation employees would commute approximately 2,300,000 km (1,400,000 mi) per year of facility operation. The New Mexico 2010 vehicle accident rates result in 51.73 injuries and 1.73 fatal accidents per 160 million vehicle km (100 million vehicle mi) traveled (UNM, 2010). Based on these rates, statistically there would be one injury (risk of less than 0.7 injury crashes) per year and no fatalities (risk of less than 0.02 fatal crashes) per year due to the Phase 1 operations traffic.

The transportation of nonradiological materials would include the delivery of routine supplies and equipment and the removal of nonradiological wastes (including hazardous wastes). The transport of hazardous waste is subject to EPA and DOT regulations. Nonradiological deliveries and waste removal would require an estimated 1,950 truck round-trips per year, or approximately 8 round-trips per day (IIFP, 2011a). As with the commuter traffic, the NRC staff finds that this increase in traffic volume would have a SMALL impact on the current traffic and the carrying capacity of the affected roads would not be challenged.

Assuming a round-trip distance of 113 km (70 mi), the round-trip distance to the furthest nonradiological waste disposal facility likely to be used by IIFP, these trucks would travel approximately 220,480 km (137,000 mi) per year of operation, therefore, no injuries (risk <0.07), and no fatalities (risk <0.002) would be expected per year of Phase 1 operation. The NRC staff concludes that impacts from accidents involving the shipment of nonradiological materials would be SMALL.

4.1.2.9.2 Incident-free Radiological Transportation

Operation of the proposed IIFP facility would require shipment of full DUF_6 cylinders from commercial enrichment facilities, empty DUF_6 cylinders back to the commercial enrichment facilities, DUO_2 to waste disposal facilities, and other process and miscellaneous LLW to waste disposal facilities. Data for the analysis came from IIFP (IIFP 2011a; IIFP, 2011b) unless specified otherwise. More detail on the analysis can be found in Appendix E.

Full DUF_6 Cylinders: The NRC staff selected all current or proposed U.S. commercial enrichment facilities as representative origins for shipments of DUF_6. These are (1) URENCO USA, just east of Eunice, New Mexico, (2) the GE-Hitachi Global Laser Enrichment Facility north of Wilmington, North Carolina, and (3) the AREVA Eagle Rock Enrichment Facility west of Idaho Falls, Idaho. The cylinders would be shipped one per 18-wheel truck. The radiation dose rate at 1 m (3.28 ft) from a cylinder is 0.0046 mSv per hour (0.46 mrem per hour) (Biwer et al., 2001). There would be 293 shipments per year of full DUF_6 cylinders for Phase 1 operations.

Empty DUF_6 Cylinders: Although it is possible that some cylinders would not be shipped back to their origin, NRC staff has assumed, for purposes of analysis, that all cylinders would be returned. In the event that cylinders are not returned, they could be disposed empty as LLW or filled with DUO_2 and disposed as LLW. The returned cylinders would have a heel of less than 23 kg (50 lb) and, thus, contain radioactive material. The cylinders are conservatively assumed to be shipped one per truck, consistent with IIFP data; however, two per truck is a likely scenario. Radiation dose rates from empty cylinders are slightly higher than from full cylinders due to the concentration of uranium daughter products and loss of self-shielding. The estimated radiation dose rate 1 m (3.28 ft) from an empty cylinder is 0.01 mSv per hour (1 mrem per hour). Conservatively, there would be 293 shipments per year of empty cylinders.

DUO_2 Waste: The DUO_2 is assumed to be waste and not sold. It would be packaged into 55-gallon drums and loaded 40 per truck (subject to weight limitations). Shipment destinations selected for analysis are the Energy*Solutions* Clive, Utah facility and the WCS facility on the Texas-New Mexico border west of Andrews, Texas (immediately east of the URENCO USA facility). Less probable destinations, such as the U.S. Ecology Washington disposal facility on the Hanford Site near Richland, Washington and the Nevada National Security Site, are represented by these analyses. The radiation dose rate 1 m (3.28 ft) from a drum would be approximately 6×10^{-4} mSv per hour (0.06 mrem per hour). IIFP estimates that there would be as many as 155 DUO_2 waste shipments per year.

Process and Miscellaneous LLW: This volume of LLW would be small compared to the DUO_2 waste. The radioactivity in most of this waste would likely be less concentrated than the DUO_2 waste. There would be 31 shipments per year, each with 40 55-gal drums. The dose rate is conservatively selected to be the same as the DUO_2 shipments, 6×10^{-4} mSv per hour (0.06 mrem per hour).

NRC staff used the TRAGIS (Johnson and Michelhaugh, 2003) transportation routing computer modeling code and the RADTRAN5 (Neuhauser and Kanipe, 2003) transportation risk assessment computer modeling code to calculate radiological impacts (collective dose) to members of the public living near the transportation route, drivers and passengers sharing the highways, persons at fueling or rest stops, the truck drivers, and package handlers. Results of that analysis are provided in Table 4-22.

Assuming a scenario in which DUF_6 shipped from the enrichment facility results in the greatest collective dose and DUO_2 waste shipped to the disposal facility results in the greatest collective dose, and summing for all receptors (Appendix E), one arrives at 0.18 person-sievert per year (18 person-rem per year). This is for receipt and return of cylinders to the GLE facility in Wilmington, North Carolina and disposal of low-level waste at the Energy*Solutions* Clive, Utah facility. Multiplying the collective dose by the Interagency Steering Committee on Radiation Standards conversion factor of 6×10^{-4} latent cancer fatalities (LCFs) per person-rem (ISCORS, 2002), estimates the transportation-related latent cancer fatalities for one year of incident-free exposure as 0.01 LCF.

The Centers for Disease Control and Prevention (CDC, 2010) estimated the age-adjusted cancer death rate in the U.S. was 178.4 deaths per 100,000 people in 2007. Similarly, 23.2 percent (23,200 per 100,000) of all deaths in the U.S. in 2007 were cancer related. Although these results are from two different studies and difficult to compare, both studies show

Table 4-22. Annual Collective Doses to Various Receptors from Radiological Transportation

Description	General Public		Drivers and Passengers		Persons at Stops		Truck Drivers		Package Handlers	
	Person-Sv	Person-Rem	Person-Sv	Person-Rem	Person-Sv	Person-Rem	Person-Sv	Person-Rem	Person-Sv	Person-Rem
Full DUF_6 cylinders from URENCO USA	2.3×10^{-5}	2.3×10^{-3}	2.6×10^{-4}	0.02.6	a	a	4.3×10^{-4}	0.043	3.9×10^{-3}	0.39
Full DUF_6 cylinders from GLE Facility	6.8×10^{-3}	0.68	0.07	4.7	0.075	7.5	0.029	2.9	3.9×10^{-3}	0.39
Full DUF_6 cylinders from AREVA Eagle Rock	3.9×10^{-3}	0.39	0.039	3.9	0.075	7.5	0.026	2.6	3.9×10^{-3}	0.3
Empty DUF_6 cylinders to URENCO	5.1×10^{-5}	5.1×10^{-3}	5.6×10^{-4}	0.056	a	a	9.3×10^{-4}	0.093	8.4×10^{-3}	0.84
Empty DUF_6 cylinders to GLE Facility	0.015	1.5	0.10	10	0.16	16	0.06.3	6.3	8.4×10^{-3}	0.81
Empty DUF_6 cylinders to AREVA Eagle Rock	8.5×10^{-3}	0.85	0.084	8.4	0.16	16	0.056	5.6	8.4×10^{-3}	0.84
DUO_2 and Miscellaneous LLW to Energy Solutions	3.0×10^{-4}	0.03	3.3×10^{-3}	0.33	5.3×10^{-3}	0.53	2.0×10^{-3}	0.20	5.8×10^{-3}	0.5
DUO_2 and Miscellaneous LLW to Waste Control Specialists	2.0×10^{-6}	2.0×10^{-4}	2.1×10^{-5}	2.1×10^{-3}	a	a	3.6×10^{-5}	3.6×10^{-3}	5.8×10^{-3}	0.5

Source: See Appendix E
[a] A stop was not assumed because the route was short

that cancer fatalities are significant in normal life. Given these cancer fatality rates, the addition of 0.01 LCF from the proposed action is considered by NRC staff to be a SMALL impact. While mitigation measures are not required, IIFP would be required by NRC regulation 10 CFR 20 to maintain all radiation doses As Low As Reasonably Achievable (ALARA).

Estimates of radiological exposure to the workforce and public from facility operations other than radiological transport are discussed in Section 4.1.2.12.

4.1.2.10   Noise Impacts

As discussed in Section 3.11, noise from the operation of the proposed facility would be minimal, occur mostly inside the buildings, and be attenuated by distance. The proposed facility is in a relatively remote location, surrounded by other industrial facilities, and far from lands uses that could be adversely affected by increases in noise levels. Noise at the nearest residences and recreational areas would not increase due to operation of the proposed IIFP facility. Therefore, the NRC staff concludes that impacts from noise of operations would be SMALL.

4.1.2.11   Public and Occupational Health Impacts

Normal operations at the proposed IIFP facility have the potential to impact the health of workers and the public due to exposures from permitted chemical and radiological gaseous emissions and liquid effluents. Additionally, workers could be impacted from direct radiation exposures and occupational hazards. This section discusses these potential impacts. Although normal operations at the proposed IIFP facility create the potential for radiological and nonradiological impacts, plant design would incorporate features to minimize gaseous and liquid effluent releases and to keep them well below regulatory limits. These features include the following (IIFP, 2011a):

- DUF_6 cylinders would be moved only when cool and when DUF_6 is in solid form, which minimizes the risk of inadvertent release due to mishandling.

- Process off-gas from DUF_6 purification and other operations would be solidified to reclaim as much DUF_6 as possible. Remaining gases pass through high-efficiency filters and chemical absorbers, which remove HF and uranium compounds.

- Liquid and solid waste handling systems and techniques would be used to control wastes and effluent concentrations.

- Gaseous emissions would pass through pre-filters, high efficiency filters, and carbon filters, all of which greatly reduce the radioactivity in the final discharged emission to very low concentrations.

- Uranium-bearing liquid waste would be routed to the Decontamination Building for removal of uranium and the treated water would be evaporated or reused in the Decontamination Building.

> **10 CFR 20 Exposure Limits**
>
> The NRC exposure limits place annual restrictions on the total dose equivalent exposure (1 mSv [100 mrem]), which includes external plus internal radiation exposures, and the dose equivalent rate (0.02 mSv [2 mrem]) in any 1 hour in unrestricted areas that are accessible by members of the public who are not employees, but who may be present during the year at the facility.
>
> Source: 10 CFR 20.1301

- Effluent paths would be monitored and sampled to assure compliance with regulatory discharge limits.

Radiological Impacts

The general public could be impacted by radiation and radioactive material from the IIFP facility via controlled releases of gas associated with the uranium process lines during routine operations and from decontamination and maintenance of equipment, or direct radiation exposure associated with transportation and storage of DUF_6 cylinders and wastes.

The radiation exposure limits for the general public have been established by the NRC in 10 CFR 20. Routine operations would be conducted to ensure that public exposure at off-site locations would be within these limits. Annual exposure to the public would be maintained \ALARA through effluent controls and monitoring.

The potential radiological impacts to the public from operations at the proposed IIFP facility are those associated with chronic exposure to low levels of radiation and not the immediate health effects associated with acute radiation exposure. The major sources of potential radiation exposure (chronic or acute) are the gaseous discharges from the plant scrubber systems for the DUF_4 and fluorine extraction processes and the dust collector scrubber system. It is estimated that the total amount of uranium released to the air from the proposed IIFP facility would be less than 0.5 kg (1.1 lb) per year. Due to the low volume of contaminated liquid waste anticipated by the applicant, no liquid effluent discharges are expected to contain radiological waste. Therefore, there would be no dose pathway and no significant radiological impact to the public or the environment from liquid effluent discharges. The radiological impacts associated with direct radiation from indoor operations are not expected to be a significant contributor to dose to the public because the low-energy gamma-rays associated with the uranium would be absorbed almost completely by the process lines, equipment, cylinders, and building structures (IIFP, 2011a). Routine radiological gaseous releases from the proposed IIFP facility are listed in Table 4-23.

Table 4-23.   Estimated and Bounding Radiological Releases from the Stacks

Radionuclide	DUF_6 to DUF_4 Stack		SiF_4 and BF_3 Production Stack	
	kBq/yr	Ci/yr	kBq/yr	Ci/yr
Estimated Releases				
^{234}U	461	1.25×10^{-5}	42.2	1.14×10^{-6}
^{235}U	44.5	1.20×10^{-6}	4.08	1.10×10^{-7}
^{238}U	3,500	9.46×10^{-5}	321	8.66×10^{-6}
Total	**4,005.5**	**1.08×10^{-4}**	**367.3**	**9.91×10^{-6}**
Bounding Releases				
^{234}U	922	2.49×10^{-5}	84.5	2.28×10^{-6}
^{235}U	89.1	2.41×10^{-6}	8.16	2.21×10^{-7}
^{238}U	7,000	1.89×10^{-4}	641	1.73×10^{-5}
Total	**8,010**	**2.16×10^{-4}**	**734**	**1.98×10^{-5}**

Source: IIFP, 2011a
kBq = kilobecquerel (2.7×10^{-7} curies)

There are three primary exposure pathways associated with plant effluent: inhalation; direct radiation due to deposited radioactivity on the ground surface ("ground plane exposure"); and ingestion of contaminated food products. Of these three exposure pathways, inhalation exposures are expected to be the predominant pathways at site boundary locations and also at off-site locations that are relatively close to the site boundary. Because airborne concentrations decrease with the distance from the discharge point, for gaseous releases from the proposed IIFP facility, the highest off-site airborne concentrations (and, hence, the greatest radiological impacts) are expected at locations near the site boundary. Beyond those locations, the concentrations of airborne radioactive material would decrease continuously because of dispersion of the material and depletion processes.

The critical populations for determining dose impacts include the resident nearest to the proposed IIFP facility (at the northwest boundary) and the maximally exposed individual (MEI). The MEI is a hypothetical person living at the point of highest projected total uranium concentrations. The impact due to gaseous releases was evaluated for the dose from the three primary exposure pathways identified above. Because there is no pathway for contamination of drinking water, no radiological contamination of drinking water was considered in the analysis. The analysis included dose equivalent assessments for four age groups (i.e., adults, teens, children, and infants) for these pathways.

IIFP calculated doses using GENII (version 2.08), which is a dose assessment model developed for EPA for calculating radiation dose and risk from radionuclides released to the environment. Dose equivalents for the MEI and the nearest resident due to gaseous releases were calculated by pathway for the total body in adults, teens, children, and infants, and are presented in Tables 4-24 and 4-25, respectively. For the MEI, the highest committed effective dose equivalent (CEDE) from the proposed IIFP facility emissions was calculated to be 1.40×10^{-7} Sv (1.40×10^{-5} rem) per year. For the adult fulltime resident nearest to the facility, the highest CEDE from the IIFP facility was calculated to be 9.46×10^{-8} Sv (9.46×10^{-6} rem) per year.

> **Committed Effective Dose Equivalent (CEDE)**
>
> Committed effective dose equivalent is the sum of the products of the weighting factors applicable to each of the body organs or tissues that are irradiated and the committed dose equivalent to these organs or tissues.
>
> Source: 10 CFR 20.1003.

In its environmental report (IIFP, 2009a), IIFP calculated direct dose rates for the MEI and the nearest resident. These doses rates were extremely small (e.g., less than 1.04×10^{-2} mSv per year [1.04 mrem per year]). The CEDE and the direct dose equivalent were totaled to determine the total effective dose equivalent (TEDE) for the MEI. The highest TEDE was determined to be 0.21 mSv per year (20.8 mrem per year), which is approximately one-fifth of the NRC exposure limit of 1 mSv (100 mrem). Doses for public receptors at other sites of interest (e.g., schools and hospitals) would be lower than those of the MEI because the airborne

> **Total Effective Dose Equivalent (TEDE)**
>
> Total effective dose equivalent is the sum of the effective dose equivalent or the deep-dose equivalent (for external exposures) and the committed effective dose equivalent (for internal exposures).
>
> Source: 10 CFR 20.1003.

concentrations of uranium would be lower at these more distant locations. Therefore, NRC staff anticipates that radiological impacts to off-site receptors from routine combined effluent releases and direct radiation would to be SMALL.

Table 4-24. Annual and Committed Dose Equivalents for Exposures to the MEI from Gaseous Effluents

Source	Units	Adult EDE	Teen EDE	Child EDE	Infant EDE
Cloud Immersion	Sv	5.77×10^{-16}	5.77×10^{-16}	5.77×10^{-16}	5.77×10^{-16}
	rem	5.77×10^{-14}	5.77×10^{-14}	5.77×10^{-14}	5.77×10^{-14}
Inhalation	Sv	3.06×10^{-8}	3.67×10^{-8}	6.19×10^{-8}	1.30×10^{-7}
	rem	3.06×10^{-6}	3.76×10^{-6}	6.19×10^{-6}	1.30×10^{-5}
Ingestion	Sv	1.30×10^{-9}	1.96×10^{-9}	2.35×10^{-9}	9.79×10^{-9}
	rem	1.30×10^{-7}	1.96×10^{-7}	2.35×10^{-7}	9.79×10^{-7}
Ground Plane Exposure	Sv	2.08×10^{-10}	2.08×10^{-10}	2.08×10^{-10}	2.08×10^{-10}
	rem	2.08×10^{-8}	2.08×10^{-8}	2.08×10^{-8}	2.08×10^{-8}
Total Dose	**Sv**	3.21×10^{-8}	3.88×10^{-8}	6.45×10^{-8}	1.40×10^{-7}
	rem	3.21×10^{-6}	3.88×10^{-6}	6.45×10^{-6}	1.40×10^{-5}

Source: IIFP, 2011a

Table 4-25. Annual and Committed Dose Equivalents for Exposure to the Nearest Resident from Gaseous Effluents

Source	Units	Adult EDE	Teen EDE	Child EDE	Infant EDE
Cloud Immersion	Sv	4.40×10^{-17}	4.40×10^{-17}	4.40×10^{-17}	4.40×10^{-17}
	rem	4.40×10^{-15}	4.40×10^{-15}	4.40×10^{-15}	4.40×10^{-15}
Inhalation	Sv	2.20×10^{-8}	2.65×10^{-3}	4.44×10^{-8}	9.38×10^{-6}
	rem	2.20×10^{-6}	2.65×10^{-6}	4.44×10^{-6}	9.38×10^{-6}
Ingestion	Sv	9.91×10^{-11}	1.49×10^{-10}	1.79×10^{-10}	7.43×10^{-10}
	rem	9.91×10^{-9}	1.49×10^{8}	1.79×10^{-8}	7.43×10^{-8}
Ground Plane Exposure	Sv	1.59×10^{-11}	1.59×10^{-11}	1.59×10^{-11}	1.59×10^{-11}
	rem	1.59×10^{-9}	1.59×10^{-9}	1.59×10^{-9}	1.59×10^{-9}
Total Dose	**Sv**	2.21×10^{-8}	2.66×10^{-8}	4.46×10^{-8}	9.46×10^{-8}
	rem	2.21×10^{-6}	2.66×10^{-6}	4.46×10^{-6}	9.46×10^{-6}

Source: IIFP, 2011a

Potential doses to the total population within an 80-km (50-mi) radius of the proposed IIFP facility were also determined. The local area population distribution was derived from U.S. Census Bureau 2000 data for counties in New Mexico and Texas (IIFP, 2011a) that fall all or in part within the 80-km (50-mi) radius of the proposed IIFP site. A standard 16-sector compass rose was centered on the IIFP site and divided into annular rings out to a distance of 80 km (50 mi) (see Figure 4-1). Using census data, significant population groups, typically towns or cities, within the 80-km (50-mi) area were identified in those sectors. Table 4-26 and Table 4-27 present the total population doses expected in units of person-sieverts and person-rem,

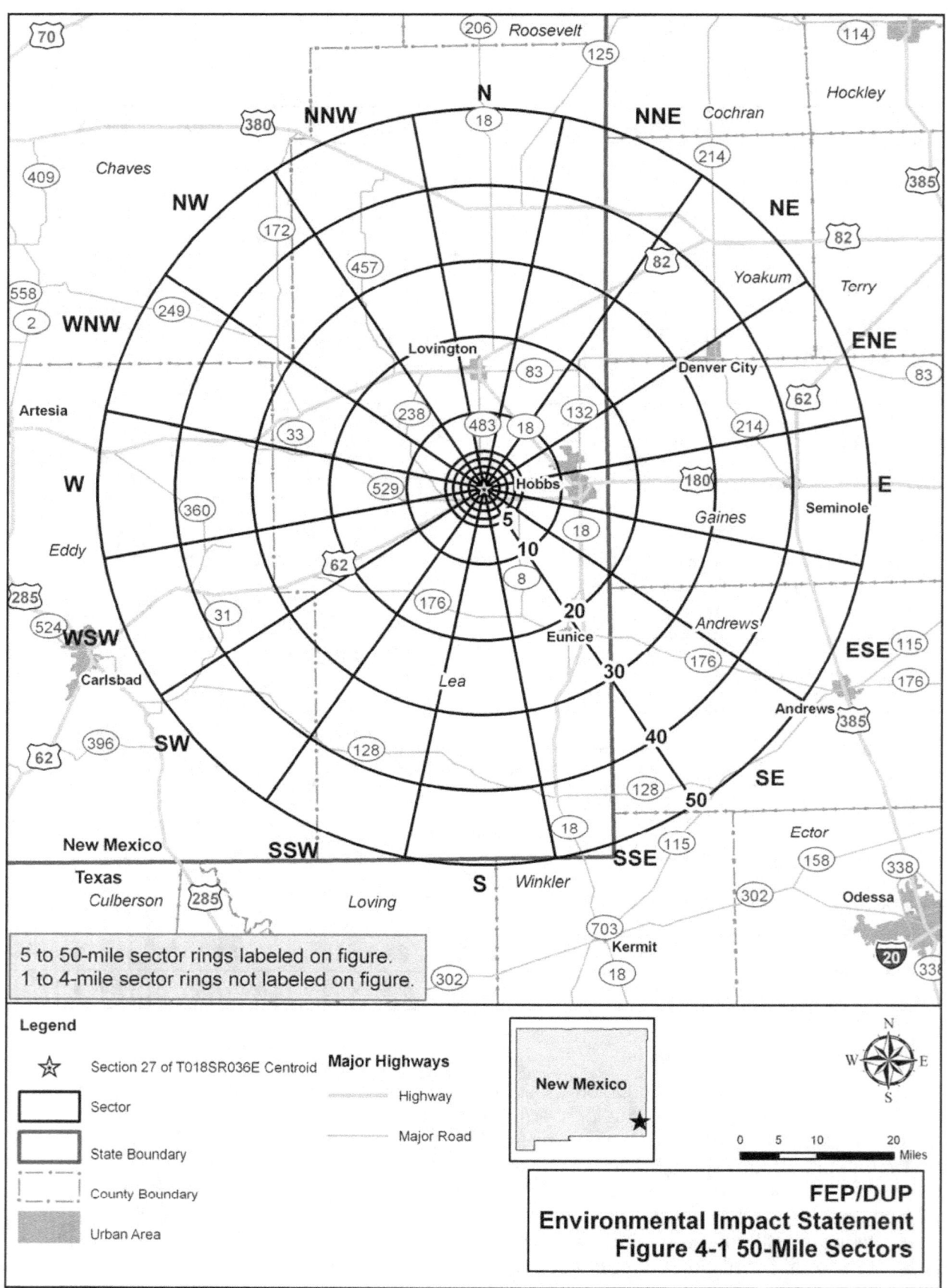

5 to 50-mile sector rings labeled on figure.
1 to 4-mile sector rings not labeled on figure.

Legend

☆   Section 27 of T018SR036E Centroid

▢   Sector

▢   State Boundary

▢   County Boundary

▨   Urban Area

Major Highways

——— Highway

——— Major Road

New Mexico

FEP/DUP
Environmental Impact Statement
Figure 4-1 50-Mile Sectors

Table 4-26. Collective Dose Equivalents to All Ages Population (person-Sv) (gas release pathways)

Vector	0-1 mi	1-2 mi	2-3 mi	3-4 mi	4-5 mi	5-10 mi	10-20mi	20-30 mi	30-40 mi	40-50 mi	Total
E	0.00	0.00	0.00	0.00	0.00	0.00	1.82×10^{-4}	4.03×10^{-6}	9.53×10^{-6}	5.15×10^{-6}	2.01×10^{-4}
ENE	0.00	0.00	0.00	0.00	0.00	3.72×10^{-5}	4.08×10^{-5}	0.00	1.16×10^{-5}	4.77×10^{-5}	9.44×10^{-5}
NE	0.00	0.00	0.00	0.00	0.00	0.00	0.00	0.00	0.00	3.25×10^{-6}	3.25×10^{-6}
NNE	0.00	0.00	0.00	0.00	0.00	0.00	6.54×10^{-6}	0.00	0.00	0.00	6.54×10^{-6}
N	0.00	0.00	0.00	0.00	0.00	0.00	9.55×10^{-5}	3.56×10^{-6}	3.56×10^{-6}	0.00	1.03×10^{-4}
NNW	0.00	0.00	0.00	0.00	0.00	0.00	9.89×10^{-6}	0.00	0.00	0.00	9.89×10^{-6}
NW	0.00	0.00	0.00	0.00	0.00	0.00	0.00	0.00	0.00	0.00	0.00
WNW	0.00	0.00	0.00	0.00	0.00	0.00	0.00	0.00	0.00	0.00	0.00
W	0.00	0.00	0.00	0.00	0.00	0.00	0.00	0.00	0.00	0.00	0.00
WSW	0.00	0.00	0.00	0.00	0.00	0.00	0.00	0.00	0.00	0.00	0.00
SW	0.00	0.00	0.00	0.00	0.00	0.00	0.00	0.00	0.00	0.00	0.00
SSW	0.00	0.00	0.00	0.00	0.00	0.00	0.00	0.00	0.00	0.00	0.00
S	0.00	0.00	0.00	0.00	0.00	0.00	0.00	0.00	0.00	3.28×10^{-6}	3.28×10^{-6}
SSE	0.00	0.00	0.00	0.00	0.00	0.00	0.00	8.68×10^{-6}	0.00	0.00	8.68×10^{-6}
SE	0.00	0.00	0.00	0.00	0.00	0.00	0.00	0.00	0.00	0.00	0.00
ESE	0.00	0.00	0.00	0.00	0.00	0.00	4.43×10^{-6}	0.00	0.00	6.05×10^{-7}	5.03×10^{-6}
Ring	0.00	0.00	0.00	0.00	0.00	3.72×10^{-5}	3.39×10^{-4}	1.63×10^{-5}	2.47×10^{-5}	1.71×10^{-5}	4.34×10^{-4}
Cumulative	0.00	0.00	0.00	0.00	0.00	3.72×10^{-5}	3.76×10^{-4}	3.92×10^{-4}	4.17×10^{-4}	4.34×10^{-4}	

Source: IIFP, 2011a

Table 4-27. Collective Dose Equivalents to All Ages Population (person-rem) (gas release pathways).

Vector	0-1 mi	1-2 mi	2-3 mi	3-4 mi	4-5 mi	5-10 mi	10-20 mi	20-30 mi	30-40 mi	40-50 mi	Total
E	0.00	0.00	0.00	0.00	0.00	0.00	0.0182	4.03×10^{-4}	9.53×10^{-4}	5.15×10^{-4}	0.0201
ENE	0.00	0.00	0.00	0.00	0.00	3.72×10^{-3}	4.08×10^{-3}	0.00	1.16×10^{-3}	4.77×10^{-4}	9.44×10^{-3}
NE	0.00	0.00	0.00	0.00	0.00	0.00	0.00	0.00	0.00	3.25×10^{-4}	3.25×10^{-4}
NNE	0.00	0.00	0.00	0.00	0.00	0.00	6.54×10^{-4}	0.00	0.00	0.00	6.54×10^{-4}
N	0.00	0.00	0.00	0.00	0.00	0.00	9.55×10^{-3}	3.56×10^{-4}	0.00	0.00	0.0103
NNW	0.00	0.00	0.00	0.00	0.00	0.00	9.89×10^{-4}	0.00	0.00	0.00	9.89×10^{-4}
NW	0.00	0.00	0.00	0.00	0.00	0.00	0.00	0.00	0.00	0.00	0.00
WNW	0.00	0.00	0.00	0.00	0.00	0.00	0.00	0.00	0.00	0.00	0.00
W	0.00	0.00	0.00	0.00	0.00	0.00	0.00	0.00	0.00	0.00	0.00
WSW	0.00	0.00	0.00	0.00	0.00	0.00	0.00	0.00	0.00	0.00	0.00
SW	0.00	0.00	0.00	0.00	0.00	0.00	0.00	0.00	0.00	0.00	0.00
SSW	0.00	0.00	0.00	0.00	0.00	0.00	0.00	0.00	0.00	0.00	0.00
S	0.00	0.00	0.00	0.00	0.00	0.00	0.00	0.00	0.00	3.28×10^{-4}	3.28×10^{-4}
SSE	0.00	0.00	0.00	0.00	0.00	0.00	0.00	8.68×10^{-4}	0.00	0.00	8.68×10^{-4}
SE	0.00	0.00	0.00	0.00	0.00	0.00	0.00	0.00	0.00	0.00	0.00
ESE	0.00	0.00	0.00	0.00	0.00	0.00	4.43×10^{-4}	0.00	0.00	6.05×10^{-7}	5.03×10^{-4}
Ring	0.00	0.00	0.00	0.00	0.00	3.72×10^{-3}	0.0339	1.63×10^{-3}	2.47×10^{-3}	1.71×10^{-3}	**0.0434**
Cumulative	0.00	0.00	0.00	0.00	0.00	3.72×10^{-3}	0.0376	0.0376	0.0417	0.0434	

Source: IIFP, 2011a

respectively. As shown on those tables, the total population dose would be 4.34×10^{-4} person-Sv/yr (4.34×10^{-2} person-rem/yr). Multiplying the total population dose by the Interagency Steering Committee on Radiation Standards conversion factor of 6×10^{-4} LCFs per person-rem (ISCORS, 2002), yields approximately 2.6×10^{-5} LCFs expected in the 80-km (50-mi) population for every for one year of operation of the IIFP facility. To put this population dose into perspective, based on statistics, the proposed IIFP facility would need to operate for approximately 38,400 years to produce 1 LCF in the 80-km (50-mi) population. Therefore, NRC staff anticipates that radiological impacts to the 80-km (50-mi) population would be SMALL.

Workers at the IIFP plant would be subject to higher potential exposures than members of the public because they would be involved directly with handling uranium cylinders, uranium processes, and decontamination and maintenance of equipment. During routine operations, workers at the plant potentially could be exposed to radiation from uranium via inhalation of airborne particles and direct exposure to equipment and components containing uranic materials. The radiation protection program at the IIFP facility would require routine radiation surveys and air sampling to ensure that worker exposures are maintained ALARA. Exposure-monitoring techniques at the plant would include personal dosimeters worn by workers, personnel breathing zone air sampling, and annual whole-body counting.

Potential doses to workers were estimated based on analyses conducted for similar DUF_6 deconversion operations at the DOE Piketon (Portsmouth) Ohio and Paducah, Kentucky facilities. For those facilities, the TEDE for workers was conservatively estimated to be about 0.75 mSv per year (75 mrem per year) for involved workers in the deconversion facility. The average TEDE for workers at the cylinder yards was estimated to range from 4.3 mSv per year (430 mrem per year) to 6.9 mSv per year (690 mrem per year) (DOE, 2004a; DOE, 2004b). These doses would be well below the regulatory limit of 50 mSv (5 rem) codified in 10 CFR 20.1201.

Annual radiation exposure for an employee would be controlled, monitored, and maintained ALARA through the Radiation Protection Program at the IIFP plant. The Radiation Protection Program would comply with all applicable NRC requirements established in 10 CFR 20, Subpart B. The radiation exposure of involved workers is estimated to be well within public health standards and the NRC staff finds that radiological impacts to facility workers would be SMALL. Section 4.1.2.9.2 discusses the potential impacts to workers associated with radiological transportation.

Nonradiological Impacts

Routine nonradiological gaseous fluoride effluents from the plant are listed in Table 4-28. For Phase 1 operations, approximately 52.7 kg/yr (116 lb/yr) of HF would be released from the IIFP process stacks. Additionally, approximately 64 kg (141 lb) of BF_3 and 3.7 kg (8.2 lb) of SiF_4 would be released through the stack annually. Emissions of regulated air pollutants would come predominately from the operating natural gas-fired boiler that would be used to provide steam for the plant heating and autoclave feed systems (the facility would have two boilers for redundancy, but only one would operate at any given time. Emission data estimated for the boiler indicates that it would not emit more than 13.2 metric tons (14.5 tons) per year of any regulated air pollutants. At 100 percent power, the boiler would emit 0.93 metric tons (1.03 tons) per year of CO, and 0.11 metric tons (0.12 tons) per year of NO_x. IIFP would determine if the boilers would require an air quality permit from the State of New Mexico (IIFP, 2011a).

Table 4-28.   Estimated Annual Nonradiological Gaseous Fluoride Emissions

| Emission | Estimated Releases | |
	DUF$_6$ Dust Collector Stack	SiF$_4$ & BF$_3$ Dust Collector Stack
SiF$_4$	N/A	3.7 kg/yr (8.19 lb/yr)
BF$_3$	N/A	62.5 kg/yr (137.75 lb/yr)
HF	49.3 kg/yr (108.72 lb/yr)	0.3 kg/yr (0.77 lb/yr)

Source: IIFP, 2011a

Nonradiological effluents would not exceed criteria in 40 CFR 50, 59, 60, 61, 122, 129, or 141 (IIFP, 2009a). The primary chemical hazard is HF. HF is a clear, colorless, corrosive, fuming liquid with a very acrid odor. A release can form dense white vapor clouds. Both liquid and vapor can cause severe burns to all parts of the body. Exposure to skin, eyes and inhalation or ingestion can cause severe health consequences, including death.

The facility would not discharge any industrial effluents to natural surface waters or soil, and there is no plant facility tie-in to a public waste water treatment facility. All effluents would be contained on the IIFP site via collection tanks. No routine liquid effluent discharge is expected; therefore, there would be no public impact.

The NRC staff finds that impacts from routine releases (Phase 1 operations) to the public would be SMALL.

No worker exposures exceeding the OSHA Standards for Toxic and Hazardous Substances (29 CFR 1910, Subpart Z) are anticipated (IIFP, 2011a). Additionally, handling of all chemicals and wastes would be conducted in accordance with the site Environment, Health, and Safety Program which would conform to 29 CFR 1910 OSHA standards and specify the use of appropriate engineered controls, as well as personnel protective equipment, to minimize potential chemical exposures (IIFP, 2011a).

In addition to the radiological hazards associated with uranium, workers may be potentially exposed to the chemical hazards associated with uranium. UF$_6$ is hygroscopic (moisture absorbing) and, in contact with water, would chemically breakdown into UO$_2$F$_2$ and HF. When released to the atmosphere, gaseous UF$_6$ combines with humidity to form a cloud of particulate UO$_2$F$_2$ and HF fumes. The reaction is very fast and is dependent on the availability of water vapor. Consequently, an inhalation of UF$_6$ is typically an internal exposure to HF and UO$_2$F$_2$. In addition to the radiation dose, a worker would be subjected to two other primary toxic effects: the uranium in the uranyl complex acts as a heavy metal poison that can affect the kidneys, and the HF can cause acid burns to the skin and lungs if concentrated. Because of low specific activity values, the radiotoxicity of UF$_6$ and its products are smaller than their chemical toxicity (IIFP, 2011a).

Because of the containment systems for gasses used or created in the plant process, and the personal protective equipment that would be used in areas where exposure could occur, worker exposure to in-plant gaseous releases would be minimal, and no exposures exceeding 29 CFR 1910, Subpart Z are anticipated (IIFP, 2009a). Laboratory and maintenance operations involving hazardous gaseous or respirable effluents would be conducted with ventilation control (i.e., fume hoods, local exhaust, or similar) and with the use of respiratory protection, as required. All regulated gaseous effluents would be below regulatory limits as specified by the New Mexico Air Quality Bureau (IIFP, 2011a). The NRC staff finds that impacts from routine releases within the facility (Phase 1 operations) to workers would be SMALL.

The proposed action involves a major industrial activity with the potential to cause temporary injuries, long-term injuries and/or disabilities, and even fatalities to workers. Common occupational accidents at facilities similar to the proposed IIFP plant typically involve hand and finger injuries, tripping accidents, minor burns and impacts due to striking objects or falling objects. To estimate the number of potential fatal and nonfatal occupational injuries from operation of the proposed IIFP facility, data on fatal and nonfatal occupational injuries per worker per year were collected from the U.S. Department of Labor's Bureau of Labor Statistics. Nonfatal and fatal occupational injury rates for the manufacturing industry were used to calculate the estimated fatal and nonfatal injuries associated with operation of the proposed IIFP facility. As shown in Table 4-29, less than four nonfatal injuries and less than one fatality are expected annually during operation of the proposed IIFP facility. The NRC staff finds that the impacts to human health from occupational injuries during operation would be SMALL.

Table 4-29. Annual Nonfatal and Fatal Occupational Injuries Projected for Operation of the IIFP Facility

Category	Injury Rate	Expected Occurrences
Nonfatal Injuries	2.3 per 100 workers[a]	3.2
Fatal Injuries	2.2 per 100,000 workers[b]	less than 1 (3.1×10^{-3})

[a] The expected nonfatal injury rate (total recordable cases) is from BLS (2010a).
[b] The fatal injury rate is from BLS (2010b)
[c] Expected occurrences are based on 140 workers during Phase 1 operations.

Worker health and safety at the proposed IIFP facility would be protected by its Chemical Safety Program, the Radiation Protection Program, and the Industrial Safety Program. These programs would comply with applicable State, NRC (10 CFR 20), and OSHA (29 CFR 1910) requirements. Work environments that present the potential for exposure to chemical, biological, or physical agents (e.g., radiation, noise, heat/cold, vibration) would be evaluated, and appropriate safety controls would be implemented and/or safety equipment would be assigned to workers. Personal protective equipment requirements would be based on the nature of the work and chemical and/or radiological hazards present and would be a key component to minimizing exposure to chemical and radiological agents. Exposure monitoring would be conducted on radiation workers to evaluate their personal exposure; if personal monitoring is not feasible, work area monitoring would be used to represent personal exposure.

The NRC staff finds that the impacts to human health from occupational injuries during operation would be SMALL.

4.1.2.12   Waste Management Impacts

Waste generation during facility operation would be minimized through reduction, reuse, and recycling, as applicable to specific waste streams. The proposed IIFP facility would incorporate waste minimization systems in its operational procedures and design with the goal of conserving materials, recycling important compounds, and preventing the spread of contamination. Good work practices would be used to collect and sort the wastes generated during operation for recycling or disposal (e.g., using designated roll-off containers and collection areas for different types of wastes) (IIFP, 2011a).

There would be no permanent onsite disposal of any waste; only temporary storage. Wastes generated at the proposed IIFP facility would be disposed of at licensed facilities designed to accept the various waste types. The management of stormwater and wastewater at the proposed IIFP facility is discussed in Section 4.1.2.6.

Solid waste, including sanitary waste, miscellaneous trash, vehicle air filters, empty cutting oil cans, miscellaneous scrap metal, and paper would be shipped offsite for recycling or minimization, if appropriate, or transported offsite to an approved local landfill (IIFP, 2011a).

The radioactive DUO_2 waste from the deconversion process would be shipped to an offsite LLW disposal facility licensed to accept DUO_2. Other LLW, including dust collector bags, ion exchange resin, crushed contaminated drums, contaminated trash, contaminated coke, and carbon trap material, would be collected in labeled containers in each Restricted Area and transferred to the Radioactive Waste Storage Area for inspection. Waste would be volume-reduced, if appropriate, and disposed of at a licensed LLW disposal facility.

Hazardous wastes and some mixed wastes would be collected at the point of generation in approved containers, transferred to the onsite Waste Storage Area, inspected, classified, and shipped by a licensed transporter to a hazardous waste treatment or disposal facility. The majority of the projected hazardous waste is the potential waste CaF_2. As described in Section 2.1.6.4.2, the KOH regeneration process results in CaF_2 that would be packaged and stored for sale. If a market for this material is not identified, the CaF_2 would be sent to a licensed hazardous waste disposal facility. Any mixed waste would be treated in its original collection container prior to shipment for offsite disposal, or shipped directly to a mixed waste processor (IIFP, 2011a).

Tables 4-30, 4-31, and 4-32 provide information on the types and estimated annual quantities of solid, hazardous, and LLW, respectively, generated from Phase 1 operations at the proposed IIFP facility.

As described in Section 4.1.1.12, nonhazardous solid wastes from the proposed IIFP facility would likely be transported to the Lea County landfill for disposal. The Lea County landfill receives approximately 82,500 tons of solid waste annually (NMED, 2009). Nonhazardous, industrial waste generated from operation of the proposed facility (up to 46 tons per year as shown in Table 4-30) would result in an increase of approximately 0.06 percent in the waste that the Lea County Landfill receives annually from all other sources. The NRC staff finds that this quantity of nonhazardous waste material would result in SMALL impacts that could be managed effectively.

Hazardous wastes would be packaged and shipped offsite to licensed hazardous waste treatment and disposal facilities in accordance with Federal and State regulations (IIFP, 2011a). Table 4-31 shows that the quantity of hazardous waste generated by operations could be as much as 154 tons per year if a market for the CaF_2 cannot be identified. The projected annual hazardous waste generation would likely classify the proposed IIFP facility as large quantity generator (over 1,000 kg/mo [2,200 lb/mo]). As discussed in Section 4.1.1.12, hazardous waste generators in New Mexico produced 1,078,672 tons of hazardous waste in 2009 (EPA, 2010b).

4-54

The maximum IIFP generation rate would result in an increase of less than 0.02 percent in the hazardous waste generated annually in the State of New Mexico. Therefore, the NRC staff finds that the quantity of operations hazardous waste material would result in SMALL impacts that could be managed effectively.

Table 4-30. Solid Waste Generation – Operations

Material	Estimated Annual Amount
Clothing	45 – 90 kg (100 – 200 lbs)
Molecular sieve	140 – 230 kg (300– 500 lbs)
Municipal trash waste	27,000 – 41,000 kg (60,000 – 90,000 lbs)
Safety gear	90 – 180 kg (200 – 400 lbs)
Waste Glass	23 – 90 kg (50 – 200 lbs)
Total	27,500 – 41,400 kg (60,650 – 91,300 lbs)

Source: IIFP, 2011a

Table 4-31. Hazardous Waste Generation – Operations

Material	Estimated Annual Amount
Aerosol cans, paints cans, bulbs	450 – 1400 kg (1,000 – 3,000 lbs)
CaF_2[a]	90,000 – 136,000 kg (200,000 – 300,000 lbs)
Lab chemicals	90 – 180 kg (200 – 400 lbs)
Oil sorbent	900 – 2,300 kg (2,000 – 5,000 lbs)
Total[a]	92,000 – 140,000 kg (203,200 – 308,400 lbs)

Source: IIFP, 2011a
[a] Includes CaF_2 that would not be waste if sold.

Table 4-32. Low Level Radioactive Waste Generation – Operations

Material	Estimated Annual Amount
Activated alumina	900 – 1,800 kg (2,000 – 4,000 lbs)
Air ventilation filters	23 – 45 kg (50 – 100 lbs)
Carbon	11,000 – 14,000 kg (25,000 – 30,000 lbs)
DUF_4 clinkers	2,300 – 4,500 kg (5,000 – 10,000 lbs)
Coke	3,600 – 5,400 kg (8,000 – 12,000 lbs)

Table 4-32.   Low Level Radioactive Waste Generation – Operations (Continued)

Material	Estimated Annual Amount
Crushed drums	450 – 1,400 kg (1,000 – 3,000 lbs)
Dust collector bags	230 – 1,400 kg (500 – 3,000 lbs)
Ion exchange resin	450 – 900 kg (1,000 – 2,000 lbs)
Oxide for burial (plus drums)	1,270,000 – 2,800,000 kg (2,800,000 – 6,200,000 lbs)
Radioactive waste trash	16,000 – 25,000 kg (35,000 – 55,000 lbs)
Scrap metal	1,800 – 3,600 kg (4,000 – 8,000 lbs)
Sintered metal tubes	450 – 900 kg (1,000 – 2,000 lbs)
Sodium fluoride	900 – 1,800 kg (2,000 – 4,000 lbs)
Spent blasting sand	45 – 90 kg (100 – 200 lbs)
Wood trash (pallets)	450 – 1,800 kg (1,000 – 4,000 lbs)
Total	**1,309,000 – 2,875,000 kg** **(2,885,650 – 6,337,300 lbs)**

Source: IIFP, 2011a

Depleted uranium is classified as Class A low level waste; however, a specific disposal site may place additional limits on concentration, volume or waste form. Disposal options, including waste form, would be determined after licensing and may change over the operating life of the facility; however, licensed LLW disposal facilities, including the U.S. Ecology site in Richland, Washington; Energy*Solutions* site in Clive, Utah, DOE's site in Area 5 of the Nevada National Security Site (formerly known as the Nevada Test Site), and the WCS facility in Andrews, Texas are potentially viable options, provided regulatory and contractual conditions can be satisfied. The U.S. Ecology facility is in the Pacific Northwest Compact, which has an agreement with Rocky Mountain Compact, of which New Mexico is a member, to dispose of waste but the U.S. Ecology facility would need a revision in the allowable total uranium inventory. Energy*Solutions* accepts shipments from all states. Shipment to the Nevada National Security Site would require DOE to accept possession of the LLW (consistent with Section 13 of the USEC Privatization Act of 1996).

The WCS facility is 42 km (26 mi) southeast of the proposed site but is currently limited to waste from the Texas Compact and therefore, would have to establish approval mechanisms for out-of-compact waste to be disposed. Furthermore, the Rocky Mountain Compact would have to approve shipment outside the compact. The analysis in this EIS is not intended to support selection of the LLW disposal facility for the DUO_2.

Decisions regarding the disposal location for DUO_2 and other LLW would be made based on economic and other considerations. For analysis purposes, the radioactive wastes were assumed to be shipped to the Energy*Solutions* site in Clive, Utah. As shown in Table 4-32, up to 3,170 tons per year of LLW could be sent for disposal. Most of the LLW generated (approximately 97 percent) would be the DUO$_2$ produced by the deconversion process. The DUO$_2$ and other LLW generated would be Class A waste (IIFP, 2009a). The projected quantities of DUO$_2$ and other Class A LLW generated by the proposed IIFP facility operations would have little effect on the available disposal capacity for such material. The projected volume of DUO$_2$ waste (up to 6,200 55-gal drums or 1,300 m^3/yr) represents approximately 0.04 percent of the 3.1 million m^3 disposal volume of the Class A cell at the Clive facility (DOE 2000). The Clive facility accepts most of the United States' Class A waste and is estimated to have capacity to accept this waste at current volume levels for more than 20 years (GAO, 2004). The NRC staff finds that the potential impact of proposed IIFP facility operations on LLW disposal capacity would be SMALL.

4.1.2.13   Impacts of Postulated Accidents

4.1.2.13.1   Facility Accidents

The operation of the proposed IIFP facility would involve risks to workers, the public, and the environment from potential accidents. The facility would be licensed under 10 CFR 40, Domestic Licensing of Source Material, and would also be subject to consideration of 10 CFR 70, Subpart H, Additional Requirements for Certain Licensees Authorized to Possess a Critical Mass of Special Nuclear Material, as part of the licensing basis for the application review of certain new source material facilities as an interim measure pending the completion of 10 CFR 40 rulemaking. NRC regulation 10 CFR 70 requires that each applicant or licensee evaluate, in an Integrated Safety Analysis (ISA), its compliance with certain performance requirements. As part of the safety review, the NRC staff would conduct a confirmatory analysis, which independently evaluates the consequences of potential accidents identified in IIFP's ISA plans. The accidents evaluated are a representative selection of the types of accidents that are possible at the proposed facility.

The analytical methods used in the NRC staff's consequence assessment are based on NRC guidance for analysis of nuclear fuel-cycle facility accidents (NRC, 1990; NRC, 1991; NRC, 1998) and regulatory guidance cited by IIFP (EPA, 1999). The consequence assessment considered the available information regarding the facility prior to final design. The NRC staff analyzed accidents involving the release of HF, the primary chemical hazard at the facility. HF is a clear, colorless, corrosive, fuming liquid. In high concentrations, a release could form dense white vapor clouds. HF releases pose a chemical risk to workers, the public, and the environment. Both direct releases of HF and releases from a byproduct reaction involving

---

Acute Exposure Guideline Levels (AEGLs)

AEGLs represent threshold exposure limits for the general public and are applicable to five emergency exposure periods (10 minutes, 30 minutes, 1 hour, 4 hours, and 8 hours) and are distinguished by varying degrees of severity of toxic effects. It is believed that the recommended exposure levels are applicable to the general population including infants and children, and other individuals who may be susceptible. The three AEGLs have been defined as follows:

AEGL-1 is the airborne concentration of a substance, expressed as parts per million or milligrams per cubic meter (ppm or mg/m^3) above which it is estimated that the general population, including susceptible individuals, could experience notable discomfort, irritation, or certain asymptomatic nonsensory effects. However, the effects are not disabling and are transient and reversible upon cessation of exposure.

AEGL-2 is the airborne concentration (expressed as ppm or mg/m^3) of a substance above which it is estimated that the general population, including susceptible individuals, could experience irreversible or other serious, long-lasting adverse health effects or an impaired ability to escape.

AEGL-3 is the airborne concentration (expressed as ppm or mg/m^3) of a substance above which it is estimated that the general population, including susceptible individuals, could experience life-threatening health effects or death.

other fluoride species (DUF$_6$, DUF$_4$, SiF$_4$ and BF$_3$) could pose accident risks. NRC staff also evaluated accidents involving radioactive materials (depleted uranium bound with fluoride and/or oxide) for radiation and chemical (heavy metal toxicity) impacts.

4.1.2.13.1.1 Accidents Considered

A number of potential accidents could occur at the proposed facility. The NRC staff selected, for detailed evaluation, a subset of the potential accident scenarios that is intended to encompass the range of possible accidents. The accident sequences the staff selected vary in severity from high- to low-consequence events, and include accidents initiated by natural phenomena (seismic event), operator error, and equipment failure.

The accident scenarios evaluated were as follows:

- Seismic event causing multiple process containment failures: This scenario would occur across multiple processes. The staff evaluation of acute effects was limited to cylinder breaches in the cylinder storage area which IIFP identified as resulting in high consequences. The staff evaluation of collective effects utilized an estimate of the total facility source term.

- Liquid DUF$_6$ cylinder drop: This scenario would include a breach and release of liquid DUF$_6$.

- SiF$_4$ release: This scenario could be caused by over-pressurization of a nitrogen loop with secondary cold trap breach.

- UF$_4$ collection drum spill.

- UF$_4$ vacuum transfer line rupture: This scenario would occur outside of the building.

IIFP's ISA attributes "likelihood categories" (highly unlikely, unlikely, or not unlikely) to each accident sequence. The staff's analysis described in this section does not include an estimate of the probability of occurrence of accidents, which, in combination with consequences, would reflect the overall risk from an accident. Instead, analyzed accidents are assumed to occur and consequences of each accident reported.

4.1.2.13.1.2 Accident Consequences

The performance requirements in 10 CFR 70, Subpart H, define acceptable levels of risk of accidents at nuclear fuel cycle facilities such as the proposed facility. The regulations in Subpart H require that IIFP reduce the risks of credible high-consequence and intermediate-consequence events, with all nuclear processes being subcritical. Table 4-33 defines the accident consequence categories used for the accident analysis. Table 4-34 defines exposure thresholds, by receptor and for intermediate- and high- consequence accidents, for each chemical species analyzed, as interpreted by IIFP. Subcritical conditions are assured because the facility would work exclusively with depleted uranium materials, and the incoming materials would be assayed to ensure this condition.

The staff evaluated the consequences of the selected accidents against the threshold values for a facility worker, a site worker 100 m (328 ft) from the release point, an individual at the site boundary, and the environment at the site boundary. Table 4-35 summarizes these results.

Table 4-33. Accident Consequence Categories

Category	Workers	Off-Site Public	Environment
Category 3 High Consequences	• Individual Radiation Dose ≥100 rem • Individual Chemical Dose = endanger life (> than AEGL-3, 10 min exposure) 75 mg soluble uranium intake	• Individual Radiation Dose ≥25 rem • Chemical Dose = long-lasting health effects (> AEGL-2, 30 min exposure) 30 mg soluble uranium intake	Radiological release >5000 times values in Table 2 of 10 CFR 20
Category 2 Intermediate Consequences	• Individual Radiation Dose ≥25 rem • Individual Chemical Dose = long-lasting health effects (>AEGL-2 but <AEGL-3, 10 min exposure)	• Individual Radiation Dose ≥5 rem • Chemical Dose = mild transient health effects (>AEGL-1 but <AEGL-2, 30 min exposure)	Radiological releases lower than Category 2
Category 1 Low Consequences	Accidents of lower radiological and chemical exposures than Category 2	Accidents of lower radiological and chemical exposures than Category 2	

Source: IIFP, 2009b

Table 4-34. Chemical Consequence Exposure Thresholds

Chemical	Intermediate Consequences				High Consequences			
	Worker Exposure		Public Exposure		Worker Exposure		Public Exposure	
	Level of Concern	Concentration, mg/m³	Level of Concern	Concentration, mg/m³	Level of Concern	Concentration, mg/m³	Level of Concern	Concentration, mg/m³
Hydrogen fluoride (HF)	AEGL-2 10 min	77.8	AEGL-1 30 min	0.82	AEGL-3 10 min	139	AEGL-2 30 min	28
Silicon tetrafluoride (SiF₄)	AEGL-2 10 min	27	AEGL-1 30 min	0.21	AEGL-3 10 min	81	AEGL-2 30 min	18
Boron trifluoride (BF₃)	AEGL-2 10 min	41	AEGL-1 30 min	2.5	AEGL-3 10 min	140	AEGL-2 30 min	47
Uranium hexafluoride (UF₆)	AEGL-2 10 min	28	AEGL-1 30 min	3.6	AEGL-3 10 min	216	AEGL-2 30 min	19
Uranyl fluoride (UO₂F₂)	AEGL-2 10 min	28	AEGL-1 30 min	3.6	AEGL-3 10 min	216	AEGL-2 30 min	19
Uranium tetrafluoride (UF₄)	AEGL-2 10 min	28	AEGL-1 30 min	3.6	AEGL-3 10 min	216	AEGL-2 30 min	19
Uranium dioxide (UO₂)	ERPG-2 10 min	201	ERPG-1 30 min	0.68	ERPG-3 10 min	180	ERPG-2 30 min	32

Source: IIFP, 2009b

ERPG = Emergency Response Planning Guideline – Concentration values established by the American Industrial Hygiene Association that meet certain human response criteria similar to those for Acute exposure guideline levels (AFGLs).

Table 4-35. Summary of Accident Analysis Results

Receptor	Parameter	Worst Case DUF$_6$ Release	Seismic event causing multiple process containment failures	Fluorine Compounds Release	UF$_4$ Spill	Transfer Line Rupture
Worker (inside room, 10 min exposure)	HF concentration (mg/m^3)	1.34 x 10^6		56.5		
	UO$_2$F$_2$ concentration (mg/m^3)	5.14 x 10^6				
	Soluble U intake (mg)	7.94 x 10^5				
	Dose (rem)	686			0.052	
	SiF$_4$ concentration (mg/m^3)			73.5		
	UF$_4$ concentration (mg/m^3)				121	
Worker (outside building, 10 min exposure)	HF concentration (mg/m^3)	1.64 x 10^4	47.3	0.452		
	UO$_2$F$_2$ concentration (mg/m^3)	6.05 x 10^4	179			
	Soluble U intake (mg)	9,340	27.6			
	Dose (rem)	8.07	0.02		4.05 x 10$^-4$	3.48 x 10^{-4}
	SiF$_4$ concentration (mg/m^3)			0.588		
	UF$_4$ concentration (mg/m^3)				0.953	0.817
Public (at Site Boundary, 30 min exposure)	HF concentration (mg/m^3)	7,800	15.7	0.367		
	UO$_2$F$_2$ concentration (mg/m^3)	2.93 x 10^4	59.4			
	Soluble U intake (mg)	1.36 x 10^4	27.4			
	Dose (rem)	11.7	0.02		0.0017	3.45 x 10^{-4}
	SiF$_4$ concentration (mg/m^3)			0.478		
	UF$_4$ concentration (mg/m^3)				1.33	0.27

Table 4-35. Summary of Accident Analysis Results (Continued)

Receptor	Parameter	Worst Case DUF$_6$ Release	Seismic event causing multiple process containment failures	Fluorine Compounds Release	UF$_4$ Spill	Transfer Line Rupture
Environment (at Site Boundary, 24 hr avg)	Activity Concentration (uCi/mL)	2.7 2 x 10^{-7}	4.96 x 10^{-10}		6.67 x 10^{-12}	2.17 x 10^{-12}
Public collective exposure	Dose (person-rem)	6.1	135		0.00317	0.00192
	LCF	0.00351	0.0297		2.63 x 10^{-6}	1.59 x 10^{-6}

Source: NRC, 2011
Note: Not all accident sequences resulted in datum for the categories listed in this table. This could be because the sequence was postulated to occur outside of a building or did not involve all the chemicals or radioactive materials listed.

The most significant accident consequences are those associated with the release of liquefied UF$_6$ caused by rupturing a cylinder. The facility emergency plan addresses this type of event as well as all other lower-risk, high- and intermediate-consequence events. IIFP would reduce the likelihood of this type of event by requiring a robust cylinder design that maintains its integrity during credible drops, shocks, collisions, and thermal events, and an interlock on the autoclave which would prevent the removal of liquid or partially full cylinders during heating/feed cycles. The NRC staff concludes that through the combination of plant design, passive and active engineered controls, and administrative controls, accidents at the facility would pose an acceptably SMALL risk to workers, the environment, and the public.

NRC regulations and IIFP's operating procedures for the proposed facility would be designed to ensure that the high and intermediate accident scenarios would be highly unlikely and unlikely, respectively. The combination of responses by Items Relied on for Safety, which mitigate or prevent emergency conditions, and the implementation of emergency procedures and protective actions in accordance with the facility emergency plan would limit the consequences and reduce the likelihood of accidents that could otherwise extend beyond the proposed facility site and property boundaries.

4.1.2.13.2 Transportation Accidents

Operation of the IIFP facility would require shipment of full DUF$_6$ cylinders from commercial enrichment facilities, empty DUF$_6$ cylinders back to the commercial enrichment facilities, DUO$_2$ to waste disposal facilities, and other process and miscellaneous LLW to waste disposal facilities. Section 4.1.2.9.2 describes these shipments, which are summarized here in Table 4-36.

NRC staff used the TRAGIS (Johnson and Michelhaugh, 2003) transportation routing computer modeling code and the RADTRAN5 (Neuhauser and Kanipe, 2003) transportation risk assessment computer modeling code to calculate the radiological transportation dose-risk to the exposed population along the transportation route. Dose-risk is the product of dose and probability for small segments along the route and summed over the entire route. Accident frequencies were taken from Saricks and Tompkins (1999). Severity fractions and package/contents response characteristics were taken from NUREG-0170 (NRC, 1977).

Results of that analysis are provided in Table 4-37, with more details on the analysis provided in Appendix E. LCF risk is the product of dose-risk times the Interagency Steering Committee on Radiation Standards conversion factor of 6 x 10^{-4} LCFs per person-rem (ISCORS, 2002).

Table 4-36.   Summary of Annual Radiological Transportation Shipments

Description	Origin	Destination	Number of Shipments	Packaging
Full DUF$_6$ cylinders	URENCO USA	IIFP	293	1 cylinder per truck
Full DUF$_6$ cylinders	GLE	IIFP	293	1 cylinder per truck
Full DUF$_6$ cylinders from AREVA Eagle Rock	Eagle Rock	IIFP	293	1 cylinder per truck
Empty DUF$_6$ cylinders	IIFP	URENCO USA	293	1 cylinder per truck
Empty DUF$_6$ cylinders	IIFP	GLE	293	1 cylinder per truck
Empty DUF$_6$ cylinders	IIFP	Eagle Rock	293	1 cylinder per truck
DUO$_2$	IIFP	Energy *Solutions*, Clive Facility	155	55-gal drums, 40 per truck
DUO$_2$	IIFP	Waste Control Specialists	155	55-gal drums, 40 per truck
Miscellaneous LLW	IIFP	Energy *Solutions*, Clive Facility	31	55-gal drums, 40 per truck
Miscellaneous LLW	IIFP	Waste Control Specialists	31	55-gal drums, 40 per truck

Source: IIFP, 2011a

Table 4-37.   Annual Accident Dose-Risk and LCF-Risk from Radiological Transportation

Description	Dose-Risk (person-Sv)	Dose-Risk (person-rem)	LCF Risk
Full DUF$_6$ cylinders from URENCO USA	4.0 x 10^{-5}	0.0040	2.4 x 10^{-6}
Full DUF$_6$ cylinders from GLE Facility	0.14	14	0.0081
Full DUF$_6$ cylinders from AREVA Eagle Rock	0.10	10	0.0060
Empty DUF$_6$ cylinders to URENCO USA	1.5 x 10^{-7}	1.5 x 10^{-5}	8.7 x 10^{-9}
Empty DUF$_6$ cylinders to GLE Facility	4.9 x 10^{-4}	0.049	2.9 x 10^{-5}
Empty DUF$_6$ cylinders to AREVA Eagle Rock	3.7 x 10^{-4}	0.037	2.2 x 10^{-5}
DUO$_2$ to Energy *Solutions*, Clive	0.10	10	0.0063
DUO$_2$ to Waste Control Specialists	3.9 x 10^{-5}	0.0039	2.3 x 10^{-6}
Miscellaneous LLW to Energy *Solutions*, Clive	5.5 x 10^{-5}	0.0055 E	3.3 x 10^{-6}
Miscellaneous LLW to Waste Control Specialists	2.0 x 10^{-8}	2.0 x 10^{-6}	1.2 x 10^{-9}

Source: See Appendix E

Assuming a scenario in which DUF$_6$ is shipped from the enrichment facility and the DUO$_2$ waste is shipped to the waste disposal facility as the greatest transportation risks, one arrives at 0.24 person-sievert (24 person-rem) of accident risk annually. This is for receipt and return of cylinders to the GLE facility in Wilmington, North Carolina and disposal of low-level waste at

Energy*Solutions* Clive, Utah facility. The equivalent number of latent cancer fatalities is 0.014 LCF.

According to the Centers for Disease Control and Prevention (CDC, 2010), there were 178.4 cancer deaths per 100,000 people in 2007 with a probability of occurrence of 100 percent. Given the high rate of cancer fatalities in the U.S. from all causes, the addition of 0.014 LCF from the risk of a radiological transportation accident from the proposed facility is considered by the NRC staff to be a SMALL impact. While mitigation measures are not required, IIFP would be required by NRC and DOT regulations to package and manage the transported waste to minimize the probability of accidental release of radioactive material.

4.1.3   Decommissioning Impacts

This section summarizes the potential environmental impacts of the decommissioning of the proposed IIFP facility. Decommissioning as described in Chapter 10 of the License Application (IIFP, 2009a), would involve the decontamination of equipment and buildings and the removal and disposal of all operating fuel-cycle facility equipment. Decommissioning would be funded in accordance with a decommissioning funding plan for the proposed IIFP facility, which will be prepared by IIFP in accordance with 10 CFR 70.25(a) and NUREG-1757 (NRC, 2006).

A complete description of the actions to be taken to decommission the proposed IIFP facility at the expiration of the plant's NRC license period (if the license is granted) cannot be provided at this time. In accordance with 10 CFR 70.38, IIFP must prepare and submit a decommissioning plan (different from the decommissioning funding plan) to the NRC for review and comment at least 12 months prior to the expiration of the proposed facility's NRC license. IIFP would submit a final decommissioning plan to the NRC for review prior to the start of decommissioning. This plan would include more detail than is available at this time. All decommissioning activities would comply with the applicable Federal, State, and local regulations in effect at the time of the decontamination and decommissioning activities.

It is reasonable to expect that decommissioning would occur over the course of three years and that it would be expected to employ 40 workers for the three-year period (IIFP, 2009a).

Two possibilities exist for decommissioning the facility. One is to leave the structures and most (non-uranium-processing) support equipment in place after they are decontaminated to appropriate (unrestricted release) levels, in accordance with 10 CFR 20, for ultimate use by another industrial tenant or owner. The second is to decontaminate and raze the entire facility, restoring the site to its current use as open range land (e.g., grazing and wildlife habitat). The final disposition of the property would be determined at the time of decommissioning. The ER assumes that "…decommissioning…will involve the removal of the internal equipment, utilities, and products from the building(s); however the physical structure, associated foundations, access roads, and utility lines will likely remain intact," (IIFP, 2009a). Therefore, this section evaluates leaving structures for industrial re-use as the likely decommissioning option.

Decontamination and decommissioning of the proposed facility is described in Section 2.1.7. Regardless of the end use of the facility, decommissioning would begin with the decontamination and removal of uranium-processing equipment and other materials to be shipped offsite for licensed disposal. The number of daily truck shipments is anticipated to be similar to the average daily shipments during operations, and the total number of shipments would depend upon the volume of demolition debris and materials packaged for disposal. Radioactively-contaminated equipment and materials would be disposed of by shipping them to a licensed treatment or disposal facility in compliance with applicable NRC and DOT requirements.

Discussions of issue- and resource-specific impacts of decommissioning include the following:

LAND USE: The chain-link perimeter security fence surrounding the facility compound could be removed following decommissioning. If decommissioning included the removal of all facilities, the land could revert to its current use for grazing and wildlife habitat. If buildings are not removed, another industry could move into the facility; and the 16 ha (40 ac) would not be available for grazing; however, the undeveloped land (240 ha, or 600 ac) could be available for grazing. Land use plans and land uses surrounding the site would be unaffected by decommissioning. The NRC staff concludes that regardless of the condition (option with structures remaining for alternate uses or option with all structures removed/site restored), the impacts to local land use due to decommissioning would be SMALL.

HISTORIC AND CULTURAL RESOURCES: Decommissioning of the facility would not involve land disturbance which could affect historic properties, districts, resources or significant historic/precontact archaeological sites. No historic resources were identified within the cultural resources APEs and three isolated artifacts that are not NRHP-eligible were identified during the cultural resource survey. No Native American Tribes expressed concerns to the NRC regarding the project.

Therefore, the NRC staff concludes that any impacts to historic properties, districts, resources or significant historic/precontact archaeological sites during facility decommissioning would be SMALL.

CLIMATE, METEOROLOGY, AND AIR QUALITY: GHG emissions associated with decommissioning would result primarily from three activities: (1) the onsite consumption of fossil fuels in vehicles and equipment used to dismantle and possibly demolish existing structures or excavate buried utilities and components, (2) the transportation of waste materials and salvage materials from the proposed site to appropriate offsite disposal or recycling facilities, and (3) the commuting decommissioning workforce.

The following are conservative assumptions that can be made relative to the proposed IIFP facility decommissioning and that can be used to estimate GHG impacts associated with decommissioning activities (IIFP, 2011a):

- CO_2 emissions from shipments of DUF_6 feed materials and operational waste shipments still occurring during the initial period of decommissioning are treated as operational GHG impacts.

- Shipments of wastes or recycling materials would occur by diesel-fueled trucks averaging 23.5 liters of fuel per100 km (10 mpg).

- LLW resulting from decontamination activities would be substantially greater in volume than LLW resulting from routine IIFP facility operation.

- All non-radioactive and non-hazardous solid wastes would be delivered to the same area landfills and treatment facilities that received wastes of similar nature during IIFP facility operation. Assuming successful decontamination of the majority of IIFP facility equipment and structures, a significantly higher number of annual trips would occur throughout the 3-year decommissioning phase than would have occurred annually during IIFP facility operation, and the resulting CO_2 emissions would be at least an order of magnitude greater than the values for such waste shipments appearing in Section 4.1.2.4.1, "Greenhouse Gases".

- All non-radioactive hazardous waste generated during IIFP facility operations would already have been transported to permitted disposal facilities. The CO_2 emissions of such deliveries would be credited to the IIFP facility operational phase. The amount of non-radioactive hazardous waste generated as a result of decommissioning is expected to be small and would likely be transported to the same disposal facilities that received similar waste during IIFP facility operation. It is further assumed that an appropriately permitted disposal facility will be located within a reasonable distance from the proposed IIFP facility, resulting in limited amounts of GHG emissions from transport.

- Except for the period at the beginning of decommissioning when some operations would still be ongoing, the decommissioning workforce would decrease from 140 to 40 employees. Therefore annual releases of CO_2 related to workforce commuting would be approximately one-third of the values shown in Table 4-13 for operations. Releases of CO_2 related to workforce commuting during the time that operations are continuing as decommissioning is beginning would be approximately one-third higher than the values shown in Table 4-13.

Therefore, the NRC staff concludes that impacts to climate and air quality would be SMALL.

GEOLOGY, MINERALS, AND SOIL: The general condition of the site geologic resources would not change during or after decommissioning activities. Minerals at the site and vicinity would not be affected by decommissioning. As with construction, demolition of structures and disturbed areas would be subject to BMPs to prevent adverse impacts to soils. As a final step n decommissioning, soil testing would demonstrate that site soils meet NRC, EPA, and NMED regulations and guidelines for free release. Accordingly, the NRC staff concludes that impacts to geology, minerals, and soil during decommissioning would be SMALL.

WATER RESOURCES: No surface water is present on the site, so decommissioning would not affect surface water. The management of stormwater is not expected to change during or after decommissioning activities, unless the site is restored to its original open range conditions Groundwater would be used during decommissioning for the potable water system, and decommissioning needs such as dust suppression. Water for facility processes would no longer be used; therefore, water withdrawal during decommissioning would be less than during operations.

Accordingly, the NRC staff concludes that impacts to water resources during decommissioning would be SMALL.

SOCIOECONOMICS: Decommissioning is expected to employ 40 workers over three years. The workers would be IIFP employees or work in the construction trades. All would be residents of the ROI. No workers would migrate into the area; however, some former IIFP staff could migrate out of the area. The NRC staff finds that impacts to socioeconomic resource would be SMALL. The NRC staff finds that no disproportionately high or adverse impacts would be incurred by any minority or low-income population.

TRAFFIC AND TRANSPORTATION: Impacts to traffic would be similar to the impacts during construction and operations. The Phase 2 construction and operations workforces and the number of trucks transporting materials to/from the facility on a daily basis would be similar to the number during Phase 1 construction and operation. IIFP would ensure that all transportation of materials met NRC and DOT regulations.

Therefore, the NRC staff concludes that impacts to traffic and transportation would be SMALL.

NOISE: Impacts from noise during decommissioning would be very similar to impacts during construction. Therefore, the NRC staff concludes that impacts would be SMALL.

OCCUPATIONAL AND PUBLIC HEALTH: Impacts to occupational and public health would be similar to impacts during construction. Therefore, the NRC staff concludes that impacts would be SMALL.

WASTE MANAGEMENT: The overall strategy for decommissioning would be to remove all radioactively contaminated materials, hazardous materials and chemicals from the site. Decommissioning programs and procedures would focus on minimizing waste volumes. For example, as described in Chapter 10 of the License Application (IIFP, 2009a), IIFP would incorporate design features that would result in minimizing the radioactive waste volumes including the following:

- A washable coating on floors and walls in the Restricted Areas, which have the potential to become radioactively contaminated during operation would lower waste volumes during decontamination and simplify the decontamination process.

- Sealed, nonporous pipe insulation in areas with higher potential to become contaminated would facilitate cleaning in event of a spill and reduce the waste volume during decommissioning.

- Tanks would have access for entry and decontamination. Design provisions would be made to allow complete draining of the wastes contained in the tanks.

- Connections in the process systems would provide access during operation and maintenance and to allow for thorough purging at plant shutdown which would remove some radioactive contamination prior to disassembly.

Decommissioning activities would include cleaning to remove radioactive and hazardous contamination that could be present on materials, equipment, and structures. Wastes produced during decommissioning would be collected, handled, and disposed of in a manner similar to that described for the wastes produced during operation. These wastes would consist of industrial trash, nonhazardous chemicals and fluids, small amounts of hazardous materials, and radioactive wastes. The radioactive waste would consist primarily of piping, tanks, hoppers, and compactable trash generated during the dismantling process.

Solid wastes would be generated by decontamination activities and by the removal of used process equipment. Decontaminated used equipment would be shipped offsite to salvage or disposal facilities, as appropriate. In the event that structures would be demolished as part of the decommissioning activities, the demolition material would be shipped offsite for disposal in permitted disposal facilities. Radioactively- contaminated equipment and materials would be shipped to a licensed treatment or disposal facility (as appropriate for the material type) or disposed of in a manner authorized by the NRC. Similarly, materials constituting hazardous wastes would be shipped to a RCRA-permitted treatment and/or disposal facility or an appropriate licensed recovery facility.

A detailed estimate of the wastes produced during decommissioning would be provided in the decommissioning plan that would be submitted to the NRC prior to initiating the decommissioning of the plant (IIFP, 2009a). Approximately 56,000,000 L (2 million ft^3) of commercial LLW were disposed of in the United States in 2008 (NRC, 2010). The estimated decommissioning LLW generation from decommissioning represents less than 1 percent of the national annual disposal volume. The LLWs from the decommissioning are expected to be Class A waste. In its analysis of LLW disposal capacity, the U.S. Government Accountability Office concluded that the availability of disposal capacity in the United States for Class A LLW is not considered to be a problem for the short or long term (GAO, 2004). The NRC staff concludes that the waste management impacts resulting from decommissioning of the IIFP facility, decontamination, disposal, and closure activities would be SMALL.

4.2　　　　Cumulative Impacts

The CEQ regulations implementing NEPA define cumulative impacts, or effects, as "the impact on the environment which results from the action when added to other past, present, and reasonably foreseeable future actions regardless of what agency (Federal or non-Federal) or person undertakes such other actions" (40 CFR 1508.7). In the following analysis, cumulative impacts are assessed from the anticipated impacts of the proposed construction, operation, and decommissioning of the proposed IIFP facility when added to other identified projects, facilities, or activities in the region that have impacts that affect the same resources or human populations. Effects from the various sources may be direct or indirect and they may be additive or interactive. Such effects are assessed that, when on their own, may be minor, but in combination with other effects may produce a cumulative effect that is of greater concern.

To identify the activities in the region that could contribute to cumulative impacts, NRC staff defined an ROI for each resource that is expected to be affected by the proposed IIFP facility. An ROI for a particular resource is the size of the surrounding area within which impacts from multiple sources may be additive or interactive. The sizes of the ROIs may be different for various resources, and some resources may be remote from the proposed site, such as a waste disposal facility. Still others might cover large areas, such as a watershed or airshed. NUREG-1748 (NRC, 2003) states that the surrounding area of the proposed action can range from less than 1.6 km to 80 km (1 mi to 50 mi). Consistent with NUREG-1748, for the proposed IIFP facility, an ROI radius of 16 km (10 mi) was identified for the majority of resources. The exceptions include socioeconomics, for which an ROI radius of 80 km (50 mi) was identified (Section 3.9); and cultural and historic resources and visual resources, for which an ROI radius of 10 km (6 mi) was identified (Section 3.3.4). Additionally, in order to assess the potential cumulative impacts of radiological transportation, the analysis includes consideration of the URENCO USA/LES uranium enrichment facility and the DOE WIPP, both of which are more than 16 km (10 mi) from the proposed IIFP facility.

In order to identify projects or activities in the region that could contribute to cumulative effects, the NRC staff conducted Internet searches, reviewed news media (local newspapers and local television), and reviewed other relevant NEPA documents (such as the Environmental Impact Statement for the Proposed National Enrichment Facility in Lea County, New Mexico [NUREG-1790; NRC, 2005a], the Final Complex Transformation Supplemental Programmatic Environmental Impact Statement [DOE/EIS-0236-S4; DOE, 2008], and the Supplement Analysis for the Waste Isolation Pilot Plant Site-Wide Operations [DOE/EIS-0026-SA-07, DOE, 2009]). This cumulative impacts analysis included review of existing activities in the region that would affect the same resources as the proposed IIFP facility, known past impacts on these resources, and reasonably foreseeable proposed new projects, activities, or facilities that could impact these resources. Section 4.2.1 discusses these projects or activities.

4.2.1 Past, Present, and Reasonably Foreseeable Future Actions

Five other projects or actions are identified and described in this section:

1. Preconstruction activities on the proposed IIFP site that could occur prior to NRC issuing a license for construction, operation, and decommissioning of the proposed IIFP facility.

2. Construction, operation, and decommissioning of Phase 2 of the IIFP facility.

3. Construction, operation, and decommissioning of the URENCO USA/LES uranium enrichment facility (formerly known as the National Enrichment Facility) in Lea County, New Mexico.

4. Operation of the DOE WIPP near Carlsbad, New Mexico.

5. Construction and operations related to energy production facilities in the region.

4.2.1.1 Proposed IIFP Facility Preconstruction Activities

The preconstruction activities would be preparatory in nature and would not involve any radiological process or safety related equipment or systems. Required Federal and State permits would be obtained prior to the start of preconstruction, and preoperational baseline environmental samples would be collected. Preconstruction activities for the proposed IIFP project would include (IIFP, 2011a):

- Clearing land
- Site grading and erosion control
- Installing temporary fencing
- Installing main entrance roadbed and drainage to highway
- Installing construction trailer
- Preparing preliminary site roadways and gravel parking area
- Drilling water wells
- Constructing power substation and electric utility lines
- Stubbing in gas line to the meter
- Beginning administration building construction
- Beginning maintenance and stores building construction
- Beginning warehouse building construction
- Installing geothermal heating/cooling loops
- Installing firewater tanks
- Installing truck washing station

Based on the characteristics of the proposed IIFP site, major grading would not be required. Excavation would be required for sewer systems, roads, pads, and structure foundations. Less than 10 percent of the total 259-ha (640-ac) area would be disturbed. The area of clearing would include locations of buildings, process structures, storage pads and roads. During this pre-licensing, preconstruction phase, conventional earthmoving and grading equipment would be used. The removal of very dense soil (caliche) may require the use of heavy equipment with ripping tools. Soil removal work for foundations would be controlled to minimize excavation. In addition, loose soil and/or damaged caliche would be removed prior to installation of foundations for seismically-designed structures. Temporary silt fencing and sediment straw bales would be installed around the areas of construction to entrap silt and to prevent its migration off site. Drainage trenches and ditch checks would be installed along the entrance road to prevent run-off and silt from the site moving onto NM 483 right-of-way. Site sloping, earth berms, underground drainage pipe, and wet sediment retention basins would be installed to entrap storm water run-off from construction areas (IIFP, 2011a).

The natural gas line feeding the site would be connected to an existing, nearby line. This would minimize impacts of short-term disturbances related to the placement of the tie-in line. A new electrical distribution line is proposed for providing electrical service to the IIFP facility. There are currently 115 and 230 kV transmission lines along US 62/180 and NM 483 and crossing the site. IIFP anticipates that the additional line would be erected in an existing right(s)-of-way. In conjunction with the new electrical lines serving the site, the local electrical utility company would install an independent substation within the 16-ha (40-ac) facility to ensure service (IIFP, 2011a).

The Clean Water Act NPDES requires an NPDES(s) permit for discharges to surface waters, for stormwater from construction projects and industrial pollutant discharges. This could include construction and operation of a facility such as the proposed IIFP. A Spill Prevention, Control, and Countermeasures (SPCC) plan would also be implemented to prevent and, if necessary respond to oil spills. An SPCC plan would be completed and an NPDES Construction Stormwater Permit with the General Construction Permit would be obtained by IIFP prior to the implementation of preconstruction activities (IIFP, 2011a), if necessary.

4.2.1.2   Proposed Phase 2 of the IIFP Facility

The proposed Phase 2 project would add additional deconversion capacity at the facility and a process for the direct deconversion of DUF_6 to uranium oxide. Phase 2 construction activities are proposed to begin in early 2015 and would be completed to support operations by mid-2016 and require a maximum of 180 additional workers (IIFP, 2011a).

Prior to the proposed Phase 2 expansion, IIFP would prepare and submit an amended license application to the NRC for the Phase 2 facility, including possession of up to 2,200,000 kg (4,850,120 lb) of DUF_6 (compared to the 750,000 kg [1,653,450 lb] of DUF_6 that were requested in the Phase 1 application). IIFP plans to submit a license amendment for this plant expansion in 2013 (IIFP, 2011a).

During Phase 2 construction, additions are planned for the DUF_6 Autoclave Building, the Oxide Process Building, Direct Oxide Staging Building, and the HF Distillation Annex. The entire site clearing would occur during preconstruction and Phase 1 construction. No roads would need to be added. Minor revisions during Phase 2 construction to paved or concrete areas may be required. Hence, no major earth grading or movement would be necessary, but excavation would be required for sewer and building foundations and floors and for tie-ins for water, natural gas, and utilities. Excavation for foundations would be minimized. Loose soil and/or damaged caliche would be removed prior to installation of foundations for seismically designed structures.

Approximately 20 percent more building space would be added to the existing Phase 1 facility. Considering the total 259-ha (640-ac) area, minimal soil disturbance would occur. Silt fences and straw bales would be used to control erosion and to protect undisturbed areas (IIFP, 2009a). As part of the Phase 2 plant expansion, another major stack would be added for venting filtered exhaust gas from the oxide process dust collector system. Phase 2 construction would be accomplished with an average construction crew of 150 to 180 workers (IIFP, 2011a).

Once the Phase 2 facility is operational in mid-2016, all of the fluorides in the DUF_6 could be directly converted to AHF, and SiF_4 and BF_3 would not be produced unless warranted by market conditions for these products. Despite different internal operations, many aspects of the Phase 2 operations that would give rise to potential environmental impacts would be very similar to those in Phase 1 (IIFP, 2011a). Upon completion of Phase 2, the integrated facility would have an overall total deconversion capacity of nearly 800 DUF_6 cylinders per year; about 9.8 million kg/yr (21.7 million lb/yr) of DUF_6. Nearly 2.6 million kg/yr (5.7 million lb/yr) of AHF product is projected to be produced and sold (IIFP, 2009a).

The utilities needed to support the Phase 2 facility would be the same as those for the Phase 1 facility, although there would be an increase in overall utility usage (especially electricity and steam) with the addition of the Phase 2 facility. For example, when the Phase 2 facility becomes operational, the total steam load would increase to about 2,722 to 3,629 kg/hr (6,000 to 8,000 lb/hr) compared to 1,134 to 1,588 kg/hr (2,500 to 3,500 lb/hr) for Phase 1 operations (IIFP, 2009a). At the end of its useful life, the IIFP facility would be decommissioned consistent with the decommissioning plan that is developed.

4.2.1.3 URENCO USA/LES Uranium Enrichment Facility

In December 2003, the LES submitted a license application to the NRC to construct, operate, and decommission a facility to produce enriched U-235, up to 5 percent weight, by the gas centrifuge process. The enriched uranium would be used as fuel in commercial nuclear power plants. The NRC staff issued a Final EIS (NUREG-1790) (NRC, 2005a) and SER (NUREG-1827) (NRC, 2005b) for the facility in June 2005. In June 2006, the NRC issued LES a 30-year license to construct and operate the facility with a nominal production capacity of 3 million separative work units (SWUs) per year. On November 21, 2008, LES announced plans to expand the facility capacity to 5.7 million SWUs per year (NRC, 2010); although a license application for the facility expansion has not yet been submitted to the NRC.

The URENCO USA/LES Uranium Enrichment facility commenced initial operations on June 11, 2010. Construction of the project will continue until the plant reaches the planned 5.7 million SWU capacity and full operations are expected in 2015 (assuming a license for the additional 2.7 million SWU is granted by the NRC). The facility is located approximately 32 km (20 mi) south of Hobbs, New Mexico, 8 km (5 mi) east of Eunice, and approximately 40 km (25 mi) south of the proposed IIFP site. DUF_6 is a waste product of the uranium enrichment process, and the URENCO USA/LES Uranium Enrichment facility would be one of the likely DUF_6 suppliers to the IIFP facility.

This cumulative impacts analysis is based on information in the Final EIS (NUREG-1790) (NRC, 2005a).

4.2.1.4   DOE Waste Isolation Pilot Plant (WIPP)

The WIPP facility is the nation's first underground repository permitted to safely and permanently dispose of transuranic radioactive waste generated by defense-related activities. Waste generated at DOE sites is shipped to the WIPP and permanently disposed in an ancient salt formation 655 m (2,150 ft) below the surface. Over the planned 35-year operational lifetime ending in 2034, the WIPP is expected to receive approximately 37,000 shipments of waste from locations across the United States (DOE, 2008). The WIPP disposal site is 42 km (26 mi) east of Carlsbad, in Eddy County in the Chihuahuan Desert of southeastern New Mexico, and approximately 87 km (54 mi) from the proposed IIFP facility site.

Transuranic Waste
Transuranic waste is waste that contains alpha-emitting radionuclides with atomic numbers greater than uranium (92) and half-lives greater than 20 years, in concentrations greater than 100 nanocuries per gram of waste.

Waste disposal operations began at the WIPP in March 1999. As of August 2010, the WIPP has received 8,812 transuranic waste shipments, totaling more than 16.1 million km (10 million mi) of transport on U.S. highways of approximately 69,240 m^3 (90,566 yd^3) of transuranic waste. Based on the most recent transuranic waste inventory data, DOE estimates that approximately 140,000 m^3 (182,779 yd^3) of transuranic waste either has been disposed of or could be eligble for disposal at the WIPP (DOE, 2010).

4.2.1.5   Regional Energy Production Facilities

As shown on Figure 3-2 and described in this section, there are four energy production facilites in the vicinity of the IIFP facility that could contribute to cumulative impacts:

1. Xcel Energy Cunningham Station

2. Xcel Energy Maddox Station

3. Colorado Energy Station

4. DCP Midstream Linam Ranch Natural Gas Processing Facility

The cumulative impacts analysis is based on information in Section 4.1 of this EIS, the Environmental Report submitted by IIFP (IIFP, 2009a), Official Responses to the Environmental Report Requests for Additional Information (IIFP, 2011a), and the other references identified in Section 4.2.2.

4.2.2   Cumulative Impacts to Environmental Resources

The potential cumulative impacts are presented for each resource presented in Section 4.1.

4.2.2.1   Land Use

As described in Section 3.2, the proposed IIFP facility would be located in a sparsely populated area on undeveloped land and near four power and gas industry plants. Present land uses in the vicinity include cattle grazing and oil and gas development. The preconstruction, construction, and operation of Phase 1 would disturb less than 10 percent of the total 259-ha (640-ac) site (IIFP, 2011a). Because approximately 93 percent of Lea County (approximately 1.0 million ha [2.6 million ac]) is used as range land for grazing, the impacts resulting from restricting the current land use would be negligible due to the abundance of other nearby grazing land. There are no zoning restrictions on the property. As described in Section 4.2.1.2, during the Phase 2 expansion, no roads would be added and only minor revisions to paved or concrete areas may be required. Hence, no major earth grading and land disturbance would occur. Therefore, the NRC staff concludes that cumulative impacts on land use from the preconstruction of the proposed facility, the proposed action, and Phase 2 construction, operation, and decommissioning would be SMALL.

4.2.2.2 Historic and Cultural Resources

As described in Section 3.3, an archaeological survey of the site conducted in May 2009 identified three isolated artifacts and no archaeological sites. A review of the current listings for the New Mexico State Register of Cultural Resource Properties and the National Register of Historic Places indicate no NRHP-listed or eligible historic properties within 10 km (6 mi) of the proposed site and one State-listed property just less than 10 km (6 mi) south of the IIFP site. The archaeological consultant recommended no further work based on the survey results. The NM SHPO concurred with this determination (Appendix B). Preconstruction activities at the proposed IIFP site and Phase 2 expansion, which would occur within the same footprint as the proposed action, would have no impact on historic properties, districts, resources or significant historic/precontact archaeological sites. Therefore, the NRC staff concludes that cumulative impacts on historic and cultural resources from the proposed action, preconstruction of the proposed facility, and Phase 2 construction, operation, and decommissioning would be SMALL.

4.2.2.3 Visual Resources

As discussed in Section 3.4, the construction of the proposed facility would occur in a sparsely populated area with an existing low-quality viewshed. No regionally or locally important high quality views occur in the vicinity of the proposed IIFP facility. Consequently, the NRC staff concludes that cumulative impacts would be SMALL.

4.2.2.4 Climatology/Meteorology/Air Quality

4.2.2.4.1 Greenhouse Gases

Greenhouse gas emissions from construction vehicles and equipment were taken into account in the analysis for Phase 2. During the Phase 2 construction, it was assumed that the workforce of 180 would commute 2,900,000 km (1,800,000 mi) over the 1-year construction period (250 days). Over the course of the construction period, it was also estimated that there would be 20 deliveries each day each also traveling a distance of 64 km (40 mi). NRC staff used EPA MOVES to calculate the resulting CO_2 emissions associated with workforce commuting and construction deliveries during Phase 2 construction. The total CO_2 equivalent emissions, expected during the Phase 2 construction period would be 1,303 metric tons (1,435 tons), which are substantially less than those expected from the Phase 1 construction period.

Using calendar year 2000 as a reference point (the latest year for which New Mexico greenhouse gas emission data are available), and as shown in Table 3-1, total net CO_2 emissions for New Mexico for the year 2000 were 62 million metric tons (68 million tons) of CO_2 equivalents. For the United States for that same year, total net CO_2 emissions were 5,977 million metric tons (6,588 million tons) (EPA, 2010a). By comparison, during the Phase 2 construction phase, CO_2 emissions are projected to be 1,303 metric tons (1,435 tons), approximately 0.002 percent of the New Mexico statewide output or 0.00002 percent of the nationwide emissions for calendar year 2000. Consequently, the NRC staff concludes that potential cumulative impacts on greenhouse gas emissions would be SMALL

4.2.2.4.2   Air Quality

4.2.2.4.2.1   Air Quality (pre-construction)

Air quality impacts from the operation of construction equipment and support vehicles during the preconstruction stage were evaluated based on the construction schedules and parameters provided by IIFP (IIFP, 2011a). The proposed IIFP facility site is 16 ha (40 ac).

Activities that would take place during preconstruction are described in Section 4.2.1.1. IIFP estimates preconstruction would last for a period of approximately three months, and would be followed by approximately 12 months of Phase 1 construction (IIFP, 2011b).

During preconstruction, criteria pollutants (e.g., CO, NO_2, PM_{10}, $PM_{2.5}$, and SO_2), HAPs, and VOCs would be generated by the operation of construction vehicles and equipment (operating at 10 hours a day, 5 days a week), delivery vehicles (estimated at 20 trips a day), and workforce transport vehicles (estimated at 140 trips per day) traveling to and from the site. These emissions would include (1) fugitive dust emissions from the disturbance of unpaved surfaces, (2) combustion emissions from the operation of diesel-fired vehicles and equipment, (3) tailpipe emissions from the operation of gasoline and diesel-fired commuter and delivery vehicles, and (4) fugitive HAP and VOC emissions due to evaporative losses from diesel fuel tanks and diesel fuel transfers.

The quantities of air pollutants that would be generated from preconstruction activities at the IIFP site were estimated using the equipment list and description of planned activities provided by IIFP (2011b); and emission factors from the EPA MOVES Model (EPA, 2009a), the EPA NONROAD model (EPA, 2005), and EPA AP-42 emission factors (EPA, 1995a). Air quality impacts were evaluated using the EPA SCREEN3 (EPA, 1995b) air dispersion model.

IIFP anticipates that most of the earth moving activities would take place during preconstruction. Consequently, fugitive dust emission rates would be greater during preconstruction than during the Phase 1 construction period, however, the 3-month preconstruction period is relatively short. The estimated pollutant emissions during preconstruction would represent a very small fraction of the current emissions in Lea County.

Dispersion modeling results show that air pollutant concentrations at the IIFP site boundary during preconstruction would be similar to the concentrations during Phase 1 construction (See Appendix C). The estimated incremental increases in ambient background concentrations due to the proposed preconstruction activities would be above the NAAQS for NO_2, $PM_{2.5}$ and PM_{10} emissions. Pollutant emissions from preconstruction activities potentially could change the existing ambient air quality in the vicinity of the proposed IIFP facility temporarily. Because conservative assumptions that tend to overestimate impacts were used to produce these estimates, actual emissions from the construction activities are expected to be lower. Overall, the preconstruction impacts would be localized and short-term.

Because preconstruction and Phase 1 construction would not occur simultaneously, the impacts would not be cumulative. As discussed in Section 3.5.3, Lea County is in attainment for all criteria pollutants. The cumulative air impacts of preconstruction and other projects in the region of influence are not expected to change this attainment status. Therefore, the NRC staff concludes that the air quality impacts resulting from the preconstruction of the proposed IIFP facility would be MODERATE for NO_2, $PM_{2.5}$, and PM_{10} emissions and SMALL for other emissions. BMPs during preconstruction and construction as described in Chapter 5, Mitigation Measures and Commitments, would reduce impacts to air quality. NRC staff considers the use of BMPs to minimize impacts to air quality as an environmental commitment. Furthermore, the NRC staff finds that the BMPs committed to by IIFP would be sufficient to ensure that pre-construction impacts of the proposed IIFP facility to air quality would be MODERATE for NO_2 and particulate emissions; and SMALL for other emissions.

4.2.2.4.2.2   Air Quality (Phase 2 Construction and Operation)

During Phase 2 construction, the process area would be expanded approximately 28 percent to add a 33.5 m x 33.5 m (110 ft x 110 ft) area next to the Phase 1 process buildings. Less than 1 percent of the 16-ha (40-ac) site area would be disturbed during the Phase 2 construction period of approximately 1 year (IIFP, 2011a).

Pollutant emissions and diesel fuel consumption attributable to Phase 2 construction activities were estimated using the equipment list and description of planned activities provided by IIFP (2011b); and emission factors from the EPA MOVES Model (EPA, 2009a), the EPA NONROAD model (EPA, 2005), and EPA AP-42 emission factors (EPA, 1995a). Air quality impacts were evaluated using the EPA SCREEN3 (EPA, 1995b) air dispersion model.

Heavy earth-moving equipment (e.g. dozers, excavators, and graders) would not be required for Phase 2 construction, so annualized Phase 2 emissions would be approximately 25 percent less than annualized Phase 1 construction emissions (See Appendix C). The estimated pollutant emissions during Phase 2 construction represent a very small fraction of the current emissions in Lea County.

Dispersion modeling results show that air pollutant concentrations at the IIFP site boundary during Phase 2 construction would be much lower than the concentrations during Phase 1 construction (See Appendix C). The estimated incremental increases in ambient background concentrations due to the proposed preconstruction activities would be above the NAAQS for NO_2 emissions over a 1-hr averaging time. All other pollutant concentrations were estimated to be below NAAQS. Pollutant emissions from Phase 2 construction activities potentially could change temporarily the existing ambient air quality in the vicinity of the IIFP facility with respect to NO_2. Because conservative assumptions that tend to overestimate impacts were used to produce these estimates, actual emissions from the construction activities are expected to be lower. Overall, the Phase 2 construction impacts would be localized and short-term.

Because Phase 2 and Phase 1 construction would not occur simultaneously, the impacts would not be cumulative. As discussed in Section 3.5.3, Lea County is in attainment for all criteria pollutants. The cumulative air impacts of Phase 2 construction and other projects in the region of influence are not expected to change this attainment status. Therefore, the NRC staff finds that the air quality impacts resulting from the construction of the IIFP Phase 2 facility would be MODERATE for NO_2 emissions and SMALL for other air emissions. BMPs used during construction would reduce the impact of construction activities on air quality. These BMPs are described in Chapter 5, Mitigation Measures and Commitments. The NRC staff finds that the BMPs committed to by IIFP for the proposed facility would be sufficient to maintain impacts to air quality from Phase 2 construction as MODERATE to SMALL.

Greenhouse gas emissions from construction vehicles and equipment were taken into account in the analysis for Phase 2. During the Phase 2 construction, it was assumed that the workforce of 180 would commute 2,900,000 km (1,800,000 mi) over the 1-year construction period (250 days). Over the course of the construction period, it was also estimated that there would be 20 deliveries each day each also traveling a distance of 64 km (40 mi). NRC staff used EPA MOVES to calculate the resulting CO_2 emissions associated with workforce commuting and construction deliveries during Phase 2 construction. The total CO_2 equivalent emissions, expected during the Phase 2 construction period would be 1,303 metric tons (1,435 tons), which are substantially less than those expected from the Phase 1 construction period.

Greenhouse gas emissions from operation of the IIFP facility would be insignificant (less than 0.1 percent) when compared to the greenhouse gas emissions from the regional energy facilities. In 2008, the total CO_2 emissions from the Cunningham Station, Maddox Station, Colorado Energy Station, and DCP Midstream Linam Ranch Natural Gas Processing Facility were more than 1.3 million metric tons (1.43 million tons) (NMED, 2010).

For Phase 2 operations, criteria pollutant emissions attributable to operations are well below Title V and Class II PSD thresholds. IIFP evaluated regional impacts with SCREEN3 based on frequency-weighted site-specific meteorological data. Pollutant concentrations at the site boundary were determined to be well below the NAAQS (IIFP, 2011a). The cumulative air impacts of Phase 2 operations of the IIFP facility and other projects in the region of influence, including IIFP Phase 1 operations, are not expected to change the attainment status of Lea County. Consequently, the NRC staff concludes that potential cumulative impacts on air quality would be SMALL.

4.2.2.5 Geology, Minerals, and Soil

Preconstruction would occur within about 16 ha (40 ac) of the 259-ha (640-ac) proposed site (IIFP 2009a; IIFP 2011a); and construction, operation, and decommissioning for Phase 2 would occur within the previously disturbed 16-ha (40-ac) footprint of the Phase 1 IIFP facility. Therefore, these actions would have little or no additional impacts on geology, minerals, seismology, and soil beyond those of the proposed action.

During all preconstruction and Phase 2 construction activities, BMPs would be employed to limit soil loss and mitigate these impacts. These would include:

- Soil stabilization (e.g. temporary and permanent seeding),

- Structural controls (e.g. hay bales and sediment fences),

- Drainage trenches and ditch checks would be installed along the entrance road to prevent run-off and silt from the site onto NM 483 right-of-way, and

- Management practices (e.g. construction sequencing, materials delivery sequencing, physical delineation of disturbed areas) (IIFP, 2011a).

Once the Phase 2 facility is constructed, no additional impacts to geology, minerals, seismicity, and soil are expected. Thus, the NRC staff concludes that cumulative impacts from preconstruction of the proposed IIFP facility, the proposed action, and Phase 2 construction, operation, and decommissioning would be SMALL.

4.2.2.6   Water Resources

Preconstruction activities are not expected to require any use of on-site of groundwater. During the preconstruction period, up to two new wells would be installed, and capped at the wellheads for connections to the facility water distribution systems after possible NRC license approval. For dust control during preconstruction activities, IIFP would bring in tanker trucks of water from the City of Hobbs municipal system. The City of Hobbs groundwater allocation is included as part of Lea County's 40-Year Water Development Plan and preconstruction activities would not result in cumulative impacts to groundwater use. Site sloping, earth berms, underground drainage pipe, and wet sediment retention basins would be installed to entrap storm water run-off from construction areas. As discussed in Section 3.7.2, no permanent surface water or jurisdictional waters are present on the proposed IIFP site and, therefore, there would not be any cumulative impacts to surface water.

Approximately 3.79 m^3/day (1,000 gal/day) of groundwater would be required during Phase 2 construction, mainly for dust suppression control, fill compaction, and concrete formation. Average and peak site water requirements for Phase 2 operations are expected to be approximately 11.36 m^3/day (3,000 gal/day) and 37.85 m^3/day (10,000 gal/day), respectively.

Phase 2 facility operation would require relatively low volumes of water because it would recycle process water and re-circulate cooling water. Groundwater use during operation is projected to be less than 37,854 L (10,000 gal) per day (IIFP, 2011a), and would be below the water allotment set aside by Lea County. Therefore, the NRC staff concludes that cumulative impacts to groundwater use from preconstruction of the proposed IIFP facility, the proposed action and Phase 2 construction and operation would be SMALL.

As summarized in Section 3.13.5, there are four energy production facilities in the vicinity of the proposed IIFP facility. Each of these facilities uses groundwater from the Ogallala aquifer, as would the proposed action. The Xcel Energy Cunningham Station, which is adjacent to the IIFP site, is a zero discharge plant, meaning no process waters are discharged from the plant site. The cooling water from the Cunningham Station is reused to irrigate pecan orchards. The groundwater rights and use for the four facilities were allocated prior to the development of the Lea County 40-Year Water Development Plan and, thus, are not reliant on Lea County's assigned unappropriated 4,215.2 ha-m (34,173 ac-ft) per year of water rights. In 2005, the four energy plants were factored into the Lea County annual groundwater withdrawals of 2,293,700 ha-m/yr (185,952 ac-ft/yr) (McCoy and Perry, 2004). The Lea County 40-Year Water Development Plan includes an assessment of groundwater use impacts from existing and future beneficial uses of groundwater.

The National Enrichment Facility operations are expected to use on an average approximately 87,600 million m^3 (23.1 million gal) of water annually. For the life of the facility, the National Enrichment Facility could use up to 263,000 m^3 (695 million gal) of the Ogallala waters, encompassing both construction and operations use. This constitutes a small portion, 0.004 percent, of the 60 billion m^3 (49 million ac-ft or 16 trillion gal) of Ogallala reserves in the State of New Mexico territory. Water use during decontamination and decommissioning would be less than or equal to the water consumption during operations (NRC 2005a).

As discussed in Section 4.1.2.6, Lea County has allocated up to 175 ac-ft/yr of water rights to the proposed IIFP site in their 40-year plan, which takes into account existing groundwater users. As discussed above, the IIFP site's groundwater use would be much less than this allotment. In accordance with regulations of the New Mexico Office of the State Engineer for wells installed in a non-Critical Management Area, the two site wells are not expected to create drawdowns that exceed the limit of 2.4 m (8 ft) over 40 years, or 0 06 m/yr (0.20 ft/yr). Therefore, the NRC staff concludes that the cumulative groundwater use impact related to the Lea County unappropriated water rights from the operation of the four existing energy production facilities, the National Enrichment Facility, and the activities associated with the proposed IIFP facility would be SMALL.

With respect to groundwater quality the Xcel Energy Cunningham Station, which is the closest energy facility to the proposed IIFP Facility, operated with an unlined cooling tower and boiler cleanout pond for a number of years. The pond has recently been lined. Xcel Energy monitoring wells along the western boundary of the proposed IIFP site were installed to monitor contaminants in groundwater that potentially originated from cooling water pond and/or agricultural fields. Data since 2004 from these monitoring wells indicate that concentrations of sulfate, chloride, and total dissolved solids have exceeded New Mexico Water Quality Control Commission Standards for Groundwater (IIFP, 2011a).

During preconstruction, operations, and decommissioning of the proposed IIFP facility, control of surface water runoff would be required by the NPDES permit. As a result, no impacts are expected to surface or groundwater bodies. Stormwater and effluent sampling would be conducted as required by the NPDES permit to protect surface water quality. In addition, site-wide groundwater levels would continue to be monitored routinely, and samples from the groundwater monitoring-well and pumping-well networks would continue to be analyzed to confirm that cumulative impacts to groundwater quality would be SMALL (IIFP, 2011a). Therefore, the NRC staff finds that groundwater quality impacts would be SMALL.

4.2.2.7 Ecological Resources

Most of the impacts to ecological resources would occur during the preconstruction activities. Land clearing would occur within the 16 ha (40 ac) facility area and would destroy the Western Great Plains Shortgrass Prairie and Apacherian-Chihuahuan Mesquite Upland Scrub vegetation communities. The amount of vegetation cleared would be limited, to the extent practicable, to the land area needed for the proposed IIFP facility's operational, security, and utility requirements (IIFP, 2011a). However, neither of these vegetation communities provides unique habitat in the area. The existing natural habitats on the proposed IIFP site and the region surrounding the proposed site have been previously impacted by domestic livestock grazing, wildfires, oil/gas pipeline rights-of-way and access roads (IIFP, 2011a). The total area to be disturbed for the facility (16 ha [40 ac]) represents less than one-tenth of the total site area. Therefore, the NRC staff finds that the loss of 16 ha (40 ac) of either habitat type, for both direct and onsite cumulative impacts, would have a SMALL impact on native vegetation in the vicinity of proposed action.

During preconstruction, an access roadway off of northbound NM 483 would be built to support construction and delivery of materials to the site during construction. Roadway preconstruction activities would have a SMALL effect on ecological resources, due to the limited amount of area involved.

Noise, dust, and air emissions associated with site clearing would be short-lived and represent only a temporary adverse impact to the biota of the IIFP site (IIFP, 2011a). Removal of the vegetation and the soil disturbance that would occur during preconstruction activities would likely destroy nesting substrates for many of the potential breeding bird species found in this area (see Table 3-17). However, the impacts are not likely to have population-level impacts to the affected species (SORA, 2011). NMGF has suggested a minimization measure, for preconstruction to take place outside of the nesting season of migratory birds, which, if instituted, would impact few nesting activities in the affected habitat. Accordingly, preconstruction site clearing activities would have a SMALL effect on ecological resources.

Construction of Phase 2, which will occur on recently disturbed land adjacent to the Phase 1 facility, would not affect ecological resources. Accordingly, Phase 2 construction would have a SMALL effect on ecological resources.

Therefore, the NRC staff concludes that cumulative impacts to ecological resources would be SMALL.

4.2.2.8 Socioeconomic Resources

Preconstruction activities are assumed to begin in 2011 and to conclude prior to the end of 2011. Initially 35 and later as many as 70 workers would be involved in preconstruction activities. During preconstruction, the work force would consist of heavy equipment operators and structural crafts, most of which are expected to come from the ROI. Preconstruction activities are expected to result in impacts that would be approximately one-fourth to one-half the impacts presented in Section 4.1.8 for Phase 1 construction. As such, the NRC staff finds that there would be a correspondingly SMALL impact on housing, taxes, infrastructure and community services (IIFP, 2011a).

Phase 2 would use a construction crew of 150 to 180 workers. IIFP estimates approximately 27 workers of the construction work force are expected to move into the vicinity as new residents (15 percent of 180 workers). The increases in area population during Phase 2 construction, therefore, would be approximately the same as Phase 1 construction and the NRC staff finds that those increases would have SMALL impacts to socioeconomic resources.

The Phase 2 operations of the IIFP facility would require a maximum of 40 additional workers (IIFP, 2009). Using the same assumptions for the Phase 1 operations workforce, the NRC staff assumed that 32 workers would already reside in the area, and that 8 would in-migrate. Given the excess housing, public utilities and capacity in local schools, as described in Section 3.9, the NRC staff concludes that socioeconomic impacts from Phase 2 operations would be SMALL.

No disproportionately high or adverse impacts would occur to environmental justice populations in the ROI. The NRC staff finds that the cumulative impacts of preconstruction, the proposed action and Phase 2 construction and operation on socioeconomic resources would be SMALL.

The URENCO USA/LES Uranium Enrichment facility is expected to employ a maximum of 210 people annually and would indirectly create an additional 173 jobs (NRC, 2005a).

Overall, the NRC staff concludes that the cumulative impacts from the proposed IIFP project and the UNENCO facility Phase 2 construction and operation are expected to be SMALL.

4.2.2.9 Traffic and Transportation

The peak preconstruction workforce is estimated to be 70 employees (INIS, 2011). The construction work force would predominantly use NM 483 and US 62/180 to access the IIFP site. The existing AADT of both of these roadways is within the general capacity of 3,400 personal cars per hour for two-lane highways. There would be an increase of a maximum of 140 trips per day, two trips per potential employee, plus up to 40 additional trips associated with preconstruction equipment or supply deliveries (IIFP, 2009a). During preconstruction, the roadways would still operate well within their capacity. There would be no radiological transportation during preconstruction. The NRC staff finds that the impacts from increased traffic during preconstruction would be SMALL and temporary; therefore, the NRC staff concludes that no cumulative impacts would occur.

Latent Cancer Fatality (LCF)

A latent cancer fatality (LCF) is a fatality associated with acute or chronic environmental exposures to chemicals or radiation. The fatality may occur many years after the exposure.

An average construction crew of 150 to 180 workers would be required during the approximately 15-month Phase 2 construction period. Once operational, the workforce at the IIFP facility would increase from approximately 120-138 for Phase 1 operations to 145-160 for Phase 2 operations. If all the construction traffic used the access road off NM 483 this would result in a 75 percent increase during Phase 2 construction (including construction and operations traffic). The vast majority of this increase is expected to be on the 2.4 km (1.5 mi) section between the access road and US 62/180. Compared with the traffic count for the various highways from 2006 through 2008 and the transportation commuting statistics in Lea County from the 2000 census data, the impact of this temporary increase in traffic during Phase 2 construction is considered to be MODERATE for the peak construction period on NM 483. During construction of Phase 2 mitigation could include staggering the construction and operations shifts, encouraging carpooling or providing vans to transport construction workers from remote locations. Mitigation would reduce the impacts from MODERATE to SMALL.

After Phase 2 is operational, there would be a maximum of 40 additional round trips per day due to operation workers, resulting in an additional 80 vehicles on the area highways per day which would not exceed the design capacity of the roadways. The NRC staff finds that operational traffic would have a SMALL impact on the local transportation pattern.

During Phase 2 operations, the number of radiological shipments (including DUO_2 and LLW) per year would increase from 145 -155 shipments of DUO_2 (IIFP, 2011b) during Phase 1 with a total of approximately 700 radiological shipments (IIFP, 2011b) total, to 450-500 shipments of DUO_2 (IIFP, 2011b) during Phase 2 with a total of approximately 2,150 radiological shipments (IIFP, 2011b). The number of non-radiological shipments is not expected to change from 1,950 shipments. Therefore during Phase 2 operations, a total of 4,100 shipments are estimated annually or approximately 16 round trips per day. Compared with the transportation commuting statistics in Lea County from the 2000 census data and the AADT on the specific highways, the NRC staff finds that this increase in traffic from operational deliveries and waste removal would be SMALL for Phase 2 operations. One mitigation measure to be considered by IIFP is to schedule operations worker shift changes and truck shipments for off-peak traffic periods, when practical.

The URENCO USA/LES Uranium Enrichment facility truck shipments of feed, product, and waste materials (including DUF_6) could result in 2 LCFs to the general population over the life of the facility due to vehicle emissions and fewer than 0.03 LCF due to direct radiation. All rail shipments of feed, product, waste materials, and empty cylinders were estimated to result in fewer than 0.08 LCF to the general population over the life of the facility, and 0.1 LCF from direct radiation (NRC, 2005a).

Some adverse transportation impacts are expected as a result of moving the transuranic wastes from sites across the country to the WIPP. One of the official WIPP routes is US 62/180, which runs along the southern boundary of the proposed IIFP facility site. DOE estimated that the non-radiological impacts of transportation related to WIPP operations would result in approximately one traffic fatality and less than one death from pollution health effects. Radiological impacts associated with WIPP-related accident-free transportation are expected to be much less than 1 LCF (DOE, 2009).

The radiological impacts associated with combined Phase 1 and Phase 2 operations would result in a total population dose of 1.7 person-Sv (170 person-rem) annually. Statistically, this dose could result in 0.10 LCFs annually. When combined with the radiological transportation impacts from operation of the LES (0.1 LCFs over the facility life) and radiological transportation impacts from the WIPP (less than 1 LCF annually), the NRC staff finds that the cumulative radiological impacts from transportation would be SMALL (less than 1 LCF annually) (IIFP, 2009a).

4.2.2.10 Noise

As discussed in Section 3.11.2, there are no noise sensitive receptors in close proximity to the proposed IIFP facility. The nearest commercial facility is the Xcel Energy Cunningham Generating Station, approximately 1.6 km (1 mi) west of the proposed site. The nearest residence is approximately 2.6 km (1.6 mi) northwest of the site and there are no recreational facilities areas within 8.0 km (5.0 mi) of the proposed site. Because of the absence of any sensitive noise receptors, no noise impacts are anticipated during preconstruction activities and Phase 2 construction activities and no cumulative impacts would occur.

Cumulative impacts from all site noise sources would remain at or below HUD guidelines of 65 dBA Ldn (24 CFR 51), and the EPA guidelines of 55 dBA Ldn, (EPA, 1974) at the site boundary during IIFP facility construction and operation. Therefore, the NRC staff concludes that the cumulative noise of all site construction and operation activities, even when considered in conjunction with surrounding regional energy production facilities would have a SMALL impact and to only those receptors closest to the site boundary.

4.2.2.11 Public and Occupational Health

The preconstruction activities have the potential to cause industrial accidents, material-handling accidents, falls, etc., that could result in temporary injuries, long-term injuries and/or disabilities, and even fatalities. The proposed activities are not anticipated to be any more hazardous than the construction activities discussed in Section 4.1.1.11. The preconstruction workforce would be smaller than the construction workforce and the duration of preconstruction would be less than that of construction. Less than six nonfatal injuries and no fatalities (less than one fatality) are expected during preconstruction activities. Therefore, the NRC staff concludes that preconstruction health and safety impacts would be SMALL.

The Phase 2 construction activities have the potential to cause industrial accidents, material-handling accidents, falls, etc., that could result in temporary injuries, long-term injuries and/or disabilities, and even fatalities. The proposed activities are not anticipated to be any more hazardous than the construction activities discussed in Section 4.1.1.11. The Phase 2 construction workforce would be slightly larger than the Phase 1 construction workforce, and less than 13 nonfatal injuries and no fatalities (less than 1 fatality) are expected during Phase 2 construction activities.

Once operational, the workforce at the IIFP facility would increase from 140 for Phase 1 operations to 180 for Phase 1 and Phase 2 operations. Statistically, this would increase the potential number of both nonfatal and fatal occupational injuries by approximately 10 - 15 percent. Overall, less than seven nonfatal injuries and no fatalities (less than 1 fatality) are expected annually during the proposed operation of Phase 1 and Phase 2.

The NRC staff finds that radiological impacts associated with operation of the Phase 2 facility would be SMALL. The differential in the total population dose between the integrated Phase 1 and Phase 2 operations and Phase 1 operations alone would be an increase of 2.33×10^{-4} person-Sv/yr (2.33×10^{-2} person-rem/year). The differential in the dose to the MEI would be 1.62×10^{-8} person-Sv/yr (1.62×10^{-6} rem/yr). The differential between the two operational phases for the dose to the nearest resident would be 1.18×10^{-8} Sv/yr (1.18×10^{-6} rem/year) (IIFP, 2011a). The difference, therefore, between operational phases is very low.

The types of postulated accidents and release scenarios for the Phase 2 facility would not differ from those already addressed in Phase 1 operations because Phase 2 operations only add inventory and capacity; no new types of chemical or radiological risks would be added. As such, the types of accidents and the description of postulated accidents for the Phase 1 facility would be representative of the range of credible accidents associated with the Phase 2 facility.

Therefore, the NRC staff concludes that the cumulative impacts to occupational and public health from preconstruction, the proposed action, and Phase 2 construction and operations would be SMALL.

4.2.2.12 Waste Management

As discussed in Section 4.1.1.12, the NRC staff finds that the quantities of wastes generated during construction of the proposed IIFP facility would result in SMALL impacts that could be managed effectively. Approximately 300-500 kg (650-1,100 lbs) of solid waste and 270–820 kg (600-1,800 lbs) of hazardous waste would be generated (INIS, 2011). Preconstruction activities are expected to generate waste types similar to and with volumes less than those estimated for construction (IIFP, 2011a). No radiological wastes would be generated during preconstruction.

As a point of comparison, the operation of the National Enrichment Facility would generate approximately 172,500 kg (380,400 lbs) of solid nonradioactive waste annually, including approximately 1,900 L (500 gal) of hazardous liquid wastes (NRC, 2005a). Approximately 87,000 kg (191,800 lbs) of radiological and mixed waste would be generated annually, of which approximately 50 kg (110 lbs) would be mixed waste. When added to the wastes from other waste generators, such as the National Enrichment Facility, the NRC staff finds that the impacts and cumulative impacts of disposal of hazardous and solid (nonhazardous) wastes from preconstruction activities of the proposed IIFP facility would be SMALL.

Phase 2 construction would necessitate connections to existing Phase 1 facilities and installation of additional autoclaves (IIFP, 2011b). Radiological materials would not be used in the construction of the Phase 2 facility. However, Phase 2 construction involving connections to Phase 1 facilities could result in generation of radioactive wastes. The construction waste types and volumes would be similar to those during Phase 1 construction. Tables 4-38 through 4-40 provide the estimated annual quantities of solid, hazardous, and radioactive wastes generated during Phase 2 construction.

Table 4-38.   Phase 2 Construction Solid Waste Generation

Waste Type	Estimated Annual Amount
Air filters(vehicle)	23 – 45 kg (50 – 100 lbs)
Cardboard / packing	136 – 227kg (300 – 500 lbs)
Fiber drums	136 – 318 kg (300 – 700 lbs)
Total	295 – 590 kg (650 – 1,300 lbs)

Source: IIFP, 2011a

Table 4-39.   Phase 2 Construction Hazardous Waste Generation

Waste Type	Estimated Annual Amount
Adhesives, resins, caulking residues	54 – 109 kg (120 – 240 lbs)
Lead (batteries)	45 –113 kg (100 – 250 lbs)
Oil filters	45 – 91 kg (100 – 200 lbs)
Paints, thinners, solvents, organic residues	45 – 227 kg (100 – 500 lbs)
Pesticides	45 – 68 kg (100 – 150 lbs)
Petroleum products, oils, lubricants residues	45 – 227 kg (100 – 500 lbs)
Total	281 – 835 kg (620 – 1,840 lbs)

Source: IIFP, 2011a

Table 4-40.   Phase 2 Construction Radioactive Waste Generation

Material	Estimated Annual Amount
Scrap metal	1,800 – 2,700 kg (4,000 – 6,000 lbs)
Spent blasting sand	45 kg (100 lbs)
Wood trash (pallets)	450 – 680 kg (1,000 – 1,500 lbs)
Total	2,300 – 3,400 kg (5,100 – 7,600 lbs)

Source: IIFP, 2011a.

As described in Section 4.1.1.12, all construction wastes would be transferred offsite to licensed waste disposal facilities with adequate disposal capacity for the estimated volumes. Thus, it is also anticipated by NRC staff that the waste management impacts from Phase 2 construction would be SMALL.

The URENCO USA/LES Uranium Enrichment facility commenced initial operations on June 11, 2010 and full operations are expected in 2015. Projected waste volumes from the enrichment facility operations include 173,000 kg/yr (380,400 lb/yr) of solid waste; 1,890 kg/yr (4,165 lb/yr) of hazardous and mixed waste; and 87,000 kg/yr (191,800 lb/yr) of LLW (NRC, 2005b). DUF_6 is a waste product of the uranium enrichment process, and the URENCO USA/LES Uranium

Enrichment facility would be one of the likely DUF_6 suppliers to the proposed IIFP facility. The enrichment facility will produce depleted uranium at a rate of 627 cylinders or 7,800 metric tons/yr (NRC, 2005b). During Phase 1, the proposed IIFP facility would process 266 cylinders annually of DUF_6 as feed to the deconversion process.

Solid waste from the enrichment facility would be disposed of at the Lea County Landfill along with waste from the proposed IIFP facility. The solid waste generated by the enrichment facility would potentially increase the volume of wastes received at the landfill by less than 0.03 percent (NRC, 2005b). That increase in combination with the highest IIFP annual solid waste generation rate (during Phase 1 and Phase 2 operations) would result in less than 0.1 percent change in the waste received by the Lea County Landfill. Hazardous waste generated by the enrichment facility (less than 1, 814 kg [2 tons] per year) and the proposed IIFP facility (up to 154 tons/yr during Phase 1 operations) represents less than 0.02 percent of the hazardous waste managed in the state of New Mexico (more than 1 million tons in 2009). The NRC staff finds that the combined impacts of managing the solid and hazardous wastes generated by both facilities on the available capacity would be SMALL.

In the final EIS for the URENCO USA/LES Uranium Enrichment facility (NUREG-1790; NRC, 2005a), NRC staff considered the impacts of conversion of the DUF_6 from the enrichment process (up to 15,727 cylinders over the operating life) to depleted U_3O_8 and disposal of the resulting Class A LLW in a licensed disposal facility. The NRC staff concluded that both the environmental impacts of shallow land disposal such as the Energy*Solutions* site in Clive, Utah, and the effect on national disposal capacity for Class A LLW would be SMALL. The deconversion of DUF_6 by the proposed IIFP facility and disposal of the resulting DUO_2 as Class A LLW represents a subset of the impacts previously considered in NUREG-1790 (NRC, 2005a) (the oxide form of the converted depleted uranium waste, whether U_3O_8 or UO_2, would not materially change the consequences). Therefore, the NRC staff concludes that the cumulative effects of the management of depleted uranium wastes from the proposed IIFP facility and the URENCO USA/LES Uranium Enrichment facility would be SMALL.

The wastes from Phase 2 construction would generate much less than 1 percent of the annual wastes from the National Enrichment Facility (172,500 kg [380,400 lbs] of solid nonradioactive waste and approximately 87,000 kg [191,800 lbs] of radiological and mixed waste). Based on available capacities at hazardous, solid, and radioactive waste treatment and disposal sites, and the expectation that there would be no large developments in the Hobbs area that would cause a significant increase in municipal waste disposal volume, the NRC staff finds that the cumulative impacts from hazardous, solid, and radioactive waste generation would be SMALL.

As described in Section 4.1.2.12, the NRC staff finds that the impact of disposal of hazardous, solid, and radioactive wastes from operation of the proposed Phase 1 IIFP facility at the appropriate offsite facilities would be SMALL. Phase 2 operations would generate waste types similar to those during Phase 1 operations. The hazardous waste volumes are expected to be lower and LLW volumes higher than from the Phase 1 facility.

The cumulative LLW generation rate during combined Phase 1 and 2 operations would be about three times higher than from Phase 1 alone. Most of that increase would result from tripling the production of DUO_2. The generation rate of other LLW streams (e.g., trash, waste drums and pallets) would also increase with the expanded Phase 2 facility. Tables 4-41 through 4-43 provide the estimated annual waste quantities generated during combined Phase 1 and 2 operations, for solid, hazardous, and radioactive wastes, respectively.

The cumulative solid waste generation (up to 49,900 kg [55 tons] per year) would be 20 percent greater than during Phase 1 operations. Cumulative Phase 1 and 2 operations would result in an increase of approximately 0.07 percent in the waste that the Lea County landfill receives annually from all other sources. The NRC staff finds that this quantity of nonhazardous waste material would result in SMALL impacts that could be managed effectively.

The quantity of cumulative hazardous waste could be as much as 46,300 kg (51 tons) per year if a market for the CaF_2 cannot be identified during Phase 2 operations. Because of the added process technology used in the expansion to the Phase 2 facility, when Phase 2 becomes operational, a large part of the fluoride-bearing spent scrubber liquids (the HF liquor portion) from the plant KOH scrubbing system can be recycled to the add-on direct oxide deconversion process and recovered rather than be treated with lime to generate CaF_2, thus reducing the amount of hazardous waste produced compared to Phase 1 operations. As discussed in Section 4.1.1.12, hazardous waste generators in New Mexico accounted for 978,554,778 kg (1,078,672 tons) of hazardous waste in 2009 (EPA, 2010b). The maximum cumulative generation rate would result in an increase of less than 0.005 percent in the hazardous waste generated in the State of New Mexico.

DUO_2 and other radiological waste would be shipped offsite to licensed disposal facilities. As shown in Table 4-43, up to 9,168,009 kg (10,106 tons) per year of LLW could be sent for disposal each year. Most of the estimated annual LLW generation (approximately 99 percent) would be the DUO_2 produced by the deconversion process. Assuming 450 kg (1,000 lbs) per oxide drum, Phase 1 and 2 operations would result in 8,700 to 20,000 drums of material being sent for disposal. This uranium oxide waste volume represents 3.1 percent to 7.2 percent of the annual commercial waste volume currently received at the EnergySolutions facility in Clive, Utah (NRC, 2010). The Clive facility accepts the majority of the United States' Class A waste and is estimated to have capacity to accept this waste at current volume levels for more than 20 years (GAO, 2004). The NRC staff finds that the estimated generation of depleted uranium oxide and other LLW from the Phase 2 deconversion process would result in SMALL impacts to LLW disposal capacity.

Table 4-41.   Cumulative Solid Waste Generation – Phase 1 and 2 IIFP Facility

Material	Estimated Annual Amount
Clothing	68 – 136 kg (150 – 300 lbs)
Molecular sieve	136 – 227 kg (300 – 500 lbs)
Municipal trash waste	32,659 – 48,988 kg (72,000 – 108,000 lbs)
Safety gear	181 – 363 kg (400 – 800 lbs)
Waste Glass	34 – 136 kg (75 – 300 lbs)
Total	33,078 – 49,850 kg (72,925 – 109,900 lbs)

Source: IIFP, 2011a

Table 4-42. Cumulative Hazardous Waste Generation — Phase 1 and 2 IIFP Facility

Material	Estimated Annual Amount
Aerosol cans, paints cans, bulbs	907 – 1,814 kg (2,000 – 4,000 lbs)
Calcium fluoride*	27,216 – 40,823 kg (60,000 – 90,000 lbs)
Lab chemicals	91 – 182 kg (200 – 400 lbs)
Oil sorb	1,361 – 3,175 kg (3,000 – 7,000 lbs)
Total*	29,574 – 45,994 kg (65,200 – 101,400 lbs)

Source: IIFP, 2011a
*Includes calcium fluoride which would not be waste if sold.

Table 4-43. Cumulative Radioactive Waste Generation – Phase 1 and 2 IIFP Facility

Material	Estimated Annual Amount
Activated alumina	907 – 1,814 kg (2,000 – 4,000 lbs)
Air ventilation filters	29 – 45 kg (65 – 100 lbs)
Carbon	11,340 – 13,608 kg (25,000 – 30,000 lbs)
DUF$_4$ clinkers	2,268 – 4,536 kg (5,000 – 10,000 lbs)
Coke	3,629 – 5,443 kg (8,000 – 12,000 lbs)
Crushed drums	907 – 3,629 kg (2,000 – 8,000 lbs)
Dust collector bags	454 – 1,361 kg (1,000 – 3,000 lbs)
Ion exchange resin	907 – 1,814 kg (2,000 – 4,000 lbs)
Oxide for burial (plus drums)	3,946,258 – 9,071,858 kg (8,700,000 – 20,000,000 lbs)
Radioactive waste trash	31,752 – 45,359 kg (70,000 – 100,000 lbs)
Scrap metal	5,443 – 7,257 kg (12,000 – 16,000 lbs)
Sintered metal tubes	907 – 1,361 kg (2,000 – 3,000 lbs)
Sodium fluoride	907 – 1,814 kg (2,000 – 4,000 lbs)
Spent blasting sand	45 – 91 kg (100 – 200 lbs)
Wood trash (pallets)	1,361 – 5,443 kg (3,000 – 12,000 lbs)
Total	4,007,115 – 9,167,702 kg (8,834,165 – 20,211,300 lbs)

Source: IIFP, 2011a

The wastes generated during cumulative Phase 1 and 2 operations would be transferred offsite to licensed waste facilities with adequate disposal capacity for the estimated volumes. Thus, the NRC staff anticipates that the waste management impacts from cumulative operations would be SMALL.

4.3   No-Action Alternative

As presented in Section 2.2 of this EIS, the no-action alternative would be to not construct, operate, and decommission the proposed IIFP facility in Lea County, near Hobbs, New Mexico. As discussed in Section 2.1, IIFP expects to carry out preconstruction activities (i.e., site preparation and non-safety related construction activities) prior to issuance of a license by NRC. If NRC does not ultimately grant IIFP a license for the proposed plant, these would be activities associated with the no-action alternative.

Preconstruction would be overseen by the NMED and Lea County, pursuant to applicable permit requirements. NMED state permits would include those listed in Section 1.4, including stormwater controls, and erosion and sedimentation controls. A New Mexico Department of Transportation right-of-way permit would be required in order to construct access to NM 483. Lea County ordinances would require adherence to applicable building codes and fire code standards. There could be additional activities at the proposed site in the future under the no-action alternative that could have (adverse or beneficial) impacts on the environment and community. The impacts associated with these activities would depend on what IIFP would decide to do with the proposed site or the improvements (e.g., access roads, buildings, etc.) already constructed on the site, should the license not be granted. The conclusions presented in this section for the no-action alternative address the impacts of denying the license, but do not include the impacts of the NRC-approved preconstruction activities, which have been discussed in Section 4.2.2, Cumulative Impacts.

Under the no-action alternative, as discussed in Section 2.2.1, commercial uranium enrichment facilities would continue to store depleted uranium. DOE, which operated three gaseous diffusion plants near Oak Ridge, Tennessee, Piketon, Ohio, and Paducah, Kentucky, as part of the process of enriching uranium for civilian and defense applications, and would continue to deconvert its stockpiles of DUF_6. In the future, DOE deconversion technology and recently constructed deconversion plants at Paducah and Piketon would become available to deconvert commercial DUF_6. However, this would only occur when all of the DOE's DUF_6 stockpiles at Oak Ridge, Portsmouth, and Paducah are deconverted which is projected to take 25 years. Therefore, the no-action alternative would involve long term storage of depleted uranium at commercial enrichment facilities until such time as DOE could accept DUF_6 from facilities other than its own.

As discussed in Section 2.2.2, other alternatives, including Alternative Sites, Alternative Technologies, Overseas Shipment, Indefinite Storage, or deconversion at the commercial enrichment plants, have been dismissed for various reasons, and it is, therefore, assumed that none of those alternatives would occur. The environmental impacts of deconversion alternatives involving other sites and technologies would have impacts similar to or greater than the impacts from the Proposed Action. Alternatives involving overseas shipment would involve unreasonable shipment costs, increased potential for adverse transportation-related impacts, and potentially unacceptable risks. Indefinite storage is not practical or reasonable compared to the benefits of added commercial fluoride product and increased safety afforded by the deconversion process. Deconversion onsite at commercial uranium enrichment facilities is not feasible, considering technological and marketing advantages of using existing DOE facilities and/or the proposed IIFP facility. Therefore, the no-action alternative is limited to the Proposed Action, and its affiliated positive or adverse effects, not occurring.

The no-action alternative assumes that preconstruction occurred within the 16-ha (40-ac) facility and that fencing was erected around the 259-ha (640-ac) proposed IIFP site. Impacts from the no-action alternative to affected resources are as follows:

LAND USE: If the fencing was not removed, it would restrict cattle grazing. Other land uses in the vicinity of the site would be unaffected. A Site Redress Plan (SRP) could be required by NRC. The SRP could require the site to be restored to original grade and condition, including removal of fencing. The NRC staff finds that impacts to local land use would be SMALL, because of the amount of land adversely affected compared to the large amount of land available in the vicinity.

HISTORIC AND CULTURAL RESOURCES: For the same reasons as described for the proposed action, the NRC staff finds that impacts to historic and cultural resources would be SMALL and would result in no effect on historic properties, districts, resources or significant historic/precontact archaeological sites.

VISUAL RESOURCES: The existing character of the area would be altered only within the preconstruction area perimeter and access road. This disturbance would be limited to less than ten percent of the Section 27 (IIFP) site, and the NRC could require that all structures be removed. Therefore, the NRC staff finds that impacts would be SMALL.

CLIMATE, METEOROLOGY, AND AIR QUALITY: Impacts to air quality from preconstruction activities would be small, localized, and temporary. Local and global atmospheric conditions would not be altered noticeably, and the NRC staff finds that impacts to air quality would be SMALL.

GEOLOGY, MINERALS, AND SOILS: Land disturbance from preconstruction clearing, grading, and excavation would have occurred. For reasons discussed in Section 4.2.2.5, the NRC staff finds that the impacts would be SMALL.

WATER RESOURCES: Consumptive use of water for preconstruction is anticipated (dust suppression and domestic use for workers). Water would be brought to the IIFP site by tanker trucks from the Hobbs municipal water system, which has adequate capacity. No surface water is present on the site, and so no impacts to surface water quality, including from sedimentation are expected. Water tanker trucks used for preconstruction would likely obtain water from groundwater wells within the region, near to, but not within the IIFP site. As indicated in Section 4.1.2.6.1, the NRC staff finds that impacts to surface water and groundwater would be SMALL.

ECOLOGICAL RESOURCES: Preconstruction would result in the loss of vegetation and terrestrial habitat. Some wildlife may be destroyed. This disturbance would be limited to less than 10 percent of the 259-ha (640-ac) site. If NRC requires an SRP, the site could be restored to near original grade and condition, including replanting or re-seeding to allow vegetation to reclaim the facility location. The NRC staff finds that impacts to ecological resources would be SMALL.

SOCIOECONOMICS AND ENVIRONMENTAL JUSTICE: Any consequences of the construction, operation, and decommissioning of the proposed IIFP facility (positive or adverse) would not occur and socioeconomic conditions in the ROI would remain unchanged. Population in the ROI would grow in accordance with current projections. The socioeconomic characteristics of the region, including housing availability, school enrollment, availability of health service resources, and law enforcement and firefighting resources, would not be affected by the proposed action. The no-action alternative would not cause any high and adverse impacts, including to low-income and minority populations. Therefore, there would not be any environmental justice concerns. The NRC staff finds that impacts of no action on socioeconomic conditions, including those of low-income and minority populations, in the region would be SMALL.

TRAFFIC AND TRANSPORTATION: There would be no increased traffic as a result of the no-action alternative. The NRC staff finds that impacts to the regional and national traffic and transportation system would be SMALL.

NOISE: Temporary, slight increases in ambient noise levels in the immediate area of the facility would occur during preconstruction, however, other than those temporary increases in local noise, the no-action alternative would not affect ambient noise levels. No changes in land use plans or traffic are expected. Therefore, based on all of these considerations, the NRC staff finds that impacts to noise would be SMALL.

PUBLIC AND OCCUPATIONAL HEALTH: Except for the potential for construction-related injuries during preconstruction, the no-action alternative would not affect public or occupational health. Therefore, the NRC staff finds that impacts to public and occupational health would be SMALL.

WASTE MANAGEMENT: The no-action alternative would not be expected to cause changes in management of solid, hazardous, or mixed waste in the region. No radiologically contaminated waste would be generated during preconstruction or to meet the requirements of the SRP, and, therefore, the NRC staff finds that impacts to waste management, would be SMALL.

ACCIDENTS: The no-action alternative would not cause accidents to occur within the IIFP facility. Therefore, the NRC staff finds that impacts from accidents would be SMALL.

4.4 References

(BEA, 2010) U.S. Department of Commerce, Bureau of Economic Analysis. 2010. RIMS II Multipliers (2002/2007) Table 1.5 Total Multipliers for Output, Earnings, Employment, and Value Added by Detailed Industry IIFP Socioeconomic Site (Type II). Economic and Statistics Administration, Washington, D.C. September, 2010. ADAMS Accession No. ML112720418.

(Biwer et al., 2001) Biwer, B. M., F. A. Monette, L.A. Nieves, and N. L. Ranek. 2001. Transportation Impact Assessment for Shipment of Uranium Hexafluoride (UF$_6$) Cylinders from the East Tennessee Technology Park to the Portsmouth and Paducah Gaseous Diffusion Plants, ANL/EAD/TM-112, Environmental Assessment Division, Argonne National Laboratory, Argonne, Il. October, 2001. Available at http://web.ead.anl.gov/uranium/pdf/ANL-EAD-TM-112.pdf.

(BLS, 2010a) U. S. Department of Labor, Bureau of Labor Statistics. 2010. "Workplace Injuries and Illnesses - 2009;" October 21, 2010. Available at http://www.bls.gov/news.release/archives/osh_10212010.pdf. Accessed on March 15, 2011. ADAMS Accession No. ML112720427.

(BLS, 2010b) U. S. Department of Labor, Bureau of Labor Statistics. 2010. National Census of Fatal Occupational Injuries, in 2010 (preliminary results). August 19, 2010. Available at http://www.bls.gov/news.release/pdf/cfoi.pdf. Accessed on March 3, 2011. ADAMS Accession No. ML 112720434.

(BMI, 1981) Battelle Memorial Institute. 1981. Migration and Residential Location of Workers at Nuclear Power Plant Construction Sites; Forecasting Methodology (Volume 1) and Profile Analysis of Worker Surveys (Volume 2), NUREG/CR-2002, PNL-3757, prepared by Pacific Northwest Laboratory for U. S. Nuclear Regulatory Commission. April 1981. ADAMS Accession No. ML 112840173.

(CDC, 2010) Centers for Disease Control and Prevention. 2010. "Deaths: Final Data for 2007," National Vital Statistics Reports, Volume 58, Number 19. May 20, 2010. ADAMS Accession No. ML112720451.

(Crone and Wheeler, 2000) Crone, A. J. and Wheeler, R. L. 2000. Data for Quaternary Faults, Liquefaction Features, and Possible Tectonic Features in the Central and Eastern United States, East of the Rocky Mountain Front. U.S. Geological Survey Open File Report 00-0260. Available at http://pubs.usgs.gov/of/2000/ofr-00-0260/ofr-00-0260.pdf. ADAMS Accession No. ML103080519.

(DOE, 2000) U.S. Department of Energy. 2000. Evaluation of the Acceptability of Potential Depleted Uranium Hexafluoride Conversion Products at the Envirocare Disposal Site. ORNL/TM-2000/355. December 2000. ADAMS Accession No. ML071130404

(DOE 2004a) U.S. Department of Energy. 2004. Final Environmental Impact Statement for Construction and Operation of a Depleted Uranium Hexafluoride Conversion Facility at the Paducah, Kentucky Site, DOE/EIS-0359. Washington, D.C., June 2004. ADAMS Accession No. ML050380331.

(DOE 2004b) U.S. Department of Energy. 2004. Final Environmental Impact Statement for Construction and Operation of a Depleted Uranium Hexafluoride Conversion Facility at the Portsmouth, Ohio Site, DOE/EIS-0360, Washington, D.C., June 2004. ADAMS Accession No. ML050380334

(DOE, 2008) U. S. Department of Energy. 2008. Final Complex Transformation Supplemental Programmatic Environmental Impact Statement, Volume II, Chapter 6 (DOE/EIS-0236-S4). Washington, D.C. Available at http://nepa.energy.gov/1017.htm. Accessed on August 16, 2010. ADAMS Accession No. ML103430383.

(DOE, 2009) U.S. Department of Energy. 2009. Supplemental Analysis for the Waste Isolation Pilot Plant Site-Wide Operations (DOE/EIS-0026-SA-07). Washington, D.C. Available at http://www.wipp.energy/gov/libraryseis/DOE_EIS-0026-SA-07_May_29_2009.pdf. Accessed August 16, 2010. ADAMS Accession No. ML103430391.

(DOE, 2010) U.S. Department of Energy. 2010. WIPP Shipment and Disposal Information. Available at http://www.wipp.energy.gov/shipments.htm. Accessed August 16, 2010 and October 29, 2010. ADAMS Accession No. ML103430396.

(EPA, 1974) U. S. Environmental Protection Agency. 1974. Information on Levels of Environmental Noise Requisite to Protect Public Health and Welfare with an Adequate Margin of Safety," EPA/ONAC 550/9-74-004. March 1974. Available online at http://www.nonoise.org/library/levels74/levels74.htm. Accessed on December 13, 2010. ADAMS Accession No. ML110110692.

(EPA, 1995a) U.S. Environmental Protection Agency. 1995. Compilation of Air Pollutant Emissions Factors, Fifth Edition and Supplements, AP-42. Office of Air Quality Planning and Standards, Research Triangle Park, North Carolina. January 1995. . Available at http://www.epa.gov/ttn/chief/ap42/index.html. Accessed May 18, 2011. ADAMS Accession No. ML112720457.

(EPA, 1995b) U.S. Environmental Protection Agency. 1995. SCREEN3 Model User's Guide, EPA-454/B-95-004, Office of Air Quality Planning and Standards, Research Triangle Park, North Carolina. September 1995. Available at http://www.epa.gov/ttn/scram/userg/screen/screen3d.pdf. Accessed May 18, 2011. ADAMS Accession No. ML071930125.

(EPA, 1999) U.S. Environmental Protection Agency. 1999. Risk Management Program Guidance for Off-Site Consequence Analysis, EPA-550-B-99-009. Washington, D.C. April 1999. Available at http://www.epa.gov/oem/docs/chem/oca-all.pdf. ADAMS Accession No. ML112720461.

(EPA, 2005) U.S. Environmental Protection Agency. 2005. EPA NONROAD Emissions Model User's Guide. EPA420-R-05-013, Office of Transportation and Air Quality. December 2005. Available at http://www.epa.gov/otaq/models/nonrdmdl/nonrdmdl2005/420r05013.pdf. Accessed May 18, 2011. ADAMS Accession No. ML112720500 and ML112720491.

(EPA, 2009a) U.S. Environmental Protection Agency. 2009. Motor Vehicle Emission Simulator (MOVES) 2010 User Guide, EPA-420-B-09-041, Office of Transportation and Air Quality. December 2009. Available at http://www.epa.gov/otaq/models/moves/420b09041.pdf. Accessed May 18, 2011.

(EPA, 2009b) U.S. Environmental Protection Agency. 2009. U.S. Environmental Protection Agency. 2009. County Air Quality Report – Criteria Air Pollutants for Eddy, Lea, and Sandoval Counties, New Mexico, and El Paso County, Texas. Washington, D.C., January 10, 2009. Available at http://www.epa.gov/oar/data/geosel.html. Accessed on May 31, 2011.

(EPA, 2010a) U.S. Environmental Protection Agency. 2010. Inventory of U.S. Greenhouse Gas Emissions and Sinks: 1990–2008. Executive Summary. April 15, 2010. http://www.epa.gov/climatechange/emissions/usgginv_archive.html. Accessed March 10, 2011. ADAMS Accession No. ML112710366.

(EPA, 2010b) U.S. Environmental Protection Agency. 2010. "The National Biennial RCRA Hazardous Waste Report (Based on 2009 Data)". Available at: http://www.epa.gov/wastes/inforesources/data/br09/index.htm. ADAMS Accession No. ML112720562.

(FAA, 1992) U.S. Department of Transportation, Federal Aviation Administration. 1992. Change 2 to Obstruction Marking and Lighting. Advisory Circular. AC No. 70/7460-1H. July 15, 1992. Available online: http://wireless.fcc.gov//antenna/documentation/faadocs/7460-1H.pdf. Accessed March 2, 2011. ADAMS Accession No. ML112720574.

(GAO, 2004) U.S. Government Accountability Office. 2004. Low-Level Radioactive Waste: Disposal Availability Adequate in the Short Term, But Oversight Needed to Identify Any Future Shortfalls. Report to the Chairman, Committee on Energy and Natural Resources, U.S.

Senate.GAO-04-604. June 2004. Available at http://www.gao.gov/new.items/d04604.pdf. ADAMS Accession No. ML042010144.

(GL Environmental, 2010) GL Environmental, Inc. 2010. Existing Groundwater Condition in Section 27, Range 18 South, Township 36 East. Prepared for International Isotopes Fluorine Products. ADAMS Accession No. ML112720073.

(IIFP, 2009a) International Isotopes Fluorine Products. 2009. Fluorine Extraction Process and Depleted Uranium De-conversion Plant (FEP/DUP) Environmental Report. ER-IFP-001, Revision A. December 27, 2009. ADAMS Accession No. ML100120758.

(IIFP, 2009b) International Isotopes Fluorine Products, Inc. 2009. Fluorine Extraction Process & Depleted Uranium De-conversion Plant (FEP/DUP) Integrated Safety Analysis Summary, Revision A. December 23, 2009. ADAMS Accession No. ML100630501.

(IIFP, 2011a) International Isotopes Fluorine Products. 2011. Fluorine Extraction Process and Depleted Uranium De-conversion Plant (FEP/DUP) Official Responses to Environmental Report RAIs, Revision A. March 31, 2011. ADAMS Accession No. ML110970481.

(IIFP, 2011b) International Isotopes Fluorine Products. 2011. Fluorine Extraction Process and Depleted Uranium De-conversion Plant (FEP/DUP) Official Responses to Talking Points Regarding the Environmental Report. May 24, 2011. ADAMS Accession No. ML111530402.

(INIS, 2011) International Isotopes, Inc. 2011. "International Isotopes Inc. Provides an Update Summary on the Progress of the Uranium Deconversion and Fluorine Extraction Processing Facility" Press Release. Idaho Falls, ID, February 7, 2011. Available online: http://www.intisoid.com/wp-content/uploads/2010/08/INISAnnouncesContinuedProgress TowardstheLicensingand-Construction.pdf. Accessed March 28, 2011. ADAMS Accession No. ML112710381.

(ISCORS, 2002) Interagency Steering Committee on Radiation Standards. 2002. Estimating Radiation Risk from Total Effective Dose Equivalent (TEDE). ISCORS Technical Report 2002-02, Final Report. Washington, D. C. ADAMS Accession No. ML112720579.

(Johnson and Michelhaugh, 2003) Johnson, P. E. and R. D. Michelhaugh. 2003. Transportation Routing Analysis Geographic Information System (TRAGIS) User's Manual. Oak Ridge National Laboratory, Oak Ridge, Tennessee. June, 2003.

(Leedshill-Herkenhoff et.al., 2000) Leedshill-Herkenhoff, Inc., John Shomaker & Associates, Inc., and Montgomery and Andrews. 2000. Region 16: Lea County Regional Water Plan. Prepared for the Lea County Water User's Association. Available at http://www.ose.state.nm.us/isc_regional_plans16.html. Accessed July 25, 2010.

(Machette et al., 2000) Machette, M.N., Personius, S.F., Kelson, K.I., Haller, K.M., and R.L. Dart. 1998. Map of Quaternary Faults and Folds in New Mexico and Adjacent Areas. U.S. Geological Survey Open File Report 98-521 (Revised 2000). ADAMS Accession No. ML103080536.

(McCoy and Perry, 2004) McCoy. A.M., and R.L. Perry. 2004. Second Draft, Lea County Deep Aquifer Study. Prepared for Lea County Water Users Association. John Shomaker & Associates, Inc. ADAMS Accession No. ML103080447.

(Neuhauser and Kanipe, 2003) Neuhauser, K. S. and F. L. Kanipe. 2003. RADTRAN 5 User Guide. SAND2000-2354. Sandia National Laboratory, Albuquerque, New Mexico. July 7, 2003.

(NMDOT, 2009) New Mexico Department of Transportation. Consolidated Highway Database (CHDB). Website no longer accessible online. Accessed April, 10, 2009. ADAMS Accession No. ML112860287.

(NMED, 2009) New Mexico Environment Department, Solid Waste Bureau. 2009. New Mexico Solid Waste Annual Report, 2009. Available at http://www.nmenv.state.nm.us/swb/documents/ 2009FinalAR.pdf. Accessed September 23, 2009. ADAMS Accession No. ML112720586.

(NMED, 2010) New Mexico Environment Department. 2010. Frequently Asked Questions, Proposed Greenhouse Gas Cap-and-Trade Rule, Updated June 8, 2010. Available at http://www,nmenv.state.nm.us/aqb/ghg/documents/FAQ_Rev06082010.pdf. Accessed August 17, 2010. ADAMS Accession No. ML 103430404.

(NMJC, 2011) New Mexico Junior College. 2011. NMJC State of the College Report. Available at http://www.nmjc.edu/assets/documents/NMJC State of the College Report 2011.pdf. Accessed March 10, 2011. ADAMS Accession No. ML112720596. (NMOSE, 2009a) New Mexico Office of the State Engineer. 2009. Lea County Underground Water Basin Guidelines for Review of Water Right Applications. Adopted September 16, 2009. Available at http://www.ose.state.nm.us/PDF/RulesRegsGuidelines/LeaCounty/ LeaCountyGuidelines-2009-09-16.pdf. Accessed July 29, 2010. ADAMS Accession No. ML103080986.

(NMOSE, 2009b) New Mexico Office of the State Engineer. 2009. Documentation of the Lea County Underground Basin Guidelines for the Review of Water Right Applications. December 1, 2009. Available at http://www.ose.state.nm.us/PDF/RulesRegsGuidelines/LeaCounty/ GuidelineBasis-LeaCo-Report-2009-12-01.pdf. Accessed March 1, 2011. ADAMS Accession No. ML112720601.

(NMOSE, 2010) New Mexico Office of State Engineer. 2010. Lea County Wells with Well Logs. Available at http://nmwrrs.ose.state.nm.us/nmwrrs/index.html. Accessed July 29, 2010.

(NRC, 1977) U. S. Nuclear Regulatory Commission. 1977. Final Environmental Impact Statement on the Transportation of Radioactive Material by Air and Other Modes, NUREG/CR-0170. Office of Standards Development. Washington, D.C. December 1977.

(NRC, 1990) U.S. Nuclear Regulatory Commission. 1990. Control Room Habitability System Review Methods, NUREG/CR-5659. Division of Safety Resolution. Washington, D.C. December, 1990.

(NRC, 1991) U.S. Nuclear Regulatory Commission. 1991. Chemical Toxicity of Uranium Hexafluoride Compared to Acute Effects of Radiation, Final Report, NUREG-1391. Office of Nuclear Regulatory Research. Washington, D.C. February 1991.

(NRC, 1998) U.S. Nuclear Regulatory Commission. 1998. Nuclear Fuel Cycle Facility Accident Analysis Handbook, NUREG/CR-6410. Office of Nuclear Regulatory Research. Washington, D.C., March 1998.

(NRC, 2003) U.S. Nuclear Regulatory Commission. 2003. Environmental Review Guidance for Licensing Actions Associated with NMSS Programs, Final Report, NUREG-1748. Division of

Waste Management, Office of Nuclear Material Safety and Safeguards. Washington, D.C., August 2003. ADAMS Accession No. ML032450279.

(NRC, 2005a) U. S. Nuclear Regulatory Commission. 2005. Final Environmental Impact Statement for the National Enrichment Facility in Lea County, New Mexico NUREG-1790. Office of Nuclear Materials Safety and Safeguards. Washington, D. C. June 2005. ADAMS Accession No. ML051730238 and ML051730292.

(NRC 2005b) U. S. Nuclear Regulatory Commission. 2005. Safety Evaluation Report for the National Enrichment Facility in Lea County, New Mexico, Louisiana Energy Services NUREG-1827. Office of Nuclear Materials Safety and Safeguards. Washington D.C. June 2005,

(NRC, 2006) U.S. Nuclear Regulatory Commission. 2006. Consolidated Decommissioning Guidance, NUREG-1757. Office of Nuclear Material Safety and Safeguards. Washington, D.C., September 2006.

(NRC, 2010) U.S. Nuclear Regulatory Commission. 2010. Low-Level Waste Disposal Statistics. Available at http://www.nrc.gov/waste/llw-disposal/licensing/statistics.html. Accessed February 26, 2011.

(NRC, 2011) U.S. Nuclear Regulatory Commission 2011, "Revised Table 4-35." Personal communication between A. Malliakos, NRC and T. Harvey, et al., Straughan Environmental. August 9, 2011.

(Saricks and Tompkins, 1999) Saricks, C. L. and M. M. Tompkins. 1999. State-level Accident Rates of Surface Freight Transportation: a Re-examination, ANL/ESD/TM-150. Argonne National Laboratory, Argonne, Illinois. April, 1999. ADAMS Accession No. ML112720605.

(SORA, 2011) SORA. 2011. Lesser Prairie-Chicken Survey on the International Isotopes Fluorine Products Project Site-2011. Prepared for: International Isotopes, Inc. May 24, 2011. ADAMS Accession No. ML112720317.

(TRB, 2000) Transportation Research Board. 2000. Highway Capacity Manual 2000, 3rd. Ed. National Academy of Sciences Research Council. Washington, D.C., 2000.

(UNM, 2010) University of New Mexico. 2010. Lea County Crash History for the New Mexico Traffic Safety Bureau, August 19, 2010. Available at http://www.unm.edu/~dgrint/CountyCrash History/cny0609/LeaCounty.html. Accessed March 11, 2011. ADAMS Accession No. ML112720608.

(USCB, 2010a). U.S. Census Bureau. 2010. American Factfinder. New Mexico Fact Sheet. 2005-2009 American Community Survey 5-Year Estimates. Available online at http://factfinder.census.gov. Accessed April 11, 2011. ADAMS Accession No. ML112720609.

(Yarger, 2009) Yarger, F. 2009. Seismic Probability in Lea County, NM: A Brief Analysis, New Mexico Institute of Mining and Technology, New Mexico Center for Energy Policy. ADAMS Accession No. ML103080942.

5.0 MITIGATION MEASURES AND COMMITMENTS

This chapter identifies possible measures to mitigate potential environmental impacts from the proposed action, as required by Appendix A to Subpart A of 10 CFR 51. CEQ's regulation for implementing NEPA at 40 CFR 1500.2 (f) requires Federal agencies to "[u]se all practicable means consistent with the requirements of the NEPA and other essential considerations of national policy to restore and enhance the quality of the human environment and avoid or minimize any possible adverse effects of their actions on the quality of the human environment." The CEQ regulations (40 CFR 1508.20) note that mitigation activities include those that "(1) avoid the impact altogether by not taking a certain action or parts of an action; (2) minimize impacts by limiting the degree or magnitude of the action and its implementation; (3) repair, rehabilitate, or restore the affected environment; (4) reduce or eliminate impacts over time by preservation or maintenance operations during the life of the action; or (5) compensate for the impact by replacing or substituting resources or environments." As such, mitigation measures are those actions or processes (e.g., process controls and management plans) that would be implemented to control and minimize potential impacts associated with the proposed IIFP facility.

IIFP must comply with applicable laws and regulations, including obtaining all required construction and operating permits, and decommissioning requirements. Chapter 5 summarizes the mitigation measures that were proposed by IIFP (IIFP, 2009). The proposed mitigation measures do not include environmental monitoring activities. Environmental monitoring activities are described in Chapter 6 (Environmental Measurements and Monitoring Programs). The NRC staff has reviewed the mitigation measures proposed by IIFP and has concluded that the mitigation measures would reduce or minimize impacts.

IIFP identified measures in its Environmental Report and in responses to Requests for Additional Information that would mitigate environmental impacts associated with the proposed action (IIFP, 2009; IIFP, 2011). Table 5-1 lists measures proposed to mitigate the impacts of construction. Table 5-2 lists measures proposed to mitigate the impacts of operations. These measures do not preclude additional mitigation that may be considered by IIFP based upon consultations with regulatory agencies other than NRC. In a letter to the NRC dated June 21, 2011, the NMGF recommended additional mitigation measures such as a noxious weed management plan, protective screening of all open stacks and vents to exclude birds or bats, and designing stormwater retention ponds to exclude wildlife or to provide a means of escape from the ponds. A copy of this letter is included in Appendix B Consultation/Coordination) of this EIS.

Table 5-1. Summary of Potential Mitigation Measures Proposed by IIFP for Construction (Including Preconstruction Activities)

Impact Area	Activity	Mitigation Measures
Land Use	Land disturbance	• The construction footprint would be minimized to the extent possible. • After construction is complete, disturbed areas of the site would be stabilized with native, drought-resistant landscaping; and areas expected to handle regular vehicular and pedestrian traffic would be stabilized with pavement or gravel. • To the extent possible utilities would be placed within existing rights-of-way.
Historic and Cultural Resources	Disturbance of historic and cultural resources eligible for listing on the National Register of Historic Places (NRHP)	• In the event that human remains or items of archaeological significance were discovered during construction, IIFP would cease work in the area around the discovery and notify the NM SHPO and Native American Tribes, so that they could determine the appropriate measures to identify, evaluate, and treat these discoveries. • Avoidance and data collection are the two most common forms of mitigation for sites considered eligible based on NRHP. When possible, avoidance is the preferred alternative because the site is preserved in place and mitigation costs are minimized. When avoidance is not possible, data collection becomes the preferred alternative. Data collection proceeds after the sites have been determined eligible. A treatment plan would be submitted to the appropriate regulatory agencies. The plan would describe the expected data content of the sites and how data would be collected, analyzed, and reported. A treatment/mitigation plan would be developed by IIFP, if necessary.
Visual Resources	Change in visual character	• Native, drought-resistant landscaping would be used to limit visual impacts. • Disturbed areas would be promptly re-vegetated or covered
Water Resources	Runoff	• Stormwater would be controlled at the proposed facility during preconstruction and construction by complying with the NPDES Construction General Permit requirements and by applying BMPs as detailed in the Stormwater Pollution Prevention Plan. • Construction equipment would be in good repair and without visible leaks of oil, grease, or hydraulic fluid, and would be periodically maintained and inspected. • BMPs would be used for dust control associated with excavation and fill operations during construction. Water conservation would be considered when deciding how often dust suppression sprays would be applied. • Stone construction pads would be placed at entrance/exits where an unpaved construction access road intersects a publicly maintained road. • Stormwater basins would be designed to facilitate the prompt, systematic sampling of runoff in the event of any special needs.

Table 5-1. Summary of Potential Mitigation Measures Proposed by IIFP for Construction (Including Preconstruction Activities) (Continued)

Impact Area	Activity	Mitigation Measures
Water Resources (continued		• A spill control program would be implemented for accidental oil spills. An SPCC Plan would be prepared prior to the start of construction or prior to the storage of oil on site in excess of *de minimis* quantities and would contain the following information: • Identification of potential significant sources of spills and a prediction of the direction and quantity of flow that would result from a spill from each source • Identification of the containment-type or diversionary structures such as dikes, berms, culverts, booms, sumps, and diversion basins used at the facility to prevent discharged oil from reaching the surrounding environment • Procedures for inspection of potential sources of spills and spill containment/diversion structures • As part of the SPCC Plan, other measures would include control of drainage of rain water from dike areas, containment of oil and diesel fuel in bulk storage tanks, above-ground tank integrity testing, and oil and diesel fuel transfer operational safeguards, as appropriate. • Sanitary wastes generated during site construction would be handled by portable systems until the plant sanitary waste treatment facility was available for use.
Air Quality	Fugitive dust and point-source releases of criteria pollutants	• Construction BMPs would minimize fugitive dust: Water or dust suppressants would be used to control dust on dirt roads. Water conservation will be considered when deciding how often dust suppression sprays would be applied. • Designate personnel to monitor dust emissions and direct increased watering where necessary. • Implement monitoring and inspection programs to identify equipment malfunction so that corrective action can be taken promptly. • Work practices would prevent or reduce air emissions releases. • Beds of open-bodied trucks transporting materials likely to give rise to airborne dust would be covered when in motion. • Construction equipment and related vehicles would be equipped with standard pollution control devices and maintained in good working order.
Geology, Minerals, and Soil	Soil disturbance	• Erosion impacts due to site clearing and grading would be mitigated by use of construction and erosion control BMPs. • The construction footprint would be minimized to the extent possible. • Disturbed soils would be stabilized by placing crushed stone on areas of concentrated runoff to reduce potential for erosion and sedimentation.

Table 5-1. Summary of Potential Mitigation Measures Proposed by IIFP for Construction (Including Preconstruction Activities) (Continued)

Impact Area	Activity	Mitigation Measures
Geology, Minerals, and Soil (Continued)		• Earthen berms, dikes, and sediment fences would be installed as necessary to limit suspended solids in runoff.
		• Cleared areas not covered by structures or pavement would be stabilized by acceptable means as soon as practical.
		• Watering or dust suppressants would be used to control fugitive dust and prevent loss of topsoil.
		• The facility would be designed and constructed to collect surface runoff in temporary detention basins.
		• Standard drilling and blasting techniques, if required, would be used to minimize impacts to bedrock, reducing the potential for over-excavation and thereby minimizing damage to the surrounding rock.
		• Drainage culverts and ditches would be stabilized and lined with rock aggregate to reduce flow velocity and trap sediments.
		• Soil stockpiles would be constructed in a manner to reduce erosion.
		• Site slopes would be limited to a horizontal-to-vertical ratio of three to one.
		• Excavated materials would be reused whenever possible.
		• An SPCC Plan would be implemented.
Waste Management	Waste generation and management	• The quantities of waste generated would be minimized by collecting and sorting waste for recycling or disposal.
		• An assessment for each onsite waste storage area would be performed to identify and prevent potential accidental releases to the environment.
		• Onsite waste storage facilities would be monitored and inspected on an established schedule to detect any leaks or releases to the environment, so that corrective action could be taken promptly.
		• Waste that requires offsite storage, treatment, or disposal would be shipped to a licensed facility appropriate for the waste type and in compliance with State and Federal requirements.
Ecological Resources	Disturbance to plant and animal habitat	• The construction footprint would be minimized to the extent possible.
		• Site stabilization practices would be implemented to reduce the potential for soil erosion and deposition of sediment into down slope wildlife and aquatic habitats.
		• Unused open areas would be left undisturbed and managed for the benefit of wildlife.

Table 5-1. Summary of Potential Mitigation Measures Proposed by IIFP for Construction (Including Preconstruction Activities) (Continued)

Impact Area	Activity	Mitigation Measures
Ecological Resources	Disturbance to plant and animal habitat (continued)	• Security lighting for all ground level facilities and equipment would be directed downward. • The use of native plant species in disturbed areas for revegetation would enhance and maximize the opportunity for native wildlife habitat to be reestablished at the site. • No herbicides would be used during construction

Transportation	Dust deposition on roadways	• To control fugitive dust production, reasonable precautions would be taken to prevent particulate matter from becoming airborne, including the following actions:
		  • Use water or dust-suppressants to control dust on dirt roads and in clearing and grading operations and construction activities. Water conservation would be considered when deciding how often dust suppression sprays would be applied.
		  • Adequate containment methods would be used during excavation.
		  • Open-bodied trucks transporting materials likely to give rise to airborne dust would be covered when in motion.
		  • Disturbed areas would be stabilized or covered promptly once earth moving activities are completed.
		  • Construction equipment and related vehicles would be operated with standard pollution control devices maintained in good working order.
		  • Designated personnel would be assigned to monitor dust emissions and increase watering or application of dust suppressants where necessary.

	Traffic	• During the course of construction, short-duration activities (e.g., concrete and other construction material deliveries) would be scheduled to minimize traffic impacts. • Work shifts would be implemented during construction to minimize impacts to traffic.
Noise	Operation of construction vehicles	• Heavy truck and earth moving equipment usage would be prohibited after twilight and during early morning hours. • Noise suppression systems (mufflers) on construction vehicles would be kept in proper operation. • When possible, quiet equipment or methods to minimize noise emissions would be utilized during an activity. • When possible and practical, equipment with internal combustion engines would be operated at the lowest operating speed to minimize noise emissions. • Engine housing doors would be closed during operation of the equipment to reduce noise emissions from the engine. • Equipment engine idling would be avoided to the extent possible.

Table 5-1. Summary of Potential Mitigation Measures Proposed by IIFP for Construction (Including Preconstruction Activities) (Continued)

Impact Area	Activity	Mitigation Measures
Public and Occupational Health	Hazardous materials worker safety	• Integrated Safety Management System program and procedures would be adhered to. • All construction personnel would be required to take safety training and IIFP and all construction contractors would ensure that OSHA practices for construction are implemented and followed.

Source: IIFP, 2009; IIFP 2011

Table 5-2. Summary of Potential Mitigation Measures Proposed by IIFP for Operations

Impact Area	Activity	Mitigation Measures
Land Use		No mitigation measures necessary
Geology, Minerals, and Soil	Materials storage	• Aboveground storage tanks would be constructed of appropriate materials according to industry standards and applicable regulations and appropriate measures for spill containment would be installed. • Tanks storing petroleum products and hazardous chemicals would be equipped with secondary containment. • Routine visual inspections and preventive maintenance would be conducted. • Spill cleanup materials would be stored in the areas of fuel line and tank hose connections and maintained in good working order. • Contaminated soils would be sampled, analyzed, and managed in accordance with NRC, State, and other Federal requirements. • An SPCC plan would be developed and implemented.
Water Resources	Runoff	• All aboveground petroleum storage tanks would be surrounded by berms to contain spills or leaks. • Routine visual inspections and preventive maintenance would be conducted. • Any hazardous materials would be handled by approved methods and hazardous wastes would be shipped offsite to licensed disposal sites. • The facility's liquid effluent collection and treatment system would provide a means to control liquid waste within the plant, including the collection, evaporation, and minimization of liquid wastes for disposal. • Radioactive liquid effluent releases to the evaporative tank would be maintained at concentrations below 10 CFR 20 uncontrolled release limits.

Table 5-2. Summary of Potential Mitigation Measures Proposed by IIFP for Operations (Continued)

Impact Area	Activity	Mitigation Measures
Water Resources (Continued)		• Control of surface water runoff would be required for activities as required by the NPDES General Permit. • Stormwater and effluent sampling would be conducted as required by the NPDES permit to protect surface water quality. In addition, groundwater would be monitored to confirm that the impacts to groundwater from the IIFP facility were minimal. • An SPCC plan would be implemented which would include: • Identification of potential significant sources of spills and a prediction of the direction and quantity of flow that would result from a spill from each source.
	Water use	• Identification of containment-type or diversionary structures such as dikes, berms, culverts, booms, sumps, and diversion basins used at the facility to prevent discharged oil from reaching the surrounding environment. • Procedures for inspection of potential sources of spills and spill containment/diversion structures. • Assigned responsibilities for implementing the plan, inspections, and reporting. • Control of drainage of rain water from dike areas. • Containment of oil and diesel fuel in bulk storage tanks. • Aboveground tank integrity testing. • Oil and diesel fuel transfer operational safeguards. • Native, drought-resistant vegetation would be planted. • Floor washing using mops and self-contained cleaning machines would be used to reduce water usage, as opposed to conventional washing with a hose. • High-efficiency washing machines would be installed. • Closed-loop cooling systems would be incorporated where possible. • Process waste water would be treated and recycled. Any small amounts of excess water from miscellaneous processes would be retained in a storage tank and sent to an evaporator.
Waste Management	Waste generation and management	• Minimize the quantities of waste by collecting and sorting waste for recycling or disposal. • Perform an assessment for each onsite waste storage area to identify and prevent potential accidental releases to the environment. • Monitor and inspect onsite waste storage facilities on an established schedule to detect any leaks or releases to the environment due to equipment malfunctions, so that corrective action could be taken promptly.
Waste Management		• Ship waste that requires offsite storage, treatment, or disposal to a licensed facility

Table 5-2.   Summary of Potential Mitigation Measures Proposed by IIFP for Operations (Continued)

Impact Area	Activity	Mitigation Measures
(Continued)		appropriate for the waste type and in compliance with State and Federal requirements.
Air Quality	Emissions	• Process design features would be developed to lower the potential for air emissions.
		• Implement monitoring and inspection programs to detect any air emissions from equipment malfunction during operations, so that corrective action can be taken promptly.
		• Work practices would be employed to prevent or reduce air emissions releases.
		• Air emissions control systems (i.e., scrubber systems and dust collectors) would be designed to collect and strip potentially hazardous gases from plant effluents prior to release into the atmosphere.
		• Emission stacks would be sampled continuously and routinely analyzed.
Ecological Resources	Removal of plant and animal habitat	• A raptor perch would be placed in an unused open area.
		• Bird feeders would be installed at the visitor's center and quail feeders would be placed in unused open areas away from buildings.
		• Unused open areas, including areas of native grasses and shrubs, would be managed for the benefit of wildlife.
		• Drought-resistant native plant species would be used to revegetate disturbed areas and to enhance wildlife habitat.
		• Netting or other suitable material would be used to ensure birds are excluded from retention (evaporation) basins that do not meet New Mexico Water Quality Control Commission surface water standards for wildlife usage.
		• Animal-friendly fencing would be used within the site so that wildlife would not be injured or entangled.
		• The number of open trenches would be minimized at any given time.
		• Air-scrubbers system liquids would be treated prior to disposal or recycled.
		• Security lighting for all ground level facilities and equipment would be directed downward.
		• Herbicides would be used in limited amounts according to government regulations and manufacturer's instructions to control unwanted noxious vegetation.
Transportation	Traffic	• Shift changes and truck shipments would be scheduled for off-peak traffic periods, when practical.
Noise	Operation of Equipment and Vehicles	• The facility would be designed so that the reaction vessel systems, valves, transformers, pumps, generators, and other equipment would generally be located inside structures; the buildings themselves would limit noise outside the facility.
		• Distance, vegetation, and site buildings and structures would mitigate noise from equipment located outside of structures.

Table 5-2. Summary of Potential Mitigation Measures Proposed by IIFP for Operations (Continued)

Impact Area	Activity	Mitigation Measures
Public and Occupational Health	Hazardous materials processing	• To protect the public and workers, the plant design would incorporate features to minimize gaseous and liquid effluent releases and to keep them well below regulatory limits. • The radiation protection program would require routine radiation surveys and air sampling to assure that worker exposures are maintained ALARA. Exposure-monitoring techniques at the plant would include use of personal dosimeters by workers, personnel breathing zone air sampling, and annual whole-body counting. • Annual radiation exposure for an employee would be controlled, monitored, and maintained ALARA through the IIFP Radiation Protection Program. • Worker health and safety would be protected by a Chemical Safety Program, a Radiation Protection Program, and an Industrial Safety Program. • Handling of all chemicals and wastes would be conducted in accordance with an Environment, Health, and Safety Program which would conform to 29 CFR 1910 and specify the use of appropriate engineered controls, and personnel protective equipment to minimize potential chemical exposures. • Laboratory and maintenance operations activities involving hazardous gaseous or respirable emissions would be conducted with ventilation control (i.e., fume hoods, local exhaust or similar) and/or with the use of respiratory protection.

Source: IIFP, 2009; IIFP 2011

5.1 References

(IIFP, 2009) International Isotopes Fluorine Products, Inc. 2009. Fluorine Extraction Process and Depleted Uranium De-conversion Plant (FEP/DUP) Environmental Report, Revision A, ER-IFP-001. December 27, 2009. ADAMS Accession No. ML100120758.

(IIFP, 2011) International Isotopes Fluorine Products, Inc. 2011. Fluorine Extraction Process and Depleted Uranium De-conversion Plant (FEP/DUP) Official Responses to Environmental Report RAI's. Revision A. March 31, 2011. ADAMS Accession No. ML110970481.

6.0 ENVIRONMENTAL MEASUREMENTS AND MONITORING PROGRAMS

This chapter describes programs that wou d be used to measure and monitor radiation, radiological materials, and chemicals associated with operation of the proposed IIFP facility. It also provides data on principal pathways of exposure to the public and biota. This chapter is organized as follows: Section 6.1 describes the radiological monitoring program; Section 6 2 describes the physicochemical (i.e., chemical and meteorological properties that affect measurements) monitoring program; and Section 6.3 describes the ecological monitoring program.

These monitoring programs would comprise soil and vegetation sampling, water/sediment sampling, continuous airborne emission particulate monitoring and measuring, groundwater monitoring, direct radiation measuring, and sampling of stack emissions and air vents within the facility. Exact sampling locations would be determined at a later date based on site information (IIFP, 2009).

The facility would have an onsite analytical environmental monitoring laboratory equipped with analytical instruments necessary to ensure that the operation of the plant activities complies with Federal. State and local regulations and requirements. Compliance would be demonstrated by monitoring/sampling at various plant and process locations, and in the environment surrounding the facility, analyzing the samples and reporting the results of these analyses to the appropriate agencies. The environmental sampling/monitoring locations would be selected by the Health, Safety and Environmental staff in accordance with facility permits and good sampling practices.

The onsite laboratory would perform analyses on air, water, soil, flora, and fauna samples obtained from designated release points and areas around the plant. In addition to its environmental and radiological capabilities, the environmental monitoring laboratory also would be capable of performing bioassay analyses when necessary. Commercial, offsite laboratories may also be contracted to perform bioassay analyses.

6.1 Radiological Monitoring Program

The proposed IIFP facility would address radiological monitoring through two programs: the Effluent Monitoring Program and the Radiological Environmental Monitoring Program. The Effluent Monitoring Program wou d monitor, record, and report data for radiological contaminants being discharged from specific emission points such as an airborne release stack. Radiological Environmental Monitoring Program would monitor radioactivity in environmental media (i.e., soil, sediment, groundwater, biota, and air) within and outside the proposed IIFP facility site boundary. The following subsections provide information on the two radiologica monitoring programs.

6.1.1 Effluent Monitoring Program

The NRC requires nuclear fuel cycle facilities such as the proposed IIFP facility to monitor and report the release of radiological airborne and liquid effluents to the environment in accordance with Title 10, "Energy," of the U.S. Code of Federal Regulations (10 CFR), 20.1501(a) and (b). Table 6-1 lists the guidance documents that apply to the radiological monitoring program.

Table 6-1. Guidance Documents Applicable to Radiological Monitoring Program

Document	Applicable Guidelines
Regulatory Guide 4.15[1]	Quality Assurance for Radiological Monitoring Programs (Inception to Normal Operations to License Termination) - Effluent Streams and the Environment. This guide describes a method acceptable to the NRC for designing a program to ensure the quality of the results of measurements for radioactive materials in the effluents and the environment outside of nuclear facilities during normal operations.
Regulatory Guide 4.16[2]	Monitoring and Reporting Radioactive Materials in Liquid and Gaseous Effluents from Nuclear Fuel Cycle Facilities. This guide describes a method acceptable to the NRC for submitting semiannual reports that specify the quantity of each principal radionuclide released to unrestricted areas to estimate the maximum potential annual dose to the public resulting from effluent releases.

[1]NRC, 2007
[2]NRC, 2010

Public exposure to radiation from routine operations at the proposed IIFP facility may occur as the result of the discharge of liquid and gaseous effluents, including controlled releases from the uranium deconversion process lines during decontamination and maintenance of equipment. In addition, radiation exposure to the public may result from the transportation and storage of DUF_6 feed cylinders. Of these potential pathways, discharge of gaseous effluent has the highest potential to introduce uranium into the environment (IIFP, 2009). Section 4.1.2.11 of this EIS presents the potential impacts from the potential release pathways.

Compliance with 10 CFR 20.1301, Dose limits for individual members of the public, would be demonstrated using a calculation of the total effective dose equivalent (TEDE) to the individual likely to receive the highest dose in accordance with 10 CFR 20.1302(b)(1) (IIFP, 2009). The determination of the TEDE pathway analysis is supported by appropriate models, codes, and assumptions that accurately represent the facility, site, and the surrounding area. The computer codes used to calculate dose associated with potential gaseous and liquid effluent from the plant follow the methodology for pathway modeling, as described in Regulatory Guide 1.109 (NRC, 1977), and have undergone validation and verification by NRC.

Administrative action levels are established for effluent samples and monitoring instrumentation as an additional check in the effluent control process. These action levels are well below regulatory limits; their purpose is to support implementation of corrective actions before releases approach regulatory limits. Effluent samples that exceed the action level are cause for an investigation into the source of elevated radioactivity. For example, radiological analyses would be performed more frequently on ventilation air filters if there is an unexplained increase in gross radioactivity, or when a process change or other circumstance change radioactivity concentrations in the effluent stream. Progressively more rigorous corrective actions would be implemented based on the radioactivity level, through means of automatic shutdown programming and operating procedures to be developed in the detailed alarm design (IIFP, 2009).

Under routine operating conditions, radioactive material in effluent discharged from the facility would comply with regulatory release criteria. Compliance would be demonstrated through effluent and environmental sampling data. Processes are designed to include, when practical, provision for automatic shutdown in the event action levels are exceeded. Appropriate action levels and actions to be taken are specified for liquid effluents and gaseous releases (IIFP, 2009).

The effluent monitoring program would be overseen by IIFP Radiation Safety Program, Quality Assurance (QA) personnel and would be subject to periodic audits. Written procedures would specify the collection of representative samples, use of appropriate sampling methods and equipment, appropriate locations for sampling points, and proper handling, storage, transport, and analyses of effluent samples. In addition, IIFP would develop written procedures for maintaining and calibrating sampling and measuring equipment, including ancillary equipment such as airflow meters, to ensure that all radiological monitoring equipment is properly maintained and calibrated at regular intervals. The effluent monitoring program procedures would include functional testing and routine checks to demonstrate that monitoring and measuring instruments are in working condition. Employees involved in implementation of this program would be trained in the program procedures (IIFP, 2009).

6.1.1.1 Gaseous Effluent Monitoring

To ensure compliance with regulatory requirements, potentially radioactive effluents from the facility would be discharged only through monitored pathways. The effluent sampling program would measure the quantities and concentrations of radionuclides discharged to the environment. Uranium isotopes and daughter products are expected to be the most common radionuclides in the gaseous effluent.

Effluents would be sampled as shown in Table 6-2. Representative samples would be collected from each release point. Because uranium in gaseous effluents may exist in a variety of compounds (e.g., UF_6, uranium oxide, UF_4, and uranyl fluoride), effluent data would be maintained, reviewed, and assessed by the facility's Radiation Protection Manager to ensure that all gaseous effluent discharges comply with regulatory release criteria for uranium. However, the gaseous effluent monitoring program for the IIFP plant would be designed to determine the quantities and concentrations of all gaseous discharges to the environment, not just uranium. The process exhaust stacks would be equipped with monitors for particulates, HF, and gross radioactivity (IIFP, 2009).

Table 6-2. Gaseous Effluent Sampling Program

Area	Type of Sample	Type of Analysis	Frequency
Dust Collector Stacks	Continuous Air Filter	Gross Alpha/Beta Isotopic	Weekly/Composite/ Quarterly
Process Stacks	Continuous Air Filter	Gross Alpha/Beta Isotopic/Fluoride	Weekly/Composite/ Quarterly
Air Vents	Continuous Air Filter	Gross Alpha/Beta Isotopic	Weekly/Composite/ Quarterly

Source: IIFP, 2009

Monitoring for uranium isotopes would be performed continuously and samples would be analyzed at least once per operating shift. If an unacceptable level of uranium is detected (i.e., if it exceeded the administrative action level), IIFP would investigate the cause and corrective action would be taken. The gaseous effluent sampling program would support the determination of quantity and concentration of radionuclides discharged from the facility and support the collection of other information required for 10 CFR 20.1501(a) and (b) (IIFP, 2009).

6.1.1.2 Liquid Effluent Monitoring

Liquids potentially contaminated with low concentrations of uranium could be generated from equipment decontamination, floor washings, and laundry. Except for discharges from the Sanitary Treatment System, liquid effluents would be contained on the proposed IIFP site via collection tanks and retention basins (IIFP, 2009).

Potentially contaminated liquid effluent would be routed to the Decontamination Area for treatment. In the Decontamination Area, radioactive material would be removed from waste water through a combination of clean-up processes that would include precipitation, filtration, and ion exchange. Representative sampling would be ensured through the use of tank agitators and recirculation lines. Collection tanks would be sampled before the contents were sent through any treatment process. Treated water would then be collected in other tanks, which would be sampled. Concentrated radioactive solids generated by the liquid treatment processes would be disposed of as LLW at an off-site licensed disposal facility (IIFP, 2009).

6.1.2 Radiological Environmental Monitoring Program

The primary objective of the Radiological Environmental Monitoring Program (REMP) would be to provide verification that IIFP operations do not result in detrimental radiological impacts to the environment. The REMP data would confirm the effectiveness of effluent controls and provide additional verification of the power of the effluent monitoring program to produce results. The REMP would establish a process for collecting data for assessing radiological concentrations in the environment, estimate the potential impacts on the public, and support the demonstration of compliance with applicable radiation protection standards and guidelines.

6.1.2.1 Sampling Program

To meet the REMP objectives, representative samples from various environmental media would be collected and analyzed for radioactivity. The types and frequency of sampling and analyses are summarized in Table 6-3. Environmental media identified for sampling consist of ambient air, groundwater, soil/sediment, and vegetation.

Environmental samples would generally be analyzed at the on-site analytical laboratory. However, samples could be shipped to a qualified independent laboratory for analyses. Monitoring and sampling activities, laboratory analyses, and reporting of radioactivity in the environment would be conducted in accordance with industry-accepted and agency-approved methodologies.

The REMP would include the collection of data during pre-operational years in order to establish baseline radiological information that would be used in determining and evaluating releases from plant operations to the local environment. The REMP would be initiated at least 12 months prior to initiation of plant operations in order to develop a sufficient database before the arrival of the first uranium hexafluoride shipment. Radionuclides in environmental media would be identified using technically appropriate, accurate, and sensitive analytical instruments.

Data collected during the operational years would be compared to the baseline generated by the pre-operational data. Such comparisons would provide a means of assessing the magnitude of potential radiological impacts on members of the public and in demonstrating compliance with applicable radiation protection standards.

Table 6-3.   Radiological Sampling and Analysis Program

Sample Type	Location	Sampling	Collection Frequency	Type of Analysis
Continuous Airborne particulate	Six locations a ong fence line and in the region of influence, including the location of the nearest resident	Continuous operation of air sampler with sample collection as necessary based on dust loading, but at least biweekly	Quarterly composite samples by location	Gross beta/gross alpha analyses each filter change. Quarterly isotopic analysis on composite sample
Vegetation/Soil Analyses	Five (including four locations along fence line and a control at an offsite location some distance away)	For each vegetation and soil sample, 1 to 2 kg (2.2 to 4.4 lbs)	Quarterly pre-operation/semi-annual during operation	Isotopic analyses/fluoride
Groundwater	Four wells	Samples [4 L (1.1 gal)]	Semiannually	Isotopic analyses
Thermoluminescent Dosimeters (TLDs)	Eight locations along fence line	Samples collected quarterly	Quarterly	Gamma and neutron equivalent
Stormwater	Site Stormwater Retention Basin, DUF$_6$ Cylinder Storage Pads, Stormwater Retention Basins	Water sample 4 L (1.1 gal). Sediment samples 1 to 2 kg (2.2 to 4.4 lbs)	Semiannually	Isotopic analyses

Source: IIFP, 2009

Over time, revisions to the REMP may be necessary and appropriate to assure reliable sampling and collection of environmental cata. The rationale and actions behind such revisions to the program would be documented and reported to the appropriate regulatory agency, as required. REMP sampling focuses on locations within 1.6 km (1 mi) of the facility, but may also include distant locations as control sites. The sampling locations may be subject to change as determined from the results of periodic review of land use.

The concentrations of radioactive material in gaseous effluent from the proposed IIFP facility are expected to be very low because of process and effluent controls. Consequently, air samples collected at locations that are close to the facility would provide the best opportunity to detect and identify plant-related radioactivity in the ambient air. Therefore, air monitoring activities would concentrate on locations close to the plant, such as the plant perimeter fence or the plant property line. Air monitoring stations would be situated along the fence perimeter, at the nearest residence, and at "control comparative" locations. In addition, an air monitoring station would be located next to the Stormwater Retention Basins to measure for particulate radioactivity that may be resuspended into the air from sediment when the basin is dry. Environmental air samplers would operate on a continuous basis with sample retrieval for a gross alpha and beta analysis occurring weekly (or more often if dust loads are heavy) (IIFP, 2009).

Vegetation and soil samples, from on and offsite locations would be collected quarterly in each compass sector during the pre-operational REMP. This would ensure the development of an adequate baseline. During the operational years, vegetation and soil sampling would be performed semiannually in five compass sectors, including the three with the highest predicted atmospheric deposition (based on the prevailing wind direction). Vegetation samples may include garden vegetables or grass, depending on availability. Soil samples would be collected in the same vicinity as the vegetation samples (IIFP, 2009).

On October 15, 2010, soil and vegetation samples were collected and shipped to analytical laboratories for analysis (GL Environmental, 2010) to establish baseline conditions. Table 6-4 presents the results of these samples.

Table 6-4.   Baseline Radiological Soil and Vegetation Samples

	Soil Sample Bq/g (µCi/g)	Vegetation Sample
U-234	0.016 to 0.022 $(4.42 \times 10^{-7}$ to $5.95 \times 10^{-7})$	Less than minimum detectable concentrations
U-235/U-236	2.06×10^{-4} to 9.62×10^{-4} $(5.58 \times 10^{-9}$ to $2.60 \times 10^{-8})$	Less than minimum detectable concentrations
U-238	0.0217 to 0.0220 $(5.86 \times 10^{-7}$ to $5.95 \times 10^{-7})$	3.85×10^{-4} (1.04×10^{-8})
Other Isotopic Uranium	Less than minimum detectable concentrations	Less than minimum detectable concentrations

Source: GL Environmental, 2010
Bq/g = becquerel/gram
µCi/g = microcurie/gram

Groundwater samples from onsite monitoring wells would be collected semiannually for radiological analysis. Two monitoring wells would be downgradient of the proposed IIFP site, one would be located downgradient of the DUF$_6$ Cylinder Storage Pads, and one (background monitoring well) would be upgradient of the site. Sediment samples would be collected semiannually from the stormwater runoff retention basins on site to analyze for any buildup of uranic material being deposited (IIFP, 2009).

Direct radiation in offsite areas from processes inside the facility buildings is expected to be minimal because the low-energy radiation associated with the uranium would be shielded by process piping, equipment, and cylinders. Because the offsite dose equivalent rate from stored DUF$_6$ cylinders is expected to be very low and difficult to distinguish from the variance in normal background radiation beyond the site boundary, demonstration of compliance would rely on a system that combines direct dose equivalent measurements and computer modeling to extrapolate the measurements. Environmental TLDs would be placed at the plant perimeter fence line or other location(s) close to the DUF$_6$ cylinders to provide quarterly direct dose equivalent information. The direct dose equivalent at offsite locations would be estimated through extrapolation of the quarterly TLD data using computer programs (IIFP, 2009).

6.1.2.2   Procedures

Monitoring procedures would employ approved analytical methods and instrumentation. The instrument maintenance and calibration program would comply with manufacturers recommendations. The onsite laboratory and any contract laboratory used to analyze the IIFP facility samples would participate in third-party laboratory intercomparison programs appropriate to the media and analyses being measured. The following are examples of these third-party programs:

- The DOE Mixed Analyte Performance Evaluation Program and DOE Quality Assurance Program

- Analytics, Inc., Environmental Radiochemistry Cross-Check Program

IIFP would require that all radiological and nonradiological laboratory vendors are certified by the National Environmental Laboratory Accreditation Program or an equivalent State laboratory accreditation agency for the analytes being tested (IIFP, 2009).

The REMP would fall under the oversight of IIFP's Quality Assurance Program. Quality assurance procedures would be implemented to ensure representative sampling, proper use of appropriate sampling methods and equipment, proper locations for sampling points, and proper handling, storage, transport, and analyses of effluent samples. In addition, written procedures would ensure that sampling and measuring equipment, including ancillary equipment such as airflow meters, would be properly maintained and calibrated at regular intervals according to manufacturer recommendations. The implementing procedures would include functional testing and routine checks to demonstrate that monitoring and measuring instruments were in working condition.

Audits would be periodically conducted as part of its Quality Assurance Program (IIFP, 2009). The quality control procedures used by the analytical laboratories would conform to the guidance in Regulatory Guide 4.15 (NRC, 2007). These quality control procedures would include the use of established standards such as those provided by the National Institute of Standards and Technology and the use of standard analytical procedures such as those established by the National Environmental Laboratory Accreditation Conference (IIFP, 2009)

6.1.2.3 Reporting

Reporting procedures would comply with the requirements of 10 CFR 70.59 and the guidance specified in Regulatory Guide 4.16 (NRC, 2010). Reports of the concentrations of principal radionuclides released to unrestricted areas in effluents would be provided and would include the minimum detectable concentration (MDC) for the analysis and the error for each data point. Each year, IIFP would submit a summary report of the environmental sampling program to the NRC, including all associated data, as required by 10 CFR 70. The report also would include the types, numbers, and frequencies of environmental measurements and the identity and concentrations of nuclides found in the environmental samples. Significant positive trends would also be noted in the report, along with any adjustment to the program, unavailable samples, and deviations from the sampling program.

6.2 Physicochemical Monitoring

6.2.1 Introduction

The primary objective of physicochemical monitoring would be to provide verification that the operations at the IIFP plant do not result in detrimental chemical impacts on the environment. Effluent controls would be in place to ensure that chemical concentrations in gaseous and liquid effluents are maintained ALARA. In addition, physicochemical monitoring would provide data to confirm the effectiveness of effluent controls.

Administrative action levels would ensure that chemical discharges remain below the limits specified in the facility discharge permits: the EPA Region 6 NPDES General Discharge Permits and the New Mexico Environment Department / Water Quality Bureau WQB) Groundwater Discharge Permit/Plan. Physicochemical monitoring would be performed for routine operations with provisions for additional evaluation in response to potential accidental releases.

Physicochemical monitoring would sample stormwater, soil, sediment, vegetation, and groundwater (Table 6-5) to confirm that chemical discharges are below regulatory limits. There are no surface waters on the site; therefore, no surface water monitoring program would be implemented. However, soil sampling would include outfall/overflow areas such as the outfall at the Site Stormwater Retention Basins. In the event of any accidental release from the facility, these sampling protocols would be initiated immediately and on a continuing basis to document the extent/impact of the release until conditions have been abated and mitigated (IIFP, 2009).

Table 6-5.   Physicochemical Sampling

Sample Type	Sample Location	Frequency	Sampling and Collections[2]
Stormwater	Stormwater Detention Basins	Quarterly	Analytes as determined by baseline program
Vegetation	5 minimum[1]	Quarterly/ Semiannually[3]	Fluoride Uptake (growing seasons)
Soils	5 minimum[1]	Quarterly/ Semiannually[3]	Metals, Organics, Pesticides, and Fluoride Uptake
Water/Sediment	2 minimum[1]	Quarterly/ Semiannually[3]	Analytes as determined by baseline program
Groundwater	Selected Groundwater Wells	Semiannually	Metals, Organics, and Pesticides

Source: IIFP, 2009
[1] Locations to be established by Health Safety & Environmental organization.
[2] Analyses will meet EPA Lower Limits of Detection (LLD), as applicable, and will be based on the baseline surveys and the sample type.
[3] Quarterly during pre-operations; semiannual during operations.

Waste liquids, solids and gases from related processes and decontamination operations would be analyzed and/or monitored for chemical contamination to determine safe disposal methods or further treatment requirements.

6.2.2   Evaluation and Analysis of Samples

Samples of liquid effluents, solids and gaseous effluents from plant processes would be analyzed in the environmental monitoring laboratory. Results of process sample analyses would be used to verify that process parameters were operating within expected performance ranges. Results of liquid effluent sample analyses would be characterized to determine if treatment is required prior to discharge or disposal.

6.2.3   Quality Assurance

Quality assurance would be achieved by following a set of formalized and controlled procedures that IIFP would create, implement and periodically review for sample collection, lab analysis, chain of custody, reporting of results, and corrective actions. Corrective actions would be instituted if an action level is exceeded for any of the measured parameters. IIFP would establish three action levels: the sample parameter is three times the normal background level, the sample parameter exceeds any existing administrative limits, or the sample parameter exceeds any regulatory limit. The third scenario represents the worst case, which is not expected, however, triggering any of the three action levels would initiate an action plan. Corrective actions would be implemented to ensure that the cause for the action level exceedance is identified and immediately corrected; applicable regulatory agencies are notified, if required; communications to address lessons learned are dispersed to appropriate personnel and applicable procedures are revised accordingly, if needed. Action plans would be commensurate with the severity of the exceedance.

IIFP would ensure that the onsite laboratory and any contract laboratory used to analyze IIFP samples participates in third-party laboratory intercomparison programs appropriate to the media and analytes being measured. The IIFP facility would require all radiological and non-radiological laboratory vendors to be certified by the National Environmental Laboratory Accreditation Conference or an equivalent State laboratory accreditation agency for the analytes being tested.

6.2.4   Lower Limits of Detection

Lower limits of detection (LLDs) for the parameters sampled for in the Stormwater Monitoring Program are listed in Section 6.2.6. LLDs for the non-radiological parameters would be based on the results of the baseline surveys and the sampled media. Minimum detectable concentrations for environmental samples are listed in Table 6-6.

Table 6-6.   Required Minimum Detectable Concentrations for Environmental Sample Analyses

Medium	Analysis	Minimum Detectable Concentrations Bq/ml (µCi/ml)
Ambient Air	gross alpha	3.7×10^{-14} (1.0×10^{-18})
Vegetation	isotopic uranium	3.7×10^{-6} (1.0×10^{-10})
Soil/Sediment	isotopic uranium	1.1×10^{-2} (3.0×10^{-7})
Groundwater	isotopic uranium	3.7×10^{-8} (1.0×10^{-12})

Source: IIFP, 2009.
Bq/ml = becquerel/milliliter
µCi/ml = microcurie/milliliter

6.2.5   Effluent Monitoring

Chemical constituents that may be discharged to the environment would be below concentrations established by State and Federal regulatory agencies as protective of the public health and the natural environment. Under routine operating conditions, no significant quantities of contaminants would be released from the facility. This would be confirmed through monitoring and collection and analysis of environmental data. The facility would not directly discharge any industrial effluents to surface waters or to offsite locations, and there would be no plant tie-in to a publicly owned wastewater treatment works. Except for discharges from the sanitary treatment system, liquid effluents would be contained in the IIFP facility in collection tanks and retention basins.

No chemical sampling is planned for sanitary wastes because no plant process related effluents would be introduced into that system.

6.2.6 Stormwater Monitoring Program

A stormwater monitoring program would be initiated during construction. Data collected from the program would be used to evaluate the effectiveness of measures taken to prevent the contamination of stormwater and to retain sediments within site boundaries. A temporary detention basin would be used as a sediment control basin during construction as part of the overall sedimentation erosion control plan.

Stormwater monitoring would continue with the same frequency upon initiation of facility operation. During plant operation, samples would be collected from the DUF_6 Cylinders Storage Pad Stormwater Retention Basin and the Site Stormwater Detention Basin to demonstrate that runoff does not contain contaminants. A list of parameters to be monitored and monitoring frequencies is presented in Table 6-7.

Table 6-7. Stormwater Monitoring Program

Parameter	Frequency	Sampling Method	Lower Limit of Detection
Oil & Grease	Quarterly	Grab	0.5 ppm
Total Suspended Solids	Quarterly	Grab	0.5 ppm
5-Day Biological Oxygen Demand	Quarterly	Grab	2 ppm
Chemical Oxygen Demand	Quarterly	Grab	1 ppm
Total Phosphorous	Quarterly	Grab	0.1 ppm
Total Kjeldahl Nitrogen	Quarterly	Grab	0.1 ppm
pH	Quarterly	Grab	0.01 units
Nitrate plus Nitrite Nitrogen	Quarterly	Grab	0.2 ppm
Metals	Quarterly	Grab	Varies[1]

Source: IIFP, 2009.
[1] Analyses will meet EPA LLD, as applicable, and will be based on the baseline surveys and the sample type.
ppm = parts per million

The monitoring program would be refined to reflect applicable requirements as determined during the NPDES permit application process. Additionally, the Site Stormwater Retention Basin would adhere to the requirements of the Groundwater Discharge Permit/Plan from the New Mexico Water Quality Board.

6.2.7 Environmental Monitoring

The purpose of this section is to describe the surveillance monitoring program, which would be implemented to measure non-radiclogical chemical impacts on the environment. The ability to detect and contain any potentially adverse chemical releases from the facility to the environment would depend on chemistry data collected as part of the effluent and stormwater monitoring programs described in the preceding sections. Data acquisition from these programs encompasses both onsite and offsite sample collections. Final constituent analysis requirements would be in accordance with permit mandates. Sampling locations would be determined based on meteorological information and current land use. The sampling locations may be subject to change as determined from the results or any significant changes in land use.

The chemical monitoring program is designed to identify chemical concentrations in the environment that could be attributed to plant operations.

Vegetation samples would include grasses and shrub brush. Soil would be collected in the same vicinity as the vegetation sample. The samples would be collected from sectors chosen based on predicted direction of the prevailing winds. Sediment samples would be collected from the discharge points of the stormwater collection basins. Groundwater samples would be collected from the series of wells described in Section 6.1.2.1. Stormwater samples collected in the DUF_6 Cylinder Storage Pad Stormwater Retention Basin would be sampled to ensure no contaminants are present.

Operational sample results would be compared to baseline data collected during preoperational sampling to identify any positive trends. On October 15, 2010, two soil and two vegetation baseline samples were collected for analysis. Tables 6-8 and 6-9 present the results of these samples.

Operational monitoring surveys would be conducted at locations and frequencies established from baseline sampling data and as determined by requirements in EPA Region 6 NPDES General Discharge Permits and the New Mexico Water Quality Board Groundwater Discharge Permit/Plan.

Annually IIFP would submit a summary of the environmental sampling program results to regulatory authorities, as required. This summary would include the types, numbers and frequencies of samples collected, analytical results, and a discussion of any observed trends. Significant positive trends would be discussed, along with any adjustments to the program, unavailable samples, or deviations from the sampling protocol.

Table 6-8.   Baseline Physicochemical Soil Sample Results

	Soil Sample 1 (mg/kg)	Soil Sample 2 (mg/kg)
Barium	88.5	109
Cadmium	0.27	0.42
Chromium	10.0	12.2
Lead	11.7	14.7
All other Resource Conservation and Recovery Act Metal Concentrations	Less than minimum detectable concentrations	Less than minimum detectable concentrations

Source: GL Environmental, 2010
mg/kg = milligrams/kilogram

Table 6-9.   Baseline Physicochemical Vegetation Sample Results

	Vegetation Sample 1 (mg/kg)	Vegetation Sample 2 (mg/kg)
Resource Conservation and Recovery Act Metal Concentrations	Less than minimum detectable concentrations	Less than minimum detectable concentrations
Barium	10.6	10.9
Benzoic acid	0.48	0.46
Bis(2-ethylhexyl) phthalate	0.26	0.19
Phenol	0.40	Less than minimum detectable concentrations

Source: GL Environmental, 2010
mg/kg = milligrams/kilogram

6.2.8   Meteorological Monitoring

Atmospheric conditions (e.g., wind speed, wind direction, temperature, precipitation, relative humidity) would be monitored by electronic sensors mounted on a 40 m (131 ft) tower located on site. Data from this monitoring program would be used to characterize the site's meteorological conditions (both normal and extreme) in order to predict patterns of radionuclide and chemical dispersion and deposition. The meteorological tower would be at the same elevation as the finished facility grade. The tower would be located at a distance at least ten times the height of any obstruction to ensure that wind flow around structures would interfere with meteorological sampling. IIFP would establish instrument maintenance and calibration schedules, keep back-up monitoring equipment on hand, and deploy redundant data recorders to ensure at least 90 percent data recovery.

6.3   Ecological Monitoring

The ecological monitoring program would be designed to characterize changes that may occur in the composition of biotic communities as a result of site preparation, construction, operation, and decommissioning of the proposed IIFP facility. The program would focus on observable changes in habitat characteristics and wildlife populations.

The ecological monitoring program would be carried out in accordance with generally accepted monitoring practices and the requirements of the USFWS and NMGF. Under the program, data would be collected and analyzed. Procedures would be established, as appropriate, for data collection, storage, analysis, reporting, and corrective actions.

6.3.1   General Ecological Conditions of the Site

Section 3.8 describes the natural environment of the proposed site and vicinity. The area is a transitional zone between the shortgrass prairie north of the Mescalero Ridge (Western Great Plains Shortgrass Prairie) and the desert communities south of the Mescalero Ridge (Apacherian-Chihuahuan Mesquite Upland Scrub). These habitat types commonly occur in the vicinity of the IIFP site (Figure 3-19). The vegetation in this area is dominated by deep sand tolerant- and extreme drought- and grazing-tolerant plant species. The natural habitats on the IIFP site and the region surrounding the site have been degraded by livestock grazing, oil and gas pipeline rights-of-way and access roads. As described in Section 3.7.2 of this EIS, there are no wetlands or stream systems on the facility footprint, and therefore, no riparian habitat.

There are no important ecological communities on site that are vulnerable to change or that contain important species habitats, such as breeding areas, nursery, feeding, or other areas important to important species (Section 3.8).

6.3.2   Monitoring Program Elements

Several ecological elements would be monitored vegetation, birds, mammals, reptiles and amphibians. Currently there are no known actions or reporting levels for any of these elements. However, discussions with the responsible agencies (NMGF and USFWS) would continue and agency recommendations would be considered when developing action and/or reporting levels for each element.

IIFP would periodically monitor the proposed site property during the construction phases, operation phases, and decommissioning to ensure the risk to wildlife is minimized.

6.3.3   Observations and Sampling Design

The monitoring program would establish site baseline data collected before commencement of preconstruction activities. The procedures to characterize the baseline plant and animal populations would also be used for the construction and operations monitoring programs. Monitoring surveys during operations would be conducted annually for vegetation and semiannually for animals using the same sampling sites established during the baseline monitoring program (IIFP, 2009).

These surveys are intended to be sufficient to characterize broad changes in the composition of the ecological community in the vicinity of the facility that could be attributed to activities at the facility.

The analyses would comprise descriptive statistics (sample size, mean, standard deviation, standard error, and confidence interval for the mean). For these studies, a significance level of 5 percent would be used, resulting in a 95 percent confidence level (IIFP, 2009).

The data collected would be analyzed by the Environment, Health, and Safety staff. Annually a report summarizing the results would be prepared (IIFP, 2009). The monitoring program for each of the ecological elements described below would be used for the duration stipulated in the terms of the NRC license agreement, if granted. The anticipated duration would most likely be the first three years of operation of the proposed IIFP facility. Following that initial monitoring period, program changes could be initiated based on operational experience and the results of the initial monitoring.

6.3.3.1   Vegetation

The following vegetation parameters would be monitored: species composition, percent ground cover, stem frequency, woody plant density, and production data. Sampling from 16 permanent sampling locations on the IIFP site would occur annually in September or October. Annual sampling is scheduled to coincide with the mature flowering stages of the dominant perennial species.

The sampling locations would be selected in areas outside of the proposed footprint of the IIFP facility. The selected sampling locations would be clearly marked (i.e., staked or flagged) on site, and the Global Positioning System (GPS) coordinates recorded. Permanent sampling locations would facilitate a long-term monitoring system designed to evaluate vegetation trends and characteristics.

Transects used for data collection would extend out 30-m (98-ft) in a given compass direction at each sampling location. Ground cover and stem occurrence frequency would be determined utilizing the line intercept method. Cover measurements would be read to the nearest 0.03-m (0.1-ft). Woody plant densities would be determined using the belt transect method. All individual shrubs and trees within 2-m (6.6-ft) of the 30-m (98-ft) transect would be counted. Productivity would be determined by estimating the production within three 0.25-m^2 (2.7-ft^2) plots and harvesting each species in one 0.25-m^2 (2.7-ft^2) plot along the transect and converting the dry weight of the plot vegetation into kg of forage per ha (lbs/ac).

6.3.3.2   Birds

Site-specific avian surveys would be conducted in both the wintering and breeding seasons to verify the presence of particular bird species. For the winter survey, the distinct habitats at the site would be identified and the bird species composition within each of the habitats described. Transects, 100-m (328-ft) in length, would be established within each distinct homogenous habitat, and data would be collected along each transect. Species composition and relative abundance would be determined based on visual observations and call counts. The spring survey would also determine the nesting and migratory status of the species observed and (as a measure of the nesting potential of the site) the occurrence and number of male territories. The area would be surveyed using the standard point count method.

All birds seen or heard by a qualified observer at each point would be recorded. Surveys would begin 15 minutes prior to sunrise and conclude by 10:00 am (or earlier on warm days) to coincide with the territorial males' peak singing times. The points would be recorded using a GPS, enabling return visits. Data would be compared with species known to exist in the area.

6.3.3.3   Mammals

All mammals observed during other ecological sampling will be noted and results compared to the species list compiled for the area.

6.3.3.4   Reptiles and Amphibians

A combination of pitfall trapping and walking transects (at trap sites) would provide data in sufficient quantity to allow statistical measurements of population trends, community composition, body size distributions and sex ratios that would reflect environmental conditions and changes at the site over time.

Each sample site would be located to maximize the total catch of reptile and amphibian species, rather than data on each individual caught. Each animal caught would be identified, sexed, snout-vent length measured, examined for morphological anomalies and released (sample with replacement design). There would be two sample periods, at the same time each year, in May and late June/early July, which would coincide with breeding activity for lizards; most snakes; and depending on rainfall, amphibians.

Because reptile and amphibian species are sensitive to climatic conditions, and to account for the spotty effects of rainfall, each sampling event would also record rainfall, relative humidity and temperatures. The rainfall and temperature data would act as a covariant in the analysis.

In addition to the monitoring plan described above, general observations would be gathered and recorded concurrently with other wildlife monitoring. The data would be compared to all the species known to exist in the area.

6.4 References

(GL Environmental, 2010) GL Environmental, Inc. 2010. 2010 Soil and Vegetation Characterization Report. Prepared for International Isotopes Fluorine Products. ADAMS Accession No. ML112140543

(IIFP, 2009) International Isotopes Fluorine Products, Inc. 2009. Fluorine Extraction Process & Depleted Uranium De-conversion Plant (FEP/DUP) Environmental Report, Revision A. December 27, 2009. ADAMS Accession No. ML100120758.

(NRC, 1977) U. S. Nuclear Regulatory Commission. 1977. Calculation of Annual Doses to Man from Routine Releases of Reactor Effluents for the Purpose of Evaluating Compliance with 10 CFR 50, Appendix I. Regulatory Guide 1.109, Revision 1. October, 1977.

(NRC, 2007) U.S. Nuclear Regulatory Commission. 2007. Quality Assurance for Radiological Monitoring Programs (Inception Through Normal Operations to License Termination)-Effluent Streams and the Environment. Regulatory Guide 4.15. Revision 2. 2007.

(NRC, 2010) U.S. Nuclear Regulatory Commission. 2010. Monitoring and Reporting Radioactive Materials in Liquid and Gaseous Effluents from Nuclear Fuel Cycle Facilities. Regulatory Guide 4.16. Revision 2. 2010.

7.0 COST-BENEFIT ANALYSIS SUMMARY

This chapter summarizes benefits and costs associated with the proposed action and the no-action alternative. Chapter 4 (Environmental Impacts) of this EIS discusses the potential impacts of the construction, operation, and decommissioning of the proposed IIFP facility.

Implementation of the proposed action would generate national, regional, and local benefits and costs. The primary national benefit of the proposed IIFP facility would be a benefit to the national uranium fuel cycle by ensuring that commercial enrichment facilities throughout the nation do not have to rely on long-term storage of DUF_6. The regional benefits of the proposed project would be increased employment, economic activity, and tax revenues in the region around the proposed site. Some of these regional benefits, such as tax revenues, would accrue specifically to Lea County and the City of Hobbs. Other benefits may extend to neighboring Eddy County. Environmental costs associated with the proposed IIFP facility are, for the most part, limited to the area immediately surrounding or on the site.

The data for this analysis are drawn largely from Chapter 4, the assessment of environmental impacts. Monetary cost data is taken from IIFP's environmental report prepared for the license application (IIFP, 2009) and subsequent responses to NRC staff's requests for additional information (IIFP, 2011). The analysis separately covers both the construction (including preconstruction) and operations phases. As described in Section 4.1.3, NRC regulation 10 CFR 40.36 requires IIFP to have a decommissioning plan and provide for funding of the decommissioning. Decommissioning costs are evaluated in this analysis only in terms of payments to a decommissioning fund.

Section 7.1 presents the costs and benefits of the no-action alternative. Section 7.2 presents costs of the proposed action. Section 7.3 presents benefits of the proposed action. Section 7.4 presents a summary of the cost-benefit analysis, including NRC staff's determination of cost-effectiveness.

7.1   Costs and Benefits of the No-Action Alternative

Under the no-action alternative, NRC would not grant a license to IIFP to construct, operate and decommission the facility. No DUF_6 would be deconverted into fluoride products (for commercial resale) and depleted uranium oxides (for disposal). Without a deconversion facility such as the proposed facility, DUF_6 would continue to be stored, primarily at commercial uranium enrichment facilities in the United States. Fluoride products would not be manufactured and sold to end users. Planned or existing commercial enrichment facilities would not be able to send their DUF_6 to the IIFP facility for deconversion. As a result, the proposed site would not be disturbed by the proposed project activities. Ecological, natural, and socioeconomic resources would remain unaffected by the proposed action, except for what occurred during preconstruction. All potential environmental impacts from the proposed action (that is, not including preconstruction) would be avoided. Similarly, all project-specific socioeconomic impacts (e.g., related to employment, economic activity, population, housing, local finance) would be avoided.

Table 2.5 of Section 2.3 summarizes and compares the external environmental costs and benefits of both the proposed action and the no-action alternatives. Section 4.1 provides details on these external environmental and socioeconomic costs and benefits for the proposed action. Section 4.3 provides details for the no-action alternative.

7.2 Costs of the Proposed Action

The costs for a project are usually presented as internal and external costs. Internal costs are those that are borne by the owner, IIFP in this instance. These costs are most easily expressed as monetary costs. External costs are those borne by others or by the environment. Such costs can be monetary, but most often include both quantitative and qualitative environmental impacts. As described in Sections 2.1 and 2.2.2.2.1, IIFP intends to develop this project in two phases, with the Phase 1 component the subject of the current license application. Because Phase 2 is closely related to Phase 1 and is a reasonably foreseeable action for which analysis of cumulative impacts is required, this section presents both Phase 1 and Phase 2 costs. Section 7.2.1 discusses costs during the construction phase, and Section 7.2.2 discusses costs during the operations phase.

7.2.1 Construction Costs

7.2.1.1 Internal Costs

Internal construction costs include capital costs and labor costs. All costs are presented in 2009 dollars.

IIFP's environmental report provides cost estimates based on the assumptions presented there. Table 7-1 of this section presents the capital costs and labor costs. Both capital and labor costs are spread out over the years of construction (2012 through 2013 for Phase 1 and 2015 through 2016 for Phase 2).

Table 7-1. Construction Capital and Labor Costs for the IIFP Facility (Millions of 2009 Dollars)

Cost Category	Phase 1 Costs[a] (in millions of dollars)	Phase 2 Costs[a] (in millions of dollars)	Total Phases 1 and 2 Costs (in millions of dollars)
Capital Costs			
Fixed Capital			
DUF$_4$ plant	$9 – $12	0	$9 – $12
FEP plant	$15 – $19	0	$15 – $19
Oxide add-on plant	0	$26 – $34	$26 – $34
Balance of Plant	$15 – $20	$1 – $1.5	$16 – $21.5
Engineering, procurement, and construction management	$7 – $11	$7 – $9	$14 – $20
Project management and programs	$2 – $3	$1 – $1.5	$3 – $4.5
Contractor fees	$2 – $3	$1 – $2	$3 – $5
Contingency	$5 – $6	$3 – $4	$8 – $10
Subtotal Fixed Capital	$55 – $74	$39 – $52	$94 – $126
Development/Startup Capital			
Regulatory, licenses, permits	$3 – $4	$1 – $1.5	$4 – $5.5
Pre-startup working capital	$9 – $12	$1 – $2	$10 – $14

Table 7-1. Construction Capital and Labor Costs for the IIFP Facility (Millions of 2009 Dollars) (Continued)

Cost Category	Phase 1 Costs[a] (in millions of dollars)	Phase 2 Costs[a] (in millions of dollars)	Total Phases 1 and 2 Costs (in millions of dollars)
Spare parts and startup inventories	$3 – $4	$1 – $1.5	$4 – $5.5
Subtotal Development/Startup	$15 – $20	$3 – $5	$18 – $25
Total Capital Costs	$70 – $94	$42 – $57	$112 – $151
Labor Costs			
Construction and installation	$22.3 – $34.1	$13.7 – $20.9	$36 – $55
Engineering, procurement, and construction management	$6.1 – $9.2	$3.7 – $5.7	$9.8 – $14.9
Project management	$1.6 – $2.3	$0.9 – $1.4	$2.5 – $3.7
Total Labor Costs	$29.9 – $45.6	$18.4 – $28.0	$48.3 – $73.6
Total Capital and Labor costs	$99.9 – $139.6	$60.4 – $85.0	$160.3 – $224.6

Source: IIFP, 2009

[a]Phase 1 and Phase 2 labor costs are estimated from the cumulative costs, based on the 62 percent-38 percent cost split for capital costs as found in the capital costs.

7.2.1.2 External Costs

External construction costs are summarized here.

Land Use: 259 ha (640 ac) of grazing land converted to industrial use

Historic and Cultural Resources: no resources expected to be affected

Visual Resources: no adverse impact expected

Climatology, Meteorology, and Air Quality: small, temporary, and local impacts to air quality; some small amount of CO_2 and other GHGs, criteria pollutants, and HAPs released

Geology, Mineral, and Soils: no prime farmland affected; 16 ha (40 ac) cleared

Water Resources: groundwater withdrawal a small percentage of that available; groundwater quality not expected to be adversely impacted; no surface water use or discharge

Ecological Resources: 16 ha (40 ac) of grassland removed; no threatened or endangered species expected to be affected

Socioeconomic Resources and Local Community Services: small decrease in available public service capacities; small increases in local tax revenues; small influx of money to the local economy; small improvement in employment rate

Traffic and Transportation: small increase in traffic near the intersection of NM 483 and US 62/180, but not sufficient to warrant mitigation

Noise: no adverse impact expected

Public and Occupational Health Impacts: construction injuries typical for industrial construction; no fatalities expected statistically

Waste Management: waste generation a small percentage of existing disposal capacities

7.2.2 Operations Costs

7.2.2.1 Internal Costs

Internal operations costs include raw materials, utilities, marketing and distribution, operations and maintenance, labor, waste disposal, and replacement capital costs. All costs are presented in 2009 dollars. The annual costs presented were estimated based on a 40-year plant operating life. The data presented here are from IIFP's environmental report (IIFP, 2009) and subsequent responses to NRC staff's requests for additional information (IIFP, 2011), and based on the assumptions presented in these documents.

Raw Materials

IIFP states (IIFP, 2009) that the proposed plant would use relatively small amounts of raw materials. This is because the primary input to the plant is a waste product from existing and proposed commercial enrichment facilities. The primary raw materials, other than the DUF_6 feedstock, are SiO_2, B_2O_3, $Ca(OH)_2$, KOH, and hydrogen gas. These materials are not expected to be procured in the region of influence (Lea and Eddy counties). The annual costs (in 2009 dollars) for raw materials are as follows:

Phase 1: $1.89 million
Phase 2 (incremental): $0.82 million
Cumulative: $2.71 million

Utilities

Utilities include electricity, natural gas, water, nitrogen, steam, and compressed air. Some of these utilities would be produced on site. However, approximately $1.5 million (2009 dollars) per year of utilities would be procured during the Phase 1 only facility operations between 2013 and the beginning of 2017. An additional $1.7 million per year of utilities for Phase 2 would be procured each year from 2017 through 2050 as a result of the expansion to the Phase 2 facility. Beginning in 2017, the cumulative utilities procured from utility companies located in the region or State would cost approximately $3.2 million each year, thereby benefiting the local and state economies.

Marketing and Distribution

IIFP reports that the marketing and distribution of FEP products would likely amount to 8 percent of the SiF_4 cost or approximately $200,000 to $250,000 annually (2009 dollars). Only SiF_4 is accompanied by any marketing and distribution costs because the other products are sold to only a few customers under contracts. This is an annual cost that would be incurred irrespective of the startup of Phase 2, because SiF_4 is generated in the Phase 1 process.

Operations and Maintenance

Operations and maintenance (O&M) costs would be those associated with purchasing materials for repair and replacement of equipment or infrastructure, and operating supplies such as office supplies, safety equipment, or laboratory chemicals. IIFP estimates that the annual O&M costs (2009 dollars) would be:

Phase 1: $2.7 million
Phase 2 (incremental): $1.6 million
Cumulative O&M cost: $4.3 million

Not all of these monies would be spent in the region of influence.

Labor

Section 4.1.2.8 presents the workforce requirements for the IIFP facility operations. In Tables 7-8 and 7-9 of IIFP's environmental report (IIFP, 2009), IIFP projects the annual labor costs for both Phase 1 and Phase 2. These are as follows, in 2009 dollars:

Phase 1: $7.9 million to $9.1 million
Phase 2 (incremental): $1.4 million to $1.7 million
Cumulative labor cost: $9.6 million to $10.5 million

Waste Disposal

The types and quantities of waste for disposal are reported in Section 4.1.2.12. The largest disposal costs would be associated with depleted uranium oxide; however, other LLW, RCRA waste, and sanitary waste would be disposed as well. The costs for Phase 1 and Phase 2 waste disposal are presented in Table 7-10 of the IIFP's environmental report (IIFP, 2009 as modified by IIFP [2011]) and are reproduced in Table 7-2.

Table 7-2.   Estimated Annual Waste Disposal Costs (millions of 2009 dollars)

Waste Type	Phase 1 (in millions of dollars)	Phase 2 (in millions of dollars)	Cumulative (in millions of dollars)
Depleted uranium oxide	$2.6 – $7.0	$5.4 – $15.5	$8.0 – $22.5
Other process low-level waste	$0.25 – $0.40	$0.01 – $0.05	$0.26 – $0.45
Miscellaneous low-level waste	$0.23 – $0.35	$0.22 – $0.30	$0.45 – $0.65
RCRA waste	$0.009 – $0.035	$0.005 – $0.010	$0.014 – $0.045
Sanitary waste	$0.002 – $0.003	negligible	$0.002 – $0.003
Total[1]	$3.1 – $7.8	$5.6 – $16	$8.7 – $24

[1]Totals rounded to two significant digits.

Replacement Capital

Replacement capital would be required to replace infrastructure and equipment over the life of the facility. IIFP estimates that replacement costs over the 40-year assumed life of the facility would be approximately $60 million to $85 million (2009 dollars); however, no replacement capital expenditures are expected for the first 7 years. The costs accumulate more heavily as the facility ages. The NRC staff calculated an average annual replacement capital cost of $1.8 million to $2.8 million over the 13 years of maximum replacement expenditures.

Table 7-3 reports the values reported by IIFP in Chapter 7 of the environmental report (IIFP, 2009) and the subsequent response to NRC staff's requests for additional information (IIFP, 2011).

Table 7-3.   Estimated Replacement Capital Expenditures (millions of 2009 dollars)

Time Period	Phase 1[1] (in millions of dollars)	Phase 2[1] (in millions of dollars)	Cumulative (in millions of dollars)
2011 – 2016	0	0	0
2017 – 2027	$4.6 – $5.6	$4.4 – $5.4	$9 – $11
2028 – 2037	$17.9 – $21.9	$17.2 – $21.1	$35 – $43
2038 – 2050	$16.3 – $19.9	$15.7 – $19.1	$32 – $39
Total 40-year period	$38.8 – $47.4	$37.3 – $45.6	$76 – $93

[1] IIFP (2011) states that 51 percent and 49 percent of the replacement capital costs would be associated with Phase 1 equipment and Phase 2 equipment, respectively.

Summary of Internal Operations Costs

Table 7-4 provides the total internal operations costs per year.

Table 7-4.   Total Annual Internal Operations Costs (millions of 2009 dollars)

Type of Internal Cost	Phase 1 (in millions of dollars)	Phase 2 (in millions of dollars)	Cumulative (in millions of dollars)
Raw materials	$1.89	$0.82	$2.71
Utilities	$1.5	$1.7	$3.2
Marketing and distribution	$0.20 – $0.25	0.0	$0.20 – $0.25
O&M	$2.7	$1.6	$4.3
Labor	$7.9 – $9.1	$1.4 – $1.6	$9.6 – $10.5
Waste disposal	$3.1 – $7.8	$5.6 – $16	$8.7 – $24
Replacement capital	$38.8 – $47.4	$37.3 – $45.6	$76 – $93
Total[1]	$56 – $71	$48 – $67	$100 – $140

[1] Totals rounded to two significant digits.

7.2.2.2   External Costs

External operations costs are summarized here.

<u>Land Use:</u> Land use would be consistent with other uses in the area

<u>Historic and Cultural Resources:</u> no resources expected to be affected

Visual Resources: no adverse impact expected

Climatology, Meteorology, and Air Quality: small and local impacts to air quality

Geology, Mineral, and Soils: no adverse impact

Water Resources: groundwater withdrawal a small percentage of that available; groundwater quality not expected to be adversely impacted; no surface water use or discharge

Ecological Resources: no adverse impact expected

Socioeconomic Resources and Local Community Services: small decreases in public service capacities; small increases in local tax revenues; small influx of money to the local economy; small improvement in employment rate

Traffic and Transportation: small increase in traffic near the intersection of NM 483 and US 62/180, but not sufficient to warrant mitigation; radiation doses to members of the public from transport of radioactive wastes and depleted uranium far less than normal background

Noise: no adverse impact expected

Public and Occupational Health Impacts: operation injuries typical for industrial plant operation; no fatalities expected statistically; radiological emissions produce immeasurably small impacts; chemical emissions small and localized

Waste Management: waste generation a small percentage of existing disposal capacities

7.3   Benefits of the Proposed Action

7.3.1   Construction

Taxes

Phase 1 construction-related activities, purchases, and workforce expenditures would require several types of tax payments, including individual income taxes, gross receipts taxes, and property taxes. Increased tax revenues are considered a benefit to the State of New Mexico, Lea County, the Hobbs Municipal School District, the New Mexico Junior College, the communities in Lea County, and other locales where plant-related spending would occur.

IIFP (2011) estimates that approximately $554,400 of fee in lieu of property taxes would be paid to the Hobbs Municipal School District and the New Mexico Junior College during the Phase 1 construction period. IIFP is exempt from any other property tax.

IIFP estimates (in 2009 dollars) that Phase 1 construction costs would be between $70 million and $94 million (Section 4.1.1.8). Some portion of those expenditures would occur within the ROI and other counties nearby. The expenditures would generate gross receipts tax revenues for both the affected counties and for the State of New Mexico (IIFP, 2011b). Because IIFP would have an industrial revenue bond with Lea County, some facility-related expenditures would be exempt from gross receipts taxes.

Regional spending on goods and services by IIFP employees would generate gross receipts tax revenues for Lea and Eddy County municipalities, Lea County, Eddy County, New Mexico, and other locales where spending occurs.

Employment

During Phase 1 construction of the IIFP facility, 80 percent of the 140 IIFP construction jobs are expected to be filled by workers that already reside within the two-county ROI (Section 4.1.1.8). The 112 residents that would fill the construction jobs would represent 0.2 percent of the June 2010 labor force within the region. If all 112 of the jobs were filled by unemployed workers, the unemployment rate in the region of influence would decrease by 0.2 percent. The remaining 28 jobs would be filled by workers that would migrate into the ROI. The in-migrating workers would increase the labor force by 0.05 percent (Section 4.1.1.8). The 12 indirect jobs that would be created during Phase 1 construction of the IIFP facility would likely be filled by regional residents. If all 12 jobs were filled by unemployed workers, those workers would represent 0.3 percent of the unemployed labor force in June 2010 (Section 4.1.1.8).

Economy

IIFP (2011b) estimates that between $9,140,000 and $13,900,000 (2009 dollars) would be infused into the economy annually during the construction period for labor and materials. Most of these values would be spent within the ROI.

7.3.2   Operations

Taxes

Phase 1 operations-related activities, purchases, and workforce expenditures would require several types of tax payments, including corporate income taxes, individual income taxes, gross receipts taxes, and property taxes. Increased tax revenues are viewed as a benefit to the State of New Mexico, Lea County, the Hobbs Municipal School District, the New Mexico Junior College, the communities in Lea County, and other locales where plant-related spending would occur.

Table 4-21 presents the estimated corporate income and gross receipts taxes that would be paid to the State of New Mexico and Lea County entities. The low estimate of corporate income and gross receipt taxes paid to the State is $144,200,000 and $6,500,000 to Lea County. The low estimate on property taxes is $8,700,000 to Lea County (IIFP, 2011b).

In addition to IIFP's income and gross receipts tax payments, plant employees would contribute state individual income and state and county gross receipts tax revenues. IIFP facility employee earnings would be taxed as individual income. Regional spending on goods and services by IIFP employees would generate gross receipts tax revenues for Lea County, Eddy County, the State of New Mexico, and other locales where their spending would occur.

Employment

Approximately 80 percent of the IIFP operation positions would be filled by people currently residing in the two-county ROI (Table 4-19). Those 112 workers would represent 0.2 percent of the June 2010 two-county labor force (Section 4.1.2.8). If all 112 of these jobs were filled by unemployed workers in the region, the unemployment rate would decrease by 0.2 percent.

Approximately 20 percent of the IIFP operation positions (28 jobs) would be filled by people migrating into the region of influence from outside the region (Section 4.1.2.8). The in-migrating workers would represent a 0.2 percent increase of the June 2010 labor force (Section 4.1.2.8).

The in-migration of 28 workers to fill operation positions would also create 51 new indirect jobs within the ROI because of the multiplier effect (Section 4.1.2.8). If unemployed workers fill the 51 indirect jobs that would be created during the Phase 1 operation of the IIFP facility, they would represent 1.3 percent of the unemployed labor force in June 2010.

Economy

The regional economy would benefit from the capital investment expenditures and recurring costs associated with the operation of the IIFP facility. IIFP has provided estimates for some of these costs. The payroll associated with Phase 1 operating wages is within the range of $7,900,000 to $9,100,000 annually (Section 4.1.2.8). Operations employees and workers in indirect positions would spend earnings on goods and services within the region of influence. Additional costs associated with operations include replacement capital, waste disposal, insurance premiums, taxes, utilities, and maintenance materials and supplies. These expenditures would range from $17,315,000 to $23,727,000 annually (Section 4.1.2.8).

National Benefits

Long-term storage of DUF_6 poses potential health risks because of the physical and chemical characteristics of DUF_6. If DUF_6 is released to the atmosphere, t reacts with water vapor in the air, forming HF fumes and a uranium-fluoride compound, UO_2F_2. These products are chemically toxic. HF is an extremely corrosive gas that can damage the lungs and cause death if inhaled.

DUF_6 has been stored at DOE sites for approximately 40 years. The Defense Nuclear Facilities Safety Board, in 1995, issued a Technical Report (DNFSB, 1995) calling for improved safety analysis, inspections, and handling procedures to ensure safe storage of DUF_6. DOE has since embarked on a program of creating deconversion capability at two locations where uranium enrichment has been performed.

The proposed IIFP facility would provide a benefit to the national uranium fuel cycle by ensuring that commercial enrichment facilities throughout the nation do not have to rely on long-term storage.

Silicon tetrafluoride is used in the electronics industry. Boron trifluoride is used for ion implantation, as a catalyst for polymer reactions, and as a gas in neutron radiation detectors. Anhydrous hydrogen fluoride has many industrial uses. These byproducts of IIFP's deconversion process are marketable. The benefit to the nation is that the IIFP plant would be an alternate source of inexpensive (because it is the byproduct of the main process) fluoride products.

7.4   Evaluation Summary of the Proposed IIFP Facility

The internal construction and operations costs for the IIFP facility are based on proprietary business analyses performed by IIFP. Given that company investors are willing to pursue the license in light of these costs, the NRC staff's concern is primarily evaluation of costs to the communities around the facility and the State of New Mexico. Implementation of the proposed action would have a SMALL positive overall economic impact on the region of influence. The implementation of the proposed action would generate national, regional, and local benefits and costs.

The primary national benefit of building the proposed IIFP facility would be improved management of the DUF_6 part of the uranium fuel cycle. The regional benefits of building the proposed IIFP facility would be increased employment, economic activity, and tax revenues in the region around the site. Some of these regional benefits, such as tax revenues, accrue specifically to Lea County. Other benefits may extend to neighboring counties in the state of New Mexico.

Costs associated with the proposed IIFP facility are, for the most part, limited to the area surrounding the site and the communities within commuting distance. These include monetary and environmental costs. As summarized above, the environmental costs are SMALL to MODERATE (for air quality). The influx of money into the State and local economies from the proposed action would appear to more than offset the small financial burdens placed on community services. The benefits to Lea County, Eddy County, the State of New Mexico, and the nation's capacity to maintain the uranium fuel cycle weigh somewhat favorably for the benefit side of this comparison.

7.5   References

(DNFSB, 1995) U.S. Defense Nuclear Facilities Safety Board. 1995. Integrity of Uranium Hexafluoride Cylinders. Technical Report DNFSB/TECH-4. May 5, 1995.

(IIFP, 2009) International Isotopes Fluorine Products. Inc. 2009. Fluorine Extraction Process and Depleted Uranium De-conversion Plant (FEP/DUP) Environmental Report, Revision A, ER-IFP-001. December 27, 2009. ADAMS Accession No. ML100120758.

(IIFP, 2011) International Isotopes Fluorine Products. 2011. Fluorine Extraction Process and Depleted Uranium De-conversion Plant (FEP/DUP) Official Responses to Environmental Report RAIs, Revision A. March 31, 2011. ADAMS Accession No. ML110970481.

8.0   SUMMARY OF ENVIRONMENTAL CONSEQUENCES

On December 30, 2009, IIFP submitted an application to the NRC for a license to construct, operate, and decommission the proposed IIFP facility (IIFP, 2009). IIFP proposes to locate the facility in Lea County, New Mexico, approximately 22.5 km (14 mi) west of Hobbs, New Mexico. If licensed, the proposed facility would deconvert DUF_6 into fluoride products (for commercial resale) and depleted uranium oxides (for disposal).

Source material licenses, such as the one requested for the proposed IIFP facility, are regulated under Title 10, Code of Federal Regulations, Part 40 (10 CFR 40), in accordance with the *Atomic Energy Act of 1954*. Section 102 of the *National Environmental Policy Act of 1969*, as amended (NEPA) (Public Law 91-190; Title 42, Section 4321 et seq., United States Code [42 U.S.C. 4321 et seq.]), directs that an Environmental Impact Statement (EIS) is required for major Federal actions that significantly affect the quality of the human environment. Section 102(2)(C) of NEPA requires that an EIS include information about the following:

- environmental impacts of the proposed action,
- any adverse environmental effects that cannot be avoided, should the proposal be implemented,
- alternatives to the proposed action,
- the relationship between local short-term uses of the environment and the maintenance and enhancement of long-term productivity, and
- any irreversible and irretrievable commitments of resources that would be involved if the proposed action is implemented.

NRC's regulations under 10 CFR 51 implement the requirements of NEPA. Because the NRC is responsible for licensing this facility, the licensing action is a Federal action, and must meet the requirements of NEPA. Based on the EIS and other information [including the original license application and responses to Requests for Additional Information (RAIs) received by NRC from the applicant] and analysis of the magnitude of potential impacts, the NRC staff will determine whether to issue a license to IIFP for the construction, operation, and decommissioning of the proposed IIFP facility.

IIFP anticipates two phases to the project, but the current license application is for the first phase only. Phase 2, under NEPA, is considered a "reasonably foreseeable future action" (40 CFR 1508.7). Therefore, Phase 2 impacts are considered cumulative impacts, and have been addressed in Section 4.2 of this EIS. IIFP expects to begin preconstruction activities in late 2011. If the license application is approved, IIFP expects to begin facility construction in 2012, which would continue for one year. Phase 2 construction would begin in 2015 and continue for one year.

As part of its license application, IIFP submitted an Environmental Report (ER). Information in the ER and supplemental environmental documentation provided by IIFP has been reviewed and independently verified by the NRC staff and used, in part, by the NRC staff in preparing this EIS. Upon acceptance of the ER, the NRC staff began the environmental review process described in 10 CFR 51 by publishing, on July 15, 2010, in the Federal Register (75 FR 42142) a Notice of Intent to prepare an EIS and conduct scoping. The purpose of the EIS scoping process was to assist in determining the range of actions, alternatives to the proposed action, and potential impacts to be considered in the EIS, and to identify significant issues related to the proposed action. Comments and information from the public and government agencies were received during the scoping period. As part of the scoping process, the NRC staff held a public scoping meeting on July 29, 2010, in Hobbs, New Mexico. NRC staff considered the public comments received during the scoping process for preparation of this EIS; the summary of the EIS scoping process is provided in Appendix A.

In addition to reviewing IIFP's ER and supplemental documentation, the NRC staff consulted with appropriate Federal, State, and local agencies and Native American Tribes.

Further comments from the public and government agencies were received after the NRC issued a Draft EIS for public review and comment on January 13, 2012, and announced its availability in the *Federal Register* (77 FR 2096) in accordance with 10 CFR 51.74. The public comment period ended on February 27, 2012. During the public comment period, the NRC held a public meeting in Hobbs, New Mexico, on February 2, 2012, where oral comments from members of the public were received on the Draft EIS. In addition to comments received at the public meetings, the NRC received written comments by postal mail and e-mail during the public comment period. The public meeting transcripts and the written comments are part of the public record for the proposed IIFP facility. These comments were considered by the NRC in preparing this EIS. Comment summaries and the NRC's responses are contained in Appendix F of this EIS.

Included in this EIS are (1) the results of the NRC staff's analyses, which consider and weigh the environmental effects of the proposed action; (2) mitigation measures for reducing or avoiding adverse effects; (3) the environmental impacts of alternatives to the proposed action; and (4) the NRC staff's assessment regarding the proposed action based on its environmental review.

Potential environmental impacts are evaluated in this EIS using the three-level standard of significance – SMALL, MODERATE, or LARGE – developed by the NRC using guidelines from the Council on Environmental Quality (CEQ) (40 CFR 1508.27). Table B-1 of 10 CFR 51, Subpart A, Appendix B provides the following definitions of the three significance levels:

- SMALL – Environmental effects are not detectable or are so minor that they would neither destabilize nor noticeably alter any important attribute of the resource.

- MODERATE – Environmental effects are sufficient to alter noticeably, but not to destabilize, important attributes of the resource.

- LARGE – Environmental effects are clearly noticeable and are sufficient to destabilize important attributes of the resource.

8.1   Unavoidable Adverse Environmental Impacts

Section 102(2)(c)(ii) of NEPA requires that an EIS include information on any adverse environmental effects that cannot be avoided, should the proposed action be implemented. Unavoidable adverse environmental impacts are those potential impacts that cannot be avoided and for which no practical means of mitigation are available.

This section summarizes the environmental consequences for the proposed action that cannot be avoided and for which no practical means of mitigation are available. Identification and description of the environmental impacts for the proposed action that would result from construction, operation, and decommissioning of the proposed IIFP facility are presented in Chapter 4, "Environmental Impacts." The mitigation measures that would be incorporated into the proposed action to control and minimize potential adverse environmental impacts are summarized in Chapter 5, "Mitigation Measures and Commitments." The monitoring programs that would be incorporated into the proposed action are listed in Chapter 6, "Environmental Measurements and Monitoring Programs."

Implementing the proposed action would result in unavoidable adverse impacts to land use, ecological resources, groundwater quantity, and air quality. Unavoidable adverse impacts to land use would occur at the initiation of the project, commencing with restricting the current land use, grazing, from the property and committing it, for the duration of the facility license, to industrial purposes. Site preparation will destroy up to 16 ha (40 ac) of Western Shortgrass Prairie or Apacherian-Chihuhuan Mesquite Upland Scrub habitat. However, both habitats are common throughout the region. Some topsoil would be lost during the grading and clearing, but this loss would be minimized with BMPs. Animal habitats would be destroyed and some mortality of individuals would occur during construction. The presence of the facility could prevent some animals from foraging or nesting in the vicinity of the facility.

During construction and operation, facility operations will consume small amounts of groundwater; the greatest groundwater use would occur during operations. The facility would use a small amount (approximately 0.5 percent) of the estimated annual 40-year planning period groundwater demand for Lea County, and 0.15 percent annually of the unappropriated water rights assigned to Lea County by the New Mexico Office of the State Engineer.

Construction and operation would release small quantities of pollutants, including radionuclides to the atmosphere. Emissions of CO_2 and other greenhouse gases, and CO and SO_2 during construction would be SMALL, however, construction could result in MODERATE impacts from NO_2, $PM_{2.5}$ and PM_{10} emissions. Construction impact to air quality would be localized and temporary. BMPs would minimize impacts to air quality during construction. Plant design would minimize emissions of radiological and chemical pollutants to levels well below regulatory limits; concentrations higher than background will not be detectable beyond the site boundary, and the releases will not adversely affect local or regional air quality.

8.2   Irreversible and Irretrievable Commitments of Resources

Environmental Review Guidance for Licensing Actions Associated with NMSS Programs [NUREG-1748 (NRC, 2003)], defines an "irreversible" commitment and an "irretrievable" commitment as follows:

- "Irreversible" refers to the commitment of environmental resources that cannot be restored.

- "Irretrievable" refers to the commitment of material resources that once used cannot be recycled or restored for other uses by practical means.

The implementation of the proposed action as described in Section 2.1 would include the commitment of land, water, energy, raw materials, and other natural and manmade resources. Approximately 16 ha (40 ac) on the 259-ha (640-ac) site would be affected by the construction, operation, and decommissioning of the proposed IIFP facility.

It is likely that, once the land has been committed to an industrial use, it will remain in industrial use in perpetuity, so this should be considered an irreversible commitment.

Groundwater use by the facility during both construction and operation would be consumptive. Groundwater withdrawn from the Ogallala aquifer will not be returned to the aquifer. Some will be lost to evaporation in the process, and the treated sanitary wastewater used to irrigate landscaping will transpire to the atmosphere through the process of photosynthesis. The depth to groundwater at the site is approximately 30 ft, so it is unlikely any landscape water will return to the groundwater.

Energy consumption will be in the form of gasoline and diesel fuel for construction equipment and generators, and coal or natural gas to generate electricity to power the facility. Some natural gas will be consumed in the production of hydrogen at the facility. These represent irretrievable uses of those resources.

The construction and operation of the proposed IIFP facility would require commitments of significant quantities of concrete, steel, nonferrous metals, plastics, and other material resources. At decommissioning, certain building materials and equipment could be recycled, however some materials would not be recyclable, and some materials would have been consumed by the deconversion process. Resources used in the construction and operation of the facility that could not be reused or recycled at the end of their useful life would represent an irreversible commitment. Materials consumed during the deconversion process would be irreversible commitments of resources. Hazardous and radioactive waste streams would be irreversible commitments of resources, as would the land needed to properly dispose of those waste streams.

No other irreversible or irretrievable commitments of resources were identified for the construction, operation, and decommissioning of the proposed IIFP facility.

8.3 Relationship between Local Short-Term Uses of the Environment and the Maintenance and Enhancement of Long-Term Productivity

Consistent with the CEQ definition in 40 CFR 1502.16 and the definition provided in NUREG-1748 (NRC, 2003), this EIS defines short-term uses and long-term productivity as follows:

- Short-term uses generally affect the present quality of life for the public (i.e., the 40-year license period for the proposed IIFP facility).

- Long-term productivity affects the quality of life for future generations on the basis of environmental sustainability (i.e., long-term is the period after license termination for the proposed IIFP facility).

Construction, operation, and decommissioning of the proposed IIFP facility would necessitate short-term commitments of resources. The short-term commitment of resources would include land, water and energy sources, and materials which could be recovered or recycled. Impacts would be minimized by mitigation measures and resource management. The short-term use of these resources would result in potential long-term socioeconomic benefits to the local area and the region, such as improvements to the local economy and infrastructure supported by worker income and tax revenues and the maintenance and enhancement of a skilled worker base.

Workers, the public, and the environment would be exposed to slightly elevated concentrations of radioactive and hazardous materials over the short term from the operation of the proposed IIFP facility due to process emissions and the transport and disposal of hazardous and radioactive waste.

Upon expiration of the license, IIFP would decommission the facility, recycle some equipment and restore the facility for another use. The use of the site and the buildings for other industrial purposes would constitute a long-term benefit to the community and would increase long-term productivity. Continued employment, expenditures, and tax revenues generated during preconstruction, construction, and operation of the proposed IIFP facility and from future site uses after the facility is decommissioned would directly benefit the local, regional, and State economies and would be considered a long-term benefit.

8.4   References

(IIFP, 2009) International Isotopes Fluorine Products, Inc. 2009. Fluorine Extraction Process & Depleted Uranium De-conversion Plant (FEP/DUP) Environmental Report, Revision A, ER-IFP-001. December 27, 2009. ADAMS Accession No. ML100120758.

(NRC, 2003) U.S. Nuclear Regulatory Commission. 2003. Environmental Review Guidance for Licensing Actions Associated with NMSS Programs, NUREG-1748, Washington D.C., 2003. ADAMS Accession No. ML032450279.

9.0 LIST OF PREPARERS

9.1   U.S. Nuclear Regulatory Commission Contributors

Matt Bartlett; NMSS Project Manager
BS, Physics, Bob Jones University, Greenville, SC, 1997
MS, Physics, Clemson University, Clemson, SC, 2000
PhD, Physics, Clemson University, Clemson, SC, 2004
Years of Relevant Experience: 6

Greg Chapman P.E., C.H.P; Analyst, Accidents
BEE, Georgia Tech 1987
ME, University of Florida, 1993
Graduate Certificate, University of Tennessee, 2010
Years of Relevant Experience: 18

Diana Diaz-Toro; Chief, Environmental Review Branch-A; General Reviewer
BS, Chemical Engineering, University of Puerto Rico, 2002
MBA, American University, 2008
Years of Relevant Experience: 9

Nathan Goodman; Reviewer, Ecological Resources
BS, Biology, Muhlenberg College, 1998
MS, Environmental Science, Johns Hopkins University, 2000
Years of Relevant Experience: 12

Christopher Grossman; Reviewer, Waste Management
BS, Civil Engineering, Purdue University, 1998
MS, Environmental Engineering & Science, Clemson University, 2001
Years of Relevant Experience: 9

Kellee L. Jamerson; Reviewer, Environmental Measurements and Monitoring Programs
BS, Environmental Science, Tuskegee University, 2006
Years of Relevant Experience: 3

Stephen Lemont, Ph.D.; Technical Reviewer, Statutory and Regulatory Requirements,
Alternatives, Reasonably Foreseeable Actions, Land Use, Historic and Cultural Resources,
Water Resources, Waste Management, and Geology, Minerals, and Soil
BS, Chemistry, Brooklyn College of the City University of New York, 1971
PhD, Physical Chemistry, Columbia University, 1976
Years of Relevant Experience: 30

Asimios Malliakos, Ph.D.; EIS Project Manager; General EIS Reviewer
BS, Physics, University of Thessaloniki, Greece, 1975
MS, Nuclear Engineering, Polytechnic Institute of New York, 1977
PhD, Nuclear Engineering with a Minor Degree in Probability and Statistics, University of
Missouri-Columbia, 1980
Years of Relevant Experience: 30

Christopher A. McKenney; Contributor and Technical Reviewer, Waste Management
BS, Nuclear Engineering, Oregon State University, 1991
Years of Relevant Experience: 20

Johari Moore; Reviewer, Public and Occupational Health, Traffic and Transportation
BS, Physics, Florida A & M University, 2003
MSE, Nuclear Engineering and Radiological Sciences, University of Michigan, 2005
Years of Relevant Experience: 3

Ashley Waldron; Reviewer, Ecological Resources, Air Quality and Environmental Monitoring
BS, Biology, Frostburg State University, 2009
Years of Relevant Experience: 3

Jean Trefethen; Reviewer, Traffic and Transportation, Noise, and Public and
Occupational Health
BA, Biology, Carroll College, 1987
Years of Relevant Experience: 2

9.2   Straughan Environmental and Tetra Tech Contributors

Lawson Bailey; Contributing Author, Affected Environment
BS, Biology, Virginia Polytechnic Institute & State University, 1979
Years of Relevant Experience: 32

Greg M. Beachley, Ph.D.; Contributor, Air Quality
BS, Chemistry and BS, Biology, James Madison University, 2001
PhD, Analytical & Environmental Chemistry, University of Maryland, 2009
Years of Relevant Experience: 6

Jacqueline K. Boltz; Technical Editor, Public Outreach Manager
BA, French Language and Literature, Boston University, 1991
MBA, General Business, Boston University, 1991
Years of Relevant Experience: 19

Steven J. Connor; Analyst, Proposed Action and Alternatives and Cost-Benefit Analysis
Summary; Technical Reviewer
BS, Physics, Georgia Institute of Technology, 1973
MS, Physics, Georgia Institute of Technology, 1974
Years of Relevant Experience: 34

Charles Conrad; Analyst, Air Quality and Traffic
BS, Chemical Engineering, University of Wisconsin – Madison, 2002
Years of Relevant Experience: 5

Krista Dearing; Analyst, Geology, Minerals, Soils and Groundwater
BS, Geology, University of Cincinnati, 1991
MS, Geology, University of Cincinnati, 1993
Years of Relevant Experience: 16

Amanda J. Deering; Contributor and Technical Reviewer
BS, Environmental Studies, Towson University, 2010
Years of Relevant Experience: 1

Eric T. Duce; Contributor and Technical Reviewer, Land Use and Ecology
BS, Natural Resources Management, University of Maryland, 1998
MS, Environmental Science and Policy, Johns Hopkins University, 2005
Years of Relevant Experience: 11

Alverna R. (A.J.) Durham, Jr.; Analyst and Technical Reviewer, Air Quality
BS, Industrial Technology, North Carolina Agricultural and Technical State University, 1999
Years of Relevant Experience: 6

Kristin Fusco Rowe, P.E.; Analyst, Noise Abatement Specialist
BS, Engineering Science, Loyola College, 2002
Years of Relevant Experience: 9

Kristi Hagood; Analyst, Climatology, Meteorology, and Air Quality
BS, Chemistry, Converse College, 1998
MS, Astrophysical, Planetary and Atmospheric Sciences, University of Colorado, 2000
Years of Relevant Experience: 9

Tim J. Harvey, PMP; Program Manager
BS, Natural Resources, Cornell University, 1983
Years of Relevant Experience: 25

Justin M. Haynes; Technical Reviewer
BS, Integrated Science & Technology, Environmental Concentration, James Madison
University, 2003
Years of Relevant Experience: 7

Richard L. Heimbach, II, AICP; Project Manager
BS, Environmental Resource Management, Pennsylvania State University, 1989
Years of Relevant Experience: 20

Chimere M. Lesane-Matthews; Analyst, Land Use, Socioeconomics, Traffic and Transportation
BS, Civil Engineering, Morgan State University, 2001
Years of Relevant Experience: 10

Anne Lovell; Contributor, Air Quality
BS, Chemical and Petroleum Refining Engineering, Colorado School of Mines, 1985
Years of Relevant Experience: 25

Lisa Matis; Analyst, Waste Management
MS, Mechanical Engineering, Stevens Institute of Technology, 1989
BS, Chemical Engineering, Stanford University, 1984
Years of Relevant Experience: 26

Sarah M. Michailof; Analyst, Historic and Cultural Resources
BA, Anthropology and Biology, University of North Carolina, Chapel Hill, 1994
MA, Historic Preservation, Goucher College, 2007
Years of Relevant Experience: 16

Phil Moore; Technical Editor
BA, English, University of South Carolina, 1975
MS, Wildlife Biology (Fisheries Emphasis), Clemson University, 1983
Post Graduate Study, Zoology, Clemson University, 1977 1979
Years of Relevant Experience: 25

Ellen K. Mussman; Analyst, Land Use, Ecology, Mitigation and Monitoring
BS, Conservation of Soil, Water, and the Environment, University of Maryland,
College Park, 2002
MS, Forest Resources, University of Washington, 2006
Years of Relevant Experience: 6

Karen Patterson; Task Manager and Senior Reviewer
BA, Biology, Randolph-Macon Woman's College, 1973
MA, Biology, Wake Forest University, 1977
MLIS, University of South Carolina, 1999
Years of Relevant Experience: 37

Nikki J. Radke; Contributor, Graphics
BS, Biology and Wildlife Management, University of Wisconsin-Stevens Point, 1999
MS, Wildlife Management, Texas Tech University, 2005
Years of Relevant Experience: 7

Noreen Raza; Contributor, Environmental Justice
BS, English, Towson University, 2003
Years of Relevant Experience: 5

Jay Rose; Analyst, Cumulative Impact Assessment, Health and Safety, Accidents, and
Executive Summary
BS, Ocean Engineering, U.S. Naval Academy, 1983
JD, Catholic University, Columbus School of Law, 1996
Years of Relevant Experience: 27

Kenneth R. Scarlatelli; Technical Reviewer, Ecology
BS, Wildlife Biology, University of Massachusetts at Amherst, 1983
MA, Environmental Studies, Environmental Management Concentration, 1994
Years of Relevant Experience: 27

Alan Toblin; Analyst, Accidents
BE, Cooper Union, 1968
MS, University of Maryland, 1970
Years of Relevant Experience: 39

9.3   Center for Nuclear Waste Regulatory Analysis Contributors

James Durham; Principal Investigator, Contributor, Comment Responses
PhD, Nuclear Engineering, University of Illinois at Urbana-Champaign, 1987
Years of Relevant Experience: 25

Amy Hester; Contributor, Socioeconomics
BA, Environmental Studies, University of Kansas, 1998
Years of Relevant Experience: 14

Robert Lenhard; Project Manager
PhD, Soil Physics, Oregon State University, 1984
Years of Relevant Experience: 28

Lauren Mulverhill; Technical Editor
BS, Radio-Television, Southern Illinois University – Carbondale, 1984
Years of Relevant Experience: 28

John Stamatakos; Program Manager, Contributor, Geophysics
PhD, Geology, Lehigh University, 1990
Years of Relevant Experience: 25

10.0 Glossary

Abatement: Diminution in amount, degree, or intensity.

Activity: A measure of the rate at which a material emits nuclear radiation, usually given in terms of the number of nuclear disintegrations occurring in a given length of time. The common unit of activity is the curie (Ci), which amounts to 37 billion disintegrations per second. The international unit of activity is the becquerel (Bq) and is equal to one disintegration per second.

Air pollutant: Any substance in air which could, if present in high enough concentration, harm humans, animals, vegetation, or material. Pollutants may include almost any natural or artificial substance capable of being airborne.

Air quality: A measure of the concentrations of pollutants, measured individually, in the air. These concentrations are often compared to regulatory standards.

Air quality standards: The concentration of a pollutant in air prescribed by regulations that may not be exceeded during a specified time in a defined area. Air quality standards are used to provide a measure of the health-related and visual characteristics of the air.

ALARA: Acronym for "as low as (is) reasonably achievable." An approach to keep radiation exposures (both to the workforce and the public) and releases of radioactive material to the environment at levels that are as low as social, technical, economic, practical, and public policy considerations allow. ALARA is not a dose limit; it is a practice in which the objective is the attainment of dose levels as far below applicable limits as possible.

Alluvium: Clay, silt, sand, and/or gravel deposits found in a stream channel or in low parts of a stream valley that is subject to flooding. Ancient alluvium deposits frequently occur above the elevation of present-day streams.

Alternative site: A ranked site, other than the proposed site, that was evaluated in the fine-screening step.

Ambient air: The surrounding atmosphere, usually the outside air, as it exists around people, plants, and structures. It is not the air in immediate proximity to emission sources.

Ambient Air Quality Standards: Standards established on a State or Federal level, that define the limits for airborne concentrations of designated "criteria" pollutants (nitrogen dioxide, sulfur dioxide, carbon monoxide, total suspended particulates, ozone, and lead), to protect public health with an adequate margin of safety (primary standards) and to protect public welfare, including plant and animal life, visibility, and materials (secondary standards).

Ambient Noise Level: A sound level that represents the background noise from community or environmental sound sources.

Anhydrous: Without water (H_2O).

Anthropogenic: Caused or influenced by humans.

Aqueous: Related to water.

Aquifer: Geologic unit sufficiently permeable to conduct groundwater.

Area of potential effect (APE): The geographic area or areas within which an undertaking may directly or indirectly cause alterations in the character or use of historic properties, if any such properties exist. The area of potential effects is influenced by the scale and nature of an undertaking and may be different for different kinds of effects caused by the undertaking.

Assay: The qualitative or quantitative analysis of a substance; often used to determine the proportion of isotopes in radioactive materials.

Asymptomatic: Without symptoms.

Atmosphere: The layer of air surrounding the earth.

Atomic Energy Act of 1954 as amended: A Federal law that created the Atomic Energy Commission, which later split into the Nuclear Regulatory Commission and the Energy and Research and Development Administration (ERDA). ERDA became part of the Department of Energy in 1977. This act encouraged development and the use of nuclear energy for the general welfare and the security of the United States. This act authorized the Nuclear Regulatory Commission to regulate and license fuel fabrication facilities that seek to receive, possess, use, or transfer special nuclear material.

Attainment area: A region that meets the U.S. EPA National Ambient Air Quality Standards (NAAQS) for a criteria pollutant under the Clean Air Act.

Autoclave: A strong, pressurized, steam-heated vessel, as for laboratory experiments, sterilization, or cooking.

Background radiation: Radiation from: (1) naturally occurring radioactive materials, as they exist in nature prior to removal, transport, or enhancement or processing by man; (2) cosmic and natural terrestrial radiation; (3) global fallout as it exists in the environment; (4) consumer products containing nominal amounts of radioactive material or emitting nominal levels of radiation; and (5) radon and its progeny in concentrations or levels existing in buildings or the environment that have not been elevated as a result of current or past human activities.

Baghouse: A large chamber or room for holding bag filters used to filter gas streams.

Berms: A level space, shelf, or raised barrier separating two areas.

Baseline: A quantitative expression of conditions, costs, schedule, or technical progress to serve as a base or standard for measurement during the performance of an effort; the established plan against which the status of resources and the progress of a project can be measured.

Basin: A topographic or structurally low area or the area drained by a stream system.

Basalt: A fine-grained dark igneous (volcanic) rock that is low in silica content and has congealed from a molten (magma) state.

Best Management Practices (BMP): Structural, nonstructural, and managerial techniques recognized to be the most effective and practical means to reduce surface water and groundwater contamination while still allowing the productive use of resources.

Beta particle: A charged particle emitted from a nucleus during radioactive decay, with a mass equal to 1/1837 that of a proton. A negatively charged beta particle is identical to an electron. A positively charged beta particle is called a positron. Large amounts of beta radiation may cause skin burns, and beta emitters are harmful if they enter the body. Beta particles may be stopped by thin sheets of metal or plastic.

Bioassay analyses: A method for quantitatively determining the concentration of a substance by its effect on the growth of a suitable animal, plant, or microorganism under controlled conditions.

Biomass: The dry mass of living matter, expressed in terms of a given area or volume.

Bollard: A strong wooden or metal post mounted on a wharf, quay, etc. to protect the stationary structure from, and stop, a moving craft or vehicle.

Boom: As used in this EIS, a temporary floating barrier launched on water to contain material such as an oil spill.

Boron: Semi-metallic chemical element, with atomic number 5, which has the chemical symbol B.

Bounding: That which represents the maximum reasonably foreseeable event or impact. All other reasonably foreseeable events or impacts would have fewer and/or less severe environmental consequences.

Buffer area: A designated area of land that is designed to permanently remain vegetated in an undisturbed and natural condition in order to protect an adjacent aquatic or wetland site from upland impacts and to provide habitat for wildlife.

Byproduct: A product from a manufacturing process that is not considered the principal material.

Candidate species: A species of plants or animals considered as a candidate for possible listing as endangered or threatened by a government agency.

Carbonaceous: Consisting of, containing, relating to, or yielding the element carbon (carbon is element with atomic number 6, and has the chemical symbol C).

Carbon monoxide: An odorless, colorless, poisonous gas produced by incomplete burning of carbon in fuels. Exposure to carbon monoxide reduces the delivery of oxygen to the body's organs and tissues. Elevated levels can cause impairment of visual perception, manual dexterity, learning ability, and performance of complex tasks.

Caliche: Calcium carbonate (chemical symbol $CaCO_3$) deposited in the soils of arid or semiarid regions.

Clarifier: A piece of equipment that removes suspended impurities or solid matter by settling, heating gently, or filtering.

Clean Air Act: A Federal law that requires the EPA to set and enforce air pollutant emissions standards for stationary sources and motor vehicles.

Climatology: The science devoted to the study, over time, of the conditions of the natural environment (rainfall, daylight, temperature, humidity, air movement) prevailing in specific regions of the earth.

Code of Federal Regulations (CFR): All Federal regulations in force are published in codified form in the Code of Federal Regulations.

Coke: The solid residue of impure carbon obtained from bituminous coal and other carbonaceous materials after removal of volatile material by destructive distillation.

Cold traps: A device that condenses all vapors except the permanent gases into a liquid or solid.

Committed dose equivalent: The predicted dose equivalent to a tissue or organ over a 50-year period after an intake of a radionuclide into the body. It does not include dose contributions from radiation sources external to the body. Committed dose equivalent is expressed in units of rem (or sievert) (1 rem = 0.01 sievert).

Committed effective dose equivalent: The sum of the committed dose equivalents to various organs or tissues in the body from radioactive material taken into the body, each multiplied by the tissue-specific weighting factor. Committed effective dose equivalent is expressed in units of rem (or sievert).

Community: A group of people (or animals) within a defined area that could be exposed to health risks from industrial pollutants or disturbed by noise, dust, and traffic associated with development of an industrial facility but that could also benefit from improved employment opportunities, higher land values, and infrastructure improvements associated with the project.

Concentration: The amount of a substance contained in a unit quantity (mass or volume) of a sample.

Conservative: When used with predictions or estimates, leaning on the side of pessimism. A conservative estimate is one in which the uncertain inputs are used in the way that provides a reasonable upper limit of the estimate of an impact.

Containment: Retention of a material or substance within prescribed boundaries.

Contamination: The presence of an unwanted chemical or radiological constituent in or on a material, person, property, or structure.

Cooling water: Water circulated through a nuclear reactor or processing plant to remove heat.

Cost-benefit analysis: A formal quantitative procedure comparing costs and benefits of a proposed project or act under a set of pre-established rules.

Council on Environmental Quality: The President's Council on Environmental Quality (CEQ) was established by the enactment of National Environmental Policy Act (NEPA). The CEQ is responsible for developing regulations to be followed by all Federal agencies in developing and implementing their own specific NEPA implementation policies and procedures.

Criteria pollutants: Six pollutants (ozone, carbon monoxide, total suspended particulates, sulfur dioxide, lead, and nitrogen oxide) known to be hazardous to human health and for which the EPA sets National Ambient Air Quality Standards under the Clean Air Act.

Critical habitat: The specific areas within the geographical area occupied by a species at the time it is listed as threatened or endangered on which are found those physical or biological features that are essential to the conservation of the species and that may require special management considerations or protection. It also includes specific areas outside the geographical area occupied by the species at the time it is listed if these areas are determined to be essential for the conservation of the species.

Cryogenic: Of, or relating to low temperatures; or requiring low temperatures for storage.

Cultural resources: Archaeological sites, architectural features, traditional use areas, and Native American sacred sites or special use areas.

Cumulative impacts: Cumulative impacts are those impacts on the environment that result from the incremental impact of the action when added to other past, present, and reasonably foreseeable future actions regardless of what agency (Federal or non-Federal) or person undertakes such other actions. Cumulative impacts can result from individually minor but collectively significant actions taking place over a period of time.

Curie: A unit of radioactivity equal to 37 billion (3.7×10^{10}) disintegrations per second.

Daughter products: The remaining nuclide left over from radioactive decay.

Decibel (dB): A standard unit for measuring sound-pressure levels based on a reference sound pressure of 0.0002 dyne per square centimeter. This is the smallest sound a human can hear. In general, a sound doubles in loudness with every increase of slightly more than 3 decibels.

Deciduous: Falling off at maturity or tending to fall off and is typically used in reference to trees or shrubs that lose their leaves seasonally.

Decommissioning: The removal of a facility from active service.

Decontamination: The reduction or removal of an unwanted chemical or radiological constituent from a structure, area, object, or person. Decontamination of radiological contamination may be accomplished by (1) treating the surface to remove or decrease the contamination, (2) letting the material stand so that the radioactivity is decreased as a result of natural radioactive decay, or (3) covering the contamination to shield or attenuate the radiation emitted.

Deconversion: As used in this EIS, the process by which uranium hexafluoride (UF_6) is chemically converted to uranium oxide (UO_2) producing anhydrous hydrogen fluoride (HF) and other marketable fluoride byproducts.

Degradation: The process by which organic substances are broken down by living organisms.

Delaware Basin: An area in southeastern New Mexico and the adjacent parts of Texas where the Permian sea deposited a large thickness of evaporites some 220 to 280 million years ago. It is partially surrounded by the Capitan Reef.

Depleted uranium: Uranium having a percentage of uranium-235 smaller than the 0.7 percent found in natural uranium. In the context of this EIS, it is the residue or tails from the uranium enrichment process.

Depleted uranium hexafluoride (DUF$_6$): A compound of uranium and fluorine from which most of the uranium-235 isotope has been removed.

Diffusion: Movement of atoms, ions, or molecules of one substance into or through another as a result of thermal or concentration gradients.

Dike: A barrier (typically, an embankment for controlling or holding back water; or, in geology, a type of sheet intrusion that cuts discordantly across the geologic body).

Dispersion: The occurrence in which particles are dispersed in air, water, soil, or other another medium.

Dose equivalent: The product of absorbed dose in rad (or gray) in tissue and a quality factor. Dose equivalent is expressed in units of rem (or sievert).

Dose rate: The radiation dose delivered per unit time (e.g., rem per hour).

Ecology: The science dealing with the relationship of all living things with each other and with the environment.

Ecoregion: A classification of land based on similar climate, vegetation, and topography.

Effective dose equivalent: The sum of the products of the dose equivalent received by specified organs or tissues of the body and a tissue-specific weighting factor. The effective dose equivalent is expressed in units of rem (or sievert).

Effluent: A gas or fluid discharged into the environment, treated or untreated. Most frequently, the term applies to wastes discharged to surface waters.

EIS: Environmental impact statement; a document required by the National Environmental Policy Act for proposed major Federal actions involving potentially significant environmental impacts.

Emissions: Substances that are discharged into the air.

Endangered species: Plants and animals that are threatened with extinction, serious depletion, or destruction of critical habitat. Requirements for declaring a species endangered are contained in the Endangered Species Act.

Endangered Species Act of 1973: An act requiring Federal agencies, with the consultation and assistance of the Secretaries of the Interior and Commerce, to ensure that their actions will not likely jeopardize the continued existence of any endangered or threatened species or adversely affect the habitat of such species.

Enrichment (process): Increasing the concentration of the uranium isotope U^{235} to more than that which exists in natural uranium ore, for use in atomic energy.

Environment: The sum of all external conditions and influences affecting the life development and, ultimately, the survival of an organism.

Environmental justice: The fair treatment of people of all races, cultures, incomes, and educational levels with respect to the development, implementation, and enforcement of environmental laws, regulations, and policies. Fair treatment implies that no population of people should be forced to shoulder a disproportionate share of the negative environmental impacts of pollution or environmental hazards due to a lack of political or economic strength.

Environmental monitoring: The act of measuring, either continuously or periodically, some quantity of interest, such as radioactive material in the air.

Ephemeral stream: A stream channel that carries water only during part of the year, immediately after periods of rainfall or snowmelt.

Equilibrium: A state of rest in a chemical or mechanical system.

ER: Environmental Report required as part of an environmental assessment, which identifies, describes and evaluates the likely significant effects on the environment of implementing a plan or program.

Erosion: Removal and transport of materials by wind, ice, or water on the earth's surface.

Escarpment: A long, nearly continuous cliff or relatively steep slope facing in one general direction, breaking the continuity of the land by separating two level or gently sloping surfaces, and produced by erosion or faulting.

Exposure limit: The level of exposure to a hazardous chemical (set by law or a standard) at which or below which adverse human health effects are not expected to occur.

Exposure pathways: A route or sequence of processes by which a radioactive or hazardous material may move through the environment to humans or other organisms. Each exposure pathway includes a source or release from a source, an exposure point, and an exposure route.

Fault: A fracture or a zone of fractures along which there has been displacement parallel to the fracture.

Fauna: The animal life of any particular region or time.

Floodplain: Low-lying areas adjacent to rivers and streams that are subject to natural inundations typically associated with precipitation.

Flora: The plant life occurring in a particular region, generally the naturally occurring or indigenous plant life.

Fluorocarbon: A halocarbon in which some hydrogen atoms have been replaced with fluorine.

Fluorine: The chemical element with atomic number 9, represented by the chemical symbol F.

Formation: A mapable geologic body of rock identified by lithic characteristics and stratigraphic position. Formations may be combined into groups or subdivided into members.

Fuel cycle: The series of steps involved in supplying fuel for nuclear power reactors. It can include mining, milling, isotopic enrichment, fabrication of fuel elements, use in a reactor, chemical reprocessing to recover the fissionable material remaining in the spent fuel, re-enrichment of the fuel material, re-fabrication into new fuel elements, and waste disposal.

Fugitive dust: Any solid particulate matter (PM) that becomes airborne, other than that emitted from an exhaust stack, directly or indirectly as a result of the activities of man. Fugitive dust may include emission from haul roads, wind erosion of exposed soil surfaces, and other activities in which soil is either removed or distorted.

Gamma: Short-wavelength electromagnetic radiation (high-energy photons) emitted In the radioactive decay of certain nuclides. Gammas are the same as gamma rays or gamma waves.

Gaussian plume: The distribution of material (a plume) in the atmosphere resulting from the release of pollutants from a stack or other source. The distribution of concentrations about the centerline of the plume, which is assumed to decrease as a function of its distance from the source and centerline (Gaussian distribution), depends on the mean wind speed and atmospheric stability.

Geology: The science that deals with the earth; the materials, processes, environments, and history of the planet, especially the lithosphere, including the rocks, their formation, and structure.

Geology and Soils: Those Earth resources that may be described in terms of landforms, geology, and soil conditions.

Greenhouse gas: A gas in an atmosphere that absorbs and emits radiation within the thermal infrared range.

Gross beta: The total rate of emission of beta particles from a sample, without regard to energy distributions or source nuclides.

Groundwater: All subsurface water, especially that contained in the saturated zone below the water table.

Habitat: The part of the physical environment in which a plant or animal lives.

Hazardous chemical: Under 29 CFR 1910, Subpart Z, "hazardous chemicals" are defined as "any chemical, which is a physical hazard or a health hazard." Physical hazards include combustible liquids, compressed gases, explosives, flammables, organic peroxides, oxidizers, pyrophorics, and reactives. A chemical is a health hazard when there is good evidence that acute or chronic health effects occur in exposed individuals. Hazardous chemicals include carcinogens, toxic or highly toxic agents, reproductive toxins, irritants, corrosives, sensitizers, hepatotoxins, nephrotoxins, agents that act on the hematopoietic system, and agents that damage the lungs, skin, eyes or mucous membranes.

Hazardous waste: According to the Resource Conservation and Recovery Act, a waste that, because of its characteristics, may (1) cause or significantly contribute to an increase in mortality or an increase in serious irreversible illness, or (2) pose a substantial hazard to human health or the environment when improperly treated, stored, transported, disposed of, or otherwise managed. Hazardous wastes possess at least one of the following characteristics: ignitability, corrosivity, reactivity, or toxicity. Hazardous waste is nonradioactive.

Historic Resources: The sites, districts, structures, and objects associated with historic events, persons, or social or historic movements.

Historic and Cultural Resources: Cultural resources include any prehistoric or historic district, site, building, structure, or object resulting from, or modified by, human activity. Historic properties are cultural resources listed in, or eligible for listing in, the National Register of Historic Places.

Homogenous: Describing a substance or population with uniform composition.

Hopper: A (usually funnel-shaped) container in which materials, such as chemicals, are stored in readiness for dispensing.

Hydraulic conductivity: A quantity that describes the rate at which water flows through an aquifer. It has units of length/time and is equal to the hydraulic transmissivity divided by the thickness of the aquifer.

Hydrofluorocarbons: An organic chemical containing hydrogen, fluorine, and carbon; emitted as a byproduct of industrial manufacturing.

Hydroperiod: The number of days per year that an area of land is inundated with water; or the length of time that there is standing water at a location.

Indirect jobs: Jobs generated or lost in related industries within a regional economic area as a result of a change in direct employment.

Ingestion: To take in by mouth. Material that is ingested enters the digestive system.

Inhalation: To take in by breathing. Material that is inhaled enters the lungs.

Integrated Safety Analysis (ISA): A formalized and documented process that identifies potential accident sequences in a plant's operations, designates items relied on for safety to either prevent such accidents or mitigate their consequences to an acceptable level, and describes management measures to provide reasonable assurance of the availability and reliability of items relied on for safety.

Intermittent: As used in this EIS, a drainage feature that contains water for only part of the year, typically during wet seasons. An intermittent stream often lacks the biological and hydrological characteristics commonly associated with the conveyance of water.

Ionizing radiation: Radiation capable of displacing electrons from atoms or molecules to produce ions.

Isotope: An atom of a chemical element with a specific atomic number and atomic weight. Isotopes of the same element have the same number of protons but different numbers of neutrons. Isotopes are identified by the name of the element and the total number of protons and neutrons in the nucleus. For example, uranium-235 is an isotope of uranium with 92 protons and 143 neutrons and uranium-238 is an isotope of uranium with 92 protons and 146 neutrons.

Kilovolt (kV): A unit of electrical potential equal to a thousand volts.

Kilovolt-ampere (kVA): A unit of electrical power equal to 1000 volt-amperes.

Land use: The way land is developed and used in terms of the kinds of anthropogenic activities that occur (e.g., agriculture, residential areas, industrial areas).

Latent cancer fatalities (LCFs): Deaths resulting from cancer that has become active after a latent period following radiation exposure. For radiation exposure, latent cancer fatalities can be calculated from collective dose using the risk conversion factor of 6×10^{-4} LCFs per person rem.

Lithic: Made of stone.

Load factor: The ratio of the average electric load to the peak load over a period of time.

Loam: A rich, friable soil containing a relatively equal mixture of sand and silt, clay, and humus.

Low-level mixed waste: Low-level radioactive waste that also contains hazardous chemical components regulated under the Resource Conservation and Recovery Act.

Low-level radioactive waste: Wastes containing source, special nuclear, or by-product material are acceptable for disposal in a land disposal facility. For the purposes of this definition, low-level waste has the same meaning as in the Low-Level Radioactive Waste Policy Act, that is, radioactive waste not classified as high-level radioactive waste, transuranic waste, spent nuclear fuel, or by-product material as defined in section 11e.(2) of the Atomic Energy Act (uranium or thorium tailings and waste).

Low-income population: A population where 25 percent or more of the population is identified as living in poverty.

Magnitude (earthquake): A measure of the total energy released by an earthquake. It is commonly measured in numerical units on the Richter scale. Each unit is different from an adjacent unit by a factor of 30.

Maim: To injure, disable or disfigure, usually by depriving of the use of a limb or other part of the body.

Maximally exposed individual (MEI): A hypothetical person who—because of proximity, activities, or living habits—could receive the highest possible dose of radiation or of a hazardous chemical from a given event or process.

Meteorological tower: An individual data acquisition point for weather and air related information (e.g., wind speed, wind direction, precipitation, opacity, etc.)

Meteorology: The science dealing with the atmosphere and its phenomena, especially as relating to weather.

Migration: The natural travel of a material through the air, soil, or groundwater.

Millirem (mrem): One thousandth of a rem (0.001 rem).

Mitigation: An action or actions implemented to lessen or alleviate impacts to a resource from a proposed action or activity. The purpose of mitigative actions is to avoid, minimize, rectify, or compensate for any adverse environmental impact.

Mixed waste: Waste that contains both "hazardous waste" and "radioactive waste" as defined in this glossary.

Modified Mercalli Intensity: A measurement of earthquake intensity based on the effects to people and structures. Ranges from I (low) to XII (total destruction), as opposed to the Richter scale, which measures the energy of the earthquake. Mercalli scale is often used to classify earthquakes that were not recorded on modern seismographs.

National Ambient Air Quality Standards (NAAQS): Air quality standards established by the Clean Air Act, as amended. The primary NAAQS are intended to protect the public health with an adequate margin of safety, and the secondary NAAQS are intended to protect the public welfare from any known or anticipated adverse effects of a pollutant.

National Emission Standards for Hazardous Air Pollutants (NESHAP): Emission standards for the control of releases of specified hazardous air pollutants, including radionuclides. These were implemented in the Clean Air Act Amendments of 1977.

National Environmental Policy Act (NEPA) of 1969: A Federal law constituting the basic national charter for protection of the environment. The act calls for the preparation of an environmental impact statement (EIS) for every major Federal action that may significantly affect the quality of the human or natural environment. The main purpose is to ensure that environmental information is provided to decision makers so that their actions are based on an understanding of the potential environmental and socioeconomic consequences of a proposed action and the reasonable alternatives.

National Historic Preservation Act (NHPA): A Federal law providing that property resources with significant national historic value be placed on the National Register of Historic Places. It does not require permits; rather, it mandates consultation with the proper agencies whenever it is determined that a proposed action might impact a historic property.

National Pollutant Discharge Elimination System (NPDES): Federal permitting system mandated by the Clean Water Act required for any discharges to waters of the United States.

National Register of Historic Places: A list maintained by the National Park Service of architectural, historic, archaeological, and cultural sites of local, state, or national importance.

Native vegetation: Plants that have evolved in a particular region and environment.

Nocturnal: Of, relating to, or occurring in the night.

Nonattainment areas: An area that has been designated by the EPA, or the appropriate State air quality agency, as exceeding one or more national or State Ambient Air Quality Standards.

Nonferrous: Not composed of or containing iron.

NO_x : Oxides of nitrogen, primarily nitrogen oxide and nitrogen dioxide. These are produced primarily by combustion of fossil fuels, and can constitute an air pollution problem.

Offgas treatment: An array of technologies to discharge, collect (filter), or destroy (catalyze, react, or combust) the vapors removed from soils or other media.

Order of magnitude: A multiple of ten. When a measurement is made with a result such as 3×10^7, the exponent of 10 is the order of magnitude of that measurement. To say that this result is known to within an order of magnitude is to say that the true value lies between 3×10^6 and 3×10^8.

Organic compounds: Of or designating carbon compounds. (Some simple compounds of carbon, such as carbon dioxide, are frequently classified as inorganic compounds.)

Oxide: A compound consisting of an element combined with oxygen.

Ozone: A molecule of oxygen in which three oxygen atoms are chemically attached to each other.

Package: In the regulations governing the transportation of radioactive materials, the packaging together with its radioactive contents as presented for transport.

Packaging: A shipping container without its contents.

Particulate matter: Materials such as dust, dirt, soot, smoke, and liquid droplets that are emitted into the air by sources such as factories, power plants, automobiles, construction activity, fires, and naturally by wind.

Peak ground acceleration: The maximum acceleration experienced by the particle on the ground during the course of the earthquake motion.

Permeability: The capability of a soil or rock to transmit a fluid.

Perennial: A drainage feature that contains water year-round during a year of normal rainfall. A perennial stream exhibits the typical biological, hydrological, and physical characteristics commonly associated with the continuous conveyance of water.

Personnel monitoring: The use of portable survey meters to determine the amount of radioactive contamination on individuals; or, the use of dosimetry to determine an individual's occupational radiation dose.

Person-rem: A measure of the radiation dose to a given population; the sum of the individual radiation doses received by that population.

pH: A measure of the hydrogen ion concentration in aqueous solution. Pure water has a pH of 7, acidic solutions have a pH less than 7, and alkaline solutions have a pH greater than 7.

Photosynthesis: The process in green plants and certain other organisms by which carbohydrates are synthesized from carbon dioxide and water using light as an energy source.

Physiographic: Geographic regions based on geologic setting.

Playa lake: A temporary lake, or its dry often salty bed, in a desert basin.

Plume: The elongated pattern of contaminated air or water originating at a point source, such as a smokestack or a hazardous waste disposal site.

PM_{10}: Particulate matter with a 10-micron (micrometer, μm) or less aerodynamic diameter. PM_{10} includes $PM_{2.5}$.

$PM_{2.5}$: Particulate matter with aerodynamic diameter of 2.5 micron or less. Since it is very small, $PM_{2.5}$ is important because it can be inhaled deep into the lungs.

Point source: A source of effluents that is readily identifiable and can be treated as if it were a point. This includes stacks, pipes, conduits, and tanks. A point source can be either a continuous source or a source that emits effluents only intermittently.

Pollutant: Any material entering the environment that has undesired effects.

Pollution: The addition of an undesirable agent to the environment in excess of the rate at which natural processes can degrade, assimilate, or disperse it.

Population dose: The sum of the radiation doses received by the individual members of a population.

Porosity: Percentage of void space in a material.

Potable water: Water that is safe for human consumption.

Potash: A potassium compound often used in agriculture and industry.

Prehistoric: Predating written history, in North America, also predating contact with Europeans.

Production well: A well used to retrieve water, petroleum, or gas from underground.

Purge gas: Inert gases used in chemical processes to flush a system of other gases.

Quaternary: Noting or pertaining to the present period of Earth's history, forming the latter part of the Cenozoic era, originating about 2 million years ago and including the Recent and Pleistocene epochs.

Radiation: Ionizing radiation; e.g., alpha particles, beta particles, gamma rays, X-rays, neutrons, protons, and other particles capable of producing ion pairs in matter. As used in this document, radiation does not include nonionizing radiation.

Radiation standards: Exposure standards, permissible concentrations, rules for safe handling, regulations for transportation, regulations for industrial control of radiation, and control of radioactive material by legislative means.

Radioactive waste: Materials from nuclear operations that are radioactive or are contaminated with radioactive materials and for which there is no practical use or for which recovery is impractical.

Radioactivity: The property or characteristic of radioactive material to undergo spontaneous transformations ("disintegrations" or "decay") with the emission of energy in the form of radiation. It means the rate of spontaneous transformations of a radionuclide. The unit of radioactivity is the curie (or becquerel). (1 curie = 3.7×10^{10} becquerel).

Radionuclide: A nuclide that emits radiation by spontaneous transformation.

Radon: A colorless, radioactive, inert gaseous element formed by the radioactive decay of radium.

Reactant: A substance participating in a chemical reaction, especially a directly reacting substance present at the initiation of the reaction.

Recharge: The downward vertical flow of groundwater to an aquifer. Recharge may be from seepage through the unsaturated zone (for unconfined aquifers) or downward flow from overlying layers (for confined aquifers).

Region of influence (ROI): The physical area that bounds the environmental, sociological, economic, or cultural features of interest for the purpose of impact analysis. A site-specific geographic area that includes the counties where approximately 90 percent of the site's current employees reside.

Rem: A common (or special) unit of dose equivalent, effective dose equivalent, or committed dose equivalent.

Resource Conservation and Recovery Act (RCRA): This Act was designed to provide "cradle to grave" control of hazardous chemical wastes.

Restricted area: Any area to which access is controlled for the protection of individuals from exposure to radiation and radioactive materials.

Riparian: Associated with stream banks or margins.

Risk: The likelihood of suffering a detrimental effect as a result of exposure to a hazard. In accident analysis, the probability weighted consequence of an accident, defined as the accident frequency per year multiplied by the consequence.

Risk assessment (chemical or radiological): The qualitative and quantitative evaluation performed in an effort to define the risk posed to human health and/or the environment by the presence or potential presence and/or use of specific chemical or radiological materials.

Rotary calciner: An industrial processing kiln or oven and a drum using indirect heating and mixing.

Runoff: The portion of rainfall that is not absorbed by soil, evaporated, or transpired by plants, but finds its way into streams directly or as overland surface flows.

Sanitary/industrial waste: Nonhazardous, nonradioactive liquid and solid waste generated by normal housekeeping activities.

Scrubber: An apparatus for purifying a gas.

Sediment: Eroded soil particles that are deposited downhill or downstream by surface runoff.

Seismic: Pertaining to any earth vibration, especially an earthquake.

Seismicity: All of the earthquakes that may occur in a region, regardless of magnitude.

Semi-conductor: Any of various solid crystalline substances having electrical conductivity greater than insulators but less than good conductors.

Shielding: Any material or obstruction that absorbs radiation and thus tends to protect personnel or materials from the effects of ionizing radiation.

Sievert (Sv): A unit of radiation dose used to express a quantity called equivalent dose. This relates the absorbed dose in human tissue to the effective biological damage of the radiation by taking into account the kind of radiation received, the total amount absorbed by the body, and the tissues involved. Not all radiation has the same biological effect, even for the same amount of absorbed dose. One sievert is equivalent to 100 rem.

Silicon: A nonmetallic element occurring extensively in the earth's crust in silica and silicates.

Silt: A sedimentary material consisting of fine mineral particles intermediate in size between sand and clay.

Sink: A natural or artificial means of absorbing or removing a substance or a form of energy from a system.

Slurry pump: A machine composed of an impeller, casing, shaft/bearing assembly, shaft sea and sleeve, and drive; to increase the pressure of a liquid and solids mixture (slurry) through rotational/centrifugal force and convert electrical energy into kinetic energy; which drives the mixture from one location to another.

Soil association unit: A landscape or soil grouping that has a distinctive proportional pattern of soils; it normally consists of one or more major soils and at least one minor soil, and is named for the major soil(s).

Solidification: To make solid, compact, or hard.

Source material: Uranium or thorium ores containing 0.05 percent uranium or thorium regulated under the Atomic Energy Act. In general, this includes all materials containing radioactive isotopes in concentrations greater than natural and the by-product (tailings) from the formation of these concentrated materials

Source term: The kinds and amounts of radionuclides in an assumed release of radioactive material.

State Historic Preservation Officer (SHPO): The State officer charged with the identification and protection of prehistoric and historic resources in accordance with the National Historic Preservation Act.

Stormwater: The flow of water that results from precipitation and that occurs immediately following rainfall or as a result of snowmelt.

Subcritical: Incapable of sustaining a nuclear fission chain reaction.

Succulents: Having thick, fleshy, water-absorbing leaves or stems.

Sumps: A hole at the lowest point of a building or facility into which water is drained in order to be pumped out.

Surface water: A creek, stream, river, pond, lake, bay, sea, or other waterway that is directly exposed to the atmosphere.

Surge tank: A tank used to absorb surges in flow.

Tails: In the uranium enrichment process, tails refers to uranium hexafluoride with a reduced concentration of the uranium-235 isotope.

Tectonic activity: Movement of the earth's crust, produced by internal forces, such as uplift, subsidence, folding, faulting, and seismic activity.

Teragram: 10^{12} grams or a million metric tons ("tera" represents a factor of 10^{12}).

Terrestrial: Living or growing on land; not aquatic.

Tertiary: The first period of the Cenozoic era (after the Cretaceous period of the Mesozoic era and before the Quaternary period), thought to have covered the span of time between 65 million years and 3 to 2 million years ago. The Tertiary period is divided into five epochs: the Paleocene, Eocene, Oligocene, Miocene, and Pliocene.

Threatened Species: Any species likely to become an endangered species within the foreseeable future throughout all or a significant portion of its range. Requirements for declaring a species threatened are contained in the Endangered Species Act.

Title V: Title V of the 1990 Clean Air Act Amendments requires all major sources and some minor sources of air pollution to obtain an operating permit. A title V permit grants a source permission to operate. The permit includes all air pollution requirements that apply to the source, including emission limits and monitoring, record keeping, and reporting requirements. It also requires that the source report its compliance status with respect to permit conditions to the permitting authority.

Topography: The shape of Earth's surface or the geometry of landforms in a geographic area.

Top soil: The fertile, surface portion of a soil; usually dark colored and rich in organic material.

Total effective dose equivalent (TEDE): The sum of the effective dose equivalent from radiation sources external to the body during the year plus the committed effective dose equivalent from radionuclides taken into the body. A 50-year time interval is assumed for determining committed dose.

Toxic Substances Control Act (TSCA): A Federal law authorizing the U.S. Environmental Protection Agency to secure information on all new and existing chemical substances and to control any of these substances determined to cause unreasonable risk to public health or the environment. This law requires that the health and environmental effects of all new chemicals be reviewed by the EPA before such chemicals are manufactured for commercial purposes.

Transient species: Traveling nonresident, individuals of distinct animal species; migrating between seasonal breeding habitat, and overwintering or feeding habitat.

Transuranic waste: Waste containing more than 100 nanocuries of alpha-emitting transuranic (atomic number greater than 92) isotopes per gram of waste with half-lives greater than 20 years.

Unconfined aquifer: An aquifer that is not confined by a less-permeable confining unit. An aquifer where the water table elevation represents the hydraulic potential.

Unincorporated area: An area that is not located within the jurisdiction of any local government. Such unincorporated areas are governed and taxed by county-level government.

Uranium: A radioactive element with the atomic number 92 and, as found in natural ores, an atomic weight of approximately 238. The two principal natural isotopes are uranium-235 (0.7 percent of natural uranium), and uranium-238 (99.3 percent of natural uranium). Natural uranium also includes a minute amount of uranium-234.

Viewscape: Those features which provide a range of sight that can be identified as providing a community asset such as, but not limited to, pleasing vistas, scenes and views that provide a sense of place and character.

Viewshed: The area on the ground that is visible from a specific location.

Venturi scrubber: A "wet" scrubber, using gas atomizing spray ejection technology to control fine (under 10 micrometers diameter) particulate matter.

Volatile organic compound: Any compound containing carbon and hydrogen in combination with any other element that has a vapor pressure of 77.6 millimeters of mercury (1.5 pounds per square inch) absolute or greater under actual storage conditions.

Waste management: The planning, coordination, and direction of functions related to generation, handling, treatment, storage, transportation, and disposal of waste. It also includes associated pollution prevention and surveillance and maintenance activities.

Water deluge system: A sprinkler system employing open sprinklers that are attached to a piping system that is connected to a water supply through a valve that is opened by the operation of a detection system installed in the same areas as the sprinklers; when this valve opens, water flows into the piping system and discharges from all sprinklers attached thereto; deluge systems are used where large quantities of water are needed quickly to control a fast-developing fire; deluge valves can be electrically, pneumatically or hydraulically operated.

Water resources: This term includes both freshwater and marine systems, wetlands, floodplains, and ground water.

Wetlands: Land or areas exhibiting the following characteristics: hydric soil conditions; saturated or inundated soil during some part of the year and plant species tolerant of such conditions; also, areas that are inundated or saturated by surface or groundwater at a frequency and duration sufficient to support, under normal circumstances, a prevalence of vegetation typically adapted for life in saturated soil conditions. Wetlands generally include swamps, marshes, bogs, and similar areas.

Wildlife corridor: An area of habitat connecting wildlife populations otherwise separated by human activities.

Wind rose: A plot of wind direction and speed showing the distribution of directions that the wind blows from at a measurement site. The proportion of the time that a wind blows from any given direction is indicated by the length of the "petal" on the wind rose.

Wind speed: The speed of air movement measured for a set height above ground level (agl) at a meteorological observing site. This height may vary depending on the location. Typically, anemometers at National Weather Service stations are placed at 32 ft 10 inches (10 m) agl; however, some are still found at 20 ft (6 m) agl.

APPENDIX A

SCOPING SUMMARY

ENVIRONMENTAL IMPACT STATEMENT SCOPING PROCESS

SCOPING SUMMARY REPORT

PROPOSED INTERNATIONAL ISOTOPES FLUORINE PRODUCTS, INC. (IIFP) FLUORINE EXTRACTION PROCESS AND DEPLETED URANIUM DE-CONVERSION PLANT TO BE LOCATED IN LEA COUNTY, NEW MEXICO

A.1 INTRODUCTION

On December 30, 2009, International Isotopes Fluorine Products, Inc. (IIFP) submitted an application to the U.S. Nuclear Regulatory Commission (NRC) for a license to construct and operate a proposed Fluorine Extraction Process (FEP) and Depleted Uranium De-conversion Plant (FEP/DUP) to be located at a site 22.5 kilometers (km) (14 miles [mi]) west of the City of Hobbs in Lea County, New Mexico. An Environmental Report was also submitted by IIFP at that time. If licensed, the FEP/DUP facility would be used for the deconversion of commercially-generated depleted uranium hexafluoride (DUF_6) inventories into depleted uranium oxide and other deconversion products.

In accordance with NRC regulations in Title 10 of the Code of Federal Regulations (10 CFR) Part 51 (10 CFR 51), which implement the National Environmental Policy Act of 1969, as amended (NEPA), the NRC staff is preparing an Environmental Impact Statement (EIS) for the proposed FEP/DUP facility as part of its decision-making process. The EIS will examine the potential environmental impacts associated with the proposed facility. The NRC staff has not identified any cooperating agencies for the preparation of this EIS. In addition to the EIS, the NRC staff will prepare a Safety Evaluation Report (SER) which will document the staff's review of safety and security issues associated with the proposed facility.

On July 15, 2010, NRC published in the *Federal Register* (FR) a Notice of Intent to prepare an EIS and to conduct the public scoping process (75 FR 41242). The public scoping comment period ended on August 30, 2010. Scoping is an early part of the NEPA process designed to help determine the range of actions, alternatives, and potential impacts to be considered in the EIS, and to identify significant issues related to the proposed action. In addition to the public scoping process, the NRC staff solicits input from State, local and other Federal agencies, and potentially affected Native American Tribes in order to focus on issues of genuine concern.

On July 29, 2010, the NRC staff held a public scoping meeting in Hobbs, New Mexico, to receive oral and written comments from interested parties. The public scoping meeting began with NRC staff providing a description of the NRC's roles, responsibilities, and mission. A brief overview of the licensing process was followed by a description of the environmental review process and a discussion of how the public can participate. The majority of the meeting was reserved for the public to ask questions and make comments on the scope of the environmental review.

As part of the environmental review process, the NRC staff has requested information regarding the scope of its environmental review from several sources. The NRC staff initiated consultation with the New Mexico State Historic Preservation Officer (SHPO), in accordance with the procedures in 36 CFR 800 to meet the requirements of Section 106 of the National Historic Preservation Act. In accordance with 36 CFR 800.3(f), the NRC staff has requested information from Native American Tribal members identified by the SHPO and the NRC staff. The NRC staff has also consulted with representatives of the U.S. Fish and Wildlife Service (USFWS) as required by Section 7 of the Endangered Species Act. The National Park Service was contacted and indicated that no parks would be affected by the project.

This scoping summary report addresses only comments received through the public scoping process and will be included as an Appendix of the EIS. Input from consulting agencies and potentially affected Native American Tribes will also be used as a basis for the impact assessments performed for each resource area. Correspondence with the SHPO and potentially-affected Native American Tribes are included in Appendix B of this EIS.

Correspondence with the USFWS, the National Park Service, and New Mexico Environment Department (NMED) are also included in Appendix B of this EIS.

This report has been prepared to summarize the comments received during the scoping process as required in 10 CFR 51.29(b). After publication of the draft EIS, the public will be invited to submit comments on the draft EIS. Availability of the draft EIS, the dates of the public comment period, and information about a public meeting to discuss the draft EIS will be announced in the Federal Register, on the NRC's website (http://www.nrc.gov/public-involve.html), and in the local news media. After evaluating comments on the draft EIS, the NRC staff will issue a final EIS that will serve as the basis for the NRC's consideration of potential environmental impacts in its decision on whether to license the proposed facility.

This report is organized into four main sections. Section 1 provides an introduction and background information on the environmental review process. Section 2 summarizes the comments and concerns expressed by government officials, agencies, and the public. Section 3 identifies the issues that the draft EIS will address and Section 4 describes those issues that are not within the scope of the draft EIS. Where appropriate, Section 4 also identifies other places in the decision-making process where issues that are outside the scope of the draft EIS may be considered.

A.2   ISSUES RAISED DURING THE SCOPING PROCESS

A.2.1   Overview

The public scoping process is an important component in determining the major issues that the NRC staff should address in the draft EIS. The comments provided by the public addressed several subject areas related to the IIFP proposed facility and the development of the draft EIS. Members of the public were able to submit comments on the scope of the IIFP proposed facility draft EIS by e-mail, postal mail, and by speaking and/or submitting written comments at the public scoping meeting held in Hobbs, New Mexico, on July 29, 2010. The scoping period ended on August 30, 2010.

Approximately 60 individuals not affiliated with the NRC staff attended the July 29, 2010, public scoping meeting in Hobbs, New Mexico. During the meeting, one individual asked a specific question about the licensing process. Ten individuals offered specific oral comments related to the proposed FEP/DUP facility. Including the comments received in the scoping meeting, a total of 28 oral and written comments were received from various individuals during the public scoping period, which ended on August 30, 2010. The scoping meeting transcript and the scoping comment letters received by the NRC are available on the NRC website, electronic reading room, at http://www.nrc.gov/reading-rm/adams/web-based.html. The ADAMS accession number for the scoping meeting transcript is ML102210424.

In addition to private citizens, the commenters included:

- A representative of Senator Tom Udall
- A Lea County Commissioner
- A Hobbs City Commissioner
- The Mayors of the Cities of Hobbs and Eunice
- The City Manager of Eunice
- State Senator Carroll Leavell (Letter read on his behalf)

Individuals providing oral and written comments addressed several subject areas related to the environmental review process of the proposed FEP/DUP facility. The following general topics categorize the comments received during the public scoping period:

- General support or opposition
- Socioeconomics
- Waste Management
- Water Resources
- Geology and Seismicity
- Transportation
- Public and Occupational Health
- Out of Scope

In addition to raising issues about the potential environmental impacts of the proposed facility, some commenters offered opinions and concerns that typically would not be included in an EIS. Although noted by the NRC in this summary document, comments of this type are not within the scope of environmental issues to be analyzed.

Other statements may be relevant to the proposed action, but have no direct bearing on the evaluation of alternatives or on the decision-making process regarding the proposed action. For instance, general statements of support for or opposition to the proposed action fall into this category. Comments of this type have been noted but are not used in defining the scope and content of the EIS.

A.2.2   Summary of Issues Raised

Several individuals provided comments regarding the beneficial potential socioeconomic impacts of the proposed facility on the local community. Other comments addressed potential impacts or risks posed by the facility due to seismic concerns, availability of water sources, transportation and disposal of waste, and possible health impacts associated with nuclear facilities. The following summary groups the comments received during the scoping period by technical area and issue.

A.2.2.1   General Support or Opposition

Several commenters expressed general support for the FEP/DUP facility. One commenter expressed opposition to locating the FEP/DUP facility, or any facility that deals with nuclear byproducts, in an area with a history of earthquakes and over an aquifer.

A.2.2.2   Socioeconomics

Three commenters expressed support for the project, specifically for the jobs that will be created by construction and operation of the facility and the positive economic impact it will have on the region.

A.2.2.3   Waste Management

Two commenters supported the project as a way to use uranium 'tails' that will be generated at the nearby URENCO USA uranium enrichment plant. One commenter stated that a disposal path for waste from the FEP/DUP facility to the Andrews County, Texas, nuclear waste disposal facility is an unsafe disposal path. This commenter also requested that the EIS include disposal site suitability requirements, as described in 10 CFR 61.50.

A.2.2.4 Water Resources

One commenter stated that the EIS should include the aquifer map that has been prepared by Mesa Water Company. The same commenter also stated that Lea County lacks an adequate water supply for a nuclear project. This commenter expressed concern about a site that may potentially be used for disposal of waste from the FEP/DUP facility being located over the Ogallala Aquifer. The commenter also stated that the water supply of Hobbs, Eunice, and Jal risks being polluted by allowing a nuclear project in the area.

A.2.2.5 Geology and Seismicity

One commenter stated that the EIS should include the seismic hazards that have been indicated for Lea County by the U.S. Geological Survey. This commenter also stated that the Lea County site should not have been selected due to its seismic history. The commenter also expressed concerns about possible contamination of the Ogallala Aquifer by nuclear waste released during an earthquake.

A.2.2.6 Transportation

One commenter expressed concerns about the transportation of waste from the facility in Lea County (New Mexico) to the Andrews County, Texas, nuclear waste disposal facility just across the state line.

A.2.2.7 Public and Occupational Health

One commenter submitted a New Mexico Department of Health report showing elevated cancer rates in Lea County compared to other parts of the state and stated concern that allowing nuclear industry in the area will raise cancer rates.

A.2.2.8 Out of Scope

One commenter stated that the New Mexico Environment Department's denial of his request to set up offsite radiation monitors should be included in the EIS. One commenter stated that employees of various federal agencies should waive their liability immunity through the Federal Tort Claims Act and be fully liable for any damages, pollution to the water table, and loss of livelihood and health of Lea County citizens caused by any future earthquakes.

A.3 SCOPE OF THE ENVIRONMENTAL IMPACT STATEMENT

The NEPA (42 U.S.C. 4321, et seq., as amended), and the NRC's implementing regulations for NEPA (10 CFR 51), specify in general terms what should be included in an EIS prepared by the NRC staff. Regulations established by the Council on Environmental Quality (40 CFR 1500-1508), while not binding on the NRC, provide useful guidance. Additional guidance for meeting NEPA requirements associated with licensing actions can be found in NUREG-1748, "Environmental Review Guidance for Licensing Actions Associated with NMSS Programs."

Pursuant to 10 CFR 51.71(a), in addition to public comments received during the scoping process, the EIS will also consider matters discussed in the IIFP Environmental Report. In accordance with 10 CFR 51.71(b), the EIS will consider major points of view and objections concerning the environmental impacts of the proposed action raised by other Federal, State, and local agencies, by any affected Indian Tribes/Pueblos, and by other interested persons. Pursuant to 10 CFR 51.71(c), the EIS will list all Federal permits, licenses, approvals, and other entitlements that must be obtained in implementing the proposed action, and will describe the status of compliance with these requirements. Any uncertainty as to the applicability of these requirements will be addressed in the EIS.

In accordance with 10 CFR 51.71(d), the draft EIS will include a preliminary analysis that considers and weighs the environmental effects of the proposed action, the environmental impacts of the alternatives to the proposed action, and alternatives available for reducing or avoiding adverse environmental effects. In the analysis, due consideration will be given to compliance with environmental quality standards and regulations that have been imposed by Federal, State, regional, and local agencies having responsibilities for environmental protections. The environmental impact of the proposed action will be evaluated in the EIS with respect to matters covered by such standards and requirements, regardless of whether a certification or license from the appropriate authority has been obtained. Compliance with applicable environmental quality standards and requirements does not negate the requirement for the NRC to weigh all environmental effects of the proposed action, including the degradation, if any, of water quality, and to consider alternatives to the proposed action that are available for reducing adverse effects.

While satisfaction of the NRC standards and criteria pertaining to radiological effects is necessary to meet the licensing requirements of the Atomic Energy Act, the EIS will also, for the purposes of NEPA, consider the radiological and nonradiological effects of the proposed action and alternatives. The development of the EIS is closely coordinated with the SER prepared by the NRC staff to evaluate the potential health and safety impacts of the proposed action. The EIS will also contain a discussion of the potential cumulative impacts of the proposed action.

Pursuant to 10 CFR 51.71(f), the draft EIS will include a preliminary recommendation by the NRC staff with respect to the proposed action. Any such recommendation will be reached after considering the environmental effects of the proposed action and reasonable alternatives, and after weighing the costs and benefits of the proposed action.

One goal in writing the EIS is to present the impact analyses in a manner that makes it easy for the public to understand. This EIS will provide the basis for the NRC decision with regard to potential environmental impacts. Those resources with potential significant impacts will be discussed in greater detail in the EIS than resources with potential minor or no impacts. This should allow readers of the EIS to focus on issues that were determined to be important in reaching the conclusions supported by the EIS. The following topical areas and issues will be addressed in the EIS.

Alternatives. The EIS will describe and assess the no-action alternative and other reasonable alternatives to the proposed action. Other alternatives may include alternative sites or alternative processes to the proposed chemical process.

Need for the Facility. The EIS will provide a discussion of the need for the proposed FEP/DUF facility.

Compliance with Applicable Regulations. The EIS will list relevant permits and regulations that apply to the proposed FEP/DUP facility. These include air, water, and solid waste disposal permits.

Land Use. The EIS will discuss the potential land use impacts associated with the proposed site preparation, construction, and operating activities. As appropriate, the assessment will include an analysis of mitigation measures to address potential adverse impacts.

Transportation. The EIS will discuss the potential impacts associated with the transportation of the construction materials, feed material, product, and waste during both normal transportation and under credible accident scenarios. The potential impacts on local transportation routes due to workers, delivery vehicles, and waste removal vehicles will be evaluated. As appropriate, the assessment will include an analysis of mitigation measures to address potential adverse impacts.

Geology and Soils. The EIS will assess the potential impacts to the geology and soils of the proposed FEP/DUP facility. The potential for earthquakes or any other major ground motion considerations will be addressed in the SER and potential environmental impacts of those phenomena will be evaluated in the EIS. As appropriate, the assessment will include an analysis of mitigation measures to address potential adverse impacts.

Water Resources. The EIS will assess the potential impacts on surface water and groundwater quality and water use due to the proposed action. As appropriate, the assessment will include an analysis of mitigation measures to address potential adverse impacts.

Ecological Resources. The EIS will assess the potential environmental impacts on ecological resources, including plant and animal species. Threatened and endangered species and critical habitats that may occur in the area will be discussed. The outcomes of consultations with resource protection agencies, as required by Section 7 of the *Endangered Species Act* of 1973 (16 U.S.C. Section 1536(a)(2)), will be discussed. As appropriate, the assessment will include an analysis of mitigation measures to address potential adverse impacts.

Air Quality. The EIS will make determinations concerning the meteorological conditions of the site location, the ambient air quality, the contributions of other sources to air quality, and the potential impacts of site preparation, construction, and operation of the proposed FEP/DUP facility on local air quality. In addition, the EIS will consider the impact of the proposed facility on climate change. As appropriate, the assessment will include an analysis of mitigation measures to address potential adverse impacts.

Noise. The EIS will discuss the potential impacts associated with noise from site preparation, construction, operation, and decommissioning of the proposed FEP/DUP facility. As appropriate, the assessment will include an analysis of mitigation measures to address potential adverse impacts.

Historic and Cultural Resources. The EIS will address the potential impacts of the proposed FEP/DUP facility on the historic and archaeological resources of the area. The outcomes of consultations with historic and cultural resource protection agencies, consistent with Section 106 of the *National Historic Preservation Act* of 1966 (36 CFR 800) will be discussed. As appropriate, the assessment will include an analysis of mitigation measures to address potential adverse impacts.

Visual and Scenic Resources. Potential impacts to the overall visual and scenic character of the area will be addressed. As appropriate, the assessment will include an analysis of mitigation measures to address potential adverse impacts.

Socioeconomics. The EIS will address demography, economic base, labor pool, housing, utilities, public services, education, and recreation potentially affected by the proposed action and alternatives. The hiring of new workers from outside the area could lead to potential impacts on regional housing, public infrastructure, and economic resources. Potential population changes leading to changes in the housing market and demands on the public infrastructure will be assessed. As appropriate, the assessment will include an analysis of mitigation measures to address potential adverse impacts.

Costs and Benefits. The EIS will compile in one place the costs and benefits of the proposed project so that a determination can be made of any net positive benefit to Lea County, the region, and the Nation. The EIS will compare the potential environmental and monetary costs and benefits of constructing and operating the proposed FEP/DUP facility.

Resource Commitments. The EIS will identify the potential for any unavoidable adverse impacts and irreversible and irretrievable commitments of resources. It will also address the relationship between local, short-term uses of the environment and the maintenance and enhancement of long-term productivity. Associated mitigative measures and environmental monitoring requirements will be presented, as applicable.

Public and Occupational Health. The EIS will include a determination of potentially adverse effects on human health that result from chronic and acute exposures to ionizing radiation and hazardous chemicals, and from physical safety hazards. Potentially adverse effects on human health might occur during site preparation, construction, operation, or decommissioning. Potential impacts associated with the implementation of the proposed action will be assessed under normal operation and credible accident scenarios. As appropriate, the assessment will include an analysis of mitigation measures to address potential adverse impacts.

Waste Management. The EIS will discuss the management of wastes, including by-product materials, generated from the site preparation, construction, and operation of the proposed FEP/DUP facility to assess the potential impacts of generation, storage, and disposal.

Decommissioning. The EIS will provide a discussion of facility decommissioning and associated potential impacts.

Cumulative Impacts. The EIS will address the potential cumulative impacts from past, present, and reasonably foreseeable future activities at and near the site, including preconstruction activities and a proposed facility expansion.

Environmental Justice. The EIS will address any potential disproportionately high and adverse environmental impacts of the proposed FEP/DUP facility on low-income and minority populations.

A.4   ISSUES CONSIDERED TO BE OUTSIDE THE SCOPE OF THE ENVIRONMENTAL IMPACT STATEMENT

The purpose of an EIS is to assess the potential environmental impacts of a proposed action in order to assist in an agency's decision-making process – in this case, NRC's licensing process.

As noted in Section 2.1, some issues and concerns raised during the scoping process are not relevant to the EIS because they are not directly related to the assessment of potential environmental impacts or the decision-making process. The lack of in-depth discussion in the EIS, however, does not mean that an issue or concern lacks value. Issues beyond the scope of the EIS either may not yet be at the point where they can be resolved or are more appropriately discussed and decided in other venues.

Some of the issues raised during the public scoping process for the proposed facility are outside the scope of the EIS, but are analyzed in the SER. For example, health and safety issues are considered in detail in the SER prepared by the NRC staff for the proposed action and are summarized in the EIS. The EIS and the SER are related in that they may cover some of the same topics and may contain similar information, but the analysis in the EIS is focused on the assessment of potential environmental impacts. In contrast, the SER deals primarily with safety evaluations and procedural requirements or license conditions to ensure the health and safety of workers and the general public. The SER also covers other aspects of the proposed action such as demonstrating that the applicant will provide adequate funding for the proposed facility in compliance with the NRC's financial assurance regulations.

Some of the issues raised during the public scoping process are not addressed in the EIS as they are not appropriate for resolution in the EIS. Other issues, including support of or opposition to nuclear facilities and the liability of federal workers under the Federal Tort Claims Act, are also beyond the scope of the EIS. The mission of the NRC is to license and regulate the Nation's civilian use of byproduct, source, and special nuclear materials in order to protect public health and safety, promote the common defense and security, and protect the environment. The NRC's regulations are designed to protect both the public and workers against radiation hazards from industries that use radioactive materials. The NRC's scope of responsibility includes regulation of commercial nuclear power plants; research, test, and training reactors; nuclear fuel cycle facilities; medical, academic, and industrial uses of radioactive materials; and the transport, storage, and disposal of radioactive materials and wastes. Activities not within the jurisdiction of the NRC are not subject to NRC regulations nor appropriate for consideration in the NRC's decision making process.

APPENDIX B

CONSULTATION/CORRESPONDENCE

June 29, 2010

The Honorable Louis Maynahonah Sr.
Chairman
Apache Tribe of Oklahoma
P.O. Box 1220
Anadarko, OK 73005

SUBJECT: INITIATION OF THE NATIONAL HISTORIC PRESERVATION ACT SECTION
           106 PROCESS FOR INTERNATIONAL ISOTOPES FLUORINE PRODUCTS,
           INC. PROPOSED FLUORINE EXTRACTION PROCESS & DEPLETED
           URANIUM DE-CONVERSION PLANT

Dear Chairman Maynahonah:

International Isotopes Fluorine Products, Inc. (IIFP), a wholly owned subsidiary of International
Isotopes, Inc. (INIS), has submitted a license application to the U.S. Nuclear Regulatory
Commission (NRC) to construct, operate, and decommission a proposed uranium processing
facility. The facility is proposed to be located within a 640-acre section near Hobbs, New Mexico
in Lea County (see enclosed map), of which approximately 40 acres would be developed. The
40-acre site would be fenced in and contain process-related buildings and an administrative
office building. The proposed facility would provide services to the uranium enrichment industry
for de-conversion of depleted uranium hexafluoride (DUF_6) into uranium oxides for long-term
stable disposal. The proposed facility would also produce high-purity inorganic fluorides for
applications in the electronic, solar panel, and semiconductor markets and anhydrous
hydrofluoric acid for various industrial applications.

As established in Title 10 Code of Federal Regulations Part 51 (10 CFR Part 51), the NRC
regulation that implements the National Environmental Policy Act of 1969, as amended, the NRC
is preparing an Environmental Impact Statement (EIS) for the proposed action. The NRC
process includes an opportunity for public and intergovernmental participation in the
environmental review. We want to ensure that you are aware of our efforts and pursuant to
10 CFR 51.28(b), the NRC invites you to provide input to the scoping process for this EIS. In
addition, as outlined in 36 CFR 800.8(c), the NRC plans to coordinate compliance with
Section 106 of the National Historic Preservation Act of 1966 through the requirements of the
National Environmental Policy Act of 1969. In accordance with Section 106 of the National
Historic Preservation Act, the EIS will include an analysis of potential impacts to historic and
cultural properties. To support the environmental review, the NRC is requesting information to
facilitate the identification of tribal historic sites or cultural resources that may be affected by the
proposed facility. Any input you provide will be used to enhance the scope and quality of our
review in accordance with 10 CFR 51 and 36 CFR 800. After assessing the information you
provide, the NRC will determine what additional actions are necessary to comply with Section
106 of the National Historic Preservation Act.

We would also like to invite you to attend a public meeting that we will be holding on Thursday,
July 29, 2010, at the Lea County Event Center, 5101 Lovington Highway in Hobbs, New Mexico,
from 5:30 p.m. until 8:30 p.m. The purpose of this meeting is to solicit comments from
stakeholders and members of the public on the scope of the EIS review.

L. Maynahonah   2

The IIFP license application is publicly available in the NRC Public Document Room (PDR) located at One White Flint North, 11555 Rockville Pike, Rockville, Maryland, 20852, or from the NRC's Agencywide Documents Access and Management System (ADAMS). The ADAMS Public Electronic Reading Room is accessible at http://www.nrc.gov/reading-rm/adams.html. The accession number for the license application is ML100630503. Persons who do not have access to ADAMS or encounter problems, should contact the NRC's PDR reference staff by telephone at 1-800-397-4209, or 301-415-3747, or by e-mail at pdr@nrc.gov.

Please submit any comments you may have to offer on the environmental review within 30 days of receipt of this letter. If you have any questions, please contact Asimios Malliakos of my staff by telephone at 301-415-6458 or by email at Asimios.Malliakos@nrc.gov. Thank you for your assistance.

Sincerely,

/RA/

Diana Diaz-Toro, Branch Chief
Environmental Review Branch A
Environmental Protection and Performance
   Assessment Directorate
Division of Waste Management
   and Environmental Protection
Office of Federal and State Materials
   and Environmental Management Programs

Docket No.: 40-9086

Enclosure:
Figure 1, Proposed IIFP Site
   Location

Identical letters, as the one sent to Apache Tribe of Oklahoma, presented in pages B-3 and B-4, were also sent on June 29, 2010, to the following addresses:

Mr. Michael Burgess
Chairman
Comanche Indian Tribe
P.O. Box 908
Lawton, OK 73502

Mr. Donald G. Tofpi
Tribal Chairman
Kiowa Tribe of Oklahoma
P.O. Box 369
Carnegie, OK 73015

Ms. Holly B. E. Houghten
Tribal Historic Preservation Officer
Mescalero Apache Tribe
P.O. Box 227
Mescalero, NM 88340

Mr. Frank Paiz
Governor
Ysleta del Sur Pueblo
119 South Old Pueblo Road
El Paso, TX 79907

Mr. Samuel Cata
Tribal Liaison
New Mexico Historic Preservation Division
Bataan Memorial Building
407 Galisteo St., Suite 236
Santa Fe, NM 87501

Ms. Jodie Hayes
Tribal Administrator
Shawnee Tribe
29 South Highway, 69A
Miami, OK 74354

July 2, 2010

Ms. Jan V. Biella
Interim New Mexico State Historic
  Preservation Officer
Historic Preservation Division
Bataan Memorial Building
407 Galisteo St., Suite 236
Santa Fe, New Mexico 87501

SUBJECT: INITIATION OF THE NATIONAL HISTORIC PRESERVATION ACT SECTION
           106 PROCESS FOR INTERNATIONAL ISOTOPES FLUORINE PRODUCTS,
           INC. PROPOSED FLUORINE EXTRACTION PROCESS & DEPLETED
           URANIUM DE-CONVERSION PLANT

Dear Ms. Biella:

International Isotopes Fluorine Products, Inc. (IIFP), a wholly owned subsidiary of International
Isotopes, Inc. (INIS), has submitted a license application to the U.S. Nuclear Regulatory
Commission (NRC) to construct, operate, and decommission a proposed uranium processing
facility. The facility is proposed to be located within a 640-acre section near Hobbs, New
Mexico in Lea County (see enclosed map), of which approximately 40 acres would be
developed. The 40-acre site would be fenced in and contain process-related buildings and an
administrative office building. The proposed facility would provide services to the uranium
enrichment industry for de-conversion of depleted uranium hexafluoride (DUF_6) into uranium
oxides for long-term stable disposal. The proposed facility would also produce high-purity
inorganic fluorides for applications in the electronic, solar panel, and semiconductor markets
and anhydrous hydrofluoric acid for various industrial applications.

As established in Title 10 Code of Federal Regulations Part 51 (10 CFR Part 51), the NRC
regulation that implements the National Environmental Policy Act of 1969, as amended, the
NRC is preparing an Environmental Impact Statement (EIS) for the proposed action. In
accordance with 36 CFR 800.8(c), the NRC plans to coordinate compliance with Section 106 of
the National Historic Preservation Act of 1966 through the requirements of the National
Environmental Policy Act of 1969. In accordance with Section 106 of the National Historic
Preservation Act, the EIS will include an analysis of potential impacts to historic and cultural
properties. To support the environmental review, the NRC is requesting information to facilitate
the identification of State historic sites or cultural resources that may be affected by the
proposed facility. Any input you provide will be used to enhance the scope and quality of our
review in accordance with 10 CFR 51 and 36 CFR 800. After assessing the information you
provide, the NRC will determine what additional actions are necessary to comply with Section
106 of the National Historic Preservation Act.

We would also like to invite you to attend a public meeting that we will be holding on Thursday,
July 29, 2010, at the Lea County Event Center, 5101 Lovington Highway in Hobbs, New Mexico,
from 5:30 p.m. until 8:30 p.m. The purpose of this meeting is to solicit comments from
stakeholders and members of the public on the scope of the EIS review.

J. Biella   2

The IIFP license application is publicly available in the NRC Public Document Room (PDR) located at One White Flint North, 11555 Rockville Pike, Rockville, Maryland, 20852, or from the NRC's Agencywide Documents Access and Management System (ADAMS). The ADAMS Public Electronic Reading Room is accessible at http://www.nrc.gov/reading-rm/adams.html . The accession number for the license application is ML100630503. Persons who do not have access to ADAMS or encounter problems, should contact the NRC's PDR reference staff by telephone at 1-800-397-4209, or 301-415-3747, or by e-mail at pdr@nrc.gov .

Please subm it any comments you may have t o offer on the environmental review within 30 days of receipt of this letter. If you have any questions, please contact Asimios Malliakos of my staff by telephone at 301-415-6458 or by email at Asimios.Malliakos@nrc.gov . Thank you for your assistance.

                                          Sincerely,

                                          /RA/

                                          Diana Diaz-Toro, Branch   Chief
                                          Environmental Review Branch A
                                          Environmental Protection and Performance
                                            Assessment Directorate
                                          Division of Waste Management
                                            and Environmental Protection
                                          Office of Federal and State Materials
                                            and Environmental Management Programs

Enclosure:
Figure 1, Proposed IIFP Site
 Location

Docket No.: 40-9086

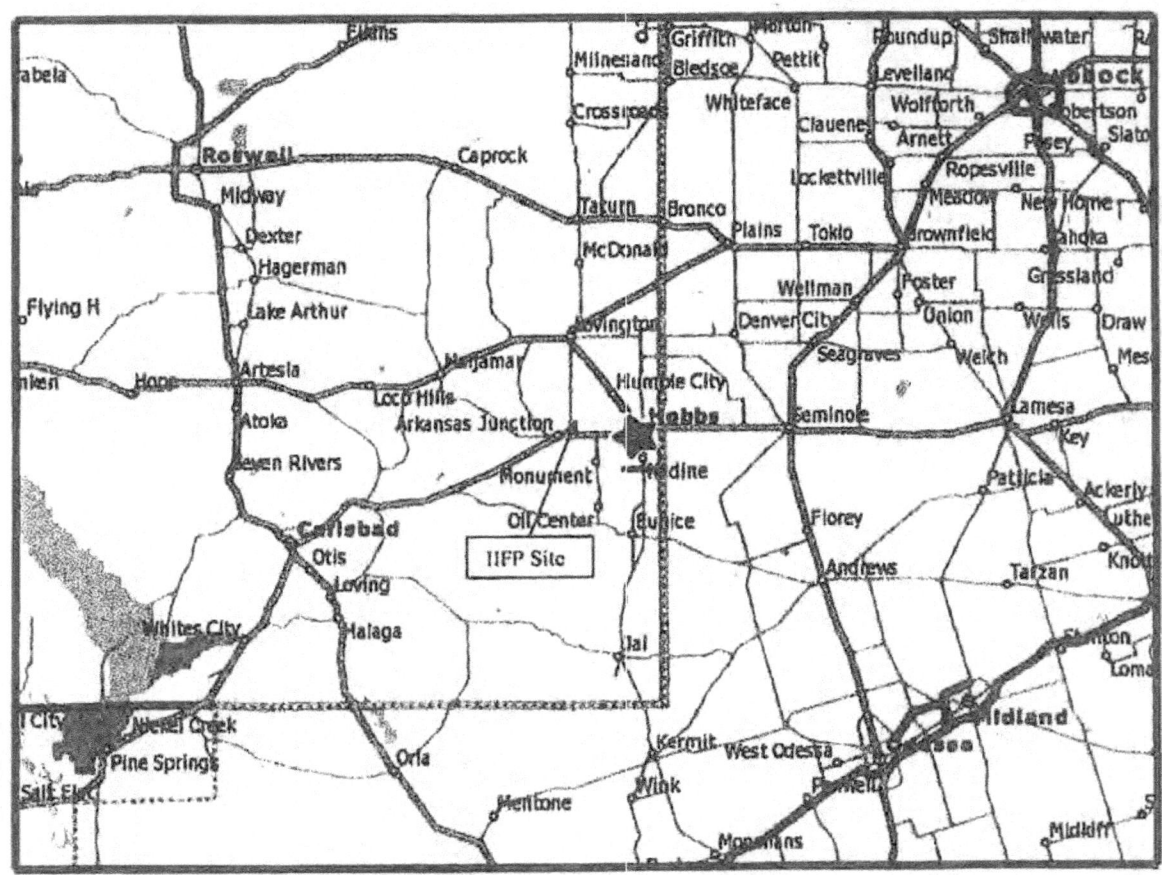

Figure 1. Proposed IIFP Site Location - The proposed site location is in Township 18S, Range 37E, Sections 26, 27, 34, and 35. The approximate center of the site is at latitude 32 degrees and 43 min North and 103 degrees and 20 min West longitude.

Ysleta del Sur Pueblo

Tribal Council

117 South Old Pueblo Road * P.O. Box 17579 * El Paso, Texas 79917 * (915) 859-8053 * Fax: (915) 859-4252

July 13, 2010

Diana Diaz-Toro
Branch Chief
Environmental Protection Office
United States Nuclear Regulatory Commission
Washington, D.C. 20555-0001

Dear Diana Diaz-Toro:

This letter is in response to the correspondence received in our office in which you provide the Ysleta del Sur Pueblo the opportunity to comment on International Isotopes Fluorine Products, Inc. (IIFP) initiation of the National Historic Preservation Act Section 106 Process, and submittal for a license application to the U.S. Nuclear Regulatory Commission (NRC) to construct, operate, and decommission a proposed uranium processing facility near Hobbs, (Lea County) New Mexico.

While we do not have any comments on the preparation of an Environmental Impact Statement (EIS) and believe that this proposed project will not adversely affect traditional, religious or culturally significant sites of our Pueblo and have no opposition to it; we would like to request consultation should any human remains or artifacts unearthed during this project be determined to fall under Native American Graves Protection and Repatriation Act (NAGPRA) guidelines. Copies of our Pueblo's Cultural Affiliation Position Paper and Consultation Policy are available upon request.

Thank you for allowing us the opportunity to comment on the proposed project.

Sincerely,

Javier Loera
War Captain/Tribal Historic and Preservation Officer
Ysleta del Sur Pueblo
E-mail: jloera@ydsp-nsn.gov

DEPARTMENT OF CULTURAL AFFAIRS
HISTORIC PRESERVATION DIVISION

BATAAN MEMORIAL BUILDING
407 GALISTEO STREET, SUITE 236
SANTA FE, NEW MEXICO 87501
PHONE (505) 827-6320   FAX (505) 827-6338

BILL RICHARDSON
Governor

July 15, 10

Diana Diaz-Toro, Branch Chief
Environmental Review Branch A
Environmental Protection and Performance Assessment Directorate
Division of Waste Management and Environmental Protection
Office of Federal and State Materials and Environmental Management Programs
U.S. Nuclear Regulatory Commission
Washington D.C. 20555-0001

Re: Proposed Flourine Extraction Process and Depleted Uranium De-Conversion
Plant

Dear Ms. Diaz-Toro:

Thank you for providing the maps, photos, and scope of work for the above referenced
project. We will need additional information in order to continue consultation on this
project under Section 106 of the National Historic Preservation Act (NHPA).

Our archaeological records show that a no cultural resource surveys to identify historic
properties have been conducted for the project area. In order to identify historic
properties within the project area, or area of potential effect (APE), as required for
compliance with Section 106, this office recommends that you engage the services of a
professional archaeologist to conduct a pedestrian archaeological survey of the property.
For federal undertakings on state lands, archaeological surveys require a contractor to
hold a state archaeological survey permit and meet the Secretary of the Interior's
standards. It is not necessary for the archaeologist to have a state permit if the property is
privately owned; however, consultants with a state permit are familiar with New Mexico
state standards for survey and reporting. A list of archaeologists and archaeological firms
with permits for state lands in New Mexico may be found at
http://www.nmhistoricpreservation.org/documents/99.DOCUMENT.pdf. The archaeologist
will write a report detailing the results of his/her work, including recommendations about
eligibility and effect of all sites in/near the project area and submit it to your office.

Compliance with Section 106 also includes consultation with Native American tribes that
may be culturally affiliated with historic properties, sacred sites and/or traditional cultural
properties (TCPs) in the project area. A list of tribes who wish to be consulted
concerning projects within Lea County is available at
http://www.nmhistoricpreservation.org/documents/21.DOCUMENT.pdf. Please contact any
other tribes that you believe would be interested in commenting on this project.

The consultation letter should provide the tribes with information about the proposed
project, funding sources, contact name and information for NRC, information on

archaeological sites in the project area, their eligibility for listing to the National Register of Historic Places or State Register of Cultural Properties, and what may happen to the archaeological sites as a result of the project. You will be able to obtain the latter information from the conclusions of the archaeological survey report. The letters should invite the tribes to comment on all the information provided and request that they provide their concerns about any TCPs that may be affected by the project.

Any information tribes report to you will be considered during our 30-day review period. Once the archaeological survey report is complete, please send us the report for review and consultation regarding any effects the proposed project may have on historic properties in the area of potential effect. The report should be accompanied by a cover letter from your office requesting a formal determination of effect for the undertaking (i.e. no effect, no adverse effect, adverse effect). Any attachments to the report that the consultant provides must also be sent to our office (i.e. NMCRIS Investigation Abstract Form (NIAF), Laboratory of Anthropology (LA) site records, etc.). At this time, you should also send a sample tribal consultation letter, along with information on who was contacted, and copies of any responses received. If you conduct follow-up telephone calls, which is encouraged, please note in your letter, or in a separate document, the results of the phone calls so that we ensure that any concerns have been addressed.

If you have any questions concerning the additional information requested, or questions on how the tribal consultation should proceed, please do not hesitate to contact me. I can be reached by telephone at (505) 827-4225 or by email at Bob.Estes@state.nm.us.

Sincerely,

Bob Estes
Bob Estes
Archaeologist

Log: 89794
cc. Asimios Malliakos

PATRICK H. LYONS
COMMISSIONER

State of New Mexico
Commissioner of Public Lands
310 OLD SANTA FE TRAIL
P.O. BOX 1148
SANTA FE, NEW MEXICO 87504-1148

COMMISSIONER'S OFFICE
Phone (505) 827-5760
Fax (505) 827-5766
www.nmstatelands.org

090631

14 October 2010

Jan Biella
Historic Preservation Division
407 Galisteo Street, Suite 236
Santa Fe, New Mexico 87501

RECEIVED
HISTORIC PRESERVATION
DIVISION

Re: Proposed Depleted Uranium Processing Facility, Active Land Sale / Exchange, International
Isotopes, Inc; Nuclear Regulatory Commission; Lone Mountain Archaeological Services Report #
1224.; New Mexico State Land Office compliance file 10DE277

Dear Ms. Biella:

I have reviewed the captioned document prepared by Lone Mountain Archaeological
Services, Inc. (LMAS) on behalf of Gordon Environmental, Inc. (GEI) under contract to
International Isotopes, Inc. (III). Enclosed, please find one copy of the report as prepared
by LMAS, together with a map I have prepared to supplement their report. Also enclosed
herewith for your reference are copies of correspondence between III and the State
Historic Preservation Officer (SHPO), and between III and tribal governments, all dating
to 2009, and my recent email communications with GEI. I submit this suite of materials
to you in support of the larger federal undertaking, but also in order to address the state
undertaking consisting of the land exchange / sale itself.

I first became aware of this project on 14 May 2009, via notification from LMAS of
impending survey in support of proposed construction of a depleted uranium deconversion and fluorine extraction processing facility on trust lands. The location
surveyed (Section 27, T18S, R36E, N.M.P.M.) is on lands whose surface and subsurface
estates are managed by the New Mexico State Land Office (SLO). You will note that
there is no mention in LMAS' survey documentation of either an intended land exchange
/ sale, the role of III, or the involvement of the Nuclear Regulatory Commission (NRC).

Until queried briefly by New Mexico Historic Preservation Division (HPD) staff on 11
August, and contacted on 07 and 08 October 2010 by GEI, I was unaware of completion
of the survey, NRC involvement, the apparently already accomplished exchange of the
land with Lea County, or the impending sale of same to III. I have not been contacted by
anyone previous to 07 / 08 October regarding the exchange / sale. Similarly, I have not
been contacted by Lea County, the NRC, or III regarding the federal undertaking. I
understand from correspondence with GEI (see email of 08 October, attached) that they
believe you have not yet received copies of the tribal consultation letters, so I have
provided the copies thereof as forwarded to me by GEI.

2

The report itself indicates that the entire area (640 acres, more or less, within Section 27) was subjected to intensive pedestrian survey using appropriate methods. The results were largely negative, identifying only three isolated occurrences. These isolated occurrences are not thought to be cultural properties worthy of further consideration and protection. The map I prepared shows the location of the parcel, the adjacent pattern of state trust and private ownerships, the areas of previous archaeological surveys, and the locations of the known archaeological sites. The gray ring surrounding the subject parcel illustrates the limits of a five-mile (8000-meter) buffer area. The current survey nearly doubles the total acres of survey that have been conducted within the overall buffer area. Note also that only four sites have been discovered and documented in that area. This area of approximately 64,000 acres has now seen an arbitrary, non-random, surveyed sample of approximately 1500 acres. It is not surprising that the current survey returned negative results, given the observed site density estimated from the findings of previous surveys.

The map also illustrates the location of all state trust lands (regardless of surface or subsurface estates) that are located within five miles of any registered cultural property. This presentation is based on a dataset derived from GIS analysis of data currently displayed by the New Mexico Cultural Resource Information System, Archaeological Records Management Section, in their on-line system. Note that the subject parcel is just outside five miles from a registered cultural property -- LA 43256 (SR #162), a site variously known as Monument Springs, Monument Springs Site, and the HAT Ranch Headquarters.

Given the situation outlined above, the SLO recommends a finding of no effect / no cultural properties / no historic properties for both undertakings. There are no documented cultural properties within the area of potential effect (APE) when considering direct effects. Similarly, there are no registered cultural properties within the assumed, five-mile APE when considering indirect effects.

As always, if any cultural materials are discovered when ground disturbance associated with construction begins, all work in the vicinity of the discovery should cease, and the SHPO should be notified. If you believe that the SLO can be of any assistance at any time, we would be happy to oblige.

If you have questions or require further information, please do not hesitate to contact me.

Sincerely,

David C. Eck
Trust Land Archaeologist
Xc: Compliance file 10DE277cd

(505) 827-5857
deck@slo.state.nm.us

Concur with recommendations as proposed.   10/25/10

nichelle Ensey

for NM State Historic Preservation Officer

NMCRIS INVESTIGATION ABSTRACT FORM (NIAF)

1. NMCRIS Activity No.: 113862	2a. Lead (Sponsoring) Agency: NM State Land Office	2b. Other Permitting Agency(ies):	3. Lead Agency Report No.:

4. Title of Report: *Cultural Resource Survey of 640 Acres for the Arkansas Junction Site, Lea County, New Mexico* Author(s) S. Daras	5. Type of Report ☒ Negative ☐ Positive

6. Investigation Type
☐ Research Design ☒ Survey/Inventory ☐ Test Excavation ☐ Excavation ☐Collections/Non-Field Study
☐ Overview/Lit Review ☐ Monitoring ☐Ethnographic study ☐ Site specific visit ☐Other

7. Description of Undertaking (what does the project entail?): The proposed project is for the construction of the International Isotopes Inc. depleted uranium de-conversion and fluorine extraction processing facility. The facility will be located within a limited footprint inside the 640 acres, with extensive buffer zones.	8. Dates of Investigation: (from: May 18, 2009 to May 25, 2009 9. Report Date: May 26, 2009

10. Performing Agency/Consultant: Lone Mountain Archaeological Services, Inc. Principal Investigator: Cathy Travis Field Supervisor: Thoras R. Dye Field Personnel Names: Richard Fransisco and Francisco Britton	11. Performing Agency/Consultant Report No.: 1224 12. Applicable Cultural Resource Permit No(s): NM State Permit: NM 09-073

13. Client/Customer (project proponent): Gordon Environmental Contact: Dacia R. Tucholke Address: 213 S. Camino del Pueblo Bernalillo, NM 87004 Phone: (432) 688-6884	14. Client/Customer Project No.:

15. Land Ownership Status (*Must* be indicated on project map):

Land Owner	Acres Surveyed	Acres in APE
State	640	640
TOTALS	640	640

16 Records Search(es):

Date(s) of ARMS File Review April 17, 2009	Name of Reviewer(s) C. Travis	
Date(s) of NR/SR File Review April 17, 2009	Name of Reviewer(s) S. Daras	
Date(s) of Other Agency File Review	Name of Reviewer(s)	Agency

17. Survey Data:

a. Source Graphics ☒ NAD 27 ☐ NAD 83
☒ USGS 7.5' (1:24,000) topo map ☐ Other topo map, Scale:
☒ GPS Unit Accuracy ☐<1.0m ☒ 1-10m ☐ 10-100m ☐>100m

b. USGS 7.5' Topographic Map Name   USGS Quad Code

Monument North, NM	32103-G8

c. County(ies): Lea

17. Survey Data (continued):

d. Nearest City or Town: Hobbs

e. Legal Description:

Township (N/S)	Range (E/W)	Section	¼	¼	¼
18 S	36 E	27	Entire section		

Projected legal description? Yes [ ] , No [X]   Unplatted [ ]

f. Other Description (e.g. well pad footages, mile markers, plats, land grant name, etc.): Barbed wire fences border the northern and western areas of project and NM State 483 extends along the western edge of the project area. The southern and eastern sides of the project area are not bounded. Two large power lines run east-to-west just outside of the southern boundary of the project area.

18. Survey Field Methods:
Intensity: ☒ 100% coverage   ☐ <100% coverage

Configuration: ☒ block survey units   ☐ linear survey units (l x w):   ☐ other survey units (specify):

Scope: ☒ non-selective (all sites recorded)   ☐ selective/thematic (selected sites recorded)

Coverage Method: ☒ systematic pedestrian coverage   ☐ other method (describe)

Survey Interval (m): 15 Crew Size: 3 Fieldwork Dates: May 18, 2009 to May 25, 2009

Survey Person Hours: 180 Recording Person Hours: 0 Total Hours: 180

Additional Narrative:

19. Environmental Setting (NRCS soil designation; vegetative community; elevation; etc.): The project area is located on a flat plain with a few shallow intermittent playas. A southeast-trending drainage is located in the far southwest quarter of the project area. The area is characterized by gently sloping terrain in the Querecho Plains, dominated primarily by the Kimbrough-Lea complex with 0 to 3 percent slopes (USDA Web Soil Survey 2009). The soil is derived from mixed alluvium and/or eolian sands. Other soil associations present are the Kimbrough gravelly loam, Portales loam, Portales-Stegall loams, and Stegall and slaughter soils.

Vegetation is characteristic of semidesert grassland (Brown 1994), and includes ringtail muhley, hairy grama, and other various forbs and grasses. Mesquite, prickly pear, horse crippler cacti, and rainbow cacti were also observed. Elevation is 3,814 ft (1,163 m) amsl in the northwest corner and 3,784 ft (1,153) amsl in the southeast corner of the project area.

a. Percent Ground Visibility: 100% in burned areas and 75-80% in grassy areas b. Condition of Survey Area (grazed, bladed, undisturbed, etc.): Numerous power lines, buried pipelines, and associated two-track roads are present throughout the project area. Approximately 45 percent of the survey area (eastern portion of the survey area) has been burned by recent grass fires. The south ½ of the southeast ¼ has been utilized as a gravel pit, crusher and hot plant site. One dry hole (abandoned well pad) is also located in the SW ¼ of the SW ¼, and it appears to have been capped in the 1980's or 1990's.

21. CULTURAL RESOURCE FINDINGS ☒ Yes, See Page 3   ☐No, Discuss Why: Three isolated occurrences were identified. A files check yielded three previous NMCRIS activities, but no previously recorded sites within 1 km of the project area. The absence of cultural resources in the project area may be explained by the presence of shallow sediments with exposed caliche (indicating a lack of lithic raw materials), and a lack of permanent water sources. This may have made the location unattractive to prehistoric peoples.

22. Required Attachments (check all appropriate boxes):
☒ USGS 7.5 Topographic Map with sites, isolates, and survey area clearly drawn
☒ Copy of NMCRIS Mapserver Map Check
☐ LA Site Forms - new sites (*with sketch map & topographic map*)
☐ LA Site Forms (update) - previously recorded & un-relocated sites (*first 2 pages minimum*)
☐ Historic Cultural Property Inventory Forms
☒ List and Description of isolates, if applicable see page 3)
☐ List and Description of Collections, if applicable

23. Other Attachments:
☐ Photographs and Log
☐ Other Attachments

(Describe):

24. I certify the information provided above is correct and accurate and meets all applicable agency standards.

Principal Investigator/Responsible Archaeologist: Cathy Travis

Signature ___Cathy Travis___ Date ___May 26, 2009___ Title (if not PI):

<table>
<tr><td>25. Reviewing Agency:
Reviewer's Name/Date

Accepted ( )   Rejected ( )

Tribal Consultation (if applicable): ☐ Yes ☐No</td><td>26. SHPO
Reviewer's Name/Date:

HPD Log #:
SHPO File Location:
Date sent to ARMS:</td></tr>
</table>

CULTURAL RESOURCE FINDINGS
[fill in appropriate section(s)]

1. NMCRIS Activity No.: 113862	2. Lead (Sponsoring) Agency: NM State Land Office	3. Lead Agency Report No.:

SURVEY RESULTS:

Sites discovered and registered: 0
Sites discovered and NOT registered: 0
Previously recorded sites revisited (*site update form required*): 0
Previously recorded sites not relocated (*site update form required*): 0
TOTAL SITES VISITED: 0
Total isolates recorded: 3   **Non-selective isolate recording?** ☒
Total structures recorded (*new and previously recorded, including acequias*): 0

MANAGEMENT SUMMARY: Three isolated occurrences were encountered during this survey. The isolated occurrences have been completely recorded in a manner consistent with current standards and do not require any additional work. Therefore, the proposed undertaking will have no effect on cultural resources.

Isolated Occurrences (UTM NAD 27, Zone 13)

IO No.	Northing	Easting	Description
IO 1	3621150	656161	A brown chert San Jose projectile fragment, distal end, reworked (35 mm x 23 mm x 7 mm)(see Figure 1)
IO 2	3621745	655564	One gray quartzite hammerstone, one end and edge battered (53 mm x 43 mm x 26 mm)
IO 3	3621263	654810	Three manganese decolorized glass body fragments, ¼ in thick

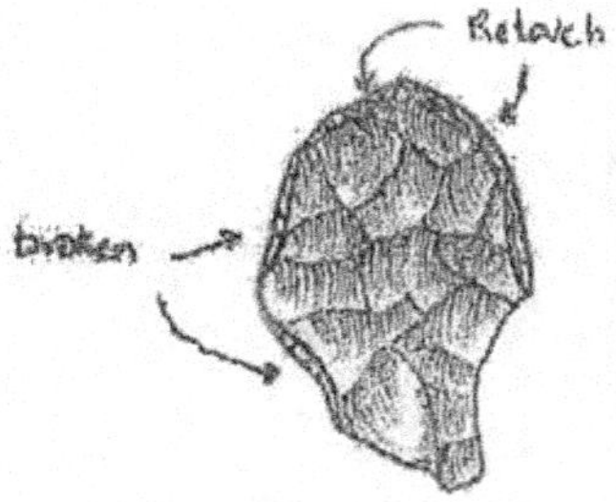

IF REPORT IS NEGATIVE YOU ARE DONE AT THIS POINT.

Figure 1: IO 1,
San Jose Projectile Point
(actual size)

SURVEY LA NUMBER LOG

Sites Discovered:

LA No.	Field/Agency No.	Eligible? (Y/N, applicable criteria)

Previously recorded revisited sites:

LA No.	Field/Agency No.	Eligible? (Y/N, applicable criteria)

From:	Malliakos, Asimios
Sent:	Wednesday, June 15, 2011 3:06 PM
To:	JimmyA@ComancheNation.com
Subject:	Historic Preservation Act Section 106 for International Isotopes Proposed De-Conversion Plant
Attachments:	Letter to the tribes.pdf; Site Location ML1011600270.pdf

Dear Mr. Jimmy Arterbery,

As we discussed in the phone attached please find the letter we sent to the tribes. Although the letter is addressed to the Honorable Louis Maynahonah Sr., in the last page of the letter shows the addresses that identical letters were sent. The list includes the name of Mr. Michael Burgess, Chairman, Comanche Indian Tribe. Attached also please find a map with the site location which is mentioned in the letter. I will appreciate any comments you may have before the end of this month, June 2011.

For your convenience the Environmental Report for the International Isotopes proposed De-Conversion plant is accessible at the web address:
http://pbadupws.nrc.gov/docs/ML1001/ML100120758.pdf

Please be aware the NRC is preparing for the proposed facility a Draft Environmental Impact Statement (DEIS) which is expected to be published on November 2011. The DEIS will include discussion on Historic and Cultural Resources. A copy of the DEIS will be send to the Comanche Indian Tribe. As you requested, I will be sending the copy of the DEIS directly to you and I will be requesting your comments.

Thank you

Asimios Malliakos
Environmental Project Manager
U. S. Nuclear Regulatory Commission
Office of Federal and State Materials and
    Environmental Management Programs
Mail Stop: T-8F5
Washington, DC 20555-0001
Telephone: 301-415-6458
Fax: 301-415-5369
Email: Asimios.Malliakost@nrc.gov

From:	Jimmy Arterberry [jimmya@comanchenation.com]
Sent:	Wednesday, June 15, 2011 3:43 PM
To:	Malliakos, Asimios
Subject:	RE: Historic Preservation Act Section 106 for International Isotopes Proposed De-Conversion Plant

Asimios,
I've had a chance to look over the document sent and have no comment at this time. I will anticipate the Draft EIS.
Thank you, jimmy

Jimmy W. Arterberry, THPO
Comanche Nation
P.O. Box 908
Lawton, Oklahoma 73502
(580) 595-9960 or 9618
(580) 595-9733 FAX

This message is intended only for the use of the individuals to which this e-mail is addressed, and may contain information that is privileged, confidential and exempt from disclosure under applicable laws. If you are not the intended recipient of this e-mail, you are hereby notified that any dissemination, distribution or copying of this communication is strictly prohibited. If you have received this e-mail in error, please notify the sender immediately and delete this e-mail from both your "mailbox" and your "trash." Thank you.

-----Original Message-----
From: Malliakos, Asimios [mailto:Asimios.Malliakos@nrc.gov]
Sent: Wed 6/15/2011 2:06 PM
To: Jimmy Arterberry
Subject: Historic Preservation Act Section 106 for International Isotopes Proposed De-Conversion Plant

Dear Mr. Jimmy Arterberry,

As we discussed in the phone attached please find the letter we sent to the tribes. Although the letter is addressed to the Honorable Louis Maynahonah Sr., in the last page of the letter shows the addresses that identical letters were sent. The list includes the name of Mr. Michael Burgess, Chairman, Comanche Indian Tribe. Attached also please find a map with the site location which is mentioned in the letter. I will appreciate any comments you may have before the end of this month, June 2011.

For your convenience the Environmental Report for the International Isotopes proposed De-Conversion plant is accessible at the web address:
http://pbadupws.nrc.gov/docs/ML1001/ML100120758.pdf

Please be aware the NRC is preparing for the proposed facility a Draft Environmental Impact Statement (DEIS) which is expected to be published on November 2011. The DEIS will include discussion on Historic and Cultural Resources. A copy of the DEIS will be send to the Comanche Indian Tribe. As you requested, I will be sending the copy of the DEIS directly to you and I will be requesting your comments.

Thank you

Asimios Malliakos
Environmental Project Manager

U. S. Nuclear Regulatory Commission
Office of Federal and State Materials and
    Environmental Management Programs
Mail Stop: T-8F5
Washington, DC 20555-0001
Telephone: 301-415-6458
Fax: 301-415-5369
Email: Asimios.Malliakost@nrc.gov<mai to:Asimios.Malliakost@nrc.gov>

From:	Malliakos, Asimios
Sent:	Wednesday, June 15, 2011 4:02 PM
To:	Samuel.Cata@state.nm.us
Subject:	Historic Preservation Act Section 106 for International Isotopes Proposed De-Conversion Plant
Attachments:	Letter to the tribes.pdf; Site Location ML1011600270.pdf; Cultural Resource Report.pdf

Dear Mr. Samuel Cata,

As we discussed in the phone attached please find the letter we sent to the tribes. Although the letter is addressed to the Honorable Louis Maynahonah Sr., in the last page of the letter shows the addresses that identical letters were sent including your name. Attached also please find a map with the site location which is mentioned in the letter. In addition attached find the cultural survey report, no findings were made but I am attaching the report for your review. I will appreciate any comments you may have before the end of this month, June 2011.

For your convenience the Environmental Report for the International Isotopes proposed De-Conversion plant is accessible at the web address:
http://pbadupws.nrc.gov/docs/ML1001/ML100120758.pdf

Please be aware the NRC is preparing for the proposed facility a Draft Environmental Impact Statement (DEIS) which is expected to be published on November 2011. The DEIS will include discussion on Historic and Cultural Resources. A copy of the DEIS will be send to you and I will be requesting your comments.

Thank you

Asimios Malliakos
Environmental Project Manager
U. S. Nuclear Regulatory Commission
Office of Federal and State Materials and
    Environmental Management Programs
Mail Stop: T-8F5
Washington, DC 20555-0001
Telephone: 301-415-6458
Fax: 301-415-5369
Email: Asimios.Malliakost@nrc.gov

From:	Cata, Samuel, DCA [samuel.cata@state.nm.us]
Sent:	Wednesday, June 15, 2011 4:17 PM
To:	Malliakos, Asimios
Subject:	RE: Historic Preservation Act Section 106 for International Isotopes Proposed De-Conversion Plant

Mr. Asimios Malliakos

I have received your E-mail correspondence and have submitted it to our staff for internal monitoring. We will reply as appropriate. Thank you very much for this information and I do appreciate that you will keep us advised on the status of this proposed activity.

Thank You

Sam

From: Malliakos, Asimios [mailto:Asimios.Malliakos@nrc.gov]
Sent: Wednesday, June 15, 2011 2:02 PM
To: Cata, Samuel, DCA
Subject: Historic Preservation Act Section 106 for International Isotopes Proposed De-Conversion Plant

Dear Mr. Samuel Cata,

As we discussed in the phone attached please find the letter we sent to the tribes. Although the letter is addressed to the Honorable Louis Maynahonah Sr., in the last page of the letter shows the addresses that identical letters were sent including your name. Attached also please find a map with the site location which is mentioned in the letter. In addition attached find the cultural survey report, no findings were made but I am attaching the report for your review. I will appreciate any comments you may have before the end of this month, June 2011.

For your convenience the Environmental Report for the International Isotopes proposed De-Conversion plant is accessible at the web address:
http://pbadupws.nrc.gov/docs/ML1001/ML100120758.pdf

Please be aware the NRC is preparing for the proposed facility a Draft Environmental Impact Statement (DEIS) which is expected to be published on November 2011. The DEIS will include discussion on Historic and Cultural Resources. A copy of the DEIS will be send to you and I will be requesting your comments.

Thank you

Asimios Malliakos
Environmental Project Manager
U. S. Nuclear Regulatory Commission
Office of Federal and State Materials and
    Environmental Management Programs
Mail Stop: T-8F5
Washington, DC 20555-0001
Telephone: 301-415-6458
Fax: 301-415-5369
Email: Asimios.Malliakost@nrc.gov

Historic Preservation Act Section 106 for International Isotopes Proposed De-Conversion Plant
From: Malliakos, Asimios [Asimios.Malliakos@nrc.gov]
Sent: Thursday, June 30, 2011 4:37 PM
To: 'holly@mescaleroapache.org'
Subject: Historic Preservation Act Section 106 for International Isotopes Proposed
De-Conversion Plant

Attachments: Letter to the tribes.pdf; Site Location ML1011600270.pdf; Cultural
Resource Report.pdf

Dear Ms Houghten,

On June 29, 2010, Diana Diaz-Toro from the U.S. Nuclear Regulatory Commission (NRC)
sent you a letter, for the International Isotopes proposed de-conversion plant, near
Hobbs, in Lea County New Mexico, pursuant to the Historic Preservation Act Section
106. Attached please find the letter we sent to several tribes, Although the
attached letter is addressed to the Honorable Louis Maynahonah Sr., in the last page
the letter shows the addresses that identical letters were sent including you.
Attached also please find a map with the site location which is mentioned in the
letter. In addition attached find the cultural survey report, no findings were made
but I am attaching the report for your review. I will appreciate any comments you
may have on the attached letter before July 15, 2011.

For your convenience the Environmental Report for the International Isotopes
proposed De-Conversion plant is accessible at the web address:

http://pbadupws.nrc.gov/docs/ML1001/ML100120758.pdf

Please be aware the NRC is preparing for the proposed facility a Draft Environmental
Impact Statement (DEIS) which is expected to be published on November 2011. The
DEIS will include discussion on Historic and Cultural Resources. A copy of the
DEIS will be send to you and I will be requesting your comments on the DEIS at that
time.

Thank you

Asimios Malliakos

Environmental Project Manager

U. S. Nuclear Regulatory Commission

Office of Federal and State Materials and

     Environmental Management Programs

Mail Stop: T-8F5

Washington, DC 20555-0001

Telephone: 301-415-6458

Fax: 301-415-5369

From: Asimios.Malliakos@nrc.gov
To: kjumper_shawneetribe@hotmail.com
Date: Fri, 1 Jul 2011 15:32:58 -0400
Subject: Historic Preservation Act Section 106 for International Isotopes Proposed De-Conversion Plant

Dear Kim Jumper,

As a follow-up to our conversation today, attached please find a letter we sent to the tribes, for the International Isotopes proposed de-conversion plant, near Hobbs, in Lea County New Mexico, pursuant to the Historic Preservation Act Section 106. Although the letter is addressed to the Honorable Louis Maynahonah Sr., in the last page of the letter shows the addresses that identical letters were sent including Ms. Jodie Hayes, of the Shawnee Tribe of Oklahoma. Attached also please find a map with the site location which is mentioned in the letter. In addition attached find the cultural survey report, no findings were made but I am attaching the report for your review. I will appreciate any comments you may have by July 15, 2011.

For your convenience the Environmental Report for the International Isotopes proposed De-Conversion plant is accessible at the web address:
http://pbadupws.nrc.gov/docs/ML1001/ML100120758.pdf

Please be aware the NRC is preparing for the proposed facility a Draft Environmental Impact Statement (DEIS) which is expected to be published on November 2011. The DEIS will include discussion on Historic and Cultural Resources. A copy of the DEIS will be send to you and I will be requesting your comments.

Thank you

Asimios Malliakos
Environmental Project Manager
U. S. Nuclear Regulatory Commission
Office of Federal and State Materials and

From:	Kim Jumper [kjumper_shawneetribe@hotmail.com]
Sent:	Wednesday, July 13, 2011 10:36 AM
To:	Malliakos, Asimios
Subject:	RE: Historic Preservation Act Section 106 for International Isotopes Proposed De-Conversion Plant

This letter is in response to the above referenced project.

The Shawnee Tribe's Tribal Historic Preservation Department concurs that no known historic properties will be negatively impacted by this project. We have no issues or concerns at this time, but in the event that archaeological materials are encountered during construction, use, or maintenance of this location, please re-notify us at that time as we would like to resume consultation under such a circumstance.

Thank you for giving us the opportunity to comment on this project.

Sincerely,
Kim Jumper, THPO
Shawnee Tribe

From: Asimios.Malliakos@nrc.gov
To: kjumper_shawneetribe@hotmail.com
Date: Fri, 1 Jul 2011 15:32:58 -0400
Subject: Historic Preservation Act Section 106 for International Isotopes Proposed De-Conversion Plant

Dear Kim Jumper,

As a follow-up to our conversation today, attached please find a letter we sent to the tribes, for the International Isotopes proposed de-conversion plant, near Hobbs, in Lea County New Mexico, pursuant to the Historic Preservation Act Section 106. Although the letter is addressed to the Honorable Louis Maynahonah Sr., in the last page of the letter shows the addresses that identical letters were sent including Ms. Jodie Hayes, of the Shawnee Tribe of Oklahoma. Attached also please find a map with the site location which is mentioned in the letter. In addition attached find the cultural survey report, no findings were made but I am attaching the report for your review. I will appreciate any comments you may have by July 15, 2011.

For your convenience the Environmental Report for the International Isotopes proposed De-Conversion plant is accessible at the web address:
http://pbadupws.nrc.gov/docs/ML1001/ML100120758.pdf

Please be aware the NRC is preparing for the proposed facility a Draft Environmental Impact Statement (DEIS) which is expected to be published on November 2011. The DEIS will include discussion on Historic and Cultural Resources. A copy of the DEIS will be send to you and I will be requesting your comments.

Thank you

Asimios Malliakos

Environmental Management Programs
Mail Stop: T-8F5
Washington, DC 20555-0001
Telephone: 301-415-6458
Fax: 301-415-5369
Email: Asimios.Malliakost@nrc.gov

July 2, 2010

Mr. Wally Murphy, Field Supervisor
New Mexico Ecological Service Field Office
U.S. Fish & Wildlife Service
2105 Osuna NE
Albuquerque, NM 87113

SUBJECT: REQUEST FOR INFORMATION REGARDING ENDANGERED OR
            THREATENED SPECIES AND CRITICAL HABITAT FOR INTERNATIONAL
            ISOTOPES FLUORINE PRODUCTS, INC. PROPOSED FLUORINE
            EXTRACTION PROCESS & DEPLETED URANIUM DE-CONVERSION PLANT

Dear Mr. Murphy:

International Isotopes Fluorine Products, Inc. (IIFP), a wholly owned subsidiary of International
Isotopes, Inc. (INIS), has submitted a license application to the U.S. Nuclear Regulatory
Commission (NRC) to construct, operate, and decommission a proposed uranium processing
facility. The facility is proposed to be located within a 640-acre section near Hobbs, New
Mexico in Lea County (see enclosed map). of which approximately 40 acres would be
developed. The 40-acre site would be fenced in and contain process-related buildings and an
administrative office building. The proposed facility would provide services to the uranium
enrichment industry for de-conversion of depleted uranium hexafluoride (DUF_6) into uranium
oxides for long-term stable disposal. The proposed facility would also produce high-purity
inorganic fluorides for applications in the electronic, solar panel, and semiconductor markets
and anhydrous hydrofluoric acid for various industrial applications.

As established in Title 10 Code of Federal Regulations Part 51 (10 CFR Part 51), the NRC
regulation that implements the National Environmental Policy Act of 1969, as amended, the
NRC is preparing an Environmental Impact Statement (EIS) for the proposed action. The EIS
will include an analysis of potential impacts to endangered or threatened species and critical
habitat in the action area. Please provide information that you may have regarding the
presence of endangered or threatened species and critical habitat in the action area. After
analyzing all the information collected, the NRC will follow up with your office regarding
compliance with the Section 7 consultation process.

We would also like to invite you to attend a public meeting that we will be holding on Thursday,
July 29, 2010, at the Lea County Event Center, 5101 Lovington Highway in Hobbs, New Mexico,
from 5:30 p.m. until 8:30 p.m. The purpose of this meeting is to solicit comments from
stakeholders and members of the public on the scope of the EIS review.

The IIFP license application is publicly available in the NRC Public Document Room (PDR) located at One White Flint North, 11555 Rockville Pike, Rockville, Maryland, 20852, or from the NRC's Agencywide Documents Access and Management System (ADAMS). The ADAMS Public Electronic Reading Room is accessible at http://www.nrc.gov/reading-rm/adams.html. The accession number for the license application is ML100630502. Persons who do not have access to ADAMS or encounter problems should contact the NRC's PDR reference staff by telephone at 1-800-397-4209, or 301-415-3747, or by e-mail at pdr@nrc.gov.

Please submit any comments you may have to offer on the environmental review within 30 days of receipt of this letter. If you have any questions, please contact Asimios Malliakos of my staff by telephone at 301-415-6458 or by email at Asimios.Malliakos@nrc.gov. Thank you for your assistance.

Sincerely,

/RA/

Diana Diaz-Toro, Branch Chief
Environmental Review Branch A
Environmental Protection and Performance
 Assessment Directorate
Division of Waste Management
 and Environmental Protection
Office of Federal and State Materials
 and Environmental Management Programs

Enclosure:
Figure 1, Proposed IIFP Site
 Location

Docket No.: 40-9086

An identical letter as the one sent to New Mexico Ecological Service Field Office, U.S. Fish & Wildlife Service, presented in pages B-25 and B-26, was also sent on July 2, 2010, to the following address:

Mr. Tod Stevenson, Director
New Mexico Department of Game and Fish
P.O. Box 25112
Santa Fe, NM 87504

United States Department of the Interior

FISH AND WILDLIFE SERVICE
New Mexico Ecological Services Field Office
2105 Osuna NE
Albuquerque, New Mexico 87113
Phone: (505) 346-2525 Fax: (505) 346-2542

AUG 10 2010

Thank you for your recent request for information on threatened or endangered species or important wildlife habitats that may occur in your project area. The New Mexico Ecological Services Field Office has posted lists of the endangered, threatened, proposed, candidate and species of concern occurring in all New Mexico Counties on the Internet. Please refer to the following web page for species information in the county where your project occurs: http://www.fws.gov/southwest/es/NewMexico/SBC_intro.cfm. If you do not have access to the Internet or have difficulty obtaining a list, please contact our office and we will mail or fax you a list as soon as possible.

After opening the web page, find New Mexico Listed and Sensitive Species Lists on the main page and click on the county of interest. Your project area may not necessarily include all or any of these species. This information should assist you in determining which species may or may not occur within your project area.

Under the Endangered Species Act of 1973, as amended (Act), it is the responsibility of the Federal action agency or its designated representative to determine if a proposed action "may affect" endangered, threatened, or proposed species, or designated critical habitat, and if so, to consult with us further. Similarly, it is their responsibility to determine if a proposed action has no effect to endangered, threatened, or proposed species, or designated critical habitat. On December 16, 2008, we published a final rule concerning clarifications to section 7 consultations under the Act (73 FR 76272). One of the clarifications is that section 7 consultation is not required in those instances when the direct and indirect effects of an action pose no effect to listed species or critical habitat. As a result, we do not provide concurrence with project proponent's "no effect" determinations.

If your action area has suitable habitat for any of these species, we recommend that species-specific surveys be conducted during the flowering season for plants and at the appropriate time for wildlife to evaluate any possible project-related impacts. Please keep in mind that the scope of federally listed species compliance also includes any interrelated or interdependent project activities (e.g., equipment staging areas, offsite borrow material areas, or utility relocations) and any indirect or cumulative effects.

B-28

Candidates and species of concern have no legal protection under the Act and are included on the web site for planning purposes only. We monitor the status of these species. If significant declines are detected, these species could potentially be listed as endangered or threatened. Therefore, actions that may contribute to their decline should be avoided. We recommend that candidates and species of concern be included in your surveys.

Also on the web site, we have included additional wildlife-related information that should be considered if your project is a specific type. These include communication towers, power line safety for raptors, road and highway improvements and/or construction, spring developments and livestock watering facilities, wastewater facilities, and trenching operations.

Under Executive Orders 11988 and 11990, Federal agencies are required to minimize the destruction, loss, or degradation of wetlands and floodplains, and preserve and enhance their natural and beneficial values. We recommend you contact the U.S Army Corps of Engineers for permitting requirements under section 404 of the Clean Water Act if your proposed action could impact floodplains or wetlands. These habitats should be conserved through avoidance, or mitigated to ensure no net loss of wetlands function and value.

The Migratory Bird Treaty Act (MBTA) prohibits the taking of migratory birds, nests, and eggs, except as permitted by the U.S. Fish and Wildlife Service. To minimize the likelihood of adverse impacts to all birds protected under the MBTA, we recommend construction activities occur outside the general migratory bird nesting season of March through August, or that areas proposed for construction during the nesting season be surveyed, and when occupied, avoided until nesting is complete.

We suggest you contact the New Mexico Department of Game and Fish, and the New Mexico Energy, Minerals, and Natural Resources Department, Forestry Division for information regarding fish, wildlife, and plants of State concern.

Thank you for your concern for endangered and threatened species and New Mexico's wildlife habitats. We appreciate your efforts to identify and avoid impacts to listed and sensitive species in your project area.

Sincerely,

Wally Murphy
Field Supervisor

New Mexico Energy, Minerals and Natural Resources Department

Susana Martinez
Governor

John H. Bemis
Cabinet Secretary - Designate

Brett F. Woods, Ph.D.
Deputy Cabinet Secretary

Tony Delfin
Acting Division Director
Forestry Division

9 June 2011

7/15/2010
75 FR 41242
(1)

RECEIVED

Chief, Rules and Directives Branch
Mail Stop T6-D59
U.S. Nuclear Regulatory Commission
Washington, DC 20555-0001

Dear Nuclear Regulatory Commission:

Energy, Minerals and Natural Resources Department – Forestry Division's Endangered Plant Program has no comments on the proposed Environmental Impact Statement for the Proposed International Isotopes Uranium Processing Facility near Hobbs, New Mexico. There are currently no known state endangered plant species or plant species of concern in Lea County.

Sincerely,

Robert Sivinski
Botanist
EMNRD-Forestry

SUNSI Review Complete
Template = ADH-013

E-RIDS = ADH-03
Add = A. Malliakos (ACM1)
M. Bartlett (mab11)

Forestry Division
1220 South St. Francis Drive • Santa Fe, New Mexico 87505
Phone (505) 476-3325 • Fax (505) 476-3330 • www.emnrd.state.nm.us/FD

B-30

GOVERNOR
Susana Martinez

DIRECTOR AND SECRETARY
TO THE COMMISSION
Tod W. Stevenson

STATE OF NEW MEXICO
DEPARTMENT OF GAME & FISH

One Wildlife Way
Post Office Box 25112
Santa Fe, NM 87504
Phone: (505) 476-8008
Fax: (505) 476-8124

Visit our website at www.wildlife.state.nm.us
For information call: (505) 476-8000
To order free publications call: (800) 862-9310

STATE GAME COMMISSIONERS

JIM McCLINTIC
Chairman
Albuquerque, NM

THOMAS "DICK" SALOPEK
Vice-Chairman
Las Cruces, NM

DR. TOM ARVAS
Commissioner
Albuquerque, NM

SCOTT BIDEGAIN
Commissioner
Tucumcari, NM

ROBERT V. HOFFMAN
Commissioner
Las Cruces, NM

GERALD "JERRY"A. MARACCHINI
Commissioner
Rio Rancho, NM

BILL MONTOYA
Commissioner
Alto, NM

June 21, 2011

Asimios Malliakos, Environmental Project Manager
US Nuclear Regulatory Commission
Mail Stop T-8F5
Washington DC 20555-0001

Re: International Isotopes Uranium Processing Facility; NMGF Project No. 13058

Dear Mr. Malliakos:

In response to your request, the New Mexico Department of Game & Fish (NMGF) has reviewed information
pertaining to the above referenced project. NRC is in the process of preparing an Environmental Impact Statement
as required by the National Environmental Policy Act of 1969. Public scoping was conducted in 2010, however
NMGF did not submit scoping comments at that time. We appreciate the additional opportunity to contribute to
development of the EIS for this project. The comments below are based mostly on information presented in the
project Environmental Report, Revision A, dated December 27, 2009.

The purpose and need for the facility is to provide services to the uranium enrichment industry for de-conversion of
depleted uranium hexafluoride (DUF6) into uranium oxide for long-term stable disposal. The company will also
include a commercial plant to produce specialty fluoride gas products for sale. High-purity silicon tetrafluoride
(SiF4) and boron trifluoride (BF3) will be manufactured in the IIFP facility by utilizing the fluorine derived from the
deconversion of DUF6. The fluoride gas products are highly valuable for applications in the electronic, solar panel,
and semi-conductor markets. In addition, anhydrous hydrogen fluoride (AHF) is a by-product of the de-conversion
process and is sold as an important chemical for various industrial applications. The project area is located in Lea
County, approximately 14 miles west of Hobbs NM. General habitat type is transitional between Southern High
Plains shortgrass prairie and Chihuahuan Desert scrub. Existing surface disturbance on the site is associated with
oil and gas development and utility corridors.

Important Habitat

The ER is not entirely correct where it concludes a lack of important habitat on the project area (Sections 3.5.9 and
3.5.13). Despite an unpredictable hydroperiod, ephemeral playa lakes (internal drainage basins) are important
breeding and nursery grounds for amphibians, and important stopovers areas for migratory waterfowl and

shorebirds. Project-related facilities should be aligned so as to avoid adverse impact to playa depressions, including excess siltation. Vegetated arroyos, such as the one running west to east across the project area, are used as wildlife movement corridors, and support a disproportionate density of nesting birds. Project-related facilities should be aligned so as to avoid adverse impact to the unnamed arroyo. In addition to black-tailed prairie dogs (a State sensitive and FWS Species of Concern), prairie dog colonies support a large number of associated species, including raptors and mammalian predators. It is unclear from the ER whether the project area includes prairie dog towns, however it is within a Natural Heritage Program of NM buffered location of an occurrence for black-tailed prairie dog, documented in 2005. Project-related facilities should be aligned so as to avoid any prairie dog colonies.

Wildlife Surveys

Presence of lesser prairie-chicken (State sensitive and FWS Candidate for listing) on the project area is possible although not likely. NMGF recommends that construction projects avoid lesser prairie-chicken leks (communal breeding grounds) by 1.5 miles. If construction will take place within 1.5 miles of a lek, no activity should be allowed between the hours of 3:00 to 9:00 am, from February 15 through June 30, to avoid interfering with auditory breeding activity. We recommend that the project area be surveyed in spring of 2012 to determine the presence or absence of this species. NMGF recommended survey protocol is available from our lesser prairie-chicken biologist Grant Beauprez, at (575) 478-2460, or grant.beauprez@state.nm.us.

To avoid violation of the federal Migratory Bird Treaty Act, clearance of vegetation should take place outside the general migratory bird nesting season (April through August). If vegetation will be cleared within the nesting season, nest surveys should be conducted, and active nests avoided until the nestlings have fledged. NMGF recommends pre-construction clearance surveys for swift fox and burrowing owl burrows. A burrowing owl survey and mitigation guideline is available on our website at http://wildlife.state.nm.us/conservation/habitat_handbook/documents/2007burrowingowlfinalfinal.pdf. If any swift fox burrows are likely to be impacted by construction, or included within the fenced area, please contact NMGF for appropriate mitigation measures.

Chapter 6 of the ER proposes an ecological monitoring program. This program does not respond to any particular regulatory requirement, but is intended "to characterize gross changes in the composition of the vegetative, avian, mammalian, and reptilian/amphibian communities of the site associated with operation of the plant." NMGF recommends the addition of a comparable nearby reference area to the study design, to control for climatic and other changes common to the surrounding area. The Wildlife Baseline Study guideline, available on our website at http://wildlife.state.nm.us/conservation/habitat_handbook/documents/WildlifeBaselineStudyGuidelinesand%20Appendix.pdf, includes information that may be useful in designing your monitoring study. NMGF requests that results of the ecological monitoring program be shared with this agency, for purposes of general information.

Best Management Practices

Consult the website of the NM Rare Plants Technical Council (http://nmrareplants.unm.edu/), or contact the NM Forestry Division, for information about plant species of concern. Conduct surveys of any suitable habitat that may be present on the project site, for rare plants which are known to occur in Lea County.

Prepare a noxious weed management plan, including a pre-construction survey, post-construction monitoring plan, steps to prevent new infestation or the spread of existing infestations, and assignment of responsibility for control of any plants on the NM Department of Agriculture Noxious Weed list.

It may not be necessary to exclude wildlife from stormwater retention ponds, unless they are expected to contain potentially harmful substances such as hydrocarbons, detergents, acids, salts, surfactants, dispersants, or heavy metals. Large wildlife will be excluded by site perimeter security fencing. If total exclusion is desired, ponds can be covered or netted to exclude flying and terrestrial animals. Extruded, knit or woven material is preferred above monofilament netting material, as it is less likely to ensnare wildlife and cause injury or death. Light colors are better (more visible) than dark. Netting should be maintained taut around the frame. If the pits will contain only water and soil, and they are not covered or netted, they should be provided with ramps to allow the escape of wildlife which may become trapped. If space allows, ramps may consist of sloping back at least one side of the pit to a 3:1 or greater horizontal:vertical ratio. Constructed ramps are commonly made from sheets of expanded metal for steel tanks, or constructed of packed earth for earthen pits. Ramps made of material with surface texture can be used in the presence of smooth liners or other slippery substrate. To be effective, the escape mechanism must be intercepted by an animal swimming around the periphery of the tank or pit at any reasonably anticipated water level. NMGF is available for consultation regarding netting or escape ramp options for any specific size and type of pit. Open above-ground tanks should also be covered, netted or provided with means of escape.

Screen all open stacks and vents, to exclude birds or bats which may seek these locations to nest or roost.

NMGF Trenching guidelines (http://wildlife.state.nm.us/conservation/habitat_handbook/documents/TrenchingGuidelines.pdf) should be included as specifications for all underground utility installation. All new electric distribution lines should be constructed in accordance with the Avian Power Line Interaction Committee (APLIC) *Suggested Practices for Avian Protection on Power Lines: The State of the Art in 2006.* This report may be ordered from APLIC at http://www.aplic.org.

Thank you for the opportunity to comment on this project. We have enclosed a list of state and federal Wildlife of Concern known to occur in Lea County, for your information. If there are any questions, please contact Rachel Jankowitz at 505-476-8159, or rjankowitz@state.nm.us.

Sincerely,

Matthew Wunder, Chief
Conservation Services Division

cc: Wally Murphy, Ecological Services Field Supervisor, USFWS
              George Farmer, SE Area Habitat Specialist, NMGF

NEW MEXICO WILDLIFE OF CONCERN
LEA COUNTY

For complete up-dated information on federal-listed species, including plants, see the US Fish & Wildlife Service NM Ecological Services Field Office website at http://www.fws.gov/southwest/es/NewMexico/SBC.cfm. For information on state-listed plants, contact the NM Energy, Minerals and Natural Resources Department, Division of Forestry, or go to http://nmrareplants.unm.edu/. If your project is on Bureau of Land Management, contact the local BLM Field Office for information on species of particular concern. If your project is on a National Forest, contact the Forest Supervisor's office for species information. E = Endangered; T = Threatened; s = sensitive; SOC = Species of Concern; C = Candidate; Exp = Experimental non-essential population; P = Proposed

Common Name	Scientific Name	NMGF	US FWS	critical habitat
Sand Dune Lizard	Sceloporus arenicolus	E	P	
Bald Eagle	Haliaeetus leucocephalus	T		
Aplomado Falcon	Falco femoralis	E	Exp	
Peregrine Falcon	Falco peregrinus	T	SOC	
Lesser Prairie-Chicken	Tympanuchus pallidicinctus	s	C	
Mountain Plover	Charadrius montanus	s	SOC	
Least Tern	Sterna antillarum	E	E	
Yellow-billed Cuckoo	Coccyzus americanus	s	SOC	
Burrowing Owl	Athene cunicularia		SOC	
Broad-billed Hummingbird	Cynanthus latirostris	T		
Loggerhead Shrike	Lanius ludovicianus	s		
Bell's Vireo	Vireo bellii	T	SOC	
Baird's Sparrow	Ammodramus bairdii	T	SOC	
Sprague's Pipit	Anthus spragueii		C	
Cave Myotis Bat	Myotis velifer	s		
Black-tailed Prairie Dog	Cynomys ludovicianus ludovicianus	s	SOC	
Swift Fox	Vulpes velox velox	s	SOC	
Black-footed Ferret	Mustela nigripes	E		
Western Spotted Skunk	Spilogale gracilis	s		
Sandhill White-tailed Deer	Odocoileus virginianus texana	s		

APPENDIX C

AIR EMISSIONS

AIR EMISSIONS

C.1   Introduction

The construction and operation of the proposed IIFP facility would result in an increase in air emissions due to construction, operations, and decommissioning workforce commuter vehicles and delivery vehicles, and, during construction, construction equipment. This Appendix presents the inputs and methodology used to estimate emission rates from vehicles in order to compare the estimated pollutant concentrations with National Ambient Air Quality criteria (NAAQS). The impacts of emissions on air quality also considered the downwind dispersion rates, and the input and methodology for those calculations are included in this Appendix.

C.2   Air Pollutant Emissions from On-Road Vehicles

This section discusses on-road vehicle air pollutant emissions, during construction, operation, and decommissioning of the proposed IIFP facility.

C.2.1   Model Input

The basic calculation to determine a pollutant emission rate is to multiply the number of vehicle miles by the pollutant's emission factor (explained below for pollutants listed in Table C-2). The number of commuter vehicles was conservatively estimated based on the size of the construction and operations workforces presented applicant's Environmental Report (IIFP, 2009). The daily mileage was estimated based on the likely residences of the workforces (see this EIS Sections 4.1.1.8 for construction and 4.1.2.8 for the methodology to estimate commuter mileage). The estimated numbers of daily deliveries and mileage was also estimated from information found in the Environmental Report. This information is summarized in Table C-1.

Emission factors were determined using the computer code MOVES (EPA, 2010a), an EPA emission inventory model. It provides an accurate estimate of emissions from mobile sources under a wide range of user-defined conditions. MOVES was used to calculate emission factors for volatile organic compounds (VOCs), carbon monoxide (CO), nitrogen oxides (NO_x), carbon dioxide equivalents (CO_2), sulfur dioxide (SO_2), particulate matter less than 2.5 microns in diameter ($PM_{2.5}$), particulate matter less than 10 microns in diameter (PM_{10}), benzene, methyl tertiary butyl ether (MBTE), 1,3 butadiene, formaldehyde, acetaldehyde, and acrolein for the years of interest. Phase 1 construction is expected to start in 2012 and be completed in 2013. Phase 2 construction is expected to begin in year 2015 and be completed in 2016 Facility operations would begin in 2013, and extend for the 40-year license term. The year 2011 was chosen as the model year.

Different emissions emanate from a vehicle depending on type of activity and time of the day. The model accounts for all emissions during normal daily activity. The types of emission processes are:

- **Running exhaust**—tailpipe emissions during highway travel.

- **Starting exhaust**—tailpipe emissions that occur as a result of starting a vehicle. These emissions are independent of running exhaust emissions. The magnitude of these emissions is dependent on how long the vehicle has been sitting prior to starting.

Table C-1. Worker and Delivery Vehicle Rates Due To Construction, Operation, and Decommissioning Activities of the IIFP Facility

	(vehicles)	(miles/day)	(days/phase)*	(vehicle miles/phase)
Preconstruction (3 months)				
workers	70	40	62.5	175,000
deliveries	10	40	62.5	25,000
equipment	2	40	62.5	5,000
Phase 1 Construction (1 year)**				
workers	140	40	250	1,400,000
deliveries	20	40	250	200,000
equipment	4.25	40	250	42,500
Phase 1 Operations (1 year)				
workers	140	40	250	1,400,000
deliveries	10.6	1512	250	4,006,800
Phase 2 Construction (1 year)				
workers	180	40	250	1,800,000
deliveries	20	40	250	200,000
equipment	2	40	250	20,000
Phase 2 Operations* (per year)				
workers	40	40	250	400,000
deliveries	17.2	1512	250	6,501,600
Decommissioning (3 years)				
workers	40	40	750	1,200,000
deliveries	0	-	750	0

*After 2016, both phases of the facility will be operational. The "Phase 1 operations" entries apply only to the years 2013 to 2016, when only Phase 1 is operation. "Phase 2 operations" entries include both Phase 1 and Phase 2 operations, beginning in year 2016.

**The work year was taken to be 250 days long.

Source: IIFP, 2011

- **Tirewear**—particulate emissions as friction between tires and the highway wear away the tire.

- **Brakewear**—particulate emissions from brake use.

- **Evaporation loss**—fuel loss through rubber and plastic components while the vehicle is sitting .

- **Crankcase exhaust**—the exhaust gases that escape around the piston rings and enter the crankcase during normal operation.

Table C-2 presents the results of all the sources of emissions as grams per mile driven, as calculated by the MOVES model using the input parameters from Table C-1.

Table C-2. MOVES Emission Factor Outputs for 2011

Pollutant	Emission Factor (gram/mile)		
	workers	deliveries	equipment
VOCs	7.37×10^{-1}	8.72×10^{-1}	1.02
CO	7.82	1.02×10	1.20×10
NO_x	1.04	4.63	1.71
SO_2	8.28×10^{-3}	1.12×10^{-2}	9.96×10^{-3}
PM_{10}*	3.53×10^{-2}	2.38×10^{-1}	5.30×10^{-2}
$PM_{2.5}$*	1.90×10^{-2}	1.97×10^{-1}	3.23×10^{-2}
CO_2 - equivalent	4.28×10^{2}	9.57×10^{2}	5.30×10^{2}
benzene	1.67×10^{-2}	1.92×10^{-2}	2.57×10^{-2}
MBTE	0.00	0.00	0.00
1,3 butadiene	2.86×10^{-3}	4.34×10^{-3}	4.56×10^{-3}
formaldehyde	6.41×10^{-3}	2.46×10^{-2}	1.15×10^{-2}
acetaldehyde	5.93×10^{-3}	1.27×10^{-2}	9.72×10^{-3}
acrolein	2.97×10^{-4}	1.20×10^{-3}	5.31×10^{-4}

*PM totals are the sum of organic carbon, elemental carbon, and sulfate particulate emissions.

C.2.2   Analysis Methods

Emission rates of the six criteria pollutants (i.e., CO, NO_x, SO_2, PM_{10}, $PM_{2.5}$ and VOCs, an ozone precursor), CO_2 equivalent, and six hazardous air pollutants (HAPs) (i.e., benzene, MBTE, 1,3 butadiene, formaldehyde, acetaldehyde, and acrolein) as calculated by MOVES for Lea County in 2011 (Table C-2) were multiplied by the worker and delivery vehicles mileage estimates (Table C-1) to arrive at total emissions.

C.2.3   Results

Pollutant emission amounts for the span of construction and operation phase are reported in this EIS Sections 4.1.1.4 for construction (Tables 4-4 and 4-5), 4.1.2.4 for operations (Tables 4-15 and 4-16), and 4.2.2.4 for the Phase 2 increment.

C.3   Air Pollutant Emissions from Construction Activities

This section discusses air pollutant emissions as a result of construction activities. This includes emissions from construction equipment, fugitive dust emissions from land disturbance from construction activities, and fugitive emissions from the onsite diesel refueling activities.

C.3.1   Analysis Methods

All emissions were calculated using the general equation for emissions estimation (EPA, 1995a):

$$E = A \times EF \times (1-ER/100)$$

where:

E = emissions
A = activity rate
EF = emission factor
ER = overall emission reduction efficiency, as %

For construction equipment the activity rate is measured as horsepower-hours. The following equation (EPA, 2005a) was used to determine the horsepower-hours:

$$HP\text{-}hr = (Max\ HP) \times (LF) \times (\#) \times (hrs)$$

where:

HP-hr = horsepower-hours
Max HP = maximum horsepower
LF = load factor
# = number of units used
hrs = hours that equipment operates

For fugitive dust emissions in the first equation, the activity rate is the number of acres that would be disturbed by construction activities. Because the applicant indicated that watering would be used to control fugitive dust emissions, an emission reduction efficiency of 50 percent was assumed.

For fugitive emissions from the onsite diesel refueling activities in the first equation, the activity rate is the number of gallons of diesel fuel used. The amount of diesel fuel used was calculated using the following equation (EPA, 2010b):

$$DB = BSFC \times TAF \times A$$

where:

DB = diesel burned
BSFC = brake specific fuel consumption
TAF = transient adjustment factor
A = activity rate (HP-hr)

Carbon dioxide equivalents were calculated using the equation (EPA, 2005b):

$$CO_2e = CO_2 + (21 \times CH_4) + (310 \times N_2O)$$

where:

CO_2e = carbon dioxide equivalents
CO_2 = carbon dioxide
CH_4 = methane
N_2O = nitrous oxide

The applicant provided equipment lists and schedules showing the hours of equipment operation per month for each construction phase (preconstruction, Phase 1 and Phase 2), and the amount of disturbed acreage (IIFP, 2011).

C.3.2   Emission Factors

Emission factors for CO_2, VOCs, CO, NO_2, SO_2, $PM_{2.5}$, and PM_{10} were determined using the computer code NONROAD (EPA, 2005b), an EPA emission inventory model. Default values for Lea County, New Mexico (i.e., climate/meteorology, equipment age, deterioration factors, fuel properties, and growth factors) were used as inputs for the model. The year 2011 was chosen at the modeling year.

Emission factors for the greenhouse gases methane (CH_4) and nitrous oxide (N_2O) were obtained from the EPA guidance document "Climate Leaders Greenhouse Gas Inventory Protocol Core Module Guidance Direct Emissions from Mobile Combustion Sources" (EPA, 2008).

Emission Factors for fugitive dust emissions were obtained from Section 13.2.3 of EPA AP-42 "Compilation of Air Emission Factors" (EPA, 1995a). Emission factors for refueling activities were provided by the applicant (IIFP, 2011).

C.3.3   Results

The input used in the calculations described in Section C.3.1, and the calculated monthly and annual emissions, and maximum emissions rates for each pollutant for each construction phase (Table C-2) are reported in this EIS Sections 4.1.1.4 for Phase 1, and 4.2.2.4 for preconstruction and Phase 2 construction.

C.4   Incremental Downwind Air Pollutant Concentration Increases

C.4.1   Model Input

Emissions from construction equipment would be dispersed downwind. Dispersion coefficients were determined using the computer code SCREEN3 (EPA, 1995b), an EPA single source Gaussian plume model. Dispersion coefficients were determined for the maximum concentration (at the construction site), the property border (at 900 meters from the construction site), and 1 mile (1,600 meters) from the construction site for Phase 1 preconstruction and construction, Phase 2 construction, and Phase 1 operations (Table C-3).

C.4.2   Analysis Methods

There is a direct correlation between the source emission rate and the dispersion coefficients (disp coeff) calculated by SCREEN3. For example, a 5-fold increase in the emission rate input to SCREEN3 results in a 5-fold increase in the resulting dispersion concentrations. Therefore, setting the source emission rate to 1.0 gram/second/square meter allows scaling of the emission rates by multiplying them by SCREEN3's dispersion coefficients. This was done using Eq. C.3-1 for the preconstruction, Phase 1 construction, and Phase 2 construction to determine the peak 1-hour concentrations at the site border (900 meters) and at 1 mile (1,600 meters). The peak 3-hour, 8-hour, 24-hour, and annual concentrations were derived by multiplying the peak 1-hour concentration by the conversion factors given in Table C-4 (EPA, 1992). The resulting concentrations are provided in Section C.4.3.

$$[(A + B) \times C] + [D \times E] = F$$ Eq. C.3-1

where:

A = Construction Equipment 1-hour Peak Emission Rate
B = Construction Vehicles 1=hour Peak Emission Rate
C = SCREEN3 Volume Dispersion Coefficient
D = Fugitive Dust 1-hour Peak Emission Rate
E = SCREEN3 Area Dispersion Coefficient
F = One-hour Peak Concentration at Site Boundary or 1.6 km (1 mi)

Table C-3. SCREEN3 Outputs: Dispersion Coefficients

		Preconstruction / Phase 1 Construction	
Volume	$(\mu g/m^3)/$ (g/s)	max (157 m)	935.9
		900 m	246.4
		1600 m	144.5
Area	$(\mu g/m^3)/$ $(g/s/m^2)$	max (223 m)	1.648×10^8
		900 m	2.492×10^7
		1600 m	1.565×10^7
		Phase 2 Construction	
Volume	$(\mu g/m^3)/$ (g/s)	max (30 m)	7352
		900 m	593.4
		1600 m	274.7
Area	$(\mu g/m^3)/$ $(g/s/m^2)$	max (35 m)	9.386×10^7
		900 m	1.753×10^6
		1600 m	7.636×10^5
		Phase 1 Operations - Utilities	
Point	$(\mu g/m^3)/$ (g/s)	max (107 m)	608.0
		900 m	145.5
		1600 m	132.9
		Phase 1 Operations - H_2 Generation	
Point	$(\mu g/m^3)/$ (g/s)	max (140 m)	666.5
		900 m	210.0
		1600 m	166.5

Table C-4. EPA Peak Hour Conversion Factors

3-Hour Conversion Factor	0.90
8-Hour Conversion Factor	0.70
24-Hour Conversion Factor	0.40
Annual Average Conversion Factor	0.10

Source: (EPA, 1992)

The 1-hour peak concentrations at site border for each construction phase and operations were determined according to Eq. C.3-2. All emission-generating units were conservatively assumed to operate continuously. The conversion factors given in Table C-4 were used to determine peak 3-hour, 8-hour, 24-hour, and annual concentrations. The resulting concentrations are provided in Section C.4.3.

$$[(G + H + J) \times K] + [L \times M] = N \qquad \text{Eq. C.3-2}$$

where:

G = Boilers 1-hour Peak Emission Rate
H = Generators 1-hour Peak Emission Rate
J = Firewater Pump 1-hour Peak Emission Rate
K = SCREEN3 Utilities Point Dispersion Coefficient
L = H_2 Generator 1-hour Peak Emission Rate
M = SCREEN3 H_2 Generation Point Dispersion Coefficient
N = One-hour Peak Concentration at Site Boundary or 1.6 km (1 mi)

C.4.3.   Results

Peak 1-hour, 3-hour, 8-hour, 24-hour, and annual concentrations at the site boundary for each construction phase and operations and their percent of the NAAQS that were calculated are reported in this EIS Sections 4.1.1.4 for construction (Table 4-6), 4.1.2.4 for operations (Table 4-17), and 4.2.2.4 for cumulative impacts.

C.5   References

(EPA, 1992) U.S. Environmental Protection Agency. 1992. Screening Procedures for Estimating the Air Quality Impact of Stationary Sources, Revised, EPA-454/R-92-019, October, 1992. Available at http://www.epa.gov/cppt/exposure/presentations/efast/usepa_1992b_sp_ for_estim_aqi_of_ss.pdf. Accessed August 9, 2011.

(EPA, 1995a) U.S. Environmental Protection Agency. 1995. Compilation of Air Pollutant Emissions Factors, Fifth Edition and Supplements, AP-42. Office of Air Quality Planning and Standards, Research Triangle Park, North Carolina. January 1995. . Available at http://www.epa.gov/ttn/chief/ap42/index.html. Accessed May 19, 2011. Relevant sections are posted in ADAMS with ADAMS Accession Nos. ML 11194A234, ML11194A236, and ML 11182C051.

(EPA, 1995b) U.S. Environmental Protection Agency. 1995. "SCREEN3 Model User's Guide." EPA-454/B-95-004, Office of Air Quality Planning and Standards, Research Triangle Park, North Carolina. September 1995. Available at http://www.epa.gov/ttn/scram/userg/ screen/screen3d.pdf. Accessed May 18, 2011. ADAMS Accession No. ML071930125.

(EPA, 2005a) U.S. Environmental Protection Agency. 2005. "User's Guide for the Final NONROAD2005 Model," EPA420-R-05-013, Office of Transportation and Air Quality. December 2005. Available at http://www.epa.gov/otaq/models/nonrdmdl/nonrdmdl2005/420r05013.pdf.

(EPA, 2005b) U.S. Environmental Protection Agency. 2005. Metrics for Expressing Greenhouse Gas Emissions: Carbon Equivalents and Carbon Dioxide Equivalents. EPA420-F-05-002, February 2005. . Available at http://www.epa.gov/oms/climate/420f05002.pdf. Accessed October 14, 2011.

(EPA, 2008) U.S. Environmental Protection Agency. 2008. Climate Leaders Greenhouse Gas Inventory Protocol Core Module Guidance Direct Emissions from Mobile Combustion Sources. EPA430-K-08-004, May 2008. Available at http://www.epa.gov/climateleaders/documents/resources/mobilesource_guidance.pdf. Last Accessed on June 2, 2011.

(EPA, 2010a) U.S. Environmental Protection Agency. 2010. MOVES – Motor Vehicle Emission Simulator, August 26, 2010. Available at http://www.epa.gov/otaq/models/moves/index.htm, last accessed June 2, 2011.

(EPA, 2010b) U.S. Environmental Protection Agency. 2010. Exhaust and Crankcase emission Factors for Nonroad Engine Modeling -- Compression-Ignition, NR-009d. EPA-420-R-10-018, July 2010. Available at http://www.epa.gov/oms/models/nonrdmdl/nonrdmdl2010/420r10018.pdf. Accessed October 14, 2011.

(IIFP, 2009) International Isotopes Fluorine Products. 2009. Fluorine Extraction Process and Depleted Uranium De-conversion Plant (FEP/DUP) Environmental Report, Revision A. ER-IFP-001, Revision A. December 27, 2009. ADAMS Accession No. ML 100120758.

(IIFP, 2011) International Isotopes Fluorine Products, 2011. Fluorine Extraction Process and Depleted Uranium De-conversion Plant (FEP/DUP) Official Responses to Environmental Report RAIs, Revision A ,Talking Points Regarding the Environmental Report. May 24, 2011. ADAMS Accession No. ML111530402.

APPENDIX D

SOCIOECONOMIC INFORMATION

SOCIOECONOMIC INFORMATION

D.1 Introduction

This Appendix presents the bases to establish the region of influence (ROI) for socioeconomic conditions, and calculations to assess impacts in the ROI. In addition, this Appendix contains the input used for the Environmental Justice analysis.

D.2 Socioeconomic Region of Influence (ROI)

The identification of a socioeconomic region of influence for a site is dependent on many factors, which can include, but are not necessarily limited to:

- Population and population densities of the counties within 50 miles of the proposed site

- Population of those counties' largest population centers

- Geographic locations of the population centers in relation to the proposed site

- Estimated travel distance or travel time from the population centers to the proposed site

- Mean travel time to work for each county

- Employment data for each county

- Worker commuting patterns from the surrounding counties to the county containing the proposed site ("host county")

In identifying the socioeconomic ROI, the initial step was to identify counties that lie primarily within the 50-mile radius or counties with only a small portion of their area within the 50-mile radius but with a large population center within the 50-mile radius . Two counties in New Mexico and three counties in Texas have these characteristics: Lea County and Eddy County, New Mexico, and Andrews, Gaines, and Yoakum Counties, Texas.   A review of the key factors for each county, determined that the proposed action has the potential to impact socioeconomic variables (employment, population, income, housing, infrastructure, and community services) in the two New Mexico counties only (Lea and Eddy). Therefore, these counties were identified as the socioeconomic ROI. For the reasons discussed below, the proposed action is unlikely to impact socioeconomic variables in the Texas counties (Andrews, Gaines, and Yoakumand these counties were not included in the socioeconomic ROI. Each county's demographics are summarized in Tables D-1 through D-5 and briefly analyzed below.

Table D-1 provides information on population, income, distances and commuting time for counties and population centers. Table D-2 provides employment characteristics by county. Table D-3 provides county-to-county worker flows. Table D-4 provides information on housing units and staffed hospital beds. Table D-5 provides hospital beds details per hospital/medical center.

Table D-1. Population, Income, Distances and Commuting Time for Counties and Population Centers.

County	Population (2000)[a]	Population Density per Square mile of Land Area (2000)[a]	Population Estimate (2009)[a]	Mean Travel Time to Work, 2000 (Minutes)[a]	Median Household Income (2008 dollars)[a]	County's Largest Population Center[b]	Population Center Population (2000)[b]	Population Center Population (2009)[b]	Driving Miles from Population Center to proposed site
New Mexico									
Lea	55,508	12.6	60,232	18.7	$45,813	Hobbs	28,657	30,838	10-15
Eddy	51,658	12.4	52,706	18.3	$43,784	Carlsbad	25,625	26,259	60-65
Texas									
Andrews	13,004	8.7	14,057	20.6	$49,043	Andrews	9,652	10,448	70-75
Gaines	14,467	9.6	15,382	17.4	$40,489	Seminole	5,910	6,251	40-45
Yoakum	7,322	9.2	7,698	15.9	$50,317	Denver City	3,985	4,140	45-50

Sources and Notes:
[a] USCB, 2010a
[b] USCB, 2010b

Table D-2. Employment Characteristics by County.

County	Number of Jobs (2008)[a]	Construction Jobs (2008)[a]	Construction Jobs as Percent of All Jobs (2008)[a]	Professional, Scientific, and Technical Services Jobs (2008)[a]	Professional, Scientific, and Technical Services Jobs as Percent of All Jobs (2008)[a]	Civilian Labor Force (2009)[b]	Annual Average Unemployment Rate (2009)[b,c] (%)	Unemployment Rate (June 2010)[b,c] (%)
New Mexico								
Lea	37,622	3,460	9.2%	1,019	2.7%	28,890	7.6	8.0
Eddy	30,692	2,597	8.5%	1,315	4.3%	28,700	5.5	6.1
Texas								
Andrews	7,337	860	11.7%	(D)	NA	7,008	7.1	6.6
Gaines	8,043	992	12.3%	150	1.9%	7,016	6.4	6.4
Yoakum	4,980	419	8.4%	67	1.3%	4,134	7.7	6.8

Sources:
[a]BEA, 2010a
[b]BLS, 2010a
[c]BLS, 2010b
(D) - Not shown (by the BLS) to avoid disclosure of confidential information, but the amount is included in the BEA's totals.
NA = Not available

Table D-3. County-to-County Worker Flows, 2000.

Res State	Res County	Res (C)MSA	Res PMSA	Residence State-County Name	Work State	Work County	Workplace State-County Name	Count	Percent from Resident County
35	025	9999	9999	Lea Co. NM	035	015	Eddy Co. NM	303	1.5%
35	025	9999	9999	Lea Co. NM	035	025	Lea Co. NM	18,566	93.6%
35	015	9999	9999	Eddy Co. NM	035	015	Eddy Co. NM	19,236	95.3%
35	015	9999	9999	Eddy Co. NM	035	025	Lea Co. NM	195	1.0%
48	003	9999	9999	Andrews Co. TX	035	025	Lea Co. NM	49	1.0%
48	003	9999	9999	Andrews Co. TX	048	003	Andrews Co. TX	3,794	77.2%
48	165	9999	9999	Gaines Co. TX	035	025	Lea Co. NM	179	3.4%
48	165	9999	9999	Gaines Co. TX	048	165	Gaines Co. TX	4,285	80.6%
48	501	9999	9999	Yoakum Co. TX	035	025	Lea Co. NM	135	4.8%
48	501	9999	9999	Yoakum Co. TX	048	501	Yoakum Co. TX	2,383	84.4%

Source: USCB, 2003

Table D-4. Housing Units and Staffed Hospital Beds

County, State	Housing Units, 2009 [a]	Percent of Total Units	Staffed Hospital Beds [b]	Percent of Total Staffed Beds
Lea Co., NM	24,837	40.1%	226*	44.5%
Eddy Co., NM	22,645	36.5%	147*	28.9%
Andrews Co., TX	5,810	9.4%	88*	17.3%
Gaines Co., TX	5,645	9.1%	25*	4.9%
Yoakum Co., TX	3,062	4.9%	22*	4.3%
Total	61,999		508	

Sources:
[a] USCB, 2010a
[b] AHA, 2007
* See Hospital Beds details per Hospital/Medical Center, in Table D.5 below.

Table D-5. Hospital Beds Details per Hospital/Medical Center

New Mexico Hospital Beds	Hospital Beds	County Total
Eddy County		147
Carlsbad Medical Center	127	
Artesia General Hospital	20	
Lea County		226
Lea Regional Medical Center	214	
NOR-Lea General Hospital	12	
Texas Hospital Beds		
Andrews County		88
Permian Regional Medical Center	88	
Yoakum County		22
Yoakum County Hospital	22	
Gaines County		25
Memorial Hospital	25	

Source: AHA, 2007

D.2.1 Lea County, New Mexico

Lea County is the host county for the proposed IIFP project. The proposed location is approximately 14 miles west of Hobbs, New Mexico. Lea County had a year 2000 population of

55,508 and an estimated 2009 population of 60,232, with 12.6 people per square land mile in 2000 (Table D-1). The county's largest population center is Hobbs, with a 2000 population of 28,657, and an estimated 2009 population of 30,838. Hobbs is the largest city within a 50-mile radius (Carlsbad, in Eddy County New Mexico, has about 26,300 residents and lies on the 50-mile perimeter). Lea County's mean commute time is 18.7 minutes.

In 2009, Lea County's civilian labor force was 28,890 persons (Table D-2). In 2008, employment in the construction industry accounted for 9.2 percent of total employment and employment in the professional, scientific, and technical services industry (the industry classification of the proposed project) accounted for approximately 2.7 percent of the jobs. In 2009, the annual average unemployment rate was 7.6 percent. The unemployment rate in June 2010 was 8.0 percent.

In 2000, Lea County's 19,828 commuting residents traveled to a worksite (USCB, 2003). Of those, 18,566 (93.6 percent) traveled to a worksite in Lea County. An additional 303 workers (1.5 percent) commuted to a worksite in Eddy County. The remaining 4.8 percent traveled to a worksite elsewhere. Of the 19,790 jobs in Lea County in 2000, 18,566 (93.8 percent) were held by residents of Lea County. Residents of Eddy County held 195 (1.0 percent) of those jobs. No other county had residents that filled at least 1 percent of the Lea County jobs (Table D-3).

Lea County, in the vicinity of the proposed site, in particular, is well served by state and county highways and roads. Sufficient community amenities and infrastructure to support additional population are in Lea County. In 2009, Lea County had 40.1 percent of the housing inventory in the five subject counties (Table D-4). Lea County had 44.5 percent of all the staffed hospital beds in the five-county area (Tables D-4 and D-5).

Based on the proximity to the proposed project site, availability of amenities including housing, and the historical county-to-county worker travel patterns, Lea County is the most likely county for project workers to reside. Also, Lea County would be the major recipient of facility-generated property taxes. Therefore, Lea County , was included in the socioeconomic ROI of the proposed project.

D.2.2   Eddy County, New Mexico

A substantial portion of Eddy County, New Mexico is within the 50-mile radius of the proposed site. Eddy County had a year 2000 population of 51,658 and an estimated 2009 population of 52,706 with 12.4 people per square land mile in 2000 (Table D-1). The county's largest population center is Carlsbad, with a 2000 population of 25,625 and an estimated 2009 population of 26,259. Carlsbad is on the perimeter of the 50-mile radius of the proposed site. Eddy County's mean commute time is 18.3 minutes. Carlsbad is approximately 60-65 driving miles from the proposed site.

In 2009, Eddy County's civilian labor force was 28,700 persons (Table D-2). In 2008, employment in the construction industry accounted for 8.5 percent of total employment and employment in the professional, scientific, and technical services industry (the industry classification of the proposed project) accounted for approximately 4.3 percent of the jobs in the county. In 2009, the annual average unemployment rate was 5.5 percent. The unemployment rate in June 2010 was 6.1 percent.

In 2000, of Eddy County's total commuting population, 19,236 (95.3 percent) traveled to a worksite in Eddy County and 195 (1.0 percent) commuted to a worksite in Lea County (Table D-3).

Eddy county is served by several state and county highways and roads. U.S. Highway 62 travels NNE from Carlsbad to the proposed site. Eddy County has sufficient community amenities and infrastructure to support its population. In 2000, Eddy County had 36.5 percent of all housing inventory in the five subject counties and 28.9 percent of all the staffed hospital beds in (Tables D-4 and D-5).

Eddy County, New Mexico, borders the host county of the proposed project. A substantial portion of the county and a portion of its largest population center is within the 50-mile radius. The county population center is accessible to the proposed site via a major U. S. Highway. Although historically few Eddy County residents have traveled to Lea County for work, commuting patterns may change with newly available employment opportunities, particularly in the professional, scientific, and technical services industry. Based on the proximity to the proposed site, easy vehicle access, and availability of amenities including housing, this analysis concludes that some project workers would likely live in Eddy County. Therefore, Eddy County, New Mexico, was included in the socioeconomic ROI of the proposed project.

D.2.3   Andrews County, Texas

A substantial portion of Andrews County, Texas, is within the 50-mile radius of the proposed site. In 2000, Andrews County had a population of 13,004 and an estimated 2009 population of 14,057 with 8.7 persons per square land mile in 2000 (Table D-1). The county's largest population center is Andrews, with a 2000 population of 9,652 and an estimated 2009 population of 10,448. Andrews is outside the 50-mile radius of the proposed site. Andrews County's mean commute time is 20.6 minutes. The proposed site is approximately 70-75 driving miles from the city of Andrews.

In 2009, Andrews County's civilian labor force was 7,008 persons. In 2008, employment in the construction industry accounted for 11.7 percent of total employment (Employment in the professional, scientific, and technical services industry was confidential and not disclosed by the Bureau of Labor Statistics). In 2009, the annual average unemployment rate was 7.1 percent. The unemployment rate in June 2010 was 6.6 percent (Table D-2).

In 2000, 3,794 (77.2 percent)of Andrews County commuting residents traveled to a workplace in Andrews County and 49 residents (1.0 percent) commuted to a worksite in neighboring Lea County (Table D-3).

The rural county is served by state and county highways and roads. In 2000, Andrews County had less than 10 percent of all housing inventory in the five subject counties and 17.3 percent of all the staffed hospital beds (Tables D-4 and D-5).

Andrews County, Texas, borders the host county of the proposed project. A substantial portion of the county is within the 50-mile radius. However, the county population center is not readily accessible to the proposed site via a major transportation artery. Historically, few Andrews

County workers commute to Lea County., Therefore, few project workers would be expected to live in Andrews County and it was not included in the socioeconomic ROI.

D.2.4 Gaines County, Texas

A substantial portion of Gaines County, Texas, is within the 50-mile radius of the proposed site. In 2000, Gaines County had a population of 14,467 and an estimated 2009 population of 15,382with 9.6 persons per square land mile in 2000 (Table D-1). The county's largest population center is Seminole, with a 2000 population of 5,910 and an estimated 2009 population of 6,251. Gaines County's mean commute time is 17.4 minutes. The proposed site is approximately 40-45 driving miles from Seminole.

In 2009, Gaines County's civilian labor force was 7,016 persons. In 2008, employment in the construction industry accounted for 12.3 percent of total employment and employment in the professional, scientific, and technical services industry accounted for 1.9 percent of total employment. In 2009, the annual average unemployment rate was 6.4 percent. The unemployment rate in June 2010 was also 6.4 percent (Table D-2).

In 2000, 4,285 (80.6 percent) of Gaines County commuting residents traveled to a worksite in Gaines County and 179 (3.4 percent) commuted to a worksite in neighboring Lea County.

The rural county is served by state and county highways and roads. In 2000, Gaines County had less than 10 percent of all housing inventory in the five subject counties, and 25 staffed hospital beds, less than 5 percent of all the staffed hospital beds (Tables D-4 and D-5).

Gaines County, Texas, borders the host county of the proposed project. A substantial portion of the county and its largest population center are within the 50-mile radius. The county population center is accessible to the proposed site via a major transportation artery. However, because historically few Gaines County workers commute to work in Lea County and the professional, scientific, and technical industry accounts for only 1.9 percent of the relatively small county workforce. Therefore, few project workers would be expected to live in Gaines County and it was not included in the socioeconomic ROI.

D.2.5 Yoakum County, Texas

A substantial portion of Yoakum County Texas is within the 50-mile radius of the proposed site. In 2000, Yoakum County had a population of 7,322 and an estimated 2009 population of 7,698with 9.2 persons per square land mile in 2000 (Table D-1). The county's largest population center is Denver City, with a 2000 population of 3,985 and an estimated 2009 population of 4,140. Yoakum County's mean commute time is 15.9 minutes. The proposed site is approximately 45-50 driving miles from Denver City.

In 2009, Yoakum County's civilian labor force was 4,134 persons. In 2008, employment in the construction industry accounted for 8.4 percent of total employment and employment in the professional, scientific, and technical services industry accounted for 1.3 percent of total employment. In 2009, the annual average unemployment rate was 7.7 percent. The unemployment rate in June 2010 was 6.8 percent (Table D-2).

In 2000, 2,383 (84.4 percent) of Yoakum County commuting residents traveled to a workplace in Yoakum County and 135 (4.8 percent) commuted to a worksite in neighboring Lea County (Table D-3).

The rural county is served by state and county highways and roads. In 2000, Yoakum County had approximately 4.9 percent of all housing inventory in the five subject counties and less than 5 percent of all the staffed hospital beds (Tables D-4 and D-5).

Yoakum County, Texas, borders the host county of the proposed project. A substantial portion of the county and its largest population center are within the 50-mile radius. The county population center is accessible to the proposed site via a major road. However, because historically few Yoakum County workers commute to work in Lea County and the professional, scientific, and technical industry accounts for only 1.3 percent of the relatively small county workforce, few project workers would be expected to live in Yoakum County. Therefore, Yoakum County, Texas, was not included in the socioeconomic ROI.

D.2.6 Workflow Patterns Summary

Historical patterns of commuting are the strongest proxy available to predict residential settlement patterns for workers migrating to an area for new employment opportunities. County-to-county worker flow patterns are established by commuters based on their demonstrated preferences for residential areas. These demonstrated preferences are thought to include commuting times, housing, amenities, and other opportunities for employment. In this analysis, workers in Lea County demonstrated a preference for working in Lea County and residents of the surrounding counties demonstrated a reluctance to drive to a worksite in Lea County. Despite the limited employment opportunities in Andrews, Gaines, and Yoakum County, few residents of those counties have elected to drive to Lea County, with its larger employment base. Eddy's County's relatively large employment in the professional, scientific, and technical service sector reflects the presence of WIPP (Waste Isolation Pilot Plant) and related industries. These variables, coupled with the availability of highway access between Carlsbad and Hobbs, indicate a strong worker exchange between Lea and Eddy Counties.

D.3 Environmental Justice

This discussion supports the identification of minority and low-income populations within 50 miles of the proposed project location, as shown in this EIS Chapter 3, Figures 3-20 through 3-25.

Procedures for the determination of minority and low-income populations are discussed in this section. Appendix C of the Environmental Review Guidance for Licensing Actions Associated with NMSS Programs (NRC, 2003), provides the current NRC guidance for identifying minority and low-income populations. The guidance was used in identifying minority and low-income populations in this EIS.

The area potentially impacted by environmental issues was determined to be within a 50-mile radius of the site, which is the area that was evaluated for impacts of potential facility accidents. Therefore, the minority populations and low-income populations were determined for all census block groups that fell entirely or partially within 50 miles of the project location. Block groups were used because census blocks (smaller than block groups) do not report income data and census tracts (larger than block groups) might not delineate minority or low-income populations within the larger general population (NRC, 2003). U.S. Census Bureau (USCB) Summary File 1 containing race data (USCB 2000a; USCB 2000b) and Summary File 3 containing household poverty data (USCB 2000c; USCB 2000d) were obtained for all block groups in New Mexico and Texas since the 50-mile radius encompasses parts of both states.

For each race/ethnicity minority category (Black or African American, American Indian and Alaska Native, Asian, Native Hawaiian and Other Pacific Islander, Other Race, Two or More Races [Multi-Racial], and Hispanic Ethnicity), and for each block group the percentage of the total population made up of the minority/ethnicity was calculated. The Aggregate category was also determined. The Aggregate is the sum of all the minorities within a block group. The percentage of low-income households was also calculated for each block group.

The Hispanic Ethnicity category is NOT included in the aggregate of minorities because the USCB considers race and Hispanic origin (ethnicity) as two separate and distinct concepts. People who are Hispanic may be of any race. People in any race group may be either Hispanic or Not Hispanic. Each person has two attributes, their race (or races) and whether or not they consider themselves Hispanic. Because each person is counted in a race category and in either the Hispanic or not Hispanic category, including the Hispanic ethnicity in the "aggregate race" category would double count a number of individuals. As such, the race categories and the Hispanic Ethnicity categories are considered separately.

The minority demographic data and low-income data were then attributed to block group spatial data in ArcGIS® 9.3 to develop a comprehensive shapefile dataset containing demographic and low-income data for every block group in the state. ArcGIS® is a geographic information system (GIS) modeling software which is used to access and query mapped demographic and low-income data (ESRI, 2008).

In order to identify whether a minority or low-income population exists, an area larger than the proposed site and immediately surrounding environs, and that encompasses the entire area of potential impact must be identified for comparative analysis (NRC, 2003). This area is called a geographic area. Because the 50-mile radius used in this analysis includes parts of New Mexico and Texas, the geographic area used as the basis for identifying individual block groups with minority or low-income populations was the states of New Mexico and Texas. Block group low-income and minority populations in New Mexico were compared to the total low-income and minority populations in New Mexico, and block groups low-income and minority populations in Texas were compared to the total Texas low-income and minority populations.

A significant minority population is considered to be present if (1) the minority population in the census block group exceeds 50 percent or (2) the minority population percentage of the block group is significantly greater (typically at least 20 percentage points) than the minority population percentage in the geographic area (NRC, 2003). A significant low-income population is considered to be present if: (1) the low-income household population in the census block group exceeds 50 percent or (2) the percentage of households below the poverty level in an environmental impact area is significantly greater (typically at least 20 percentage points) than the low-income household percentage in the geographic area (NRC, 2003).

State and county percentages for minority and low-income populations were obtained using summary statistics in ArcGIS® 9.3 and then compared to the USCB information (USCB, 2000e: USCB, 2000f). The low-income and minority populations of all block groups wholly or partially within the 50-mile radius were identified if that block group contained a significant "minority population" or a "low-income population" as defined by NRC (2003). The results of the GIS modeling are shown on Table D.7. which indicates state and county percentages of racial composition and low income status for comparison.

Table D6 provides the number of block groups entirely or partially in the 50 mile radius with minority or low-income populations.

Table D-7 contains the state and county percentages of low-income and minority populations. These data were compared to the percentages of low income households and minority populations in each block group in the 50-mile radius to arrive at the information in Table D-6.

Ninety-six block groups are within 50 miles of the project. Block groups within 50 miles of the proposed project location have Black, Some Other Race, Aggregate, Hispanic and low-income populations (Table D-6).

D.4   Construction and Operation Workforce Characteristics Calculations

The tables below present the assumptions used for construction and operation workforce assessments presented in Chapter 4 of this EIS. Table D-8 presents the construction workforce characteristics during construction of the proposed facility (IIFP, 2011) and assumptions based on NRC studies of workforces in substantially similar situations (BMI, 1981).

Table D-9 presents the operations workforce estimated number of on-site employees during the Phase I operation of the proposed IIFP facility (IIFP, 2011), and assumptions based on NRC studies of workforces in substantially similar situations (BMI, 1981).

D.5   Socioeconomic Calculations Used in Chapter 4 – Environmental Consequences

Table D-10 presents the calculations used to support the conclusions presented in Chapter 4 of this EIS related to population, employment, income, housing, public utilities, and education.

Table D-6. Race and Low-income Population Block Groups within 50 miles of the Proposed Project.

State	County	County FIPS Number	Number of Block Groups	Black	American Indian or Alaskan Native	Asian	Native Hawaiian or Other Pacific Islander	Some Other Race	Two or More Races	Aggregate	Hispanic	Low-Income Households
New Mexico	Chaves	5	2	0	0	0	0	0	0	0	1	0
New Mexico	Eddy	15	3	0	0	0	0	0	0	0	1	0
New Mexico	Lea	25	64	1	0	0	0	14	0	10	24	10
Texas	Andrews	3	3	0	0	0	0	0	0	0	0	0
Texas	Cochran	79	1	0	0	0	0	0	0	0	0	0
Texas	Gaines	165	13	0	0	0	0	0	0	1	3	0
Texas	Loving	301	1	0	0	0	0	0	0	0	0	0
Texas	Terry	445	1	0	0	0	0	0	0	0	0	0
Texas	Winkler	495	1	0	0	0	0	0	0	0	0	0
Texas	Yoakum	501	7	1	0	0	0	1	0	0	3	0
		Totals:	96	1	0	0	0	15	0	11	32	10

Source: ESRI, 2008; USCB, 2000a; USCB, 2000b; USCB, 2000c; USCB, 2000d; USCB, 2000e; USCB, 2000f; USCB, 2000g; USCB, 2000h

Table D-7. State and County Percentages of Race and Low-Income Populations

State	County	Black(%)	American Indian or Alaskan Native (%)	Asian (%)	Native Hawaiian or Other Pacific Islander (%)	Some Other Race (%)	Multi-Racial (%)	Aggregate (%)	Hispanic (%)	Low-Income Households (%)
New Mexico (state only)	NA	1.89	9.54	1.06	0.08	17.04	3.65	33.25	42.08	16.78
New Mexico	Chaves	1.97	1.13	0.53	0.06	21.25	3.12	28.05	43.83	19.12
New Mexico	Eddy	1.56	1.25	0.45	0.09	17.67	2.64	23.66	38.76	16.72
New Mexico	Lea	4.37	0.99	0.39	0.04	23.81	3.27	32.87	39.65	19.90
Texas (state only)	NA	11.53	0.57	2.70	0.07	11.69	2.47	29.03	31.99	13.98
Texas	Andrews	1.65	0.88	0.71	0.02	16.79	2.87	22.92	40.00	16.74
Texas	Cochran	4.53	0.83	0.21	0.05	27.35	2.55	35.52	44.13	21.67
Texas	Gaines	2.28	0.76	0.15	0.01	14.17	2.35	19.72	35.77	19.08
Texas	Loving	0.00	0.00	0.00	0.00	8.96	1.49	10.45	10.45	0.00
Texas	Terry	5.00	0.53	0.22	0.02	14.28	3.40	23.45	44.09	20.53
Texas	Winkler	1.85	0.45	0.20	0.00	20.35	2.34	25.19	44.00	18.58
Texas	Yoakum	1.39	0.71	0.12	0.01	25.48	1.65	29.38	45.93	18.20

Source: USCB, 2000a; USCB, 2000b; USCB, 2000c; USCB, 2000d; USBC, 2000g; USCB, 2000h

Table D-8. Workforce Characterization During IIFP Phase 1 Construction

WORKFORCE CHARACTERIZATION	
Peak number of workers on-site during construction (IIFP, 2011)	140
WORKFORCE MIGRATION	
Percent of construction workforce migrating into ROI	20%
Total of construction workers migrating into ROI during construction peak	28
FAMILIES	
Percent of construction workers who bring families (BMI, 1981)	70%
Percent of construction workers who do not bring families	30%
Average construction worker family size (worker, spouse, children) (BMI, 1981)	3.25
Number of construction workers who would move into ROI and bring families	20
Number of construction workers who would move into ROI and not bring families	8
TOTAL IN-MIGRATION - FAMILIES AND UNACCOMPANIED WORKERS	
Number of construction workers who would bring families into ROI (total new families in ROI)	20
Number of in-migrating workers' family members	44
Number of in-migrating workers accompanied by family, plus family members	64
Number of in-migrating workers who would not bring families into ROI	8
Number of in-migrating workers and family members (= new population in ROI)	72
SCHOOL-AGE CHILDREN	
Number of school-age children per construction family (BMI, 1981)	0.8
Number of in-migrating school-age children	16
POST-CONSTRUCTION WORKFORCE RETENTION	
Percent of in-migrating construction workers that would leave, post-construction (BMI, 1981)	50%
Number of in-migrating construction workers that would leave ROI, post-construction	14
Number of in-migrating construction workers and their families plus in-migrating workers without families that would leave ROI, post-construction	36
Number of school-age children of in-migrating construction workers that would migrate to ROI	16
Number of in-migrating school-age children that would leave ROI, post-construction	8
EMPLOYMENT	
Construction workforce peak	140
Number of construction workers who migrate into ROI (20% of construction workforce peak)	28
Employment multiplier for construction workers in ROI (indirect portion only) (BEA, 2010b)	0.4324
Indirect jobs resulting from in-migrating construction workers	12

Sources: BEA .2010b; BMI. 1981; IIFP. 2011

Table D-9. Workforce Characterization During IIFP Phase 1 Operation

WORKFORCE CHARACTERIZATION	
Peak number of workers on-site during operation (IIFP, 2011)	140
WORKFORCE MIGRATION	
Percent of operation workforce migrating into ROI	20%
Number of operation workers migrating into ROI during operation peak	28
FAMILIES	
Percent of operation workers who bring families (BMI, 1981)	100%
Percent of workers who do not bring families	0%
Average New Mexico family size, 2009 (USCB, 2010c)	3.23
Number of operation workers who would move into ROI and bring families	28
Number of operation workers who would move into ROI and not bring families	0
TOTAL IN-MIGRATION - FAMILIES AND UNACCOMPANIED WORKERS	
Number of operation workers who would bring families into ROI (= total new families in ROI)	28
Number of in-migrating operation worker family members	62
Number of in-migrating operation workers accompanied by family, plus family members	90
Number of operation workers who would not bring families into ROI	0
Number of operation workers and family members migrating into ROI (= new population in ROI)	90
SCHOOL-AGE CHILDREN	
Number of school-age children per family (BMI, 1981)	0.8
Number of in-migrating school-age children	22
EMPLOYMENT	
Operation workforce peak	140
Number of operation workers who migrate into ROI (20% of workforce peak)	28
Employment multiplier for operation workers in ROI (indirect portion only) (BEA, 2010b)	1.8173
Indirect jobs resulting from in-migrating operation workers	51
Number of persons unemployed in ROI, June 2010 (BLS, 2010a)	3,993

Sources: BEA, 2010b; BLS, 2010a; BMI., 1981; IIFP, 2011; USCB, 2010c.

Table D-10. Socioeconomic Calculations

	Phase 1 Construction	Phase 1 Operation
POPULATION		
2009 ROI Population (USCB, 2010e)	112,938	112,938
Total In-migration Associated with Phase 1 of the IIFP Project	72	90
Percent ROI Population Increase related to IIFP Project Phase 1	0.06%	0.08%
EMPLOYMENT AND INCOME		
June 2010 ROI Labor Force (BLS, 2010a)	56,945	56,945
Estimated Number of people, who would become IIFP Phase 1 Employees, Currently Living within the ROI (80% of workforce)	112	112
Number of In-migrating IIFP Phase 1 Workers	28	28
June 2010 ROI Labor Force Plus In-migrating IIFP Phase 1 Workers	56,973	56,973
Percent Jobs Filled by In-migrants Represent of June 2010 ROI Labor Force	0.05%	0.05%
June 2010 ROI, Unemployment Rate (BLS, 2010a)	7.0%	7.0%
June 2010 ROI, Number of People Employed (BLS, 2010a)	52,952	52,959
June 2010 ROI, Number of People Unemployed (BLS, 2010a)	3,993	3,993
Number of Indirect Jobs Created (BEA, 2010b)	12	51
Percent Indirect Jobs Represent of the June 2010 ROI Labor Force	0.02%	0.09%
HOUSING		
Vacant Housing Units in the ROI (USCB, 2010d)	5,823	5,823
Housing Units Needed for In-migrating IIFP Workers	28	28
Percent of Needed Housing Units Represent of Vacant Housing Units	0.48%	0.48%

Table D-10. Socioeconomic Calculations (Continued)

	Phase 1 Construction	Phase 1 Operation
PUBLIC UTILITIES		
People Served by Major Public Water Suppliers in 2007-2009 (NMED, 2010a)	88,643	88,643
Number of IIFP Phase 1 Workers and their Family Members Who Would Migrate into the ROI	72	90
Percent Increase of People to be Served by Major Public Water Suppliers	0.08%	0.10%
Number of People Served by Major Public Wastewater Systems, 2009 (NMED, 2010b; Artesia, 2010: Carlsbad, 2010; Appendix A; Lovington, 2010)	78,917	78,917
Percent Increase of People to be Served by Major Wastewater Systems	0.09%	0.11%
EDUCATION		
2008 Public School Enrollment (NCES, 2010)	22,347	22,847
Number of School-Aged children of IIFP In-migrants Eligible for Public School Enrollment	16	22
Percent Increase School-aged Children In-migrants Represent of 2008 ROI Public School Enrollment	0.07%	0.10%

Source: Artesia, 2010; BEA, 2010b; BLS, 2010a; Carlsbad, 2010; Appendix A, Lovington, 2010; NCES, 2010; NMED, 2010a; NMED, 2010b; USCB, 2010d; USCB, 2010e

D.6   References

(Artesia, 2010) City of Artesia Water and Wastewater Department. 2010. Personal Communication between M. Stroud, (City of Artesia Water and Wastewater Department) and P. Baxter (TtNUS). September 28, 2010. ADAMS Accession No. ML112840402.

(BEA, 2010a) U.S. Department of Commerce, Bureau of Economic Analysis. 2010. Table CA25N-Total full-time and part-time employment in NAICS industry, Eddy County New Mexico, Lea County New Mexico, Andrews County Texas, Gaines County Texas, and Yoakum County Texas. Update April. Available at http://bea.gov/. Accessed August 20, 2010. ADAMS Accession No. ML112920317.

(BEA, 2010b) U.S. Department of Commerce, Bureau of Economic Analysis. 2010. RIMS II Multipliers for Eddy, NM and Lea, NM. Economic and Statistics Administration, Washington, D.C. September, 2010. ADAMS Accession No. ML112720418.

(BLS, 2010a) U S Department of Labor, Bureau Labor Statistics. 2010.. Local Area Unemployment Statistics (LAUS), 2010 to 2011, New Mexico, Lea County and Eddy County New Mexico" Available at http://data.bls.gov/. Accessed August 23 and 31, 2010. ADAMS Accession No. ML103430031.

(BLS, 2010b) U.S. Department of Labor, Bureau of Labor Statistics. 2010. Local Area Unemployment StatisticsAndrews County Texas, Gaines County Texas, Yoakum County Texas. Available at hhtp://data.bls.gov/. Accessed August 23 and 31, 2010. ADAMS Accession No. 112710648.

(Carlsbad, 2010) City of Carlsbad Waste Water Treatment Plant. 2010. Personal Communication between A. Sena (City of Carlsbad Waste Water Treatment Superintendent) and P. Baxter (TtNUS). November 17, 2010. ADAMS Accession No. ML112840422

(ESRI, 2006) Environmental Systems Research, Inc. 2006. U.S. Census Block Groups. Tele Atlas North America, Inc. Redlands, CA.

(IIFP, 2011) International Isotopes Fluorine Products, Inc. 2011. Fluorine Extraction Process and Depleted Uranium De-conversion Plant (FEP/DUP) Official Responses to Environmental Report RAIs, Revision A. March 31, 2011. ADAMS Accession No. ML110970481.

(Lovington, 2010) City of Lovington Wastewater Treatment Plant. 2010. Personal Communication between M. De La Cruz (City of Lovington Wastewater Treatment Plant) and P. Baxter (TtNUS). September 30, 2010. ADAMS Accession No. ML112840439.

(NCES, 2010) National Center for Education Statistics. 2010. Search for Public School Districts, Lea and Eddy Counties, New Mexico. Available at http://nces.ed.gov/ccd/districtsearch/. Accessed August 30, 2010. ADAMS Accession No. ML103430054.

(NMED, 2010a) New Mexico Environment Department. 2010. "RE: FOIA Request." Personal Communication between M. Huber (New Mexico Environment Department) and N. Hill (TtNUS). September 22, 2010. ADAMS Accession No. ML112850086.

(NMED, 2010b) New Mexico Environment Department, Ground Water Quality Bureau. 2010. Ground Water Regulated Facilities, List of Current Discharge Permits, updated 07/29/2010. Available at http://www.nmenv.state.nm.us/gwb/NMED-GWQB-Permits.htm. Accessed August 31, 2010. ADAMS Accession No. ML112710484,

(NRC, 2003). U.S. Nuclear Regulatory Commission. 2003. Environmental Review Guidance for Licensing Actions Associated with NMSS Programs. Final Report. NUREG 1748. August 2003. Division of Waste Management, Washington, D.C. ADAMS Accession No. ML032450279.

(USCB, 2000a) U. S. Census Bureau 2000. *Summary File 1 for New Mexico: Census 2000.* Available at http://www2.census.gov/census_2000/datasets/Summary_File_1/New_Mexico/. Accessed July 23, 2010. ADAMS Accession No. ML112710551.

(USCB, 2000b) U. S. Census Bureau 2000. *Summary File 1 for Texas: Census 2000.* Available at http://www2.census.gov/census_2000/datasets/Summary_File_1/Texas/. Accessed July 23, 2010. ADAMS Accession No. ML112710551.

(USCB, 2000c) U. S. Census Bureau 2000. *Summary File 3 for New Mexico: Census 2000.* Available at http://www2.census.gov/census_2000/datasets/Summary_File_3/New_Mexico/. Accessed July 23, 2010. ADAMS Accession No. ML112710554.

(USCB, 2000d) U. S. Census Bureau 2000. *Summary File 3 for Texas: Census 2000.* Available at http://www2.census.gov/census_2000/datasets/Summary_File_1/Texas/. Accessed July 23, 2010. ADAMS Accession No. ML112710554.

(USCB, 2000e) U. S. Census Bureau 2000. Texas Fact Sheet. Census 2000 Demographic Profile Highlights. American Fact Finder. Available at http://factfinder.census.gov/. Accessed july 27, 2010.

(USCB, 2000f). U. S. Census Bureau 2000. New Mexico Fact Sheet. Census 2000 Demographic Profile Highlights. American Fact Finder. Available at http://factfinder.census.gov/. Accessed July 27, 2010.

(USCB, 2000g) U. S. Census Bureau 2000. P92. Poverty Status in 1999 of Households by Household Type by Age of Householder. Detailed Tables. Census 2000 Summary File 3 (SF 3) Sample Data. Available at http://factfinder.census.gov/. Accessed July 26, 2010.

(USCB, 2000h) U. S. Census Bureau 2000. P3. Race and P4. Hispanic or Latino, and Not Hispanic or Latino by Race (Total Population). Detailed Tables Census 2000 Summary Fie 1 (SF 1) 100-Percent Data. Available at http://factfinder.census.gov/. Accessed September 7, 2010.

(USCB, 2010a) U. S. Census Bureau. 2010. State and County Quick Facts, "Lea County and Eddy County New Mexico." Revised August 16. Available at http://quickfacts.census.gov/. Accessed September 21, 2010. ADAMS Accession No. ML103430121.

(USCB, 2010b) U. S. Census Bureau 2010. Table 4. Annual Estimate of the Resident Population for Incorporated Places, April 1 2000 to July 1 2009. New Mexico and Texas. Release date June 2010. Available at http://www.census.gov/. Accessed August 23, 2010.

(USCB, 2010c) U S Census Bureau. 2010. American FactFinder. New Mexico Fact Sheet. 2005-2009 American Community Survey 5-Year Estimates. Data Profile Highlights. Available online at http://factfinder.census.gov. Accessed April 11, 2011 ADAMS Accession No. ML112720609.

(USCB, 2010d) U. S. Census Bureau. 2010. American FactFinder, Fact Sheet, American Community Survey 2006 - 2008, Eddy County, New Mexico, Carlsbad City, New Mexico, Hobbs City, New Mexico, and Lea County New Mexico. Revised August 16, 2010. Available at http://quickfacts.census.gov/. Accessed September 21, 2010. ADAMS Accession No. ML112920332.

(USCB, 2010e) U. S. Census Bureau. 2010. U. S. Census Bureau. 2010. State and County Quick Facts, New Mexico and Lea County, New Mexico, Revised August 16. Available at http://quickfacts.census.gov/. Accessed September 21, 2010. ADAMS Accession No. ML103430126.

(USCB, 2003) U. S. Census Bureau. 2003. Table 3: County-To-County Worker Flow, 2000, Residence County to Workplace County Flows for New Mexico and Texas. New Mexico and Texas. Available at http://www.census.gov/. Accessed August 23, 2010. ADAMS Accession No. ML112920337.

APPENDIX E

TRANSPORTATION OF RADIOACTIVE MATERIALS

TRANSPORTATION OF RADIOACTIVE MATERIALS

E.1   Introduction

This Appendix summarizes calculations that were used in making determinations within the EIS, related to the transportation of radioactive materials. The proposed IIFP Depleted Uranium Deconversion Plant/Fluorine Extraction Process Facility would be located in Hobbs, New Mexico. The facility would receive depleted uranium (DU) in the chemical form of DUF_6 and convert it to a more stable and disposable chemical form of DUO_2. The process would recover fluorine which would be available for sale on the market. The deconversion process requires transportation of the DU cylinders (full) from current storage locations at enrichment facilities, disposal of low-level radioactive waste (LLW), and possible transportation of empty DU cylinders.

E.2   Radioactive Materials Transportation Analysis

The DUF_6 would be transported to the IIFP facility in 48Y cylinders designed for storage and transportation of DUF_6. All current or proposed U.S. commercial enrichment facilities were identified as representative origins for shipments of DUF_6. These are (1) Urenco USA facility just east of Eunice, New Mexico, (2) the GE-Hitachi Global Laser Enrichment (GLE) Facility north of Wilmington, North Carolina, and (3) the Areva Eagle Rock Enrichment Facility west of Idaho Falls, Idaho. The cylinders would be shipped one per 18-wheel truck. The empty DUF_6 cylinders would be shipped back to the location of origin. In the event that cylinders are not returned, they could be disposed as LLW or filled with DUO_2 and disposed as LLW. The empty cylinders are conservatively assumed to be shipped one per truck, consistent with IIFP data; however, two per truck is also a likely scenario.

The DUO_2 is assumed to be waste. It would be packaged into 55-gallon drums and loaded 40 per truck (subject to weight limitations). Shipment destinations selected for analysis are the Energy*Solutions* Clive, Utah facility and the Waste Control Specialists (WCS) facility on the Texas-New Mexico border west of Andrews, Texas (immediately east of the Urenco USA facility).

Process LLW (low-level waste resulting from the deconversion process) and miscellaneous LLW (low-level waste incidental to the deconversion process) volumes would be small compared to the DUO_2 waste. The radioactivity in most of this waste would likely be less concentrated than the DUO_2 waste. The process and miscellaneous LLW also would be packaged into 55-gallon drums, loaded 40 per truck, and shipped to the same disposal facilities as the DUO_2 waste. Decommissioning waste would be similar to miscellaneous LLW and would be packaged into 55-gallon drums, loaded 40 per truck, and shipped to the same disposal facilities as the LLW and DUO_2 waste.

Routing characteristics, including distances travelled, population density along the route, and stop time for crew breaks and inspecting the cargo were generated by the TRAGIS Code, Version 1.5.4 (Johnson and Michelhaugh, 2003). Radiological impacts from radioactive material shipments were calculated using the RADTRAN Code, Version 5.6 (Wiener et. al, 2006).

Input parameters for the transportation analysis were obtained from IIFP (IIFP, 2011), NUREG-0170 (NRC, 1977), and the Louisiana Energy Services (LES) Gas Centrifuge Facility License Application (REF) and are provided in Tables E-1 and E-3. The numbers of shipments and relative travel distances were provided by IIFP (IIFP, 2011a)) and accident frequency and severity were provided by NUREG–0170 (NRC, 1977). Dimensions of packages and similar information presented in Tables E-1 and E-2 were from the LES Environmental Impact Statement (NRC, 2005). State-specific accident and fatality rates are from Table 4 of the study, State-Level Accident Rates for Surface Freight Transportation: A Reexamination (Saricks and Tompkins, 1999).

The RADTRAN results and the Microsoft Excel calculations are provided in E-4 through E-9.

Table E-1A. Input Parameters for 48Y Cylinders (Part 1 of 3)

Parameter Description	Input Parameters	
Title of Project	Truck transport of Empty/Full 48Y DUF$_6$ Cylinder to Destination	
Accident Options	Incident Free, Accident	
Output Level	1	
Health Effects	Rem/Person-rem	
Package Parameters		**Source**
Package Name	48Y-Cylinder	Appendix D, Table D-4, LES EIS
Long Dimension (m)	3.73	Appendix D, Table D-4, LES EIS
Dose Rate (mrem/h)		
Full DUF$_6$ Cylinders	2.80 x 10-1	mrem/hr @ 1 meter	Appendix D, Table D-7, LES EIS
Empty DUF$_6$ Cylinders	1.00	mrem/hr @ 1 meter	Appendix D, Table D-7, LES EIS
Gamma Fraction	1	RADTRAN Default
Neutron Fraction	0	RADTRAN Default
Radionuclide Parameters		
Package Name	48Y-Cylinder	
Radionuclide	See Inventory	
Physical/Chemical Group	Powder for solids and Gas for Radon	
Curies	See Inventory	
Vehicle Parameters		**Source**
Vehicle Name	Vehicle-1	
Number of Shipments	1	User Defined Value
Vehicle Size (m)	3.73	same as package size
Vehicle Dose Rate (mrem/h)		same as package dose rate
Gamma Fraction	1	RADTRAN Default
Neutron Fraction	0	RADTRAN Default
Crew Size	2	NUREG 0170
Crew Distance	3.1	NUREG 0170
Crew Shielding Factor	1	NUREG 0170
Crew View	1.22	Appendix D, Table D-4, LES EIS
Exclusive Use	Yes	RADTRAN Default
Package	48Y-Cylinder	User Defined Value
Number of Packages	1	User Defined Value

Table E-1B. Input Parameters for 48Y Cylinders (Part 2 of 3)

Parameter Description	_Input Parameters_			
Link Parameters				**Source**
Link Name				
Vehicle Name	Vehicle-1	Vehicle-1	Vehicle-1	
Length (km)	Route specific, see TRAGIS output			TRAGIS output
Speed (km/h)	88.49	40.25	24.16	NUREG 0170
Population Density (persons/km^2)	Route specific, see TRAGIS output			TRAGIS output
Vehicle Density (Vehicles/h)	470	780	2800	NUREG 0170
Persons per Vehicle	2	2	2	NUREG 0170
Accident Rate (accidents/veh-km)	State specific values			Saricks and Tompkins, 1999, Table 4
Fatalities Per Accident	State specific values			Saricks and Tompkins, 1999, Table 4
Zone	Rural	Suburban	Urban	RADTRAN Default
Type	Primary Highway Primary Highway Primary Highway			RADTRAN Default
Farm Fraction	0	0	0	RADTRAN Default
Stop Parameters				**Source**
Stop Name	Stop-1			
Vehicle Name	Vehicle-1			
Minimum Distance	20			NUREG 0170
Maximum Distance	20			NUREG 0170
People or People/km^2	50			NUREG 0170
Shielding Factor	1			RADTRAN Default
Time (h)	4			TRAGIS output
Handling Parameters				
Handle Name	Handle-1			
Vehicle Name	Vehicle-1			
Number of Handlers	4			NUREG 0170 (2 handlers at the shipping and 2 handlers receiving end of the route)
Distance (m)	1			NUREG 0170
Time (h)	0.25			NUREG 0170 (15 minutes)

Table E-1C. Input Parameters for 48Y Cylinders (Part 3 of 3)

Parameter Description	Input Parameters									
Accident Parameters										
	Probability Parameters									
Probability Index	0	1	2	3	4	5	6	7		
Probability Fraction	0.55	0.36	0.07	0.016	0.0028	0.0011	8.50×10^{-5}	1.50×10^{-5}	*NUREG 0170*	*NUREG 0170*
	Deposition Velocity Parameters									
Physical/Chemical Group	Powder	Gas								
Deposition Velocity (m/s)	0.01	0								
	Release Parameters									
Physical/Chemical Group	Powder									
Probability Index	0	1	2	3	4	5	6	7	*NUREG 0170*	
Release Fraction	0	0.01	0.1	1	1	1	1	1	*NUREG 0170*	
	Gas									
Probability Index	0	1	2	3	4	5	6	7	*NUREG 0170*	
Release Fraction	0	1	1	1	1	1	1	1	*User defined value*	
	Aerosol Parameters									
Physical/Chemical Group	Powder and Gas									
Probability Index	0	1	2	3	4	5	6	7	*NUREG 0170*	
Aerosol Fraction	1	1	1	1	1	1	1	1	*NUREG 0170*	
	Respirable Parameters									
Physical/Chemical Group	Powder and Gas									
Probability Index	0	1	2	3	4	5	6	7	*NUREG 0170*	
Respirable Fraction	1	1	1	1	1	1	1	1	*NUREG 0170*	
Balance of RADTRAN Inputs	RADTRAN Defaults									

Table E-2A. Input Parameters for 55-Gallon Drums (Part 1 of 3)

Parameter Description	Input Parameters	
Title of Project	Truck transport of 55-Gallon-Drums of DUO$_2$/Other Waste to Destination	
Accident Options	Incident Free, Accident	
Output Level	1	
Health Effects	Rem/Person-rem	
Package Parameters		
Package Name	55-Gallon-Drum	
Long Dimension (m)	0 88	
Dose Rate (mrem/h)		
DUO$_2$ Waste	1 93 x 10-1	mrem/hr @ 1 meter	Response to RAI 5, Table RAI 5-e-1
Other Waste	3 05 x 10^{-2}	mrem/hr @ 1 meter	Response to RAI 5, Table RAI 5-e-1 (weighted average of all except DUO$_2$)
Other Waste	9 45 x 10^{-4}	mrem/hr @ 1 meter	Response to RAI 5, Table RAI 5-e-1 (Minimum dose rate)
Gamma Fraction	1		RADTRAN Default
Neutron Fraction	0		RADTRAN Default
Radionuclide Parameters		
Package Name	55_Gallon_Drum	
Radionuclide	See Inventory	
Physical/Chemical Group	Powder for solids and Gas for Radon	
Curies	See Inventory	
Vehicle Parameters		
Vehicle Name	Vehicle_1	
Number of Shipments	1	User Defined Value
Vehicle Size (m)	12.2	the length of 20 55-gallon drums (assuming the drums are arranged 20 x 2)
Vehicle Dose Rate (mrem/h)	6.00 x 10^{-2}	same as package dose rate
Gamma Fraction	1	RADTRAN Default
Neutron Fraction	0	RADTRAN Default
Crew Size	2	NUREG 0170
Crew Distance	3.1	NUREG 0170

Table E-2A. Input Parameters for 55-Gallon Drums (Part 1 of 3) (Continued)

Parameter Description	Input Parameters	
Vehicle Parameters (con't.)		
Crew Shielding Factor	1	*NUREG 0170*
Crew View	1.22	*the width of 2 55-gallon drums*
Exclusive Use	Yes	*RADTRAN Default*
Package	55 Gallon Drum	*User Defined Value*
Number of Packages	40	*User Defined Value*

Table E-2B. Input Parameters for 55-Gallon Drums (Part 2 of 3)

Parameter Description	Input Parameters			
Link Parameters				
Link Name				
Vehicle Name	Vehicle-1	Vehicle-1	Vehicle-1	
Length (km)	Route specific, see TRAGIS output			*TRAGIS output*
Speed (km/h)	88.49	40.25	24.16	*NUREG 0170*
Population Density (persons/km²)	Route specific, see TRAGIS output			*TRAGIS output*
Vehicle Density (Vehicles/h)	470	780	2800	*NUREG 0170*
Persons per Vehicle	2	2	2	*NUREG 0170*
Accident Rate (accidents/veh-km)	State specific values			*Saricks and Tompkins, 1999, Table 4*
Fatalities Per Accident	State specific values			*Saricks and Tompkins, 1999, Table 4*
Zone	Rural	Suburban	Urban	*RADTRAN Default*
Type	Primary Highway	Primary Highway	Primary Highway	*RADTRAN Default*
Farm Fraction	0	0	0	*RADTRAN Default*
Stop Parameters				
Stop Name	Stop-1			
Vehicle Name	Vehicle-1			
Minimum Distance	20			*NUREG 0170*
Maximum Distance	20			*NUREG 0170*
People or People/km²	50			*NUREG 0170*
Shielding Factor	1			*RADTRAN Default*
Time (h)	4			*TRAGIS output*
Handling Parameters				
Handle Name	Handle-1			
Vehicle Name	Vehicle-1			
Number of Handlers	4			*NUREG 0170 (2 handlers at the shipping and 2 handlers receiving end of the route)*
Distance (m)	1			*NUREG 0170*
Time (h)	0.25			*NUREG 0170 (15 minutes)*

Table E-2C. Input Parameters for 55-Gallon Drums (Part 3 of 3)

Parameter Description	Input Parameters								
Accident Parameters									
Probability Parameters									
Probability Index	0	1	2	3	4	5	6	7	NUREG 0170
Probability Fraction	0.55	0.36	0.07	0.016	0.0028	0.0011	8.50×10^{-5}	1.50×10^{-5}	NUREG 0170
Deposition Velocity Parameters									
Physical/Chemical Group	Powder	Gas							
Deposition Velocity (m/s)	0.01	0							
Release Parameters									
Physical/Chemical Group	Powder								
Probability Index	0	1	2	3	4	5	6	7	NUREG 0170
Release Fraction	0	0.01	0.1	1	1	1	1	1	NUREG 0170
	Gas								
Probability Index	0	1	2	3	4	5	6	7	NUREG 0170
Release Fraction	0	1	1	1	1	1	1	1	User defined value
Aerosol Parameters									
Physical/Chemical Group	Powder and Gas								
Probability Index	0	1	2	3	4	5	6	7	NUREG 0170
Aerosol Fraction	1	1	1	1	1	1	1	1	NUREG 0170
Respirable Parameters									
Physical/Chemical Group	Powder and Gas								
Probability Index	0	1	2	3	4	5	6	7	NUREG 0170
Respirable Fraction	1	1	1	1	1	1	1	1	NUREG 0170
Balance of RADTRAN Inputs	RADTRAN Defaults								

Table E-3. Number of Shipments

	Phase 1	Phase 2	Cumulative
Full DUF$_6$ Cylinders each from Urenco USA, GLE Facility, and Areva Eagle Rock	293	635	928
Empty DUF$_6$ Cylinders each to Urenco USA, GLE Facility, and Areva Eagle Rock	293	496	789
DUO$_2$ each to Energy Solutions and WCS	155	295	476
Miscellaneous Waste each to Energy Solutions and WCS	31	20	51
Decommissioning Waste each to Energy Solutions and WCS			64

Values from Table 3-2 of IIFP, 2011

Table E-4. Incident Free RADTRAN Output

	Incident Free Transportation Impacts (Person-Rem)					
	Crew	Off Link	On Link	Totals	Handling	Stops
Full DUF$_6$ Cylinders from Urenco USA	1.03×10^{-4}	9.95×10^{-7}	1.09×10^{-5}	1.15×10^{-4}	8.02×10^{-4}	a
Full DUF$_6$ Cylinders from GLE Facility	6.96×10^{-3}	2.88×10^{-4}	2.01×10^{-3}	9.26×10^{-3}	8.02×10^{-4}	1.01×10^{-3}
Full DUF$_6$ Cylinders from Areva Eagle Rock	6.13×10^{-3}	1.66×10^{-4}	1.63×10^{-3}	7.93×10^{-3}	8.02×10^{-4}	1.01×10^{-3}
Empty DUF$_6$ Cylinders to Urenco USA	3.68×10^{-4}	3.55×10^{-6}	3.88×10^{-5}	4.10×10^{-4}	2.86×10^{-3}	a
Empty DUF$_6$ Cylinders to GLE	2.49×10^{-2}	1.03×10^{-3}	7.18×10^{-3}	3.31×10^{-2}	2.86×10^{-3}	3.59×10^{-3}
Empty DUF$_6$ Cylinders to Areva Eagle Rock	2.19×10^{-2}	5.92×10^{-4}	5.84×10^{-3}	2.83×10^{-2}	2.86×10^{-3}	3.59×10^{-3}
DUO$_2$ to Energy *Solutions*	3.95×10^{-3}	3.32×10^{-4}	3.60×10^{-3}	7.88×10^{-3}	1.11×10^{-2}	1.86×10^{-3}
DUO$_2$ to WCS	7.09×10^{-5}	2.15×10^{-6}	2.35×10^{-5}	9.66×10^{-5}	1.11×10^{-2}	a
Miscellaneous Waste to Energy *Solutions*	1.93×10^{-5}	1.63×10^{-6}	1.76×10^{-5}	3.86×10^{-5}	5.44×10^{-5}	9.13×10^{-6}
Miscellaneous Waste to WCS	3.47×10^{-7}	1.05×10^{-8}	1.15×10^{-7}	4.73×10^{-7}	5.44×10^{-5}	
Miscellaneous Waste to Energy *Solutions*	6.24×10^{-4}	5.25×10^{-5}	5.69×10^{-4}	1.25×10^{-3}	1.76×10^{-3}	2.95×10^{-4}
Miscellaneous Waste to WCS	1.12×10^{-5}	3.40×10^{-7}	3.71×10^{-6}	1.53×10^{-5}	1.76×10^{-3}	a

a: A stop was not assumed since the route was short.
Note: The Decommissioning Waste is the same as Miscellaneous Waste.

Table E-5. Accident RADTRAN Output

	Accident Transportation Impacts (Person-Rem)				
	Ground	Inhaled	Resuspended	Cloudshine	Total
Full DUF$_6$ Cylinders from Urenco USA	2.89×10^{-8}	4.25×10^{-6}	5.37×10^{-9}	3.84×10^{-12}	4.29×10^{-6}
Full DUF$_6$ Cylinders from GLE Facility	9.84×10^{-5}	1.45×10^{-2}	1.83×10^{-5}	1.31×10^{-8}	1.46×10^{-2}
Full DUF$_6$ Cylinders from Areva Eagle Rock	7.30×10^{-5}	1.07×10^{-2}	1.35×10^{-5}	9.67×10^{-9}	1.08×10^{-2}
Empty DUF$_6$ Cylinders to Urenco USA	2.90×10^{-10}	1.54×10^{-8}	2.31×10^{-11}	3.32×10^{-12}	1.57×10^{-8}
Empty DUF$_6$ Cylinders to GLE	9.79×10^{-7}	5.19×10^{-5}	7.80×10^{-8}	1.12×10^{-8}	5.29×10^{-5}
Empty DUF$_6$ Cylinders to Areva Eagle Rock	7.31×10^{-7}	3.87×10^{-5}	5.83×10^{-8}	8.36×10^{-9}	3.95×10^{-5}
DUO$_2$ to EnergySolutions	1.39×10^{-4}	2.17×10^{-2}	2.61×10^{-5}	1.96×10^{-8}	2.19×10^{-2}
DUO$_2$ to WCS	5.18×10^{-8}	8.12×10^{-6}	9.74×10^{-9}	7.32×10^{-12}	8.18×10^{-6}
Miscellaneous Waste to EnergySolutions (Low TI)	1.32×10^{-6}	1.05×10^{-4}	8.73×10^{-7}	7.44×10^{-11}	1.07×10^{-4}
Miscellaneous Waste to WCS (Low TI)	4.95×10^{-10}	3.91×10^{-8}	3.27×10^{-10}	2.78×10^{-14}	3.99×10^{-8}
Miscellaneous Waste to EnergySolutions	1.32×10^{-6}	1.05×10^{-4}	8.73×10^{-7}	7.44×10^{-11}	1.07×10^{-4}
Miscellaneous Waste to WCS	4.95×10^{-10}	3.91×10^{-8}	3.27×10^{-10}	2.78×10^{-14}	3.99×10^{-8}

Note: The Decommissioning Waste is the same as Miscellaneous Waste

Table E-6. Phase 1 Collective Doses to Various Receptors from Radiological Transportation

	General Public		Drivers and Passengers		Persons at Stops		Truck Drivers		Package Handlers	
	Person-Sv	Person-rem	Person-Sv	Person-rem	Person-Sv	Person-rem	Person-Sv	Person-rem	Person-Sv	Person-rem
Full DUF$_6$ Cylinders from Urenco USA	2.9×10^{-6}	2.9×10^{-4}	3.2×10^{-5}	3.2×10^{-3}	a	a	3.0×10^{-4}	3.0×10^{-2}	2.3×10^{-3}	2.3×10^{-1}
Full DUF$_6$ Cylinders from GLE Facility	8.4×10^{-4}	8.4×10^{-2}	5.9×10^{-3}	5.9×10^{-1}	3.0×10^{-3}	3.0×10^{-1}	2.0×10^{-2}	2.0	2.3×10^{-3}	2.3×10^{-1}
Full DUF$_6$ Cylinders from Areva Eagle Rock	4.9×10^{-4}	4.9×10^{-2}	4.8×10^{-3}	4.8×10^{-1}	3.0×10^{-3}	3.0×10^{-1}	1.8×10^{-2}	1.8	2.3×10^{-3}	2.3×10^{-1}
Empty DUF$_6$ Cylinders to Urenco USA	1.0×10^{-5}	1.0×10^{-3}	1.1×10^{-4}	1.1×10^{-2}	a	a	1.1×10^{-3}	1.1×10^{-1}	8.4×10^{-3}	8.4×10^{-1}
Empty DUF$_6$ Cylinders to GLE	3.0×10^{-3}	3.0×10^{-1}	2.1×10^{-2}	2.1	1.1×10^{-2}	1.1	7.3×10^{-2}	7.3	8.4×10^{-3}	8.4×10^{-1}
Empty DUF$_6$ Cylinders to Areva Eagle Rock	1.7×10^{-3}	1.7×10^{-1}	1.7×10^{-2}	1.7	1.1×10^{-2}	1.1	6.4×10^{-2}	6.4	8.4×10^{-3}	8.4×10^{-1}
DUO$_2$ to EnergySolutions	5.1×10^{-4}	5.1×10^{-2}	5.6×10^{-3}	5.6×10^{-1}	2.9×10^{-3}	2.9×10^{-1}	6.1×10^{-3}	6.1×10^{-1}	1.7×10^{-2}	1.7
DUO$_2$ to WCS	3.3×10^{-6}	3.3×10^{-4}	3.6×10^{-5}	3.6×10^{-3}	a	a	1.1×10^{-4}	1.1×10^{-2}	1.7×10^{-2}	1.7
Miscellaneous Waste to EnergySolutions	1.6×10^{-5}	1.6×10^{-3}	1.8×10^{-4}	1.8×10^{-2}	9.1×10^{-6}	9.1×10^{-3}	1.9×10^{-4}	1.9×10^{-2}	5.5×10^{-4}	5.5×10^{-2}
Miscellaneous Waste to WCS	1.1×10^{-7}	1.1×10^{-5}	1.2×10^{-6}	1.2×10^{-4}	a	a	3.5×10^{-6}	3.5×10^{-4}	5.5×10^{-4}	5.5×10^{-2}
DUO$_2$ and Misc to EnergySolutions	5.3×10^{-4}	5.3×10^{-2}	5.8×10^{-3}	5.8×10^{-1}	$3.0 \times 10{-3}$	3.0×10^{-1}	6.3×10^{-3}	6.3×10^{-1}	1.8×10^{-2}	1.8
DUO$_2$ and Misc to WCS	3.4×10^{-6}	3.4×10^{-4}	3.8×10^{-5}	3.8×10^{-3}	a	a	1.1×10^{-4}	1.1×10^{-2}	1.8×10^{-2}	1.8
Greatest risk scenario	*4.4×10^{-3}*	*4.4×10^{-1}*	*3.3×10^{-2}*	*3.3*	*1.6×10^{-2}*	*1.6*	*1.0×10^{-1}*	*1.0×10*	*2.8×10^{-2}*	*2.8*

Table E-7. Phase 2 Incremental Collective Doses to Various Receptors from Radiological Transportation

	General Public		Drivers and Passengers		Persons at Stops		Truck Drivers		Package Handlers	
	Person-Sv	Person-rem	Person-Sv	Person-rem	Person-Sv	Person-rem	Person-Sv	Person-rem	Person-Sv	Person-rem
Full DUF$_6$ Cylinders from Urenco USA	6.3×10^{-6}	6.3×10^{-4}	6.9×10^{-5}	6.9×10^{-3}	A	a	6.5×10^{-4}	6.5×10^{-2}	5.1×10^{-3}	5.1×10^{-1}
Full DUF$_6$ Cylinders from GLE Facility	1.8×10^{-3}	1.8×10^{-1}	1.3×10^{-2}	1.3	6.4×10^{-3}	6.4×10^{-1}	4.4×10^{-2}	4.4	5.1×10^{-3}	5.1×10^{-1}
Full DUF$_6$ Cylinders from Areva Eagle Rock	1.1×10^{-3}	1.1×10^{-1}	1.0×10^{-2}	1.0	6.4×10^{-3}	6.4×10^{-1}	3.9×10^{-2}	3.9	5.1×10^{-3}	5.1×10^{-1}
Empty DUF$_6$ Cylinders to Urenco USA	1.8×10^{-5}	1.8×10^{-3}	1.9×10^{-4}	1.9×10^{-2}	A	a	1.8×10^{-3}	1.8×10^{-1}	1.4×10^{-2}	1.4
Empty DUF$_6$ Cylinders to GLE	5.1×10^{-3}	5.1×10^{-1}	3.6×10^{-2}	3.6	1.8×10^{-2}	1.8	1.2×10^{-1}	1.2×10	1.4×10^{-2}	1.4
Empty DUF$_6$ Cylinders to Areva Eagle Rock	2.9×10^{-3}	2.9×10^{-1}	2.9×10^{-2}	2.9	1.8×10^{-2}	1.8	1.1×10^{-1}	1.1×10	1.4×10^{-2}	1.4
DUO$_2$ to EnergySolutions	9.8×10^{-4}	9.8×10^{-2}	1.1×10^{-2}	1.1	5.5×10^{-3}	5.5×10^{-1}	1.2×10^{-2}	1.2	3.3×10^{-2}	3.3
DUO$_2$ to WCS	6.3×10^{-6}	6.3×10^{-4}	6.9×10^{-5}	6.9×10^{-3}	A	a	2.1×10^{-4}	2.1×10^{-2}	3.3×10^{-2}	3.3
Miscellaneous Waste to EnergySolutions	1.1×10^{-5}	1.1×10^{-3}	1.1×10^{-4}	1.1×10^{-2}	5.9×10^{-5}	5.9×10^{-3}	1.2×10^{-4}	1.2×10^{-2}	3.5×10^{-4}	3.5×10^{-2}
Miscellaneous Waste to WCS	6.8×10^{-8}	6.8×10^{-6}	7.4×10^{-7}	7.4×10^{-5}	a	a	2.2×10^{-6}	2.2×10^{-4}	3.5×10^{-4}	3.5×10^{-2}

E-13

Table E-8. Cumulative Collective Doses to Various Receptors from Radiological Transportation

	General Public		Drivers and Passengers		Persons at Stops		Truck Drivers		Package Handlers	
	Person-Sv	Person-rem	Person-Sv	Person-rem	Person-Sv	Person-rem	Person-Sv	Person-rem	Person-Sv	Person-rem
Full DUF$_6$ Cylinders from Urenco USA	9.2×10^{-6}	9.2×10^{-4}	1.0×10^{-4}	1.0×10^{-2}	a	a	9.6×10^{-4}	9.6×10^{-2}	7.4×10^{-3}	7.4×10^{-1}
Full DUF$_6$ Cylinders from GLE Facility	2.7×10^{-3}	2.7×10^{-1}	1.9×10^{-2}	1.9	9.4×10^{-3}	9.4×10^{-1}	6.5×10^{-2}	6.5	7.4×10^{-3}	7.4×10^{-1}
Full DUF$_6$ Cylinders from Areva Eagle Rock	1.5×10^{-3}	1.5×10^{-1}	1.5×10^{-2}	1.5	9.4×10^{-3}	9.4×10^{-1}	5.7×10^{-2}	5.7	7.4×10^{-3}	7.4×10^{-1}
Empty DUF$_6$ Cylinders to Urenco USA	2.8×10^{-5}	2.8×10^{-3}	3.1×10^{-4}	3.1×10^{-2}	a	a	2.9×10^{-3}	2.9×10^{-1}	2.3×10^{-2}	2.3
Empty DUF$_6$ Cylinders to GLE	8.1×10^{-3}	8.1×10^{-1}	5.7×10^{-2}	5.7	2.8×10^{-2}	2.8	2.0×10^{-1}	2.0×10	2.3×10^{-2}	2.3
Empty DUF$_6$ Cylinders to Areva Eagle Rock	4.7×10^{-3}	4.7×10^{-1}	4.6×10^{-2}	4.6	2.8×10^{-2}	2.8	1.7×10^{-1}	1.7×10	2.3×10^{-2}	2.3
DUO$_2$ to Energy Solutions	1.6×10^{-3}	1.6×10^{-1}	1.7×10^{-2}	1.7	8.9×10^{-3}	8.9×10^{-1}	1.9×10^{-2}	1.9	5.3×10^{-2}	5.3
DUO$_2$ to WCS	1.0×10^{-5}	1.0×10^{-3}	1.1×10^{-4}	1.1×10^{-2}	a	a	3.4×10^{-4}	3.4×10^{-2}	5.3×10^{-2}	5.3
Miscellaneous Waste to *EnergySolutions*	2.7×10^{-5}	2.7×10^{-3}	2.9×10^{-4}	2.9×10^{-2}	1.5×10^{-4}	1.5×10^{-2}	3.2×10^{-4}	3.2×10^{-2}	9.0×10^{-4}	9.0×10^{-2}
Miscellaneous Waste to WCS	1.7×10^{-7}	1.7×10^{-5}	1.9×10^{-6}	1.9×10^{-4}	a	a	5.7×10^{-6}	5.7×10^{-4}	9.0×10^{-4}	9.0×10^{-2}
Decommissioning Waste to *EnergySolutions*	3.4×10^{-5}	3.4×10^{-3}	3.6×10^{-4}	3.6×10^{-2}	1.9×10^{-4}	1.9×10^{-2}	4.0×10^{-4}	4.0×10^{-2}	1.1×10^{-3}	1.1×10^{-1}
Decommissioning Waste to WCS	2.2×10^{-7}	2.2×10^{-5}	2.4×10^{-6}	2.4×10^{-4}	a	a	7.2×10^{-6}	7.2×10^{-4}	1.1×10^{-3}	1.1×10^{-1}
DUO$_2$ and Misc to *EnergySolutions*	1.6×10^{-3}	1.6×10^{-1}	1.7×10^{-2}	1.7	9.0×10^{-3}	9.0×10^{-1}	1.9×10^{-2}	1.9	5.4×10^{-2}	5.4
DUO$_2$ and Misc to WCS	1.0×10^{-5}	1.0×10^{-3}	1.1×10^{-4}	1.1×10^{-2}	a	a	3.4×10^{-4}	3.4×10^{-2}	5.4×10^{-2}	5.4
Greatest risk scenario	1.2×10^{-2}	1.2	9.3×10^{-2}	9.3	4.7×10^{-2}	4.7	2.8×10^{-1}	2.8×10	8.4×10^{-2}	8.4

Table E-9. Annual Accident Dose-Risk and LCF-Risk From Radiological Transportation

	Dose-Risk		LCF Risk
	Person-Sv	Person-rem	
Full DUF$_6$ Cylinders from Urenco USA	1.3×10^{-5}	1.3×10^{-3}	7.5×10^{-7}
Full DUF$_6$ Cylinders from GLE Facility	4.3×10^{-2}	4.3	2.6×10^{-3}
Full DUF$_6$ Cylinders from Areva Eagle Rock	3.2×10^{-2}	3.2	1.9×10^{-3}
Empty DUF$_6$ Cylinders to Urenco USA	4.6×10^{-8}	4.6×10^{-6}	2.8×10^{-9}
Empty DUF$_6$ Cylinders to GLE	1.5×10^{-4}	1.5×10^{-2}	9.3×10^{-6}
Empty DUF$_6$ Cylinders to Areva Eagle Rock	1.2×10^{-4}	1.2×10^{-2}	6.9×10^{-6}
DUO$_2$ to EnergySolutions	6.4×10^{-2}	6.4	3.9×10^{-3}
DUO$_2$ to WCS	2.4×10^{-5}	2.4×10^{-3}	1.4×10^{-6}
Miscellaneous Waste to EnergySolutions	3.1×10^{-4}	3.1×10^{-2}	1.9×10^{-5}
Miscellaneous Waste to WCS	1.2×10^{-7}	1.2×10^{-5}	7.0×10^{-9}
Decommissioning Waste to EnergySolutions[a]	3.1×10^{-4}	3.1×10^{-2}	1.9×10^{-5}
Decommissioning Waste to WCS[a]	1.2×10^{-7}	1.2×10^{-5}	7.0×10^{-9}
Greatest Risk Scenario	*1.1×10^{-1}*	*1.1×10*	*6.4×10^{-3}*

[a] Represents total campaign—not annual

Note: latent cancer fatalities per person rem (ISCORS, 2002) = 6.00×10^{-4}

E.3 References

(IIFP, 2011) International Isotopes Fluorine Products. 2011. Fluorine Extraction Process and Depleted Uranium De-conversion Plant (FEP/DUP) Official Responses to Environmental Report RAIs, Revision A. March 31, 2011. ADAMS Accession No. ML110970481.

(ISCORS, 2002) Interagency Steering Committee on Radiation Standards. 2002. "A Method for Estimating Radiation Risk from Total Effective Dose Equivalent (TEDE)," ISCORS Technical Report 2002-02, Final Report, Washington, D.C. ADAMS Accession No. ML112720579.

(Johnson and Michelhaugh, 2003) Johnson, P. E. and R. D. Michelhaugh. 2003. Transportation Routing Analysis Geographic Information System (TRAGIS) User's Manual. Oak Ridge National Laboratory, Oak Ridge, Tennessee. June 2003.

(NRC, 1977) U.S. Nuclear Regulatory Commission. 1977. Final Environmental Statement on the Transportation of Radioactive Material by Air and Other Modes, NUREG-0170. Office of Standards Development. December 1977.

(NRC, 2005) U.S. Nuclear Regulatory Commission. 2005. Final Environmental Impact Statement for the Proposed National Enrichment Facility in Lea County, New Mexico/ NUREG–1790, June 2005. ADAMS Accession No. ML051730238.

(Saricks and Tompkins, 1999) Saricks, C.L. and M.M. Tompkins. 1999 State-Level Accident Rates for Surface Freight Transportation: A Reexamination, ANL/ESD/TM-150, Argonne National Laboratory, April 1999. ADAMS Accession No. ML112720605.

(Wiener et. al, 2006) Wiener R. F., D. M. Osborn, G.S. Mills, D. Hinojosa, T. L. Heames, and D. J. Orcutt. 2006. "RadCat 2.3 User Guide." SAND2006-6315, Sandia National Laboratories, Albuquerque, New Mexico.

APPENDIX F

PUBLIC PARTICIPATION AND
NRC RESPONSE TO COMMENTS
ON THE DRAFT ENVIRONMENTAL IMPACT STATEMENT

PUBLIC PARTICIPATION AND
NRC RESPONSE TO COMMENTS
ON THE DRAFT ENVIRONMENTAL IMPACT STATEMENT

F.1   INTRODUCTION

This appendix summarizes the public participation process the U.S. Nuclear Regulatory Commission (NRC) staff conducted for the environmental review and preparation of the Environmental Impact Statement (EIS) for the proposed construction, operation, and decommissioning of an International Isotopes Fluorine Products, Inc. (IIFP) fluorine extraction process and depleted uranium deconversion plant. If built, the proposed IIFP facility would be located in Lea County, New Mexico. This appendix also presents all of the comments NRC received on the draft EIS and the staff's response to those comments. NRC has considered and addressed 410 comments received from 103 members of the public, government officials and agencies, and nongovernmental organizations. Comments from 28 individuals were submitted under a single cover; three members of the public each submitted two separate comment documents, and four members of the public submitted both written and oral comments. A total of 109 documents were considered, including the transcript of oral comments that six individuals provided at the public meeting the NRC staff conducted on February 2, 2012. The transcript of the public meeting is available in the NRC Agencywide Documents Access and Management System (ADAMS) database (Accession Number ML120390370) on the NRC Web site (http://www.nrc.gov/reading-rm/adams.html).

F.2   PUBLIC PARTICIPATION

This section describes the public participation process during the NRC staff's development of the EIS for the proposed IIFP facility. Public participation is an essential part of the NRC environmental review under the National Environmental Policy Act of 1969 (NEPA), as amended. NRC conducted an open, public EIS development process consistent with NEPA and the NRC's regulations under 10 CFR Part 51.

F.2.1   Initial Notification and Notice of Formal Proceeding

Upon receipt of IIFP's license application for the proposed facility and completion of an initial acceptance review, NRC published on April 5, 2010, a Notice of Opportunity to Request a Hearing in the *Federal Register* (FR) (75 FR 17170). On July 15, 2010, the NRC staff published a Notice of Intent (NOI) to Prepare an Environmental Impact Statement in the *Federal Register* (75 FR 41242). The NRC's environmental review began following acceptance and docketing of the environmental report pursuant to the requirements of 10 CFR 70.65 and 10 CFR 51.60, respectively.

F.2.2   Public Scoping

The NRC is required under 10 CFR 51.20(b)(10) to prepare an EIS, and under 10 CFR 51.26 to issue an NOI to prepare the EIS and conduct a scoping process for the EIS. The NRC's public scoping process for the EIS began on July 15, 2010, with the publication in the *Federal Register* (75 FR 41242) of an NOI to prepare an EIS. The NOI summarized the NRC's plans to prepare an EIS and presented background information on the proposed IIFP facility. The NOI also invited comments on the appropriate scope of issues to be considered and announced NRC's plan to hold a public scoping meeting. The public scoping comment period ended on August 30, 2010.

On July 29, 2010, the NRC staff held a public scoping meeting in Hobbs, New Mexico, to receive oral and written comments from interested parties. In addition to the announcement of the meeting in the *Federal Register* (75 FR 41242) on July 15, 2010, the meeting was publicized on the NRC Web site, in a local newspaper, and on local radio. At the public scoping meeting, the NRC staff described NRC's roles, responsibilities, and mission; gave a brief overview of its environmental and safety review processes; discussed how the public could participate effectively in the environmental review process; and solicited input from the public on environmental concerns related to the proposed IIFP facility. Approximately 60 members of the public participated in the meeting. The ADAMS Accession Number for the scoping meeting transcript is ML102210424. Most of the meeting was reserved for attendees to ask questions and provide comments on the scope of the environmental review. Prior to the public scoping meeting, the NRC staff hosted an informal "open house" for those who wished to attend. The open house provided members of the public with an opportunity to speak informally with individual NRC staffers. Appendix A (Scoping Summary) includes the scoping summary report that summarizes the comments received during the scoping process.

F.2.3   Issuance and Availability of the Draft EIS

Pursuant to 10 CFR 51.74, on January 13, 2012, the NRC staff published a Notice of Availability (NOA) of the draft EIS in the *Federal Register* (77 FR 2096). In this notice, the NRC staff described how to access and obtain a copy of the draft EIS. The U.S. Environmental Protection Agency issued a Notice of Availability on the same day (77 FR 2060). The NRC provided a 45 day public comment period which ended on February 27, 2012. The NRC distributed the draft EIS to approximately 55 individuals including Federal, State, and local government agencies, tribal governments, and members of the public. Electronic versions of the draft EIS and supporting information were made accessible through the NRC's project-specific Web site (http://www.nrc.gov/materials/fuel-cycle-fac/inisfacility.html) and through the NRC ADAMS database Web site (http://www.nrc.gov/reading-rm/adams.html). The public also had the opportunity to examine and request a copy of the draft EIS and other related publicly available documents from the NRC Public Document Room. A copy of the draft EIS was also made available at the Hobbs Public Library.

F.2.4   Draft EIS Public Comment Meeting

The NRC staff conducted a public meeting on February 2, 2012, to receive comments on the draft EIS. The NRC staff selected the City of Hobbs as the location for the meeting because it is the closest city to the proposed IIFP facility. The NRC staff advertised these meetings in a local newspaper and on radio stations. Six individuals provided comments during the meeting. A court reporter recorded the oral comments and prepared a written transcript (ADAMS Accession Number ML120390370). The meeting transcripts are also available in the NRC's public Web site for the proposed IIFP facility, at http://www.nrc.gov/materials/fuel-cycle-fac/inisfacility.html. The transcript is part of the public record for the proposed project and was used in developing the applicable comment summaries contained in this appendix.

F.3   COMMENTS RECEIVED ON THE DRAFT EIS

As discussed above, the NRC staff received both oral and written comments on the draft EIS during the comment period. The 45-day public comment period ended on February 27, 2012. The NRC staff considered and addressed approximately 410 comments received from 103 members of the public, government officials and agencies, and nongovernmental organizations. A total of 109 documents were considered, including the transcript of oral comments that 6 individuals provided at the public meeting the NRC staff conducted on February 2, 2012.

F.3.1   Commenter and Comment Identification

The NRC staff reviewed the public meeting transcript, letters, and emails (documents) to identify and extract the individual comments on the draft EIS. These comments are presented in Section F.4.

The NRC staff identified each individual or entity that submitted a comment or document. The NRC staff then assigned a unique identification number to each document, to aid the readers of this appendix in locating the comments and the NRC staff's corresponding responses. The meeting transcript was given its own designation (i.e., TA is the first commenter in the meeting transcript; TB is the second commenter in the meeting transcript, and so on). Table F3–1 lists all of the commenters on the draft EIS alphabetically by last name, their affiliation (if provided), their associated document number, the ADAMS accession number(s) of the document(s) in which each comment appears, and the section(s) in this appendix that address the individual's comments. On the NRC public Web site identified in Section F.2.3, readers can use the ADAMS accession numbers provided in these tables to electronically search for specific individual's comments.

The NRC staff reviewed each comment and assigned it a unique number for identification in Section F.4. For documents submitted by only one author, each specific comment in that document is identified by a number using the format xx-yy. The first number (xx) identifies the document. The second number (yy) identifies a comment within the document. For example, comment 02-05 would identify the fifth comment (xx-05) in the second document (02-yy) NRC received. For documents submitted by multiple authors, each specific comment is identified by a number using the format xx-zz-yy. As before, the first and last numbers (xx and yy) identify the document and specific comment, respectively. The middle number (zz) identifies the author within the document. For example, 54-03-01 identifies the first comment (xx-zz-01) in the third letter (xx-03-yy) of the 54th document (54-zz-yy) that NRC received. The only comment document with more than one letter for this draft EIS was comment document 54 for which each letter was from a different author.

F.3.2   Comment Organization, Review, and Response

In addition to the numbering, each identified comment was assigned a topic category based on the content and issues raised. This allowed the NRC staff to facilitate sorting and reviewing comments on similar topics. The NRC staff sorted and reviewed all comments within specific topic categories, developed comment summaries and responses, and where appropriate, made changes to the EIS.

Table F.3–1. Public Commenter Names With Affiliation, When Provided; Comment Document Number; ADAMS Accession Number; and Subsections Containing Comments and Responses

Name	Affiliation	Document Number	ADAMS Accession Number	Subsections Containing Comments and Responses
Allgood, Lane	Partnership for Science and Technology	63	ML12060A027	F.4.1, F.4.9, F.4.12
Anonymous	Member of the Public	01	ML12019A115	F.4.6
Anonymous	Member of the Public	23	ML12038A024	F.4.1, F.4.15
Barr, Clifford	Member of the Public	02, 04	ML12019A117 ML12019A116	F.4.5, F.4.6
Barr, Phil	Member of the Public	03, 05	ML12023A053 ML12023A054	F.4.5, F.4.6, F.4.9
Beall, Troy	B&D Industries	07	ML12027A143	F.4.1, F.4.9, F.4.12
Bearden, Kathi	Tactical Security Solutions, LLC	36	ML12040A042	F.4.1, F.4.9, F.4.12
Black, Ron	Lea County Commissioner	31	ML12040A037	F.4.1, F.4.9, F.4.12
Borlenghi, Giorgio	Member of the Public	18	ML12033A165	F.4.1, F.4.9, F.4.12
Brunson, Hal	Hobbs City Commissioner	43	ML12040A049	F.4.1, F.4.9, F.4.12
Bryan, Cindy	Accounting and Consulting Group, LLC	22	ML12034A108	F.4.1, F.4.9, F.4.12
Buie, Garry	Pemco of New Mexico	28	ML12040A034	F.4.1, F.4.9, F.4.12
Burns, Marilyn J.	Mayor, Town of Tatum, New Mexico	73	ML12072A071	F.4.1, F.4.9, F.4.12
Campbell, Paul	Forrest Tire Company	27	ML12040A032	F.4.1, F.4.9, F.4.12
Chedekel, Miles	Member of the Public	TI	ML120390370	F.4.15
Cisneros, Sara B.	Hobbs Hispano Chamber of Commerce	38	ML12040A044	F.4.1, F.4.9
Clifton, Trent	Steel Depot	59	ML12059A289	F.4.1, F.4.9, F.4.12
Diaz, Regina	Member of the Public	54-18	ML12048A272	F.4.1, F.4.9, F.4.12
Dunlap, Tami	Member of the Public	54-26	ML12048A272	F.4.1, F.4.9, F.4.12
Ebel, Sylvia	Member of the Public	54-3	ML12048A272	F.4.1, F.4.9, F.4.12
Ensey, Michelle	New Mexico State Historical Preservation District	66	ML12060A283	F.4.4
Ensminger, Joe Glen	Member of the Public	54-14	ML12048A272	F.4.1, F.4.9, F.4.12
Ensminger, Sharon	Member of the Public	54-16	ML12048A272	F.4.1, F.4.9, F.4.12
Estrada, Dana	Member of the Public	54-28	ML12048A272	F.4.1, F.4.9, F.4.12
Fairman, Carl	Member of the Public	54-8	ML12048A272	F.4.1, F.4.9, F.4.12
Fisher, Darrell	Washington State University, Health Physics Society	09	ML12027A145	F.4.1, F.4.9, F.4.12
Frentzel, Chris	International Brotherhood of Electrical Workers, Local 611	46	ML12048A382	F.4.1, F.4.9, F.4.12

Table F.3–1. Public Commenter Names With Affiliation, When Provided; Comment Document Number; and ADAMS Accession Number; and Subsections Containing Comments and Responses (continued)

Name	Affiliation	Document Number	ADAMS Accession Number	Subsections Containing Comments and Responses
Gallagher II, Michael P.	Lea County Board of County Commissioners	41, TF	ML12040A047 ML120390370	F.4.1, F..4.5, F.4.7, F.4.9, F.4.12
Garcia, Maribel	Member of the Public	54-24	ML12048A272	F.4.1, F.4.9, F.4.12
Gardner, Rose	Member of the Public	TH	ML120390370	F.4.7, F.4.12, F.4.15
Green, Barbara	Sullivan Green Seavy LLC	68	ML12062A107	F.4.12
Gonzalez, Mayolo	Member of the Public	54-20	ML12048A272	F.4.1, F.4.9, F.4.12
Haddad, Lara	Member of the Public	54-5	ML12048A272	F.4.1, F.4.9, F.4.12
Hager, Douglas	Member of the Public	54-6	ML12048A272	F.4.1, F.4.9, F.4.12
Hager, Michael	Member of the Public	54-7	ML12048A272	F.4.1, F.4.9, F.4.12
Heier, Kevin	Member of the Public	54-23	ML12048A272	F.4.1, F.4.9, F.4.12
Hardison, Lisa	Economic Development Corporation of Lea County	42, TC	ML12040A048 ML120390370	F.4.1, F.4.9, F.4.12
Hayes, Robert	Member of the Public	TE, TG	ML120390370	F.4.1, F.4.11, F.4.12
Herring, J. Stephen	Member of the Public	67	ML12062A106	F.4.1, F.4.9, F.4.12
Hoyl, Mike	Western Commerce Bank	49	ML12048A386	F.4.1, F.4.9, F.4.12
Hulsey, Chris	Member of the Public	54-27	ML12048A272	F.4.1, F.4.9, F.4.12
Keane, James	Member of the Public	20	ML12033A200	F.4.1, F.4.9, F.4.12
Jadli, Jack	Member of the Public	70	ML12067A040	F.4.1, F.4.9, F.4.12
Jamison, Janice	Member of the Public	44	ML12040A050	F.4.1, F.4.9, F.4.12
Jeff, David G.	Development Corporation of Lea County	29	ML12040A035	F.4.1, F.4.9, F.4.12
Jianto, Linda	Member of the Public	54-15	ML12048A272	F.4.1, F.4.9, F.4.12
Keman, Gay G.	New Mexico State Senate	40	ML12040A046	F.4.1, F.4.9
Lea County Commissioners	Lea County Board of County Commissioners	45	ML12044A129	F.4.1, F.4.5, F.4.7, F.4.9, F.4.12
Leavell, Carroll H.	New Mexico State Senate	26	ML12040A031	F.4.1, F.4.9
Legg, Ladona	Member of the Public	54-17	ML12048A272	F.4.1, F.4.9, F.4.12
Leighton, Michael	Lovington City Manager	37	ML12040A043	F.4.1, F.4.9, F.4.12
Magette, Thomas E.	EnergySolutions	60	ML12059A290	F.4.1, F.4.9, F.4.12
Manning, Dennis	Boron Products, LLC	11	ML12032A025	F.4.1, F.4.9, F.4.12
Marschke, Kerry	Member of the Public	54-25	ML12048A272	F.4.1, F.4.9, F.4.12
Martin, Charles R.	Charles Martin Custom Homes	53	ML12048A392	F.4.1, F.4.9, F.4.12
Mayer, M	Member of the Public	16	ML12033A163	F.4.1, F.4.9, F.4.12
McCleery, Steve	New Mexico Junior College	35	ML12040A041	F.4.1, F.4.9, F.4.12
McCool, Jeff	New Mexico Junior College	21	ML12033A201	F.4.1, F.4.9, F.4.12

Table F.3–1. Public Commenter Names With Affiliation, When Provided; Comment Document Number; and ADAMS Accession Number; and Subsections Containing Comments and Responses (continued)

Name	Affiliation	Document Number	ADAMS Accession Number	Subsections Containing Comments and Responses
McKenzie-Carter, Michael A.	Member of the Public	65	ML12060A282	F.4.1, F.4.9, F.4.12
Medina, Patricia	Member of the Public	54-13	ML12048A272	F.4.1, F.4.9, F.4.12
Milner, Sandra	Member of the Public	54-2	ML12048A272	F.4.1, F.4.9, F.4.12
Moore, Jodi	Member of the Public	54-10	ML12048A272	F.4.1, F.4.9, F.4.12
Moran, Kathleen A.	Member of the Public	39	ML12040A045	F.4.1, F.4.5, F.4.9
N, Mitchell	Member of the Public	14	ML12033A161	F.4.1, F.4.9, F.4.12
Nadel, Marshall	Member of the Public	08	ML12027A144	F.4.1, F.4.9, F.4.12
Nash, Sandy	Steel Depot	33	ML12040A039	F.4.1, F.4.9, F.4.12
Nelson, Audrey	Member of the Public	06	ML12023A055	F.4.1, F.4.9, F.4.12
Nesser, Brad	NPSR Architects	56	ML12059A286	F.4.1, F.4.9, F.4.12
Neuwirth, Paul D.	Member of the Public	74	ML12072A073	F.4.1, F.4.9, F.4.12
Neuwirth, Richard	Member of the Public	52	ML12048A391	F.4.1, F.4.9, F.4.12
Newman, Monty D.	Newman & Company	25	ML12040A030	F.4.1, F.4.9, F.4.12
Newsome, William	Southwest Capital, LLC	17	ML12033A164	F.4.1, F.4.9, F.4.12
Nisula, Amanda L.	Bureau of Land Management Carlsbad Field Office	69	ML12066A143	F.4.15
Nunley, Dana	Member of the Public	54-4	ML12048A272	F.4.1, F.4.9, F.4.12
Parker, Pj	City of Jal Special Projects Administrator	48	ML12048A384	F.4.1, F.4.9, F.4.12
Pettigrew, Randy	Lovington Economic Development Corporation	30	ML12040A036	F.4.1, F.4.9, F.4.12
Pless, Rebecca	Member of the Public	54-12	ML12048A272	F.4.1, F.4.9, F.4.12
Prestwich, Erika	Member of the Public	10	ML12030A092	F.4.1, F.4.9, F.4.12
Pucio, Connie	Member of the Public	54-11	ML12048A272	F.4.1, F.4.9, F.4.12
Reagan, Gary Don	City of Hobbs	51, TB	ML12048A388 ML120390370	F.4.1, F.4.9, F.4.12
Riley, Hollis	International Brotherhood of Electrical Workers	71	ML12067A041	F.4.1, F.4.9, F.4.12
Roybal, Julie	New Mexico Environment Department	62	ML12059A292	F.4.2, F.4.5, F.4.10, F.4.12
Schubert, Gary	Member of the Public	47	ML12048A383	F.4.1, F.4.9, F.4.12
Sena, Jonathan	Hobbs City Commissioner	34, 55	ML12040A040 ML12059A285	F.4.1, F.4.9, F.4.12
Sharif, Farok	URS Washington Resolutions (WIPP)	TA	ML120390370	F.4.1, F.4.9, F.4.12
Smith, Kurt	Member of the Public	12	ML12033A072	F.4.1, F.4.9, F.4.12

Table F.3–1. Public Commenter Names With Affiliation, When Provided; Comment Document Number; and ADAMS Accession Number; and Subsections Containing Comments and Responses (continued)

Name	Affiliation	Document Number	ADAMS Accession Number	Subsections Containing Comments and Responses
Smith, Rhonda	U.S. Environmental Protection Agency	72	ML12067A096	F.4.3, F.4.4, F.4.5, F.4.6, F.4.7, F.4.9, F.4.11, F.4.12
Spence, Novlette	Member of the Public	54-9	ML12048A272	F.4.1, F.4.9, F.4.12
Spencer, Samuel S.	Lea County State Bank	32, TD	ML12040A038 ML120390370	F.4.1, F.4.9, F.4.12
Spencer, Stephen R.	U.S. Department of the Interior	61	ML12059A291	F.4.13, F.4.15
Stovall, David G.	SGA, LLP	57	ML12059A287	F.4.1, F.4.9, F.4.12
Suprenant, Jamie	Member of the Public	54-1	ML12048A272	F.4.1, F.4.9, F.4.12
Tanner, John	Member of the Public	24	ML12040A029	F.4.1, F.4.9, F.4.12
Taylor, Grant	Hobbs Chamber of Commerce	19	ML12033A167	F.4.1, F.4.9
Tomar, Sally	Member of the Public	50	ML12048A387	F.4.1, F.4.9, F.4.12
Tomborello, Justin	Member of the Public	13	ML12034A026	F.4.1, F.4.9, F.4.12
Torres, Katrina	Member of the Public	54-19	ML12048A272	F.4.1, F.4.9, F.4.12
Tuttle-Wurth, Judy	Member of the Public	54-22	ML12048A272	F.4.1, F.4.9, F.4.12
Unknown Individual	Member of the Public	15	ML12033A162	F.4.1, F.4.9, F.4.12
Wallace, Krystal	Member of the Public	54-21	ML12048A272	F.4.1, F.4.9, F.4.12
Wallach, Robert	Wallach Concrete, Inc	58	ML12059A288	F.4.1, F.4.9, F.4.12
Wunder, Matthew	New Mexico Department of Game & Fish	64	ML12060A281	F.4.8, F.4.14

The NRC staff consolidated the same or similar comments received either from a specific commenter or from multiple commenters within each topic to develop responses, as allowed by NRC regulations in 10 CFR 51.91. This grouping is also consistent with the Council on Environmental Quality's NEPA regulations at 40 CFR 1503.4(b). This approach allowed multiple comments to be addressed with a single response, avoided duplication of effort, and enhanced readability of this appendix. Each comment or group of similar comments is introduced with a brief summary by the NRC staff. The summary is followed by the comment identification number(s), commenter name(s) and then the text of the comment(s). This is then followed by the NRC response. For cases in which comments have resulted in a modification to the draft EIS, those changes are noted in the staff's response and are included in the Final EIS. All other comments resulted in no modifications to the draft EIS.

F.3.3 Major Issues and Topics

The majority of the comments received specifically addressed the scope of the environmental reviews, analysis, and issues contained in the draft EIS, including statutory and regulatory requirements, NRC outreach activities, historic and cultural resources, air quality, geology and soils (seismic hazards), water resources, ecological resources, socioeconomics and environmental justice, noise, public and occupational health, waste management, and accidents (wildfires), and cumulative impacts. However, other comments addressed topics and issues that were not part of the review process for the proposed action. Those comments included questions about the NRC's safety evaluation of the proposed IIFP facility, emergency response, and general statements of support for nuclear power.

F.3.4 Comments on Out-of-Scope Topics

The scope of the EIS analysis is defined in 10 CFR 51.71(a); 10 CFR 51.91; NUREG-1748, "Environmental Review Guidance for Licensing Actions Associated With NMSS Programs" (NRC, 2003); and the Scoping Summary Report in Appendix A of this EIS. Some comments addressed issues that were not specifically related to the NRC's environmental review of IIFP's application to construct, operate, and decommission the proposed deconversion facility. Because these comments did not directly relate to the environmental impacts of the proposed action and were outside the scope of the NEPA review, the NRC did not prepare detailed responses.

F.4 COMMENT SUMMARIES AND RESPONSES

All of the comments NRC received on the draft EIS and the NRC staff's responses to those comments are presented in this section. The comments are arranged by topic and multiple comments that address a similar issue/topic have been grouped together for a common response. If multiple individuals provided a similar comment, the comment is presented once, preceded by the names of all the commenters. Text received from commenters in several cases contained more than one issue/topic. In some cases deletion of text to separate topics would have altered the message in the text. In these cases the text has been kept intact, but has been repeated in as many sections as needed to ensure all the issues are addressed.

Written comments are reproduced in this appendix "as received" (i.e., the NRC staff did not correct spelling or grammatical errors in these comments). Also, NRC acknowledges the possibility of transcription errors by the court reporters during the public comment meetings, and regrets any oral comment text that does not exactly match what was said at the public meeting.

F.4.1 General Support for the Project

The comments addressed in this subsection express general support for the proposed IIFP facility. Some supporting comments which include topics within the scope of the EIS are not included in this subsection, but are instead included and addressed in the subsections relevant to the specific topics discussed.

Comment Summary: The following comments express general support for the proposed IIFP project.

[06-01, Audrey Nelson; 07-01, Troy Beall; 08-01, Marshall Nadel; 10-01, Erika Prestwich; 11-01, Dennis Manning; 12-01, Kurt Smith; 13-01, Justin Tomborello; 14-01, Mitchell N.; 15-01, Unknown Individual; 16-01, M. Mayer; 17-01, William Newsome; 18-01, Giorgio Borlenghi; 20-01, James Keane; 22-01, Cindy Bryan; 25-01, Monty D. Newman; 27-01, Paul Campbell; 28-01, Garry Buie; 29-01, David G. Jeff; 30-01, Randy Pettigrew; 31-01, Ron Black; 33-01, Sandy Nash; 34-01, 55-01, Jonathan Sena; 36-01, Kathi Bearden; 37-01, Michael Leighton; 38-01, Sara B. Cisneros on behalf of the Hobbs Hispano Chamber of Commerce Board of Directors; 43-01, Hal Brunson; 44-01, Janice Jamison; 46-01, Chris Frentzel on behalf of the International Brotherhood of Electrical Workers (IBEW), Local 611; 47-01, Gary Schubert; 48-01, Pj Parker; 49-01, Mike Hoyl; 50-01, Sally Tomar; 51-01, TB-01, Gary Don Raegan; 52-01, Richard H. Neuwirth; 53-01, Charles R. Martin; 54-01-01, Jamie Suprenant; 54-02-01, Sandra Milner; 54-03-01, Sylvia Ebel; 54-04-01, Dana Nunley; 54-05-01, Lara Haddad; 54-06-01, Douglas Hager; 54-07-01, Michael Hager; 54-08-01, Carl Fairman; 54-09-01, Novlette Spence; 54-10-01, Jodi Moore; 54-11-01, Connie Pucio; 54-12-01, Rebecca Pless; 54-13-01, Patricia Medina; 54-14-01, Joe Glen Ensminger; 54-15-01, Linda Jianto; 54-16-01, Sharon Ensminger; 54-17-01, LaDona Legg; 54-18-01, Regina Diaz; 54-19-01, Katrina Torres; 54-20-01, Mayolo Gonzalez; 54-21-01, Krystal Wallace; 54-22-01, Judy Tuttle-Wurth; 54-23-01, Kevin Heier; 54-24-01, Maribel Garcia; 54-25-01, Kerry Marschke; 54-26-01, Tami Dunlap; 54-27-01, Chris Hulsey; 54-28-01, Dana Estrada; 56-01, Brad Nesser; 57-01, David G. Stovall; 58-01, Robert Wallach; 59-01, Trent Clifton; 63-01, Lane Allgood on behalf of the Partnership for Science and Technology; 70-01, Jack Jacli; 71-01, Hollis Riley; and 73-01, Marilyn Burns] I support this project (the construction of a depleted uranium deconversion plant as proposed by International Isotopes, Inc.) and believe that it will be a positive contribution to Lea County.

[06-02, Audrey Nelson; 07-02, Troy Beall; 08-02, Marshall Nadel; 09-01, Darrell Fisher; 10-02, Erika Prestwich; 11-02, Dennis Manning; 12-02, Kurt Smith; 13-02, Justin Tomborello; 14-02, Mitchell N.; 15-02, Unknown Individual; 16-02, M. Mayer; 17-02, William Newsome; 18-02, Giorgio Borlenghi; 20-02, James Keane; 22-02, Cindy Bryan; 25-02, Monty D. Newman; 26-01, Carroll H. Leavell; 27-02, Paul Campbell; 28-02, Garry Buie; 29-02, David G. Jeff; 30-02, Randy Pettigrew; 31-02, Ron Black; 32-01, Samuel S. Spencer; 33-02, Sandy Nash; 34-02, 55-02, Jonathan Sena;; 36-02, Kathi Bearden; 37-02, Michael Leighton; 38-02, Sara B. Cisneros on behalf of the Hobbs Hispano Chamber of Commerce Board of Directors; 43-02, Hal Brunson; 44-02, Janice Jamison; 46-02, Chris Frentzel on behalf of the International Brotherhood of Electrical Workers (IBEW), Local 611; 47-02, Gary Schubert; 48-02, Pj Parker; 49-02, Mike Hoyl; 50-02, Sally Tomar; 51-02, TB-02, Gary Don Reagan; 52-02, Richard H. Neuwirth; 53-02, Charles R. Martin; 54-01-02, Jamie Suprenant; 54-02-02, Sandra Milner; 54-03-02, Sylvia Ebel; 54-04-02, Dana Nunley; 54-05-02, Lara Haddad; 54-06-02, Douglas Hager; 54-07-02, Michael Hager; 54-08-02, Carl Fairman; 54-09-02, Novlette Spence; 54-10-02, Jodi Moore; 54-11-02, Connie Pucio; 54-12-02, Rebecca Pless;

54-13-02, Patricia Medina; 54-14-02, Joe Glen Ensminger; 54-15-02, Linda Jianto; 54-16-02, Sharon Ensminger; 54-17-02, LaDona Legg; 54-18-02, Regina Diaz; 54-19-02, Katrina Torres; 54-20-02, Mayolo Gonzalez; 54-21-02, Krystal Wallace; 54-22-02, Judy Tuttle-Wurth; 54-23-02, Kevin Heier; 54-24-02, Maribel Garcia; 54-25-02, Kerry Marschke; 54-26-02, Tami Dunlap; 54-27-02, Chris Hulsey; 54-28-02, Dana Estrada; 56-02, Brad Nesser; 57-02, David G. Stovall; 58-02, Robert Wallach; 59-02, Trent Clifton; 60-01, Thomas E. Magette; 63-02, Lane Allgood on behalf of the Partnership for Science and Technology; 67-01, J. Stephen Herring; 70-02, Jack Jacli; 71-02, Holis Riley; 73-02, Marilyn Burns; and TA-01, Farok Sharif] The draft Environmental Impact Statement thoroughly reviews all of the potential environmental impacts from the proposed facility and I/we agree with the conclusions reached by the NRC staff that the impact to Lea County would be small/minimal.

[08-03, Marshall Nadel] Our nation needs nuclear power for energy reliability, efficiency and energy independence. We must work to ensure the nuclear fuel cycle is operated in a most efficient and environmentally clean manner. Deconverting UF6 close to its source, in this case, the Urenco facility in Eunice, is the most efficient and environmentally friendly route to pursue.

[09-02, Darrell Fisher] This document is comprehensive and well-written. It provides strong scientific content. The information presented and conclusions reached support construction and operation of the deconversion facility. Our country needs constructive solutions to the problem of uranium hexafluoride storage, and I believe that the company's proposal to build and safely operate the deconversion plant provides an excellent approach to UF6 management and long-term disposal of excess depleted uranium.

The Environmental Impact Statement shows that International Isotopes can provide a safe and environmentally sound approach for treatment of depleted uranium hexafluoride. This document thoroughly reviews all of the potential impacts. from the proposed facility, and I agree with the conclusions reached by the NRC staff that those impacts would be small, and that licenses should be granted to proceed with construction and operation.

[19-01, Grant Taylor on behalf of the Hobbs Chamber of Commerce] I write in unequivocal support of the proposed International Isotopes fluorine products facility in Lea County, New Mexico. I speak for the membership and the board of directors of the Hobbs Chamber of Commerce. We collectively and enthusiastically support the broadening base of Lea County's nuclear industry, which complements the established energy base of our communities.

International Isotopes has already proven itself to be a good corporate citizen of Lea County, and we support this venture with full confidence in the company's commitment to safety and environmental stewardship during the treatment of depleted uranium hexafluoride.

This proposed project has been thoroughly vetted and reviewed, and the Hobbs Chamber of Commerce concurs with the draft Environmental Impact Statement in regards to the minimal impacts anticipated in the MRC's staff conclusions.

Please know that this project has the full support of the Hobbs Chamber of Commerce.

[21-01, Jeff McCool] As Dean of Training and Outreach at New Mexico Junior College, I wholeheartedly support the construction and operation of the International Isotopes fluorine products facility. The construction and operation of this facility is a giant step to continue the

United States' commitment to responsible and viable projects as these projects relate to the Nuclear Fuel Cycle.

I agree with the conclusions of the Environmental Impact Statement.

[24-01, John Tanner] I support the construction of a depleted uranium deconversion plant as proposed by International Isotopes Inc., and believe that it will be a positive contribution to Lea County and will fulfill an important role in the safe and environmentally sound treatment of depleted uranium hexafluoride, as well as producing valuable fluoride byproducts. I am not aware of any harmful environmental effects caused by the uranium deconversion process.

[35-01, Steve McCleery] As President of New Mexico Junior College, I wholeheartedly support the construction and operation of the International Isotopes fluorine products facility. The construction and operation of this facility is a giant step to continue the United States' commitment to responsible and viable projects as these projects relate to the Nuclear Fuel Cycle.

I agree with the conclusions of the Environmental Impact Statement.

[22-03, Cindy Bryan; 27-03, Paul Campbell; 28-03, Garry Buie; 29-03, David G. Jeff; 30-03, Randy Pettigrew; 33-03, Sandy Nash; 34-03, 55-03, Jonathan Sena; 36-03, Kathi Bearden; 38-03, Sara B. Cisneros on behalf of the Hobbs Hispano Chamber of Commerce Board of Directors; 43-03, Hal Brunson; 48-03, Pj Parker; 49-03, Mike Hoyl; 50-03, Sally Tomar; 56-03, Brad Nesser;; 58-03, Robert Wallach; 71-03, Holis Riley; 73-03, Marilyn Burns] As a local resident(s) of Lea County, I/we fully support the International Isotopes project.

[25-03, Monty D. Newman] As a former mayor, businessman and resident of Lea County, I fully support the International Isotopes project.

[26-02, Carroll Leavell] I serve Senate District 41, Lea and Eddy Counties, New Mexico. There is strong support for the project throughout Senate District 41.

Mr. Steve Laflin, President & CEO, and his management team have done a great job of educating Southeast New Mexico of all aspects of the project. I have received no negative input from the constituents in Senate District 41. This project has my strong support.

[32-02, Samuel S. Spencer on behalf of Lea County State Bank] As a locally owned business located in Lea County, the International Isotopes project has our full support.

[39-01, Kathleen A. Moran] I also agree with the conclusion stated in the Draft EIS that any potential negative impacts will be small with regard to the thirteen (13) areas listed (land, historical and cultural, visual and scenic, etc.).
Moreover, it is important to find new ways to get more from the nuclear material that is already generated. The license should be granted and the project should be allowed to move forward.

[40-01, Gay G. Kernan] I am confident that the draft Environmental Impact Statement thoroughly reviews all of the potential environmental impacts from the proposed facility and I agree with the conclusions reached by the NRC staff. The minimal impact on our area allows me to fully support this project.

[41-01, Michael P. Gallagher II on behalf of the Lea County Board of County Commissioners] The Board of County Commissioners supports this project and believes that it will be a positive contribution to Lea County.

[42-01, TC-01, Lisa Hardison on behalf of the Economic Development Corporation of Lea County (EDCLC)] The Economic Development Corporation of Lea County (EDCLC) is in full support of the International Isotopes (INIS) planned fluorine extraction facility in Lea County.

The draft Environmental Impact Statement thoroughly reviews all of the potential environmental impacts from the proposed facility and the EDCLC agrees with the conclusions reached by the NRC staff that the impacts would be minimal.

The EDCLC supports the construction and licensure of the INIS facility in Lea County,

[45-01, Lea County Commissioners and TF-01, Michael P. Gallagher II] RESOLUTION IN SUPPORT OF THE PROPOSED INTERNATIONAL ISOTOPES FLUORINE PRODUCTS FACILITY IN LEA COUNTY

WHEREAS International Isotopes Fluorine Products, Inc. seeks to build a fluorine products facility in Lea County, and

WHEREAS, the economic benefit to Southeastern New Mexico will be stability, growth, job creation, and industry diversification; and

WHEREAS, the facility will process depleted uranium and fluorine extraction to produce important fluoride products for U.S. markets and will be an asset to U.S. chemical manufacturing capabilities; and

WHEREAS, the facility will be licensed and regulated by the Nuclear Regulatory Commission, along with appropriate state agencies; and

WHEREAS, the facility will have regulated air and water emissions at or below state and federal limits, as allowed by the NRC and New Mexico Environment Department; and

WHEREAS, Lea County Board of Commissioners approved the issuance of Lea County Taxable Industrial Revenue Bonds (International Isotopes Project) Series 2010 in the amount of $72,000,000 in August 2011.

NOW, THEREFORE, BE IT RESOLVED, that the Lea County Board of Commissioners supports locating the International Isotopes Fluorine Products Facility in Lea County.

PASSED, APPROVED AND ADOPTED IN OPEN MEETING on this 24th day of January 2012.

[46-03, Chris Frentzel on behalf of the International Brotherhood of Electrical Workers (IBEW), Local 611] Please note that I fully support the International Isotopes project and look forward to seeing it move forward.

[59-03, Trent Clifton] As a local resident of Eddy County, I fully support the International Isotopes project.

[60-02, Thomas E. Magette on behalf of EnergySolutions] EnergySolutions supports the proposed International Isotopes Fluorine Products Facility in Lea County, New Mexico.

[65-01, Michael McKenzie-Carter] I believe the potential environmental impacts of the proposed action and alternatives are reasonably and conservatively assessed. I agree with the conclusion of the Draft EIS that the potential negative impacts of the proposed action would be small or moderate at most.

I support the licensing of the IIFP facility as described in the Draft EIS, and believe that the facility can be built and operated in a safe and environmentally sound manner. The benefits considerably outweigh the small potential negative environmental impacts.

[74-01, Paul D. Neuwirth] The project will benefit Lea County.

I agree the conclusions of the NRC staff that any such impacts will not be material.

For all of the above reasons, I support the project.

[TA-02, Farok Sharif] I am in full support of the International Isotopes fluorine facility.

[TD-01, Samuel S. Spencer] I believe that the economic impact to this area will far outweigh any small environmental impact and I concur with your analysis.

[TE-01, Robert Hayes] If there does come up anything that would put a stop to this or slow it down...I'm an engineer...I would personally encourage you and all concerned, as an engineer there is a way around that, if there is anything that comes up. I don't know what it might be, I'm not going to even try to guess what it is, but believe whatever that might be, if it comes up, can be addressed. We can design and we can operate safely as long as it's done correctly.

And so just that's just the caveat: f there does come up something, please consider any kind of design changes that the vendor, International Isotopes, comes up with so that we can go forward. I echo what has been said before: as a resident, I strongly encourage the location of this plant here, we really want the jobs, it will be fantastic.

Response: The NRC staff acknowledges these commenters and appreciates the public participation. These comments express general support for the licensing of the proposed IIFP facility; however, they do not provide specific information that requires a response from the NRC. These comments are outside the scope of the EIS analysis because they do not directly relate to the content of the EIS.

F.4.2 Statutory and Regulatory Requirements

Comment Summary: The comment addressed in this subsection discusses permitting requirements.

[62-01, Julie Roybal on behalf of the New Mexico Environment Department]
Surface Water Quality Bureau
The U.S. Environmental Protection Agency (EPA) requires National Pollutant Discharge Elimination System (NPDES) Construction General Permit (CGP) coverage for storm water discharges from construction projects (common plans of development) that will result in the disturbance (or re-disturbance) of one or more acres, including expansions, of total land area

Because this project exceeds one acre (including staging areas, etc.), it may require appropriate NPDES permit coverage prior to beginning construction (small, one - five acre, construction projects may be able to qualify for a waiver in lieu of permit coverage - see Appendix D). Among other things, this permit requires that a Storm Water Pollution Prevention Plan (SWPPP) be prepared for the site and that appropriate Best Management Practices (BMPs) be installed and maintained both during and after construction to prevent, to the extent practicable, pollutants (primarily sediment, oil & grease and construction materials from construction sites) in storm water runoff from entering waters of the U.S. This permit also requires that permanent stabilization measures (revegetation, paving, etc.), and permanent storm water management measures (storm water detention/retention structures, velocity dissipation devices, etc.) be implemented post construction to minimize, in the long term, pollutants in storm water runoff from entering these waters. In addition, permittees must ensure that there is no increase in sediment yield and flow velocity from the construction site (both during and after construction) compared to pre-construction, undisturbed conditions (see Subpart 10.E.1.b).

You should also be aware that EPA requires that all "operators" (see Appendix A) obtain NPDES permit coverage for construction projects. Generally, this means that at least two parties will require permit coverage. The owner/developer of this construction project who has operational control over project specifications, the general contractor who has day-to-day operational control of those activities at the site, which are necessary to ensure compliance with the storm water pollution plan and other permit conditions, and possibly other "operators" will require appropriate NPDES permit coverage for this project.

The CGP was re-issued effective June 30, 2008. The CGP, Notice of Intent (NOI), Fact Sheet, and Federal Register notice can be downloaded at http://cfpub.epa.gov/npdes/stormwater/cp.cfim.

In addition, EPA requires NPDES Storm Water Multi-sector General Permit (MSGP) coverage for facilities that engage in "industrial activities" as defined at 40 Code of Federal Regulations Part 122.26(b)(14). Although the type of business to be operated is not entirely clear in the submittal, if this business meets the definition of regulated industrial activity, it will require appropriate NPDES permit coverage prior to beginning operations.

Among other things, this permit also requires that a SWPPP be prepared for the site and that appropriate Best Management Practices (BMPs) be installed and maintained to prevent, to the extent practicable, pollutants in storm water runoff from entering waters of the U.S. A SWPPP should include such things as:

A description of potential pollutant sources—includes such things as a site map, an identification of the types of pollutants that are likely to be present in storm water discharges, an inventory of the types of materials handled at the site that potentially may be exposed to precipitation, a list of significant spills and leaks of oil, toxic or hazardous pollutants, sampling data, a narrative description of the potential pollutant sources from specific activities at the facility (i.e., pumping operations, road construction, raw material storage and handling, material transportation, fueling and other equipment maintenance), and identification of specific potential pollutants (i.e., dust, total suspended solids, total dissolved solids, turbidity, pH; nitrates, oil, grease, ethylene glycol, heavy metals, radionuclides, and others); and

A description of appropriate measures and controls—includes the type and location of existing and proposed non-structural and structural best management practices (BMPs) selected for each of the areas where industrial materials or activities are exposed to storm water. Non-structural and structural BMPs to be described and implemented include such things as good housekeeping, preventive maintenance, spill prevention and response procedures, periodic inspections, employee training, record keeping, non-storm water evaluations and certifications, sediment and erosion control, as well as implementation/maintenance of traditional storm water management practices (i.e., sediment/settling ponds, check dams, silt fences, straw bale barriers, perimeter berms, runon diversion structures), where appropriate. The MSGP also requires preparation and implementation of a reclamation plan for the site.

The NPDES Storm Water Multi-Sector General Permit for Industrial Activities (MSGP) was re-issued effective September 29, 2008 (see Federal Register/Vol. 73, No. 189/Monday, September 29, 2008, p. 56572). The MSGP, Notice of Intent (NOD, Fact Sheet, and *Federal Register* notice can be downloaded at: http://cfpub.epa.gov/npdes/stormwater/msgp.cfm).

Finally, EPA requires individual NPDES permit coverage for discharges of process wastewaters. These permits typically contain both technology and water quality based effluent limits, sampling requirements, etc. NPDES regulations at 40 CFR Part 122.44(d) require that NPDES permits include effluent limits necessary to achieve water quality standards established under §303 [33 U.S.C. 1313 - Water Quality Standards and Implementation Plans] of the federal Clean Water Act (CWA), including State narrative criteria for water quality. 40 CFR Part 122.4(i) requires that a discharge not "cause or contribute to the violation of water quality standards." The New Mexico Water Quality Control Commission (WQCC) has adopted surface water quality standards under authority of the New Mexico Water Quality Act [Chapter 74, Article 6 NMSA] pursuant to CWA § 303, which are codified as *Standards for Interstate and Intrastate Surface Waters, 20.6.4 NMAC*.

Regardless of whether or not an NPDES permit has been issued, state surface water quality standards must be met at all times and violation of these standards are enforced by the New Mexico Environment Department under authority of the New Mexico Water Quality Act.

Ground Water Quality Bureau
New Mexico Environment Department (NMED) Ground Water Quality Bureau (GWQB) staff reviewed the above-referenced letter as requested, focusing specifically on the potential effect to ground water resources in the area of the proposed project.

The letter from the U.S Nuclear Regulatory Commission (NRC) provides notification and requests comments on the Draft Environmental Impact Statement (EIS) for the proposed International Isotopes Fluorine Products Facility to be located in Lea County, NM. The proposed facility would deconvert depleted uranium hexafluoride to fluorine gas products and uranium oxide compounds for long-term disposal offsite. This memo should be regarded as a preliminary response to the EIS notification. Formal comments to the EIS will be provided in a separate letter.

If constructed, International Isotopes Fluorine Products, Inc. (IIFP) will be required to obtain a ground water discharge permit for this facility. GWQB met with IIFP on September 8, 2010 for a preliminary review of the proposed facility. On February 11, 2011, GWQB replied to an IIFP request dated January 27, 2011, with a letter providing a preliminary description of ground water monitoring requirements that would be required for the facility under a discharge permit.

The GWQB strongly recommends that IIFP submit a Discharge Permit Application at least 180 days prior to construction of the facility to allow adequate time for processing and required

F–17

public notification. During the public notice period, if a public hearing is requested and granted by the NMED Secretary, the issuance of a discharge permit may be further delayed or denied.

Please note that construction of the facility will involve the use of heavy equipment, thereby leading to a possibility of contaminant releases (e.g., fuel, hydraulic fluid, etc.) associated with equipment malfunctions. The GWQB advises all parties involved in the project to be aware of notification requirements for accidental discharges contained in 20.6.2.1203 NMAC. Compliance with the notification and response requirements will further ensure the protection of ground water quality in the vicinity of the project.

Air Quality Bureau
From the project description, it is difficult to discern what the potential is for fluorine emissions. However, fluorine is listed as a Toxic Air Pollutant (TAP) in New Mexico under 20.2.72.502 NMAC and potential emissions must be included in the NEPA analysis. Radionuclides are also identified and are subject to National Emission Standards for Hazardous Air Pollutants (NESHAP) in the Clean Air Act under 40 CFR Part 61.

Construction activities identified in this proposal have the potential to create temporary increases in emissions due to combustion-related construction activities and the use of earth-moving equipment. All asphalt, concrete, quarrying, crushing and screening facilities contracted in conjunction with the proposed project must have current and proper air quality permits. For more information on air quality permitting and modeling requirements, please refer to 20.2.72 NMAC.

Dust associated with vehicular use and earth-moving activities may also impact local air quality. However the increases should not result in non-attainment of air quality standards. Dust control measures should be considered to minimize the release of particulates due to vehicular traffic and ground disturbances. If activities result in significant ground disturbance, the project area should be reclaimed to avoid long-term problems with erosion and fugitive dust.

To further ensure air quality standards are met, applicable local or county regulations requiring noise and/or dust control must be followed; if none are in effect, controlling construction-related air quality impacts during projects should be considered to reduce the impact of fugitive dust and/or noise on community members.

Hazardous Waste Bureau
Comment #2: Depending on whether the facility is a Large Quantity Generator (LQG) or Small Quantity Generator (SQG); how much hazardous waste they intend to accumulate; and the length of time they intend to accumulate it, the facility may need a Resource Conservation and Recovery Act (RCRA) permit to store hazardous waste.

Comment #3: Table 1-3 infers that Form 8700-12 to notify NMED HW13 of EPA waste activity and obtain an EPA ID Number would be submitted in 3rd Quarter of 2011. As of January 18, 2012, there is no EPA ID Number in the RCRA Info database.

Radiation Control Bureau
The draft EIS (NUREG-2113) should meet or exceed New Mexico, Title 20: Environmental Protection Chapter 3: Radiation Protection, Part 4: Standards for the Protection against Radiation and the equivalent federal regulations in IO CFR 20, Standards for the Protection against Radiation.

Response: *By law, IIFP is required to obtain all required Federal, State, and local permits and approvals to conduct construction and operation activities. Also, 10 CFR 51.71(c) requires that the EIS provide a list of all Federal permits, licenses, approvals, and other entitlements that must be obtained in implementing the proposed action and will describe the status of compliance with these requirements. Section 1.5 describes the statutory and regulatory requirements that are applicable to the proposed IIFP facility. In addition, Table 1-1 describes the laws, regulations, and agreements that apply to the facility and the regulatory basis for these items. Finally, Section 1.5.3 was updated to provide the status of the permits for the proposed IIFP facility, should NRC grant the license.*

F.4.3   NRC Outreach Activities

Comment Summary: The following comment requests clarification of the facts associated with the public scoping meeting held in Hobbs, New Mexico, on July 29, 2010, and the public meeting held on February 2, 2012, to receive comments on the draft EIS.

[72-01, Rhonda Smith on behalf of the U.S. Environmental Protection Agency] The Draft EIS describes that approximately 50 individuals not affiliated with the NRC staff attended the July 29, 2010, public scoping meeting in Hobbs, New Mexico. A news article from the Hobbs News-Sun stated about a dozen citizens and elected officials attended the July 29th meeting. The DEIS does not clearly explain what outreach efforts were made to announce the Scoping meeting and the open house held July 29, 2010, in Hobbs, New Mexico. Please clarify in the DEIS.

The International Isotopes Inc., Web site announced that the NRC was holding a public meeting in Hobbs, New Mexico, on February 2, 2012, and stated comments will be taken until February 27, 2012. The website does not mention a location of this public meeting nor additional information regarding the submission of comments. Through internet surfing a NRC News flyer, dated January 9, 2012, explains the NRC is seeking public comments on DEIS for a proposed Uranium Deconversion Facility in New Mexico. It is unclear where the general public will see a NRC News bulletin such as this. Please clarify in the FEIS.

Response: *Public participation is an essential part of the NRC's environmental review process under NEPA. As indicated in Sections F.2.2 and F.2.3 of this appendix, the NRC staff conducted an open, public EIS development process consistent with NEPA and the NRC's NEPA implementing regulations under 10 CFR Part 51.*

The NRC's public scoping process began on July 15, 2010, with the publication in the Federal Register of a Notice of Intent (NOI) to prepare an EIS (75 FR 41242). This notice also invited members of the public to provide comments on issues to be considered in the EIS and announced the NRC's public scoping meeting and open house on July 29, 2010. The notice described the different vehicles members of the public could use to provide scoping comments as well as information on how to access the applicant's ER. A copy of the applicant's ER was also made available at the Hobbs Public Library for public inspection. An electronic version of the ER and supporting information was made available through the NRC's project-specific Web site at http://www.nrc.gov/materials/fuel-cycle-fac/inisfacility.html and through the NRC's ADAMS database at http://www.nrc.gov/reading-rm/adams.html.

The NRC staff also published a meeting notice on June 29, 2010, in its Web site at http://www.nrc.gov/public-involve/public-meetings/index.cfm announcing the July 29, 2010, public scoping meeting and open house. An NRC press release was issued on July 23, 2010,

which can be found in the NRC's Web site at: http://www.nrc.gov/reading-rm/doc-
collections/news/2010/.

*The NRC staff advertised the scoping meeting and open house in a regional daily newspaper
(The Hobbs News-Sun) on July 20 and July 25, 2010, and on local radio station (KYKK) from
July 27 through July 29, 2010.*

*During the week of July 26, 2010, the NRC staff met with the Lea County Sheriff's Department,
the Monument Volunteer Fire Department, the City of Hobbs Fire Department, the City of Hobbs
Planner and Manager, the Lea County Interim Manager, Lea County Assessor, Lea County
Attorney, the Economic Development Corporation of Lea County, and the NMED Drinking Water
Bureau to discuss the NRC's roles and responsibilities, provide an overview of the
environmental review process, and to gather information as part of the scoping process in
support of the environmental review for the proposed IIFP facility.*

*Based on the attendance cards completed during the scoping meeting held July 29, 2010, in
Hobbs, New Mexico, NRC staff determined that 62 members of the public attended the scoping
meeting.*

*The draft EIS public comment period began with the Environmental Protection Agency's Notice
of Availability (NOA) published in the FR on January 13, 2012 (77 FR 2060). The NRC staff's
NOA announcing the issuance of the draft EIS for public comment and providing notice of the
public meeting was also published on January 13, 2012 (77 FR 2096). The NRC staff
advertised the draft EIS public meeting in The Hobbs News-Sun on January 26 and February 2,
2012, and on a local radio station (KIXN) on January 26 and February 2, 2012.*

The NRC staff also published a meeting notice on January 4, 2012, in its Web site at
http://www.nrc.gov/public-involve/public-meetings/index.cfm, *announcing the February 2, 2012,
draft EIS public meeting and open house. An NRC press release was issued on January 9,
2012, which can be found in the NRC's Web site at* http://www.nrc.gov/reading-rm/doc-
collections/news/2012/.

*Copies of the draft EIS were mailed to Federal, Tribal, State, and local government officials as
well as members of the public. Additionally, an electronic version of the draft EIS was
made available through the NRC's project-specific Web site (* http://www.nrc.gov/materials/fuel-
cycle-fac/inisfacility.html*) and through the NRC's ADAMS database (*http://www.nrc.gov/
reading-rm/adams.html*). No change was made to the EIS as a result of this comment.*

Comment Summary: The following comment requests information on how the public will be
informed and educated about potential dangers during operation of the proposed IIFP facility.

[72-02, Rhonda Smith on behalf of the U.S. Environmental Protection Agency] The FEIS
should provide information as to how the IIFP or the NRC will disseminate or reach out to the
community keeping residents informed and educated about any potential dangers during the
operation of this facility after the license is approved. Please explain in the FEIS.

*Response: The applicant, IIFP, submitted an emergency plan as part of the license application
that describes the actions IIFP will take in response to emergencies. IIFP also coordinated the
development of the emergency plan with offsite response organizations expected to respond in
the event of an accident at the proposed IIFP facility. Emergency response plans are not within
the scope of the EIS, but are addressed in the Safety Evaluation Report (SER) (NRC, 2012).*

The methods that IIFP will use to disseminate information to or educate community residents about any potential dangers during facility operations are discussed in the SER (NRC, 2012), which presents the NRC's staff safety review of the license application. These methods are specifically addressed in Section 8.3.8, "Responsibilities," Section 8.3.9, "Notification and Coordination," and Section 8.3.10, "Information to be Communicated." In the ER, IIFP identified and described the emergency notification procedures that enable the organization to correctly classify emergencies, notify emergency response personnel, and initiate or recommend appropriate actions in a timely manner.

In addition, if a license is issued to IIFP, the NRC staff will also implement oversight (inspection) and enforcement programs during construction, operation, and decommissioning of the proposed IIFP facility to assure safe functions and compliance with NRC requirements. The NRC's oversight program for fuel cycle facilities includes inspections focused on reviews of safety, safeguards, and environmental protection. Inspections at fuel cycle facilities occur several times a year and typically cover activities such as chemical safety, emergency preparedness, fire safety, and radiation safety. The results of the inspections are documented in reports that are generally publicly available through the NRC's ADAMS library at: http://www.nrc.gov/reading-rm/adams.html. No change was made to the EIS as a result of this comment.

F.4.4   Historical and Cultural Resources

Comment Summary: The following comment is a concurrence of a letter sent to the New Mexico State Historic Preservation Officer by the NRC staff that provided a copy of the draft E S and information in response to the New Mexico State Historic Preservation Officer's recommendation that a cultural resources survey of the IIFP property be conducted and that consultations be held with Native American Tribes.

[66-01, Michelle Ensey on behalf of the New Mexico State Historic Preservation Officer] Concur. No Historic Properties Affected.

Response: The NRC accepts the concurrence from the New Mexico State Historic Preservation Division that the proposed action will not adversely affect any historic properties. No change was made to the EIS as a result of this comment.

Comment Summary: The following comment requests that a copy of the Final EIS be sent to a government office in Texas because of Texas' close proximity to the IIFP site.

[72-03, Rhonda Smith on behalf of the U.S. Environmental Protection Agency] With regard to historic preservation and close proximity to the State of Texas, EPA recommends that a copy of the FEIS be sent to Texas State Historical Office for their review and comment.

Response: NRC intends to send a copy of the Final EIS to the Texas State Historical Office when the Final EIS is published. No change was made to the EIS as a result of this comment.

F.4.5   Air Quality

Comment Summary: The following comments express concerns that offsite air monitoring is not planned for the proposed IIFP facility.

[03-01, Phil Barr] Note that the State of Texas wanted an health survey filled out by citizens in Lea county nm because of a nuclear waste dump in Andrews county Texas. But as I understand it the New Mexico environmental department will not install rad monitors in Eunice and Hobbs; New Mexico. One Congressman has told me that it is NMED's responsibility to monitor the air in regards to nuclear facilities in Lea county. I don't believe any nuclear facility should be allowed in Lea county NM with this kind of approach by NMED.

[04-01, Clifford Barr and 05-01, Phil Barr] Requested several times that the New Mexico Environmental Department install radiation monitors in Hobbs and eunice, new mexico. For safety..... Got turned down each time. Got turned down by a Lea county commissioner as well. But one U.S. Congressman has told me by email that NMED has the responsibility of monitoring the air here. Contrast this to the Texas commission of environmental quality who wanted Lea county nm citizens to participate in a health survey because of a nuclear waste dump 20 miles away in Texas. Its my belief that NMED and the State of New Mexico are more interested in avoiding liability from nuclear industry than safety to Lea county citizens. With this approach I don't believe any nuclear industry should be located here in Lea county nm.

Response: *Potential radiological releases during facility operation are analyzed in Section 4.1.2.11. In addition, IIFP is responsible for complying with the provisions in 10 CFR Part 20, "Standards for Protection Against Radiation," which requires monitoring for airborne radioactivity both in the workplace and at the site boundary downwind from the facility in the predominant wind direction. Sections 6.1.1 and 6.1.2 discuss the gaseous effluent monitoring and radiological environmental monitoring programs, respectively. IIFP must comply with all applicable Federal, State, and local laws and regulations, including obtaining all appropriate construction and operating permits. A discussion of applicable laws and regulations is included in Section 1.5. Under New Mexico State regulations, IIFP is required to satisfy all air quality regulatory and permitting requirements. No change was made to the EIS as a result of these comments.*

Comment Summary: The following comment expresses an opinion that NRC overestimated the environmental impact of the IIFP on air quality.

[39-02, Kathleen A. Moran] As someone who is familiar with the amount of dust that gets blown across the region during one of the frequent windy days in southeast New Mexico, I seriously question whether the dust generated during construction of the facility fourteen miles west of the nearest town will be noticed by anyone or cause any air quality issues, and I would characterize the impact on existing Air Quality as being small rather than moderate.

Response: *The NRC staff takes a reasonable and conservative approach to assessing the environmental impacts from nuclear facilities to ensure that impacts are not underestimated. Based on independent verification of the information the applicant provided, the NRC staff has determined that the environmental impact on air quality from the proposed IIFP facility is SMALL to MODERATE. No change was made to the EIS as a result of this comment.*

Comment Summary: The following comment supports air monitoring at the proposed IIFP facility project.

[45-02, Lea County Commissioners and TF-02, Michael P. Gallagher, II] ... WHEREAS, the facility will have regulated air and water emissions at or below state and federal limits, as allowed by the NRC and New Mexico Environment Department; ...

Response: The NRC staff acknowledges these commenters and appreciates the public participation. This comment expresses general support for the licensing of the proposed IIFP facility; however, they do not provide specific information that requires a response from the NRC. This comment is outside the scope of the EIS analysis because it does not directly relate to the content of the EIS.

Comment Summary: The following comment describes those regulations that pertain to dust control.

[62-02, Julie Roybal on behalf of the New Mexico Environment Department]

Air Quality Bureau
Dust associated with vehicular use and earth-moving activities may also impact local air quality. However the increases should not result in non-attainment of air quality standards. Dust control measures should be considered to minimize the release of particulates due to vehicular traffic and ground disturbances. If activities result in significant ground disturbance, the project area should be reclaimed to avoid long-term problems with erosion and fugitive dust.

To further ensure air quality standards are met, applicable local or county regulations requiring noise and/or dust control must be followed; if none are in effect, controlling construction-related air quality impacts during projects should be considered to reduce the impact of fugitive dust and/or noise on community members.

Response: By law, IIFP is required to obtain all applicable Federal, State, and local permits and approvals to conduct construction and operations activities. Also, 10 CFR 51.71(c) requires that the EIS provide a list of all Federal permits, licenses, approvals, and other entitlements for the proposed action and describe the status of compliance with these requirements, including requirements for dust control (air quality). Section 1.5 of this EIS describes the statutory and regulatory requirements that are applicable to the proposed IIFP facility. In addition, Table 1-1 describes laws, regulations, and agreements that apply to the facility; the regulatory basis and requirements for these items. . Finally, Section 1.5.3 was updated to provide the status of the permits, licenses, authorizations, and approvals for construction and operation of the proposed IIFP facility, should NRC grant the license.

The NRC staff acknowledges the importance of using BMPs. As noted in Chapter 5, IIFP identified the use of BMPs as mitigation measures for minimizing air quality impacts during road construction, land clearing, and building construction. Also, construction equipment and related vehicles would be equipped with standard pollution control devices and maintained in good working order.

Comment Summary: The following comment outlines EPA recommendations that involve vehicular air emissions.

[72-04, Rhonda Smith on behalf of the U.S. Environmental Protection Agency] Any demolition, construction, rehabilitation, repair, dredging or filling activities have the potential to emit air pollutants. EPA recommend best management practices be implemented to minimize the impact of any air pollutants. Furthermore, construction and waste disposal activities should be conducted in accordance with applicable local, State and Federal statutes and regulations.

EPA recommends the use of clean, lower-emissions equipment and technologies to reduce pollution. EPA's final Highway Diesel and Nonroad Diesel Rules mandate the use of lower-sulfur fuels in nonroad and marine diesel engines beginning in 2007. Please address these concerns in the FEIS.

Response: By law, IIFP is required to obtain all applicable Federal, State, and local permits and approvals to conduct construction and operations activities. Also, 10 CFR 51.71(c) requires the EIS to provide a list of all Federal permits, licenses, approvals, and other entitlements for the proposed action and describe the status of compliance with these requirements. Section 1.5 of the EIS describes the statutory and regulatory requirements that are applicable to the proposed IIFP facility. In addition, Table 1-1 describes laws, regulations, and agreements that apply to the facility and the regulatory basis and the requirements for these items. Finally, Section 1.5.3 was updated to provide the status of the permits, licenses, authorizations, and approvals for construction and operation of the proposed IIFP facility, should NRC grant the license.

The NRC staff acknowledges the importance of using BMPs. As noted in Chapter 5, IIFP identified the use of BMPs as mitigation measures for reducing or minimizing air quality impacts during road construction, land clearing, and building construction. Also, construction equipment and related vehicles would be equipped with standard pollution control devices and maintained in good working order.

The NRC does not regulate implementation of the Highway Diesel and Nonroad Diesel Rules. However, the NRC staff informed IIFP on the EPA's final Highway Diesel and Nonroad Diesel Rules that mandate the use of lower-sulfur fuels in nonroad and marine diesel engines beginning in 2007 as one of the issues raised in the EPA letter (ADAMS Accession Number ML12067A096).

F.4.6 Seismicity

Comment Summary: The following comments express concern that the hazards associated with seismic activity were not adequately addressed in the draft EIS.

[01-01, Anonymous] Lea County sits in an seismic hazard zone. Locating any nuclear industry is unsafe (http://earthquake.usgs.gov/earthquakes/states/texas/hazards.php).

[02-01, Clifford Barr] USGS says Lea County is in earthquake hazard area/here is a copy of file the geology should be looked at again http://earthquake.usgs.gov/earthquakes/states/texas/hazards.php. I think anything nuclear located over an earthquake zone is unsafe

[04-02, Clifford Barr and 05-02, Phil Barr] See attachment labeled UTIG from the Jackson school of geosciences showing hazard zone under nuclear waste dump site in Andrews county Texas. Which would indicate an unsafe disposal path for any nuclear waste produced locally.

Response: The NRC staff acknowledges that there is some risk of seismic activity in all regions of the United States. Regions are identified by the potential to experience earthquakes based on the probability and severity of the earthquake. The U.S. Geological Survey (USGS) and Jackson School of Geosciences maps (as provided by the commenters) both identify Lea and Andrews Counties to be in a seismically quiet region, with local earthquakes of relatively small magnitude (moment magnitudes less than 2 on the Modified Mercalli-Revised 1931 scale [MM]). Section 3.6.2 of the EIS discusses seismicity in New Mexico. Additional information on seismic hazards of the IIFP site are discussed in Section 1.3.3.5 "Site Geology" of the SER for the proposed IIFP facility, (NRC, 2012. As summarized in the SER, the applicant provided adequate and appropriate seismic description of the region and seismic design basis information for the proposed facility. No change was made to the EIS as a result of these comments.

Comment Summary: The following comment, which is applicable to multiple sections including seismicity, requests clarification on issues identified during the scoping meeting.

[72-05, Rhonda Smith on behalf of the U.S. Environmental Protection Agency] Page A–7, Paragraph A.2.2, Summary of Issues Raised: ... several comments from citizens express concerns regarding impacts or risks posed by the facility due to seismic concerns.... It was unclear where in the DEIS these specific concerns were addressed. Please address these concerns in the FEIS.

Response: Appendix A includes a summary of all the issues raised during the scoping process. The NRC staff ensured that the EIS includes relevant sections and text discussing the concerns raised from citizens during the scoping meeting. Seismic issues are discussed in Section 3.6.2. No change was made to the EIS as a result of this comment.

F.4.7   Water Resources

Comment Summary: The following comments provide support for the IIFP project and indicate that air and water emissions will be regulated.

[41-05, Michael P. Gallagher II on behalf of the Lea County Board of County Commissioners] The Board of Commissioners acknowledge the facility will have regulated air and water emissions at or below state and federal limits, as allowed by the NRC and New Mexico Environmental Department.

[45-03, Lea County Commissioners and TF-03, Michael P. Gallagher II] ... WHEREAS, the facility will have regulated air and water emissions at or below state and federal limits, as allowed by the NRC and New Mexico Environment Department; ...

Response: The NRC staff acknowledges these commenters and appreciates the public participation. These comments express general support for the licensing of the proposed IIFP facility; however, they do not provide specific information that requires a response from the NRC. These comments are outside the scope of the EIS analysis because they do not directly relate to the content of the EIS.

Comment Summary: The following comment, which is applicable to multiple sections including water resources, requests clarification on issues identified during the scoping meeting.

[72-06, Rhonda Smith on behalf of the U.S. Environmental Protection Agency] Page A–7, Paragraph A.2.2, Summary of Issues Raised: ... several comments from citizens express concerns regarding impacts or risks posed by the facility due to ... availability of water sources.... It was unclear where in the DEIS these specific concerns were addressed. Please address these concerns in the FEIS.

Response: Appendix A includes a summary of all the issues raised during the scoping process. The EIS includes relevant sections and text discussing the topics raised by citizens during the scoping meeting. Water sources are discussed in Section 3.7 and comments addressing concerns about availability of water sources are addressed in Section F.4.7. No change was made to the EIS as a result of this comment.

Comment Summary: The following comment expresses concern about groundwater wells in the area.

[TH-01, Rose Gardner] Are there any groundwater wells in that area at all that feed the Lea County communities or the ranchers or cattle in the area?

Response: There are no wells within the proposed site to provide water to the Lea County citizens, other than those used for IIFP operations. Within a mile of the site there is one domestic well and 6 industrial wells for Abbott Bros. These wells are not on the IIFP site. No change was made to the EIS as a result of this comment.

F.4.8 Ecological Resources

Comment Summary: The following comment relates to the environmental impact of the proposed IIFP facility on ecological resources.

[64-01, Mathew Wunder on behalf of the State of New Mexico Department of Game & Fish] NMGF concurs with the DEIS conclusion of small impact to ecological resources. We stand by the recommendations made in our scoping comment letter, many of which have been included as recommendations in the DEIS.

Response: This comment expresses general support for the licensing of the proposed IIFP facility; however, it does not provide specific information that requires a response from the NRC.

Comment Summary: The following comment relates to the environmental impact of the proposed IIFP facility on ecological resources.

[64-02, Mathew Wunder on behalf of the State of New Mexico Department of Game & Fish] We recommend that, prior to initiating clearing and grading, pre-construction clearance surveys should be conducted for the following species of concern which may occur on the project site: lesser prairie-chicken, burrowing owl, and swift fox.

Response: IIFP conducted surveys for Vegetation (GL Environmental, 2010c), Animal (SORA, 2011) and (NMGF, 2010a) and Cultural Resource (Daras, 2009) at the proposed facility. NRC does not have the authority to require a licensee to perform a preconstruction clearance survey; imposition of such a requirement may be within the authority of the State of New Mexico. Further, IIFP is required to obtain all required Federal, State, and local permits and approvals to conduct construction and operations activities. If the State of New Mexico requires a preconstruction clearance survey, NRC staff would expect the licensee to comply with that requirement. No change was made to the EIS as a result of this comment.

F.4.9 Socioeconomics and Environmental Justice

Comment Summary: The following comment expresses a concern that the proposed IIFP facility will negatively impact the Ogallala aquifer and result in a socioeconomic impact.

[03-02, Phil Barr] Here are 2 aquifer maps; one from T Boone Pickens Mesa Water site and the other from the Red River Authority site of Texas. The 05 map didn't have a disclaimer when I downloaded it from the Texas site and is remarkably similar to the later one from Mesa Water. No matter how the water supply is spun to make it work for nuclear projects; Lea County is running out of water and when it gets in short supply; prices will go up and the average citizen will suffer economically.

Response: Groundwater at the proposed IIFP site and in its vicinity is discussed in detail in Section 3.7.1.2.2, and groundwater use is discussed in Section 3.7.1.3. Section 2.1.5.2 discusses how the facility would use groundwater resources, stating that "the proposed facility would require relatively low volumes of process water because it would recycle process water and re-circulate cooling water. IIFP estimates that the total water supply requirement is less than 38,000 liters (L) [10,000 gallons (gal)] per day. Sanitary water requirements for showers, lavatories, drinking, toilets, and the laboratory would be 11,000 L to 17,000 L [3,000 to 4,500 gal] per day of the total. Treated sanitary waste water would be used for landscape watering. Boiler blow-down would be sent to the environmental protection process (Section 2.1.6.4.2) for treatment, if needed, and evaporation." Further, Section 2.1.5.2 states that "it is anticipated that there would be at least one but no more than two groundwater wells to supply water for the facility." As stated in Section 4.1.1.6.1.1, the proposed IIFP site has been included in Lea County's 40-Year Water Development Plan. Based on this information, NRC staff has determined that the impact on groundwater resources from the proposed IIFP facility would be SMALL and that the facility's use of groundwater would have a SMALL socioeconomic impact. No change was made to the EIS as a result of this comment.

Comment Summary: The following comments discuss potential beneficial socioeconomic impacts of the proposed IIFP facility.

[06-03, Audrey Nelson; 07-03, Troy Beall; 08-04, Marshall Nadel; 10-03, Erika Prestwich; 11-03, Dennis Manning; 12-03, Kurt Smith; 13-03, Justin Tomborello; 14-03, Mitchell N.; 15-03, Unknown Individual; 16-03, M. Mayer; 18-03, Giorgio Borlenghi; 20-03, James Keane; 22-04, Cindy Bryan; 25-04, Monty D. Newman; 27-04, Paul Campbell; 28-04, Garry Buie; 29-04, David G. Jeff; 30-04, Randy Pettigrew; 31-03, Ron Black; 32-03, Samuel S. Spencer; 33-04, Sandy Nash; 34-04, 55-04, Jonathan Sena; 36-04, Kathi Bearden; 37-03, Michael Leighton; 41-02, Michael P. Gallagher II on behalf of the Lea County Board of County Commissioners; 43-04, Hal Brunson; 44-03, Janice Jamison; 46-04, Chris Frentzel on behalf of the International Brotherhood of Electrical Workers (IBEW), Local 611; 47-03, Gary Schubert; 48-04, Pj Parker; 49-04, Mike Hoyl; 50-04, Sally Tomar; 51-03, TB-03, Gary Don Reagan; 52-03, Richard H. Neuwirth; 53-03, Charles R. Martin; 54-01-03, Jamie Suprenant; 54-02-03, Sandra Milner; 54-03-03, Sylvia Ebel; 54-04-03, Dana Nunley; 54-05-03, Lara Haddad; 54-06-03, Douglas Hager; 54-07-03, Michael Hager; 54-08-03, Carl Fairman; 54-09-03, Novlette Spence; 54-10-03, Jodi Moore; 54-11-03, Connie Pucio; 54-12-03, Rebecca Pless; 54-13-03, Patricia Medina; 54-14-03, Joe Glen Ensminger; 54-15-03, Linda Jianto; 54-16-03, Sharon Ensminger; 54-17-03, LaDona Legg; 54-18-03, Regina Diaz; 54-19-03, Katrina Torres; 54-20-03, Mayolo Gonzalez; 54-21-03, Krystal Wallace; 54-22-03, Judy Tuttle-Wurth; 54-23-03, Kevin Heier; 54-24-03, Maribel Garcia; 54-25-03, Kerry Marschke; 54-26-03, Tami Dunlap; 54-27-03, Chris Hulsey; 54-28-03, Dana Estrada; 56-04, Brad Nesser; 57-03, David G. Stovall; 59-04, Trent Clifton; 63-03, Lane Allgood on behalf of the Partnership for Science and Technology; 67-02, J. Stephen Herring; 70-03, Jack Jacli; 71-04, Holis Riley; 73-04, Marilyn Burns; and TA-03, Farok Sharif] I also believe the facility will produce important fluoride products for U.S. markets and be an asset to U.S. chemical manufacturing capability.

[09-03, Darrell Fisher] The Environmental Impact Statement shows that International Isotopes can provide a safe and environmentally sound approach for treatment of depleted uranium hexafluoride.

[12-04, Kurt Smith] It sounds like a win-win situation to me, creating jobs and saving the taxpayer money.

[17-03, William Newsome] I also believe that the synergies with other companies in the energy sector in Lea County are an important positive consideration.

[19-02, Grant Taylor on behalf of the Hobbs Chamber of Commerce] The resulting products will fill an important market niche for fluoride products across the country and around the world, while simultaneously creating a new center for U.S. chemical manufacturing.

[21-02, Jeff McCool] I sincerely believe the facility will provide a viable and important product.

[22-05, Cindy Bryan; 25-05, Monty D. Newman; 27-05, Paul Campbell; 28-05, Garry Buie; 29-05, David G. Jeff; 30-05, Randy Pettigrew; 33-05, Sandy Nash; 34-05, 55-05, Jonathan Sena; 36-05, Kathi Bearden; 38-04, Sara B. Cisneros on behalf of the Hobbs Hispano Chamber of Commerce Board of Directors; 43-05, Hal Brunson; 48-05, Pj Parker; 49-05, Mike Hoyl; 50-05, Sally Tomar; 56-05, Brad Nesser; 57-04, David G. Stovall; 58-04, Robert Wallach; 59-05, Trent Clifton; 71-05, Holis Riley; 73-05, Marilyn Burns; TA-04, Farok Sharif] I (we) support this project and believe that it will be a positive contribution to Lea County by providing safe, quality jobs and economic diversification to the area.

[23-01, Anonymous] This will be a great opportunity for persons to get employment if this facility does get built in Lea County.

[24-02, John Tanner] I believe that it will be a positive contribution to Lea County and will fulfill an important role in the safe and environmentally sound treatment of depleted uranium hexafluoride, as well as producing valuable fluoride byproducts.

[26-03, Carroll H. Leavell] The impact of this project will be very positive as it concerns employment and good paying jobs.

[32-04, Samuel S. Spencer on behalf of Lea County State Bank] Our institution is in favor of the project moving forward as we believe that it will make a positive contribution to the economic diversification of our economy.

[35-02, Steve McCleery] I sincerely believe the facility will provide a viable and important product,.

[40-02, Gay G. Kernan] I am convinced that the project will have a positive impact on the quality of life in southeast New Mexico by providing quality jobs in a safe environment.

[41-03, Michael P. Gallagher II on behalf of the Lea County Board of County Commissioners] The Board of County Commissioners support this project and believe that it will be a positive contribution to Lea County by providing safe, quality jobs and economic diversification to the area.

[42-02, TC-02, Lisa Hardison on behalf of the Economic Development Corporation of Lea County (EDCLC)] We are confident that this industry will help continue to diversify our economy and create quality jobs for Lea County residents.

It is though the due diligence process of our organization and that of the Nuclear Regulatory Commission that EDCLC has been assured that the INIS facility will have a positive impact on the economy of Lea County with little environmental concern.

[45-04, Lea County Commissioners and TF-03, Michael P. Gallagher II] ... WHEREAS, the economic benefit to Southeastern New Mexico will be stability, growth, job creation, and industry diversification; and

WHEREAS, the facility will process depleted uranium and fluorine extraction to produce important fluoride products for U.S. markets and will be an asset to U.S. chemical manufacturing capabilities; and ...

[46-05, Chris Frentzel on behalf of the International Brotherhood of Electrical Workers (IBEW), Local 611] This is a positive step forward for Lea County in terms of providing safe, worthwhile jobs and economic diversification to the area.

[60-03, Thomas E. Magette on behalf of Energy Solutions] This facility will provide a positive economic contribution to Lea County.

[65-02, Michael McKenzie-Carter] The potential benefits of the proposed action-in particular the production of industrial fluoride products-are very positive.

This project represents a positive contribution to our country by safely treating stored DUF6 much more quickly than other alternatives, and doing so in a safe and commercially-viable way.

[67-03, J. Stephen Herring] The proposed facility makes an important contribution ..., converting the DUF6 to depleted uranium oxides and producing valuable fluorine in the process.

[72-07, Rhonda Smith on behalf of the U.S. Environmental Protection Agency] In the DEIS, the summary of the costs and benefits of the proposed action explains that during operation of the plant about $56 million to $71 million (in 2009 dollars) in wages, benefits, goods and services would be spent annually. Approximately $554,400 offered in lieu of property tax would be paid to the Hobbs Municipal School District and the New Mexico Junior College during the construction period. Construction and operation of the facility would have additional indirect economic impacts by creating additional employment and economic activity within the region. Over the lifetime of operations, the low estimate of gross receipts taxes is $6,500,000 (in 2009 dollars) to Lea County. With these summary of costs and benefits, it appears the project will be good for the local economy.

[74-02, Paul D. Neuwirth] Because the facility will produce important fluoride products for markets in the United States, it will add to U.S. chemical manufacturing capability.

Response: The NRC staff acknowledges these commenters and appreciates the public participation. These comments express general support for the licensing of the proposed IIFP facility; however, they do not provide specific information that requires a response from the NRC. These comments are outside the scope of the EIS analysis because they do not directly relate to the content of the EIS.

Comment Summary: The following comment requests information on lifestyles of the Native Americans living in the region around the proposed IIFP facility that may lead to a disproportionate impact from environmental impacts on Native Americans living in the region around the proposed IIFP facility compared to non-Native Americans.

[72-08, Rhonda Smith on behalf of the U.S. Environmental Protection Agency] The proposed facility is not in close proximity to any Tribal reservation or major native/tribal population or land base. According to demographic data in the DEIS, there are approximately eight hundred (800) American Indians living near Hobbs, New Mexico. There is no information presented in the DEIS as to whether American Indians who are living in this area are associated with any particular Tribe(s). There is no information provided upon which to form an opinion about whether the lifeways of American Indians in the area could include traditional or cultural practices that create additional or unique risk pathways. Also, there is no information provided upon which to determine whether potential environmental contamination, emergency events, increased traffic or other changes associated with the proposed facility could result in disproportionate impacts due to other factors that might make these persons more vulnerable or place them at greater risk. Please clarify in the FEIS.

Many Tribes in the United States do have traditional cultural practices that may include ingestion or uses of plants, animals and other natural resources that are not common to non-Indians. In addition to traditional and cultural practices, diabetes and other health conditions may make some Native Americans more susceptible to risks or sensitive to environmental conditions. The FEIS should provide appropriate data and analysis of potential impacts to American Indians living in the area.

Response: The Native Americans living in close proximity to the proposed IIFP facility are not in close proximity to any Tribal reservation or major native/tribal population or land base. However, most of the Native Americans within the region are associated with the Navaho, Cherokee, Mexican-American Indian, and Choctaw tribes. Many of the Native Americans in the region are associated with more than one tribe.

As stated in Section 4.1.2.11, there will be no liquid radiological effluent releases and the air radiological releases from the proposed IIFP facility are projected to be extremely low. The exposure through ingestion or uses of plants, animals and other natural resources would be even lower. Therefore, preconstruction, construction, operation, and decommissioning of the proposed IIFP facility is not expected to result in disproportionately high and adverse impacts to Native Americans. No change was made to the EIS as a result of this comment.

Comment Summary: The following comment expresses concern that the EIS underestimates the positive socioeconomic impacts of and the potential for improving environmental justice from the proposed IIFP facility.

[39-03, Kathleen A. Moran] Also, I believe the positive economic and socio-economic and environmental justice effects will be more significant than stated in the report. The opportunity to develop cutting edge technology in the nuclear-medicine field will bring highly desirable jobs to the area and foster the development of the human resources needed to operate such a facility. The local population, which includes a high percentage of Hispanic peoples, will benefit from such opportunities. The schools and the community will benefit by the diversity that a new industry will bring. The community has seen that with the location of the National Enrichment Facility nearby as well. The impact of that development has been extremely positive on the community.

Response: *The NRC staff acknowledges this commenter and appreciates the public participation. This comment expresses general support for the licensing of the proposed IIFP facility; however, it does not provide specific information that requires a response from the NRC. This comment is outside the scope of the EIS analysis because it does not directly relate to the content of the EIS.*

Comment Summary: The following comment requests information on the methods used to collect information on migrant worker populations in the region around the proposed IIFP facility to evaluate environmental justice.

[72-09, Rhonda Smith on behalf of the U.S. Environmental Protection Agency] The DEIS states that NRC staff used the ESRI ArcGIS 9.3 software that accessed the 2000 decennial census to identify block groups; yet the section under Minority Populations states that the number of farms which fall either wholly or partially within the 50-mile radius is unknown. There is no additional information describing any other measures taken to get information about the migrant worker population on these farms. Please clarify in the Final EIS.

Response: *The NRC staff searched for public sources to obtain migrant farm worker population statistics.   The U.S. Department of Agriculture (USDA) Census of Agriculture information showed that the number of farms that reported migrant workers in 2007 for operations located in counties either wholly or partially within 80 km [50 mi] of the proposed IIFP site (USDA, 2007). Section 3.9.1.2.2 explains that seasonal agricultural migrant workers may make up a portion of the minority population, but that the U.S. Census Bureau (USCB) does not produce a count of migrant workers. The USDA also reports temporary farm labor (workers working for less than 150 days). Temporary laborers are not characterized by USDA the same as migrant workers, but represent a worker population that depends on seasonal crops and may not have permanent employment or residence. Therefore, temporary farm labor could make up a portion of the migrant population.*

In response to the comment, the following additional information and clarifications for temporary labor in the region of influence (ROI) is provided. Seasonal agricultural (migrant) workers may make up a portion of the minority population within the 80-km (50-mi) radius. Although migrant worker population counts are not available from the USCB, the U.S. Department of Agriculture has collected information on farms that employ migrant labor and farms that report workers that work for less than 150 days per year. The number of farms that reportedly employ migrant laborers, the number of farms that employ workers for less than 150 days, and the number of workers that work less than 150 days, respectively, in each county that falls wholly or partially within the 80-km (50-mi) radius of the IIFP can be obtained from the USDA Web site (USDA, 2007). The counties that fall wholly or partially within the 80-km (50-mi) radius of the IIFP facility are: New Mexico Lea County, New Mexico; Eddy County, New Mexico; Chaves County, New Mexico; Loving County, Texas; Winkler County, Texas, Andrews County, Gaines County, Yoakum County, Terry County, and Cochran County, Texas. No change was made to the EIS as a result of this comment.

F.4.10 Noise

Comment Summary: The following comment describes those regulations that pertain to noise control.

[62-03, Julie Roybal on behalf of the New Mexico Environment Department]

Air Quality Bureau
To further ensure air quality standards are met, applicable local or county regulations requiring noise and/or dust control must be followed; if none are in effect, controlling construction-related air quality impacts during projects should be considered to reduce the impact of fugitive dust and/or noise on community members.

Response: By law, IIFP is required to obtain all applicable Federal, State, and local permits and approvals to conduct construction and operations activities. Also, 10 CFR 51.71(c) requires that the EIS provide a list of all Federal permits, licenses, approvals, and other entitlements for the proposed action and describe the status of compliance with these requirements, including requirements for noise control. In Section 1 of the environmental report (ER), the licensee lists all necessary permits from Federal, State, and local entities. Section 1.5 of this EIS describes the statutory and regulatory requirements that are applicable to the proposed IIFP facility. In addition, Table 1-1 describes laws, regulations, and agreements that apply to the facility; the regulatory basis and requirements for these items. Finally, Section 1.5.3 was updated to provide the status of the permits, licenses, authorizations, and approvals for construction and operation of the proposed IIFP facility, should NRC grant the license.

F.4.11 Public and Occupational Health

Comment Summary: The following comment discusses the chemical toxicity of uranium.

[TE-02, Robert Hayes] I do have something that just kind of bugs me, a technical issue, and that's that for uranium it's a heavy metal and as a heavy metal it's toxic, chemically toxic, just like lead or mercury, and as lead and mercury are toxic, you don't want to eat lead, you can get lead poisoning. Until you've enriched uranium to about 20 percent, the radiotoxicity does not outweigh the chemical toxicity, and so it's really the chemical toxicity that dominates any kind of hazard here.

Response: Toxicity of uranium is discussed in Sections 3.3.2.2, 3.3.2.3, 3.3.2.4, 6.3.1.8, and A.1.1 of the SER (NRC, 2012). The NRC determined that the applicant identifies the chemical hazards at the proposed facility and has developed a safety program that is adequate to assure compliance with the regulations, as discussed in Section 6.2 and in SER Section 6.4. No change was made to the EIS as a result of this comment.

Comment Summary: The following comment, which is applicable to multiple sections including public and occupational health, requests clarification on issues identified during the scoping meeting.

[72-10, Rhonda Smith on behalf of the U.S. Environmental Protection Agency] Page A–7, Paragraph A.2.2, Summary of Issues Raised: ... several comments from citizens express concerns regarding impacts or risks posed by the facility due to... possible health impacts associated with nuclear facilities. It was unclear where in the DEIS these specific concerns were addressed. Please address these concerns in the FEIS.

Response: Appendix A includes a summary of all the issues raised during the scoping process. The NRC staff ensured that the EIS included relevant sections and text discussing the concerns raised from citizens during the scoping meeting. Public and occupational health impacts are discussed in Sections 4.1.2.11 and 4.2.2.11. No change was made to the EIS as a result of this comment.

F.4.12 Transportation/Waste Management

Comment Summary: The following comments discuss potential positive national waste management impacts of the proposed IIFP facility.

[06-04, Audrey Nelson; 07-04, Troy Beall; 08-05, Marshall Nadel; 09-04, Darrell Fisher; 10-04, Erika Prestwich; 11-04, Dennis Manning; 12-05, Kurt Smith; 13-04, Justin Tomborello; 14-04, Mitchell N.; 15-04, Unknown Individual; 16-04, M. Mayer; 17-04, William Newsome; 18-04, Giorgio Borlenghi; 20-04, James Keane; 22-06, Cindy Bryan; 25-06, Monty D. Newman; 27-06, Paul Campbell; 28-06, Garry Buie; 29-06, David G. Jeff; 30-06, Randy Pettigrew; 31-04, Ron Black; 32-05, Sam S. Spencer; 33-06, Sandy Nash; 34-06, 55-06, Jonathan Sena; 36-06, Kathi Bearden; 37-04, Michael Leighton; 43-06, Hal Brunson; 44-04, Janice Jamison; 47-04, Gary Schubert; 48-06, Pj Parker; 49-06, Mike Hoyl; 50-06, Sally Tomar; 51-04, TB-04, Gary Don Reagan; 52-04, Richard H. Neuwirth; 53-04, Charles R. Martin; 54-01-04, Jamie Suprenant; 54-02-04, Sandra Milner; 54-03-04, Sylvia Ebel; 54-04-04, Dana Nunley; 54-05-04, Lara Haddad; 54-06-04, Douglas Hager; 54-07-04, Michael Hager; 54-08-04, Carl Fairman; 54-09-04, Novlette Spence; 54-10-04, Jodi Moore; 54-11-04, Connie Pucio; 54-12-04, Rebecca Pless; 54-13-04, Patricia Medina; 54-14-04, Joe Glen Ensminger; 54-15-04, Linda Jianto; 54-16-04, Sharon Ensminger; 54-17-04, LaDona Legg; 54-18-04, Regina Diaz; 54-19-04, Katrina Torres; 54-20-04, Mayolo Gonzalez; 54-21-04, Krystal Wallace; 54-22-04, Judy Tuttle-Wurth; 54-23-04, Kevin Heier; 54-24-04, Maribel Garcia; 54-25-04, Kerry Marschke; 54-26-04, Tami Dunlap; 54-27-04, Chris Hulsey; 54-28-04, Dana Estrada; 56-06, Brad Nesser; 57-05, David G. Stovall; 58-05, Robert Wallach; 59-06, Trent Clifton; 63-04, Lane Allgood on behalf of the Partnership for Science and Technology; 67-04, J. Stephen Herring; 70-04, Jack Jacli; 71-06, Holis Riley; 73-06, Marilyn Burns; TA-05, Farok Sharif] I believe that [the project] will fulfill an important role in the safe and environmentally sound treatment of depleted uranium hexafluoride.

[08-06, Marshall Nadel] By not converting depleted UF6 at Paducah and Portsmouth in a timely manner our Department of Energy left a terrible environmental legacy in the form of thousands of cylinders of depleted UF6- a most unstable uranium compound.

[21-03, Jeff McCool , 35-03, Steve McCleery] I sincerely believe the facility will ... create an opportunity to produce a safe environment for the treatment of depleted uranium hexafluoride.

[24-03, John Tanner] I believe that it ... will fulfill an important role in the safe and environmentally sound treatment of depleted uranium hexafluoride.

[41-04, Michael P. Gallagher II on behalf of the Lea County Board of County Commissioners] The INIS facility will fulfill an important role in the safe and environmentally sound treatment of depleted uranium hexafluoride.

[42-03, TC-03, Lisa Hardison on behalf of the Economic Development Corporation of Lea County] This business will compliment Lea County's uranium enrichment processing plant by providing the safe and environmentally sound treatment of depleted uranium hexafluoride that will ultimately produce important fluoride products for U.S. markets.

[46-06, Chris Frentzel on behalf of the International Brotherhood of Electrical Workers (IBEW), Local 611] I am also impressed with the approach to protecting our environment via the prudent treatment of depleted uranium hexafluoride.

[60-04, Thomas E. Magette on behalf of Energy Solutions] This facility will ... provide a commercial entity to address the depleted uranium issue.

[65-03, Michael McKenzie-Carter] The potential benefits of the proposed action-in particular decreasing the amount of stored DUF6 -are very positive.

[67-05, J. Stephen Herring] The proposed facility makes an important contribution to the safe and environmentally responsible treatment of the tails from uranium enrichment, converting the DUF6 to depleted uranium oxides...As ceramic solids, the uranium oxides are suitable for the long-term storage or disposal of the depleted uranium, whereas the DUF6 is not a suitable chemical form for disposal.

[74-03, Paul D. Neuwirth] I believe that the proposed International Isotopes fluorine products facility in Lea County, New Mexico will play an important role in the environmentally sound and safe treatment of depleted uranium hexafluoride.

Response: The NRC staff acknowledges some commenters foresee a potential benefit to waste management from the construction and operation of the proposed IIFP facility, including the safe and environmentally sound treatment of depleted uranium hexafluoride. These comments express general support for the licensing of the proposed IIFP facility; however, they do not provide specific information that requires a response from the NRC.

Comment Summary: The following comment requests clarification on waste management data supplied in the draft EIS.

[62-04, Julie Roybal] Table 4-31 states that if CaF_2 (produced at the IIFP facility) is not sold, total hazardous waste generation during operations would be 92,000–140,000 kg annually. However, Section 4.2.2.12 Waste Management, states: "The quantity of cumulative hazardous waste could be as much as 46,300 kg (51 tons) per year if a market for the CaF_2 cannot be identified." This discrepancy should be corrected in the final report.

Response: As stated in Table 4-30 of the EIS, the total hazardous waste generation during operations would be 92,000 to 140,000 kg annually. The total hazardous waste in Table 4-40 is for Phase 1 of the operations and includes 90,000 to 136,000 kg of CaF_2. The potential future Phase 2 facility expansion would provide additional deconversion capability. Expansion and operation of the expanded facility during Phase 2 would be a reasonably foreseeable action and is evaluated as a cumulative impact. Because of the added process technology used in the expansion to the Phase 2 facility, a large part of the fluoride-bearing spent scrubber liquids (the HF liquor portion) can be recycled to the add-on direct oxide deconversion process and recovered rather than be treated with lime to generate CaF_2. When this scrubber liquor waste (weak aqueous HF solution) is recycled during Phase 2, it eventually enters the process final product stream where it is distilled and becomes valuable anhydrous HF product for sale to customers (IIFP, 2011). If a market for the CaF_2 cannot be identified, then the quantity of cumulative hazardous waste, as stated in Section 4.2.2.12, could be as much as 46,300 kg (51 tons) per year. Since this aqueous HF recycle/distillation capability does not exist within the Phase 1 process technology; a larger amount of scrubber spent liquors has to be treated with lime thus generating larger amounts of CaF_2 during Phase 1 operation than the cumulative *hazardous waste generated by Phase 1 and Phase 2. An explanation has been added to Section 4.2.2.12 of the EIS.*

Comment Summary: The following comment outlines requirements that the applicant must follow should the applicant decide to transfer waste to the Waste Control Specialists (WCS) disposal facility in Texas or the U.S. Ecology waste disposal site in Washington.

[68-01, Barbara Green on behalf of the Rocky Mountain Low-Level Radioactive Waste Compact] The following comments on the above-referenced INIS DEIS are being submitted on behalf of the Rocky Mountain low-level Radioactive Waste Compact ("Compact."). The Compact is an agreement among the states of Colorado, New Mexico, and Nevada, and was created pursuant to the federal low-level Radioactive Waste Act, and approved by Congress. Thus, the Compact is both an agreement among its member states, and federal law. The Compact has jurisdiction over the import to, export from, and disposal of low-level radioactive waste within the Compact Region.

Import and Export of Low-level Radioactive Waste to and from the Facility

Pages 2-24 and 4-56 state that the Compact must approve shipments from the facility of depleted uranium to the WCS disposal facility in Texas. This statement is true because the facility is located in New Mexico, a member state of the Compact, and the Texas facility is not within the Compact region.

The DEIS notes that the U.S. Ecology facility in Richland, Washington, may be a disposal option for depleted uranium. Please add that Compact approval is required for the export of any and all low-level radioactive waste to this facility.

Please add that the export of any and all low-level radioactive waste (as defined by the Compact) from the Compact region requires Compact approval.

Also add that the import of any low-level radioactive waste (as defined by the Compact) to the INIS facility from outside the Compact will require Compact approval, unless the material is depleted uranium oxide as described in the Declaratory Order issued by the Board dated September 14, 2010 ("Order"), attached to this letter. The Order describes the conditions under which import of DUF_6 to the facility from outside the Compact region would not require approval by the Compact.

The Compact will continue to require approval for import of out-of-region low-level radioactive waste not covered by the Order.

Response: *The NRC staff acknowledges these comments and appreciates the public participation. However, they do not provide specific information that requires a response from the NRC. These comments are outside the scope of the EIS analysis because they do not directly relate to the content of the EIS.*

Comment Summary: The following comment, which is applicable to multiple sections including transportation and disposal of waste, requests clarification on issues identified during the scoping meeting.

[72-11, Rhonda Smith on behalf of the U.S. Environmental Protection Agency] Page A–7, Paragraph A.2.2, Summary of Issues Raised... several comments from citizens express concerns regarding impacts or risks posed by the facility due to ... transportation and disposal of waste... It was unclear where in the DEIS these specific concerns were addressed. Please address these concerns in the FEIS.

Response: Appendix A includes a summary of all the issues raised during the scoping process. The NRC staff ensured that the EIS included relevant sections and text discussing the concerns raised from citizens during the scoping meeting. Transportation issues are discussed in Section 3.10; waste disposal issues are discussed in Sections 4.1.1.2, 4.1.1.12, and 4.2.2.12. No change was made to the EIS as a result of this comment.

Comment Summary: The following comment expresses concern about the transportation of liquid fluorine.

[TG-01, Rob Hayes] My biggest question, which doesn't seem it got addressed, is the large liquid tanks of fluorine that would have to get transported on our roads. I would assume that there would be quite a few of them. Liquid fluorine is pretty nasty stuff, kind of like liquid chlorine, and I just wondered if and how that was taken into consideration.

Response: As stated in Sections 4.1.2.9 and 4.1.2.12 the CaF2 will be packaged and shipped. IIFP must comply with all applicable Federal, State, and local laws and regulations, including the packaging and shipping regulations of Department of Transportation and EPA. No change was made to the EIS as a result of this comment.

Comment Summary: The following comment expresses concern about the transportation of the depleted uranium and wastes.

[TH-02, Rose Gardner] Regarding the transportation of the depleted uranium to the site and then the final disposition of the, I guess, oxides wastes after the processing, how would that material be moved, on trucks, trains, or what's the plan there?

Response: Sections 2.1.3.2, 2.1.3.3 and 4.2.4.3 of the ER states that the proposed IIFP facility would receive the depleted uranium hexafluoride in 14-ton cylinders by truck. The depleted uranium oxides and all other low level waste materials generated at the facility would be transported by truck in metal containers, such as 242-(dry) liter (55 gallon) drums or 2,973-(dry) liter (105 cubic foot) boxes, that meet disposal facility's waste acceptance criteria. These contains are shipped to a licensed facility in accordance with NRC and Department of Transportation packaging and shipping regulations. Trucks would carry 20 to 25 drums per shipment. The AHF will be transported in tank-truck trailers. The SiF4 and BF3 are transported in compressed gas tube trailers via semi-truck (IIFP, 2009a). No change was made to the EIS as a result of this comment.

F.4.13 Accidents (Wildfires)

Comment Summary: The following comment requests that the impact of wildfires be addressed in the EIS.

[61-01, Stephen R. Spencer] Commenter David Herrell: Wildfire should be addressed. This area sees repeated wildfires that could threaten the safety of the proposed plant.

Response: All credible accidents at the proposed IIFP facility, including those initiated by natural events, are considered by IIFP in the Integrated Safety Analysis. As discussed in

Section 7 of the SER (NRC, 2012), wildfires can occur in areas surrounding the facility. The potential for a wild land fire to impact the site is not considered to be a credible event given the dry desert topography and lack of forestry and vegetation. Regardless, IIFP has committed to maintain vegetation controls to minimize the fire hazard at the facility. No change was made to the EIS as a result of this comment.

F.4.14 Cumulative Impacts

Comment Summary: The following comment discusses pre-construction activities.

[64-03, Mathew Wunder on behalf of the State of New Mexico Department of Game & Fish] Pre-construction clearing, grading, road-building and utility installation are treated as cumulative effects in the DEIS, because no NRC license is required. However these actions are logically integral to construction and operation of the facility.

Response: NRC regulations, including, 10 CFR 40.4, define construction to explicitly exclude certain activities, such as land clearing and erection of administrative buildings, that do not have a reasonable nexus to radiological health and safety or the common defense and security. These activities, termed "preconstruction" activities, are regarded as "non-Federal actions" that do not require NRC approval or oversight. Thus, the impacts of these actions are not considered to be direct impacts of the Federal action (i.e., issuance of a license); however, the NRC does consider preconstruction activities with respect to cumulative impacts. No change was made to the EIS as a result of this comment.

F.4.15 Miscellaneous Comments

Comment Summary: The following comment expresses concern about underground piping.

[TH-03, Rose Gardner] Are there any underground piping in the processing of the materials? In other words, under the building will there be any underground pipes carrying the dangerous materials?

Response: IIFP in the License Application in Section 1.1.1 states that the facility will not have underground piping for transporting process waters. The License Application indicates a number of underground pipes will be built for transporting water including fire mains, sanitary sewer, and well water. The IIFP facility will also have some underground piping to transfer industrial gases (e.g., natural gas) (IIFP, 2009b). No change was made to the EIS as a result of this comment.

Comment Summary: NRC has determined that the following comment does not directly relate to a specific resource area in the EIS.

[23-02, Anonymous] But it would be in International Isotopes Fluorine Products Inc. best interest to not contract Tactical Security Solutions (TSS) out of Hobbs, NM, for their security measure needs. TSS officers currently working @ URENCO USA in Eunice, NM, have been stealing things from URENCO personnel. How can you trust TSS personnel to not steal from this facility if it ever gets built???

Response: The NRC staff acknowledges this comment and appreciates the public participation. However, these comments are outside the scope of the EIS analysis because they do not directly relate to the content of the EIS. No change was made to the EIS as a result of this comment.

Note: *[61-02, Stephen R. Spencer]* A comment document was submitted with a series of comments attributed to Amanda L. Nisula related to the discussion of the No Action Alternative and Reasonable Alternatives to the Proposed Action in the draft EIS, another comment related to navigation through the draft EIS, and comments on two tables attributed to the draft EIS. *[69-01, Amanda L. Nisula]* subsequently rescinded these comments.

Comment Summary: The following comment is regarding the safety record of the applicant.

[TI-01, Miles Chedekel] I just wanted to know what the safety record is of this applicant for other facilities that they have run of this type.

Response: The NRC staff acknowledges the commenter and appreciates the public participation. This comment is outside the scope of the EIS analysis because it does not directly relate to the content of the EIS. No change was made to the EIS as a result of this comment.

F.5   REFERENCES

(Daras, 2009) Daras, S. 2009. "Cultural Resource Survey of 640 acres for the Arkansas Junction Site, Lea County, New Mexico. 2009 Report No. 1224." Prepared for New Mexico State Land Office by Lone Mountain Archeological Services, Inc. 2009. ADAMS Accession No. ML113120204.

(IIFP, 2009a) International Isotopes Fluorine Products, Inc. 2009. Fluorine Extraction Process and Depleted Uranium De-conversion Plant (FEP/DUP) Environmental Report, Revision A, ER-IFP-001. December 27, 2009. ADAMS Accession No. ML100120758.

(IIFP, 2009b) International Isotopes Fluorine Products. Inc. 2009. Fluorine Extraction Process and Depleted Uranium De-conversion Plant (FEP/DUP) License Application, Revision A, ER-IFP-001. December 23, 2009. ADAMS Accession No. ML100630503.

(IIFP, 2011) International Isotopes Fluorine Products, Inc. "Fluorine Extraction Process and Depleted Uranium De-Conversion Plant (FEP/DUP) Official Responses to Environmental Report RAIs, Revision A." ADAMS Accession No. ML100970481.

(GL Environmental, 2010c) GL Environmental, Inc. 2010. "2010 Soil and Vegetation Survey Report." Prepared for International Isotopes Fluorine Products. November 29, 2010. ADAMS Accession No. ML112720310.

(NMGF, 2010a) New Mexico Department of Game and Fish. 2010. "New Mexico Deer Harvest Survey Report for the 2009-2010 Season." Available at http://www.wildlife.state.nm.us/recreation/hunting/harvest/documents/2009-2010_DeerHarvestReport.pdf. Accessed October 14, 2010. ADAMS Accession No. ML103090088.

(NRC, 2003) U.S. Nuclear Regulatory Commission. 2003. "Environmental Review Guidance for Licensing Actions Associated With NMSS Programs," NUREG–1748.

(NRC, 2012) U.S. Nuclear Regulatory Commission. 2012. "Safety Evaluation Report for the International Isotopes Fluorine Products, Inc. Fluorine Extraction Process and Depleted Uranium Deconversion plant in Lea County, New Mexico," NUREG–2116. ADAMS Accession No. ML113140271.

(SORA, 2011) SORA. 2011. "Lesser Prairie-Chicken Survey on the International Isotope Flourine Products Project Site-2011." Prepared for International Isotopes, Inc., May 24, 2011. ADAMS Accession No. ML112720317.

(USDA, 2007) U.S. Department of Agriculture. "Table 7: Hired Farm Labor--Workers and Payroll: 2007." 2007 Census of Agriculture. Available at http://www.agcensus.usda.gov/. (Accessed March 21, 2012).

NRC FORM 335 (12-2010) NRCMD 3.7	U.S. NUCLEAR REGULATORY COMMISSION **BIBLIOGRAPHIC DATA SHEET** *(See Instructions on the reverse)*	1. REPORT NUMBER (Assigned by NRC, Add Vol., Supp., Rev., and Addendum Numbers, if any.) NUREG-2113

2. TITLE AND SUBTITLE		3. DATE REPORT PUBLISHED	
Environmental Impact Statement for the Proposed Fluorine Extraction Process and Depleted Uranium Deconversion Plant in Lea County, New Mexico Final Report		MONTH	YEAR
		August	2012
		4. FIN OR GRANT NUMBER	

5. AUTHOR(S)	6. TYPE OF REPORT
See Chapter 9	Technical
	7. PERIOD COVERED (Inclusive Dates)

8. PERFORMING ORGANIZATION - NAME AND ADDRESS (If NRC, provide Division, Office or Region, U. S. Nuclear Regulatory Commission, and mailing address; if contractor, provide name and mailing address.)
Division of Waste Management and Environmental Protection Office of Federal and State Materials and Environmental Management Programs U.S. Nuclear Regulatory Commission Washington, DC 20555-0001

9. SPONSORING ORGANIZATION - NAME AND ADDRESS (If NRC, type "Same as above", if contractor, provide NRC Division, Office or Region, U. S. Nuclear Regulatory Commission, and mailing address.)
Same as 8 above

10. SUPPLEMENTARY NOTES

11. ABSTRACT (200 words or less)

International Isotopes Fluorine Products, Inc. (IIFP), a wholly-owned subsidiary of International Isotopes, Inc., has submitted a license application to the U.S. Nuclear Regulatory Commission (NRC) to construct, operate, and decommission Phase 1 of a fluorine extraction and depleted uranium deconversion facility in Lea County, New Mexico. The proposed facility would provide services to the uranium enrichment industry, which makes fuel for nuclear power reactors. The IIFP facility would deconvert depleted uranium hexafluoride (DUF6) into fluoride products for commercial resale, and depleted uranium oxides for disposal. The license application for Phase 1 requests NRC to license the possession of up to 750,000 kilograms (827 tons) of depleted uranium under Title 10 "Energy" of the U.S. Code of Federal Regulations (10 CFR) Part 40, "Domestic Licensing of Source Material" in accordance with the Atomic Energy Act of 1954.

This Environmental Impact Statement (EIS) was prepared in compliance with the National Environmental Policy Act (NEPA) and the NRC regulations for implementing NEPA (10 CFR 51). This EIS evaluates the potential environmental impacts of the proposed action, which is to construct, operate, and decommission Phase 1 of the fluorine extraction and depleted uranium deconversion facility, and its reasonable alternatives, and describes IIFP's monitoring program and proposed mitigation measures.

12. KEY WORDS/DESCRIPTORS (List words or phrases that will assist researchers in locating the report.)	13. AVAILABILITY STATEMENT
EIS for the Proposed Fluorine Extraction Process and Depleted Uranium Deconversion Plant in Lea County, New Mexico. NUREG-2113 National Environmental Policy Act NEPA IIFP International Isotopes Depleted Uranium Hexafluoride International Isotopes Fluorine Products, Inc.	unlimited
	14. SECURITY CLASSIFICATION
	(This Page) unclassified
	(This Report) unclassified
	15. NUMBER OF PAGES
	16. PRICE

Printed
on recycled
paper

Federal Recycling Program

UNITED STATES
NUCLEAR REGULATORY COMMISSION
WASHINGTON, DC 20555-0001

OFFICIAL BUSINESS

NUREG-2113
Final

Environmental Impact Statement for the Proposed Fluorine Extraction Process and Depleted Uranium Deconversion Plant in Lea County, New Mexico

August 2012